Handbook of Ecohydrology

Handbook of Ecohydrology

Editor: Xavier Parsons

RC CALLISTO REFERENCE

www.callistoreference.com

Callisto Reference,
118-35 Queens Blvd., Suite 400,
Forest Hills, NY 11375, USA

Visit us on the World Wide Web at:
www.callistoreference.com

ISBN: 978-1-64116-296-8 (Hardback)

Cataloging-in-Publication Data

Handbook of ecohydrology / edited by Xavier Parsons.
    p. cm.
Includes bibliographical references and index.
ISBN 978-1-64116-296-8
1. Ecohydrology. 2. Hydrology. 3. Aquatic ecology. I. Parsons, Xavier.
QH541.15.E19 H36 2020

577.6--dc23

Table of Contents

Permissions

List of Contributors

Index

Preface

Ecohydrology is concerned with the study of interactions between water and ecological systems. It is an interdisciplinary scientific field and a sub-discipline of hydrology that focuses on the ecological aspects. The interactions between water and ecological systems occur within the water bodies, like lakes and rivers. Such interactions can also occur on land, such as in forests, deserts and other terrestrial ecosystems. Ecohydrology focuses on different areas including, transpiration and plant water use, effects of vegetation and benthic plants on stream flow and adaptation of organisms to their water environment. Ecohydrology studies both terrestrial and aquatic ecosystems. Its main focus in terrestrial ecosystems is on the interaction among vegetation, the vadose zone, land surface and the groundwater. In aquatic ecosystems it emphasizes on the effects of geomorphology, water chemistry and hydrology on the structure and function of the ecosystem. This book unravels the recent studies in the field of ecohydrology. Some of the diverse topics covered in this book address the varied branches that fall under this category. It will provide comprehensive knowledge to the readers.

This book is a result of research of several months to collate the most relevant data in the field.

When I was approached with the idea of this book and the proposal to edit it, I was overwhelmed. It gave me an opportunity to reach out to all those who share a common interest with me in this field. I had 3 main parameters for editing this text:

1. Accuracy – The data and information provided in this book should be up-to-date and valuable to the readers.

2. Structure – The data must be presented in a structured format for easy understanding and better grasping of the readers.

3. Universal Approach – This book not only targets students but also experts and innovators in the field, thus my aim was to present topics which are of use to all.

Thus, it took me a couple of months to finish the editing of this book.

I would like to make a special mention of my publisher who considered me worthy of this opportunity and also supported me throughout the editing process. I would also like to thank the editing team at the back-end who extended their help whenever required.

Editor

Does consideration of water routing affect simulated water and carbon dynamics in terrestrial ecosystems?

G. Tang[1]**, T. Hwang**[2]**, and S. M. Pradhanang**[3]

[1]Division of Earth and Ecosystem Sciences, Desert Research Institute, Reno, NV, USA
[2]Institute for the Environment, University of North Carolina, Chapel Hill, NC, USA
[3]Institute for Sustainable Cities, City University of New York, New York, NY, USA

Correspondence to: G. Tang (tangg2010@gmail.com)

Abstract. The cycling of carbon (C) in terrestrial ecosystems is closely coupled with the cycling of water. An important mechanism connecting ecological and hydrological processes in terrestrial ecosystems is lateral flow of water along landscapes. Few studies, however, have examined explicitly how consideration of water routing affects simulated water and C dynamics in terrestrial ecosystems. The objective of this study is to explore how consideration of water routing in a process-based hydro-ecological model affects simulated water and C dynamics. To achieve that end, we rasterized the regional hydro-ecological simulation system (RHESSys) and employed the rasterized RHESSys (R-RHESSys) in a forested watershed. We performed and compared two contrasting simulations, one with and another without water routing. We found that R-RHESSys was able to correctly simulate major hydrological and ecological variables regardless of whether water routing was considered. When water routing was considered, however, soil water table depth and saturation deficit were simulated to be greater and spatially more heterogeneous. As a result, water (evaporation, transpiration, and evapotranspiration) and C (forest productivity, soil autotrophic and heterotrophic respiration) fluxes also were simulated to be spatially more heterogeneous compared to the simulation without water routing. When averaged for the entire watershed, the three simulated water fluxes were greater while C fluxes were smaller under simulation considering water routing compared to that ignoring water routing. In addition, the effects of consideration of water routing on simulated C and water dynamics were more apparent in dry conditions. Overall, the study demonstrated that consideration of water routing enabled R-RHESSys to better capture our preconception of the spatial patterns of water table depth and saturation deficit across the watershed. Because soil moisture is fundamental to the exchange of water and C fluxes among soil, vegetation and the atmosphere, ecosystem and C cycle models therefore need to explicitly represent water routing in order to accurately quantify the magnitude and patterns of water and C fluxes in terrestrial ecosystems.

1 Introduction

The cycling of carbon (C) in terrestrial ecosystems is closely coupled with the cycling of water. Plants need water to survive, and thus, the distribution, composition, and structure of plant communities are directly influenced by spatial patterns of available water (Band, 1993; Band et al., 1993; Caylor et al., 2005; Ivanov et al., 2008). An important mechanism that connects ecological and hydrological processes in terrestrial ecosystems is lateral water flow along landscapes. Lateral water flow can redistribute water and nutrients through space, which affects plant establishment and growth (Band et al., 1993); leaf phenology (Asbjornsen et al., 2011); ecosystem structure and function (Wang et al., 2009); and soil biogeochemical processes, such as organic matter decomposition (Ju et al., 2006; Riveros-Iregui et al., 2011). For example, studies have demonstrated that lateral water flow and connectivity act as important determinants of ecological patterns and processes in heterogeneous landscapes (Band et al., 1993; Sponseller and Fisher, 2008), and contribute to changes in surface water, energy, nutrients, and C in space

(Pockman and Small, 2010). In mountainous catchments, Hwang et al. (2012) found that lateral water flow can produce important patterns in water and nutrient fluxes as well as stores, which influences the long-term spatial development of forest ecosystems. Riveros-Iregui et al. (2011) suggested that landscape-imposed redistribution of soil water is a major cause for distinct variation of growing-season soil CO_2 efflux within small subalpine watersheds.

Hydrological connectivity via lateral water flow plays important roles in the transport of water, nutrients and sediments at catchment scales (Smith et al., 2010). Correspondingly, distributed hydrology models (DHM) that simulate lateral water flow and its spatial connectivity along landscapes or among simulated grids have been developed increasingly in recent years (Lane et al., 2009). These models – such as DHSVM (Wigmosta et al., 1994) and RHESSys (Band et al., 1993; Tague and Band, 2004) – couple runoff generation and water routing mechanisms and thus are able to explicitly simulate the effects of topographic and subsurface heterogeneities on downslope redistribution of water and nutrients (Doten et al., 2006). In fact, DHMs are used widely to identify saturated areas that produce runoff and non-point source pollution (Gérard-Marchanti et al., 2006), evaluate irrigation systems (Singh et al., 2006), and examine flood potential associated with disturbances such as deforestation (Doten et al., 2006). The representation of soil moisture variability and water routing processes at grid cell level in DHMs also enables these models to account for spatial variability of runoff-generating mechanisms and infer model parameterization from distributed geospatial data such as geology, topography, soils, and land cover (Wang et al., 2011). These advantages greatly contributed to the accuracy of hydrologic forecasting (Smith et al., 2012).

Despite the fact that lateral water flow redistributes water and nutrients in space and thus affects ecosystem structure and function as well as the cycling of water and C, the representation of lateral water flow and its spatial connectivity may not be adequate in existing ecosystem and C cycle models. For example, Riveros-Iregui et al. (2011) indicated that the robust implementation of the lateral redistribution of soil water into biogeochemical models is often lacking. Chen et al. (2005) argued that most C cycle models at regional and global scales use bucket models to estimate soil moisture and ignore lateral exchanges of water among simulated units. The causes for such inadequacy are (i) lack of detailed information on how lateral water flow may affect vegetation, water, and C dynamics in terrestrial ecosystems, and (ii) increased burden of computing when water routing is included in the model's simulation (Ju et al., 2006). This inadequacy, however, is likely to hinder better quantification of the spatial heterogeneity and complex linkages of hydrological, ecological, and biogeochemical processes in terrestrial ecosystems.

Furthermore, mountain forests account for about 23 % of the earth's forest cover and play an important role in modulating global cycling of water and C (Price et al., 2011).

Given the elevational gradient in mountain forests plus gravity, lateral water flow – such as subsurface lateral flow along slopes – is common in humid mountain forests (Ridolfi et al., 2003). In semi-arid and arid ecosystems, surface lateral flow also occurs when rainfall intensity exceeds the infiltration capacity of dry soils (Kim and Eltahir, 2004) or on topographically flat ground if the presence of the vegetation patch creates a contrast in infiltration rate (Thomspon et al., 2011). The universality and significance of lateral water flow in terrestrial ecosystems suggest that it should not be overlooked by ecosystem and C cycle models. A better understanding of how lateral water flow and its spatial connectivity may affect water and C dynamics is therefore important for accurate quantification of terrestrial water and C budgets as well as sustainable management of water and forest resources (e.g., Wang et al., 2011).

The overall objectives of this study are to investigate (i) how consideration of water routing in a process-based, hydro-ecological model affects simulated water and C dynamics in terrestrial ecosystems; and (ii) if effects of consideration of water routing on simulated C and water dynamics are more remarkable in dry conditions. Toward these ends, we rasterized a regional hydro-ecological model designed to simulate integrated water, C and nutrient dynamics at watershed and regional scales. The rasterization aimed to (i) remove the model's hierarchical structure so that all hydrological and ecological processes would be simulated at the individual cell level; and (ii) add a new control interface so that the water routing algorithm built into the model could be switched on or off. These modifications allowed us to keep all model parameters and their parameterization identical between two predesigned contrasting simulations: with vs. without water routing. In turn, this helped reduce the uncertainty of model-based comparisons that can result from differences in model structure, parameters, and parameterization – as commonly encountered in model-based intercomparison studies. Based on the rasterized model, we performed two contrasting simulations for each of the two contrasting forcing scenarios: "wet" vs. "dry". We compared simulated soil water table depth and saturation deficit, evaporation, transpiration, evapotranspiration, forest productivity, and soil respiration from these simulations. Findings gained from these comparisons provide insights into the future development of ecosystem and C cycle models for terrestrial ecosystems.

2   Material and data

2.1   Study area

The Biscuit Brook (hereafter Biscuit) watershed in the Catskill Mountain region of New York State (Fig. 1) was selected as the study region. This watershed is relatively humid, with annual total precipitation of about 145 cm and annual

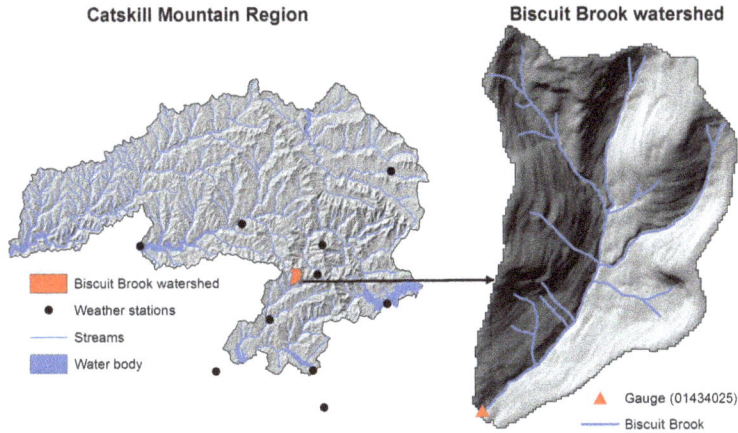

Fig. 1. The location of the Biscuit Brook watershed (red area) and the United States Geological Survey gauge station within the Catskill Mountain region of New York State. The map on the left depicts boundaries of the West of Hudson watershed and reservoirs of the New York City water supply system. The black points are 10 Cooperative Observer Program weather stations used to derive meteorological data for the watershed.

mean temperature about 4.4 °C. The slopes vary from 0.04 to 37°, and the maximum slope length is 4.73 km in a northeast to southwest direction (Fig. 1). We selected this watershed as the study region because (i) long-term historical streamflow observations from one USGS gauge station (01434025) for this watershed are available to calibrate and evaluate model simulations; (ii) this watershed is forested and thus well suited for investigating the linkages between ecological and hydrological processes; (iii) there are no human-related land use activities; and (iv) the watershed has spatially variable terrain with elevation ranging from 270 to 1270 m, providing a natural hydro-ecological laboratory to examine the effects of lateral water flow and its spatial connectivity on water, C and vegetation dynamics in terrestrial ecosystems.

2.2 Rasterizing the regional hydro-ecological simulation system

The Regional Hydro-Ecological Simulation System (RHESSys, Tague and Band, 2004) is a process-based hydro-ecological model designed for simulating integrated water, C and nutrient dynamics, as well as vegetation growth at watershed and regional scales. Although RHESSys is capable of being run in fully distributed mode, its hierarchical framework requires that some initial-state variables associated with the spatial hierarchy of basins, hillslopes, and zones be arranged per a prescribed template. In this study, we further rasterized RHESSys (version 5.12) in an attempt to remove the model's hierarchical structure. The rasterized RHESSys (hereafter R-RHESSys) adopted almost all features of its predecessor except for (i) exclusion of the hierarchical model framework of RHESSys, and (ii) modification of the user interface for controlling model simulation. The exclusion of the hierarchical structure in R-RHESSys caused the basin, hillslope, and zone hierarchical

structures existing in RHESSys to exist no longer. As a result, arrangement of some initial-state variables according to the prescribed template (i.e., the World file in RHESSys) was no longer needed. In addition, R-RHESSys excluded the TOPMODEL (Beven and Kirkby, 1979) embedded in its predecessor but retained the explicit water-routing algorithm (Wigmosta et al., 1994) for simulating surface and subsurface lateral flow as well as movement of solutes through space. The water routing algorithm in R-RHESSys can be switched on or off and thus provides users two ways (i.e., with vs. without water routing) to quantify C, water, and nutrient dynamics in terrestrial ecosystems. As in its predecessor, surface and subsurface lateral flow for stream-type patches are channelized in R-RHESSys.

Because specific algorithms for C, water, and nutrient dynamics are maintained mostly as in Tague and Band (2004), we briefly introduced calculation of subsurface and surface flow that was slightly modified for reference. In R-RHESSys, the saturated subsurface flow ($\text{SF}_{a \to b}$) (m day^{-1}) from patch a to b is calculated as follows:

$$\text{SF}_{a \to b} = \begin{cases} \delta \times \gamma \times \left(e^{-s/m} - e^{-s_{\max}/m}\right) & s \geq 0 \\ \delta \times \gamma \times \left(e^{-s/(3.5\text{m})} - e^{-s_{\max}/m}\right) & s < 0, \end{cases} \quad (1)$$

where s (m) is saturation deficit in patch a; m (dimensionless) is the decay rate of soil hydraulic conductivity with depth in patch a; $s_{\max}$ (m) is the water equivalent of soil depth; δ (dimensionless) is the empirical sensitivity parameter with a value of 1.2 when water routing is considered and a value of 0.16 when water routing is ignored. The values 1.2 and 0.16 are based on model calibrations (see below); and γ (m day^{-1}) is the percent of subsurface flow going from patch a to patch b. It is expressed as

$$\gamma = K_{\text{sat0}} \times \tan \beta_{a \to b} \times W_{a \to b}, \quad (2)$$

where K_{sat0} (m day^{-1}) is saturated hydraulic conductivity at the surface; β (degree) is the local slope from patch a to patch b; and W (dimensionless) is the flow width from patch a to patch b. The flow widths are assumed to be 0.5 times the grid size for cardinal directions and 0.354 times the grid size for diagonal directions (Quinn et al., 1991; Tague and Band, 2004).

The saturation overland flow (RF$_a$) for patch a is expressed as follows:

$$RF_a = \max(RS + U_{satS} - s, 0.0), \qquad (3)$$

where RS (m) is soil water storage in the root zone layer; and U_{satS} (m) is soil water storage in the unsaturated soil layer.

When water routing is considered in R-RHESSys, the saturated subsurface flow input from the upslope patch a (Eq. 1) is added to the downslope patch b and accounted for in patch b's local water budget. When routing is turned off, Eq. (1) is still used to calculate subsurface flow out of each patch. However, rather than being routed to downslope patches, the subsurface outflows from all patches are summed and assumed to flow out of the basin as the baseflow component of streamflow. The value of the sensitivity parameter δ in Eq. (1) for the non-routing case is reduced to reflect the change in function of this parameter from a lateral flow between patches adjustment to what is effectively a baseflow recession coefficient. The other difference between routing and non-routing is that with routing on, the surface flow generated by Eq. (3) is routed following the same topology as subsurface flow and is allowed to re-infiltrate along its flow path, whereas with no routing, the surface flow generated by Eq. (3) for all patches is summed and assumed to flow out of the basin as the runoff component of streamflow.

2.3 Meteorological data

Time series of daily maximum and minimum temperature (°C) as well as total precipitation (mm) are required to run R-RHESSys. Because there is no weather station located in the Biscuit watershed, our climate data for the period 1961–2008, a period having as long as possible available climate records and preselected for model spin-up simulation, were derived from 10 Cooperative Observer Program stations (COOP) (Fig. 1). Specifically, daily climate data for each day in each year for the watershed were estimated using the ordinary Kriging interpolation approach (Goovaerts, 1998). Before interpolation, daily records of temperatures that exceeded the long-term (1961–2008) mean of all available records from that station by four standard deviations or greater were manually removed on a case-by-case basis (e.g., Tang and Arnone III, 2013). In addition, local lapse rates of $-0.0085\,°C\,m^{-1}$ for daily maximum temperature, $-0.0054\,°C\,m^{-1}$ for daily minimum temperature, and $0.0014\,mm\,m^{-1}$ for daily precipitation were used to adjust temperature and orographic precipitation changes along the elevation gradient in the study sites. Figure S1 in the Supplement shows examples of interpolated daily maximum and minimum temperatures as well as precipitation for the Biscuit watershed in July 1994.

2.4 Land cover, soil and elevation data

The land cover data used to pre-define vegetation types for the Biscuit watershed were based on the National Land Cover Dataset 1992 (NLCD 1992; http://landcover.usgs.gov/usmap.php). The NLCD 1992 data were derived from Landsat Thematic Mapper satellite data at 30 m spatial resolution and classified land covers into 21 types for the United States (Vogelmann et al., 1998a, b). For the Biscuit watershed, only three types exist in NLCD 1992: evergreen, deciduous and mixed forests. Our soil texture data at 30 m spatial resolution were derived from the digital Soil Survey Geographic Database (http://soils.usda.gov/). We classified soil in the Biscuit watershed into four types: sandy loam, loamy skeleton, silt loam and rocky (Fig. S1d in the Supplement). Soil-texture-related parameters and their parameterization are in Table 1. The USGS National Elevation Dataset at 1 arcsec spatial resolution (about 30 m) was used in this study.

2.5 Modeling protocol, model simulation, calibration, and evaluation

Given that climate in the Biscuit watershed is relatively humid and precipitation has no distinct dry and wet cycles, we performed four simulations under two climate forcing scenarios: one wet and one dry. Under the wet scenario, time series of daily climate data for the period 1961–2008 were directly used without modification. Under the dry scenario, we set time series of daily precipitations for days in May, June, July and August in 1995 at zero while keeping others identical to those under the wet scenario. For each of the two scenarios, the two contrasting simulations (i.e., with vs. without water routing) were performed, respectively.

Our initial simulations under the wet scenario suggested that soil water table depth, leaf area index (LAI) and forest productivity tended to reach the equilibrium state after 50 simulation-years. In contrast, soil C took more than 200 simulation-years to reach the equilibrium state (Fig. S2 in the Supplement). In order to have vegetation and soil C reach equilibrium state with long-term local climate, we spun up R-RHESSys for 240 years repeatedly using 48-year (1961–2008) daily-step meteorological data. After spin-up simulations, we continued to run R-RHESSys for an additional 48 years using data from 1961 to 2008. This modeling protocol applied to all four simulations under both wet and dry forcing scenarios.

Based on results under the wet scenario, we calibrated R-RHESSys for the period 1992–1993 and evaluated it for the period 1994–1995. The period 1992–1995 was selected because observed climate records in this period from 10 COOP stations were more consistent than during other periods. This

Table 1. Major soil parameters and their parameterizations used in this study.

Variables	Unit	Soil texture			
		Sandy loam	Silt loam	Loamy skeleton	Rocky
$K^*_{sat\_0}$	m day^{-1}	89.05	48.62	48.36	109.56
m^*	DIM	0.09	0.12	0.13	0.09
Porosity	%	0.435	0.410	0.451	0.485
Porosity decay	DIM	4000	4000	4000	4000
Pore size index (PSI)	DIM (0–1)	0.204	0.189	0.186	0.228
PSI air entry	%	0.218	0.386	0.478	0.480
Soil depth	m	5.0	5.2	4.8	5.0
Active zone depth	m	10	10.0	10.0	10.0
Maximum energy capacity	°C	−10.	−10.	−10.	−10.
Albedo	DIM	0.258	0.253	0.320	0.200
Sand	%	0.70	0.20	0.80	0.75
Clay	%	0.10	0.15	0.02	0.05
Silt	%	0.20	0.65	0.18	0.20

* K_{sat_0} is saturated hydraulic conductivity at the surface; m is the decay rate of hydraulic conductivity with depth. K_{sat_0} and m were manually calibrated against observed streamflow and derived baseflow at the USGS gauge station.

can minimize the effects of the quality of atmospheric forcing data on simulated water and C dynamics. Correspondingly, model calibration and evaluation for each of the two pre-specified periods were performed for the two contrasting simulations under the wet scenario, respectively.

To investigate how consideration of water routing may affect simulated C and water dynamics, monthly average daily values of major hydro-ecological variables in July of 1994 from the two contrasting simulations under the wet scenario were compared. The July of 1994 was selected because temperature in July is generally higher than in other months and thus the effects of consideration of water routing on simulated water and C dynamics as well as vegetation growth were assumed to be more detectable. To test if effects of consideration of water routing on simulated C and water dynamics are more remarkable in dry conditions, we compared the differences in simulated monthly values of major hydro-ecological variables in 1995 between the wet and dry scenarios.

3   Results

3.1   Calibration and evaluation of simulated streamflow and baseflow

Figure 2 shows the time series of simulated daily streamflow and baseflow for the Biscuit Brook in the watershed for the calibration period 1992–1993 and the evaluation period 1994–1995. For the calibration period, the calculated Nash–Sutcliffe coefficients (NS; Nash and Sutcliffe, 1970) is 0.58 for streamflow (Fig. 2a) and 0.63 for baseflow (Fig. 2b) under the simulation that considered water routing. In contrast, the calculated NS is 0.61 for streamflow (Fig. 2c) and 0.74 for baseflow (Fig. 2d) for the simulation that neglected water routing. For the evaluation period, the calculated NS was more than 0.57 for both streamflow and baseflow regardless of whether or not water routing was considered (Fig. 2a–d). In addition, the simulated average daily streamflow for the evaluation period 1994–1995 approximated each other between the two simulations (2.54 vs. 2.50 mm day^{-1}). The difference in average daily streamflow between model simulations and observation was less than 1.25 % under both simulations. These statistics (Table S1 in the Supplement) suggested that R-RHESSys was able to accurately simulate daily streamflow and baseflow regardless of whether water routing was considered.

3.2   Comparison of simulated soil water table depth and saturation deficit

When water routing was considered, the simulated depth to the soil water table ranged from 0.15 to 2.92 m among cells and averaged 1.20 m for the entire watershed. In contrast, when water routing was ignored, the simulated depth ranged from 0.02 to 1.20 m among cells, and averaged 0.72 m for the entire watershed. In other words, the simulated water table depth was spatially more variable when water routing was simulated as indicated by the calculated standard deviations for soil water table depth among cells (Table 2 and Fig. 3a vs. b). A similar situation applied to the simulated saturation deficit, which had a wider range from 0.08 to 1.42 m under simulation with water routing but a narrower range from 0.01 to 0.54 m under simulation without water routing (Table 2). The simulated saturation deficit also was spatially more variable under simulation with water routing than that

Table 2. Comparison of simulated hydrological and ecological variables between the two contrasting simulations: with vs. without water routing.

Variables	Water routing	Minimum	Maximum	Mean	STD
Water table depth (m)	Yes	0.15	2.92	1.20	0.40
	No	0.02	1.20	0.72	0.19
Saturation deficit (m)	Yes	0.08	1.42	0.54	0.17
	No	0.01	0.54	0.33	0.08
Evaporation (mm)	Yes	0.22	3.11	0.87	0.42
	No	0.52	1.05	0.74	0.05
Plant transpiration (mm)	Yes	0.00	3.86	1.41	0.49
	No	0.92	1.95	1.35	0.13
Evapotranspiration (mm)	Yes	0.28	6.65	2.27	0.79
	No	1.44	2.99	2.09	0.18
NPP ($gC\,m^{-2}\,day^{-1}$)	Yes	0.01	5.79	3.33	0.84
	No	2.50	5.79	3.60	0.17
RA ($gC\,m^{-2}\,day^{-1}$)	Yes	0.00	0.97	0.58	0.18
	No	0.35	0.97	0.63	0.08
RH ($gC\,m^{-2}\,day^{-1}$)	Yes	0.01	1.3	0.75	0.20
	No	0.44	1.3	0.84	0.08

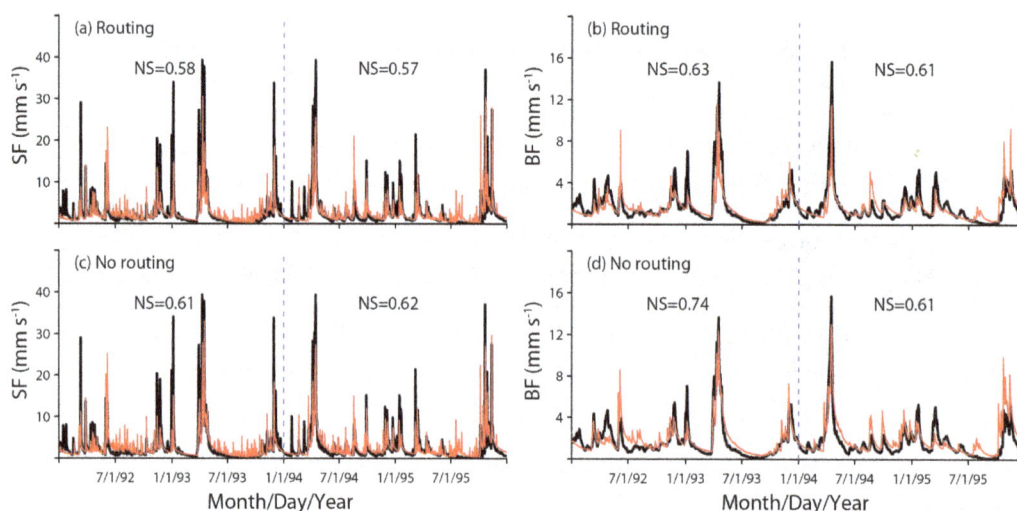

Fig. 2. Calibration (for the period 1 January 1992–31 December 1993) and evaluation (for the period 1 January 1994–31 December 1995) of R-RHESSys simulated daily streamflow (SF) and baseflow (BF) (solid red line) against observed/derived data (solid black line). Simulations in (**a**) and (**b**) considered water routing while simulations in (**c**) and (**d**) ignored water routing. NS is short for the Nash–Sutcliff coefficient. The blue-dashed line represents 1 January 1994.

without water routing (Fig. 3d vs. e), as indicated by the standard deviations for saturation deficit among cells (Table 2). Further comparison suggested that water table depth and saturation deficit were about 0.5 m (for water table) and 0.2 m (for saturation deficit) greater in the hills or ridges of the watershed when water routing was considered. In the valleys or flat areas, however, there are regions where the simulated water table depth and saturation deficit were smaller when water routing was considered compared to the simulation ignoring water routing (Fig. 3c and f). Spatially, deeper water table depth and higher saturation deficit were simulated to occur

mostly at upslope areas (Fig. 3a and d) when water routing was considered. This situation, however, did not always apply to simulations ignoring water routing, under which water table depth and saturation deficit were found to be greater at steeper slopes (Fig. 3b and e).

3.3 Comparison of simulated evaporation, transpiration, and evapotranspiration

Compared to the simulation ignoring water routing, simulated monthly average daily evaporation, transpiration, and

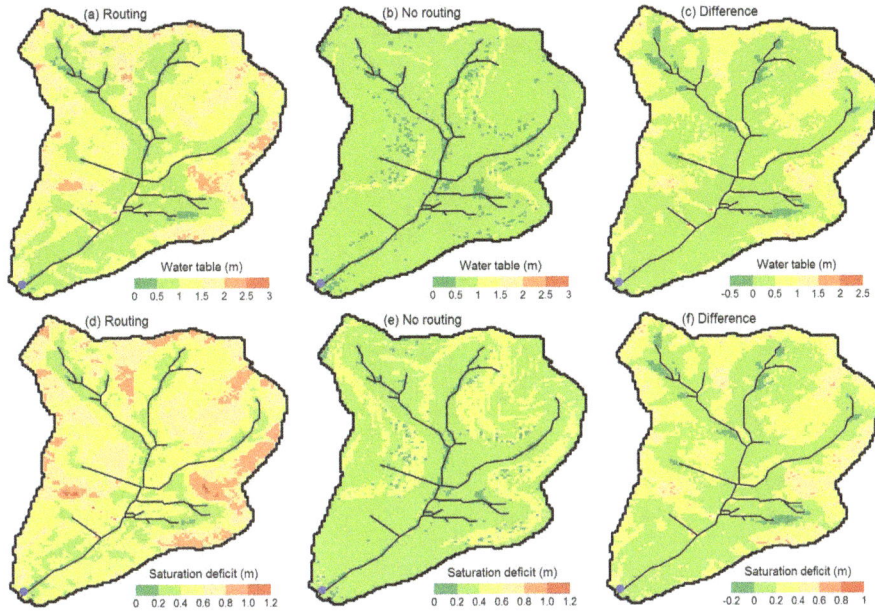

Fig. 3. Comparison of simulated monthly average daily soil water table depth and saturation deficit in July 1994 between the two contrasting simulations: (**a**) and (**d**) considered water routing while (**b**) and (**e**) ignored water routing. (**c**) and (**f**) show differences in simulated soil water table depth and saturation deficit between the two contrasting simulations.

actual evapotranspiration (ET) with water routing had a wider range among cells. For example, monthly average daily evaporation for July 1994 was simulated to vary from 0.22 to 3.11 mm day^{-1} among cells under simulation with water routing. In contrast, evaporation had a narrower range from 0.52 to 1.05 mm day^{-1} under the simulation without water routing (Table 2). When averaged for the entire watershed, monthly average daily evaporation, plant transpiration, and ET were 18 % (0.87 vs. 0.74 mm), 4 % (1.41 vs. 1.35 mm) and 9 % (2.27 vs. 2.09 mm) greater, respectively, under simulation considering water routing than that ignoring water routing (Table 2). In addition, regardless of the actual magnitudes of simulated water fluxes, the spatial patterns of evaporation, transpiration, and ET were modeled to be more variable under simulation considering water routing than that ignoring water routing, largely because extreme high and low values of evaporation, transpiration and ET were simulated to occur under the simulation with water routing (Fig. S3 in the Supplement and Fig. 4). Spatially, the effects of considering water routing on simulated evaporation, transpiration, and ET can be either positive or negative compared to the simulation neglecting water routing (Fig. S3 in the Supplement).

3.4 Comparison of simulated forest net primary productivity (NPP)

At the individual cell level, simulated monthly average daily NPP in July, 1994 (when ignoring water routing) ranged from 2.50 to 5.79 gC m^{-2}, narrower than results from the

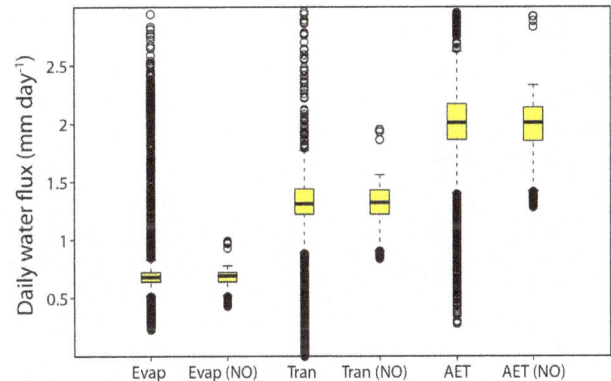

Fig. 4. Comparison of simulated monthly average daily evaporation (evap), transpiration (Tran), and actual evapotranspiration (AET) in July, 1994 between the two simulations with and without (indicated by "NO") consideration of water routing.

simulation considering water routing, which ranged from 0.10 to 5.79 gC m^{-2} among cells. In addition, although the pattern of simulated NPP was extremely similar in most areas of the watershed between the two simulations (Fig. 5a and b), simulated monthly average daily NPP among cells was spatially more variable when water routing was considered, as suggested by the calculated standard deviations for NPP among cells (Table 2). When averaged for the entire watershed, the simulated monthly average daily NPP was 8 % (3.33 vs. 3.60 gC m^{-2}) lower under simulation considering water routing than that ignoring water routing (Table 2). Nevertheless, the simulated maximum NPP between

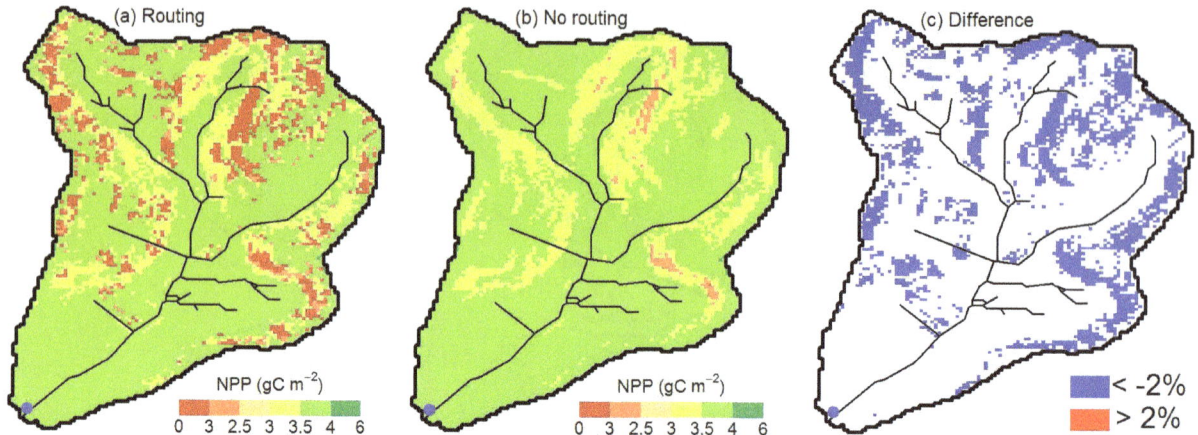

Fig. 5. Comparison of simulated monthly average daily net primary productivity (NPP) in July 1994 between the two simulations: (**a**) considering water routing and (**b**) ignoring water routing. (**c**) Shows percentage difference between (**a**) and (**b**) divided by the result from simulation (**a**) considering water routing. The white areas show no significant differences.

the two simulations was identical ($5.79\,\mathrm{gC\,m^{-2}}$), although there were regions where simulated NPP was distinctly lower ($< 3.0\,\mathrm{gC\,m^{-2}}$) under the simulation considering water routing than that ignoring water routing ($> 3.0\,\mathrm{gC\,m^{-2}}$). Overall, the simulation that neglected water routing had a tendency to overestimate forest NPP in ridges of the watershed or areas with steeper slopes (Fig. 5c).

3.5 Comparison of simulated soil autotrophic and heterotrophic respiration

Simulated monthly averaged daily soil autotrophic respiration (RA) in July 1994 ranged from 0.0 to $0.97\,\mathrm{gC\,m^{-2}}$ under the simulation with water routing. This range was slightly broader than that from the simulation without water routing, which ranged from 0.35 to $0.97\,\mathrm{gC\,m^{-2}}$ (Table 2). When averaged for the entire watershed, monthly average daily soil RA was 8 % (0.58 vs. $0.63\,\mathrm{gC\,m^{-2}}$, Table 2) lower under simulation with water routing than that without water routing. In addition, although the spatial pattern of simulated soil RA across the watershed was extremely similar in most areas between the two simulations (Fig. 6a and b), there were patches where simulated soil RA was much lower when water routing was considered (Fig. 6c). Overall, neglect of water routing has the potential to cause R-RHESSys to overestimate soil RA, while such overestimates mainly occur in areas of steeper slopes or near the ridges of the watershed (Fig. 6c). Similarly, simulated soil heterotrophic respiration (RH) had a wider range from 0.01 to $1.3\,\mathrm{gC\,m^{-2}}$ under simulation with water routing and a narrower range from 0.44 to $1.3\,\mathrm{gC\,m^{-2}}$ under the simulation without water routing (Table 2). The spatial patterns of simulated soil RH were more variable under simulation with water routing than that without water routing (Fig. 6d and e). Besides, when averaged for the entire watershed, monthly average daily soil RH was

11 % (0.75 vs. 0.84) lower under the simulation considering water routing than that ignoring water routing. Differing from soil RA, the effects of water routing on soil RH can be either positive or negative when compared to the simulation without water routing (Fig. 6f). The difference in simulated soil RH between the two simulations ranged from -0.8 to $0.12\,\mathrm{gC\,m^{-2}}$ across cells.

3.6 Comparison of the differences (with vs. without routing) in monthly values of hydro-ecological variables between the wet and dry scenarios

Figure 7 shows comparisons of the simulated differences (with vs. without water routing) in monthly values of C and water dynamics in 1995 between the wet and dry scenarios. When averaged for the entire watershed, the magnitude of the differences in monthly average water table depth and saturation deficit was not distinct for months before July between the two scenarios, while the differences diverged for months after July: greater under the wet and smaller under the dry scenario (Fig. 7a and b). For water fluxes, the absolute magnitude of the differences in monthly transpiration and AET was greater under the dry scenario for May, June, July, August, and September, and bottomed in August (Fig. 7d and e). In other months, the magnitude of the differences in monthly transpiration and ET approximated each other between the two scenarios, especially for transpiration (Fig. 7e). However, this pattern of differences in monthly transpiration and AET did not apply to evaporation (Fig. 7c). For C fluxes, the absolute magnitude of the difference in monthly average NPP, soil RA, and RH was greater under the dry scenario for May, June, July, August, and September, and bottomed in August (Fig. 7f–h). In other months, the simulated differences in the three C fluxes approximated each other between the two scenarios. These results indicated that consideration of water routing has greater effects on simulated water and C dynamics in dry conditions.

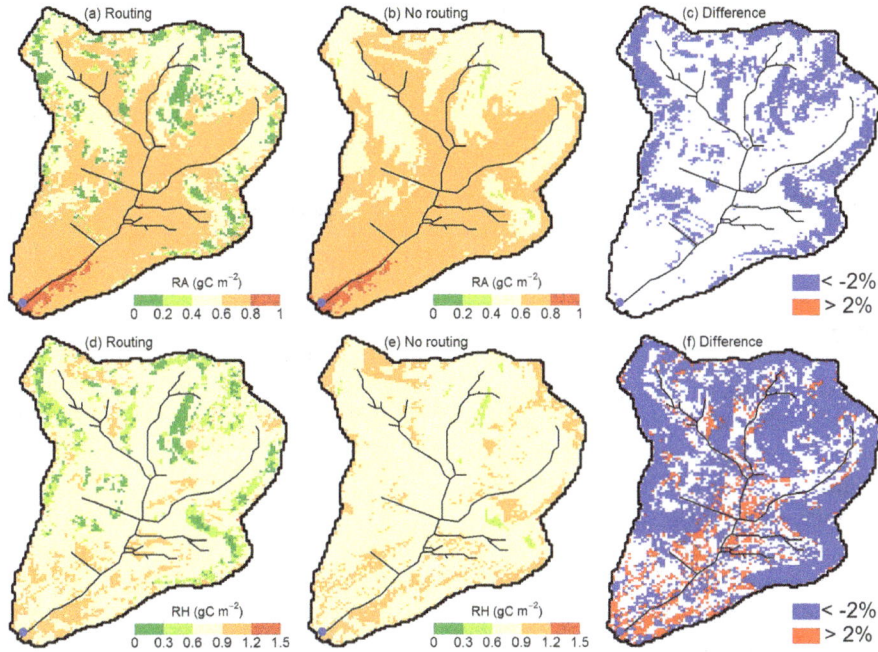

Fig. 6. Comparison of simulated monthly average daily soil autotrophic (RA) and heterotrophic respiration (RH) in July 1994 between the two simulations: (**a**) and (**d**) considering water routing while (**b**) and (**e**) ignoring water routing. (**c**) and (**f**) show percentage differences between the two simulations divided by results from the simulation considering water routing. The white areas show no significant differences.

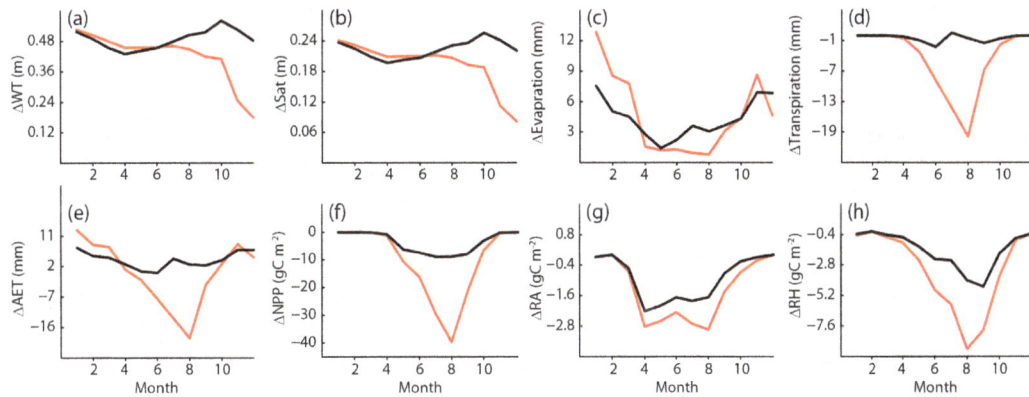

Fig. 7. Comparison of the simulated differences (with vs. without water routing) in monthly values of major hydro-ecological variables between the wet (solid black line) and dry (solid red line) scenarios.

4 Discussion

4.1 Performance and accuracy of R-RHESSys

Our model evaluation against observed streamflow and derived baseflow from the USGS gauge station indicated that R-RHESSys was able to accurately simulate river flow at watershed scales, largely because all algorithms for water, C and nutrient dynamics as well as model parameters are maintained as in RHESSys, which itself has been applied and evaluated in a number of studies (e.g., Christensen et al., 2008; Hwang et al., 2012; Tague and Band, 2001). In addition, the simulated ecological variables – such as LAI and forest NPP – all fell within the ranges of corresponding field observations. For example, modeled LAI during the growing season (May to September) averaged $3.1\,\mathrm{m^2\,m^{-2}}$ for the entire watershed and ranged from 1.2 to $3.9\,\mathrm{m^{-2}\,m^{-2}}$ across grid cells, agreeing well with observed and modeled values ranging from 2.90 to $4.5\,\mathrm{m^2\,m^{-2}}$ in mixed oak–hickory forests and northern hardwoods (Scurlock et al., 2001; Tang and Beckage, 2010), dominant forest types in the study watershed. Our modeled annual forest NPP averaged $474\,\mathrm{gC\,m^{-2}\,yr^{-1}}$, falling within the range of 391 to $574\,\mathrm{gC\,m^{-2}\,yr^{-1}}$ of field observations in oak-hickories (e.g., Pan et al., 2006; Tang et al., 2010). Nevertheless, we acknowledge that the lack of spatially distributed field measurements – such as observed soil moisture, water table depth, and forest NPP – hinder us from further evaluating the patterns of simulated major ecological and hydrological variables across the watershed. Such limitations in the model's evaluation are encountered commonly in many other distributed-model-based studies (Brooks et al., 2007) and need improvement in the future.

4.2 Effects of water routing on soil water table depth and saturation deficit

Lateral water flow and associated water redistribution across the landscape considerably influence hydrologic response in terrestrial ecosystems, including movement and storage of water in the soil (Guntner and Bronstert, 2004; Thompson and Moore, 1996). Some studies (e.g., Kim and Eltahir, 2004) indicated that topography drives lateral transport of water downslope, and water converges into concave areas or valleys through surface or subsurface runoff. As a result, water table depth tends to be significantly shallower in valleys compared to hills. However, this contrasting pattern did not occur in simulations that ignored water routing, in which the simulated water table depth and saturation deficit approximated each other between valleys and hills/ridges of the watershed (Fig. 3b and e). In other words, simulated water table depth and saturation deficit with water routing captured better our preconception of their spatial patterns across the watershed. A similar study in a humid watershed (Hotta et al., 2010) indicated that lateral flow and local infiltration descending from hillslopes often causes lower elevation sites to have a higher water table level and higher elevation sites to have a lower water table level.

A similar model-based comparison study additionally supported our findings. Sonnentag et al. (2008) compared simulated water table depth between simulations with and without considering lateral water flow in a peatland. They found that the magnitude of simulated water table depth without water routing was considerably underestimated because lateral subsurface flow moves water toward the margins of the peat body. The neglect of lateral flow resulted in the simulated water table at or very close to the ground surface, which explains why the simulated water table depth was much greater under simulation ignoring water routing (Table 2). Furthermore, Moore and Thompson (1996) found that the combination of slope curvature, microtopography, and resulting water movement produce significant variability in water table depth across the landscape. This explains why the calculated standard deviation of water table depth among cells doubled (0.40) under simulation considering water routing compared to that (0.19) ignoring water routing (Table 2).

Similar to water table depth, saturation deficit under simulation with water routing showed a distinct pattern in the watershed: higher in the valleys and lower in the hills or ridges of the watershed, which agreed better with findings from

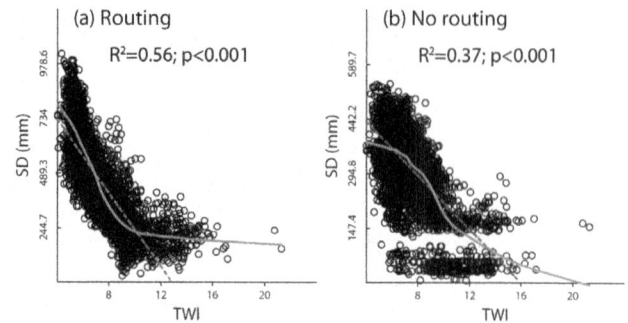

Fig. 8. Comparison of the relationships of simulated saturation deficit (SD) to topographic wetness index (TWI) across the watershed between the two simulations: (**a**) considering water routing and (**b**) ignoring water routing.

previous studies. Hopp et al. (2009) found that relatively high saturation in the soil profile occurs in the swale, and drier zone often occurs upslope and on the side ridges of hillslopes when water routing and topography were both considered in the model simulation. Crave and Gascuel-Odoux (1997) indicated that the steeper upslope parts of a watershed will be drained laterally more rapidly than the gentler downslope parts, resulting in drier slopes at the catchment scale. These patterns were captured by simulation with water routing (Fig. 3d) while not always by simulation without water routing (Fig. 3e). In addition, most previous studies indicated that the upslope contributing area, as incorporated into the TOPMODEL (Beven and Kirkby, 1979), is probably the major topographic influence on soil moisture distribution (e.g., Hotta et al., 2010; Thompson and Moore, 1996). This relationship also was captured better by simulation considering water routing as suggested by the strength of the linear relationship of simulated saturation deficit to calculated topographic wetness index (Fig. 8a vs. b) between the two simulations.

4.3 Effects of water routing on water fluxes from land to the atmosphere

Slope, aspect and surrounding topography control incident direct solar radiation, and lower-elevation regions in mountainous watersheds have more incoming longwave radiation from the surrounding landscapes plus temperature decreases as elevation increases. The highest ET values often occur in valleys, and the lowest ET in north-facing, high-elevation areas (Bertoldi et al., 2006; Christensen et al., 2008), which explains why the modeled spatial patterns of evaporation and transpiration in the watershed were generally higher in low elevations and valleys and lower in high elevations under the two contrasting simulations (Fig. S3 in the Supplement). Water routing is a major determinant of soil water table and moisture distribution, however, both of which play important roles in modulating water fluxes from land to the atmosphere.

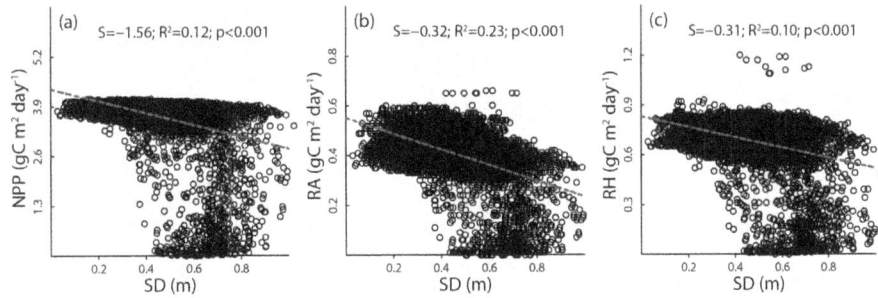

Fig. 9. The relationships of saturation deficit (SD) with (**a**) net primary productivity (NPP), (**b**) soil autotrophic respiration (RA), and (**c**) soil heterotrophic respiration (RH). Data shown here are based on the simulation considering water routing.

For example, Salvucci and Entekhabi (1995) indicated that a deeper water table typically indicates drier areas where evaporation is often suppressed. This explains why there are areas where evaporation under simulation with water routing was lower than those without water routing (Fig. S3c in the Supplement).

In addition, changes in vegetation growth resulting from moisture alteration also can affect water fluxes from land to the atmosphere due to changes in canopy leaf area. Comparing the two simulations, for cells where simulated NPP decreased (less than −2 %, Fig. 5c), 60 % experienced an increase in evaporation while 48 % experienced a decrease in transpiration due to decrease in canopy leaf area. This explains why there are areas where simulated evaporation is higher while transpiration is lower under simulation with water routing than that without water routing (Fig. S3c and f in the Supplement). At the individual cell level, because temperature, soil moisture and vegetation dynamics interact to jointly control evaporation and transpiration, differences in simulated evaporation, and transpiration can be either positive or negative (Fig. S3 in the Supplement). When averaged for the entire watershed, because evaporation showed significant increase by 18 % under simulation with water routing, the resultant AET also showed an increase by 9 % under simulation with water routing compared to that without water routing. In addition, because forest productivity is modeled to be similar in 80 % of areas between the two simulations and because transpiration accounts for two-thirds of total ET plus water is not limited, simulated transpiration and ET were extremely similar in 70 % of areas in the watershed between the two contrasting simulations, although significant differences occurred in some areas (Fig. S3f and i in the Supplement).

4.4 Effects of water routing on vegetation productivity

Changes in soil moisture condition affect canopy photosynthesis and forest productivity (Band et al., 1993). Hwang et al. (2012) found that soil moisture content has profound effects on plant growth in forested watersheds. Svoray and Karnieli (2011) indicated that plant productivity is strongly correlated with water redistribution processes. Plants in the lower physiographic units (e.g., footslope, channel) should respond well to improved water and soil conditions and, therefore, should be more productive. In contrast, the interfluve, shoulder, and backslope areas often had lower vegetative greenness values because of poor water availability. In this study, the effects of differences in simulated soil moisture condition on forest productivity were not very noticeable (defined as −2 % < NPP difference < 2 %) in 80 % of areas in the study watershed between the two contrasting simulations (Fig. 5). This is largely because incoming solar radiation and temperature are major determinants of forest productivity, and these radiative forcings were identical between the two simulations. Nevertheless, because changes in soil moisture can affect forest productivity and because the saturation deficit was simulated to be greater under the simulation with water routing, simulated forest NPP was significantly lower in steeper slope areas of the watershed when water routing was considered. In these areas where differences in NPP were less than −2 %, average soil saturation deficit (722 mm) was 45 % higher than that (498 mm) in areas where differences in NPP were not noticeable (defined as −2 % < NPP difference < 2 %) (Fig. 5). In fact, forest NPP was significantly and negatively correlated with saturation deficit in our simulation (Fig. 9a) because the deterioration of soil moisture condition can limit vegetation growth (e.g., Urgeghe et al., 2010).

4.5 Effects of water routing on soil respiration

Local topography can generate considerable spatial variability in soil temperature, incoming solar radiation, and soil water content (Running et al., 1987; Kang et al., 2004). Although each of these factors differentially affects soil respiration, soil temperature plays a major role in soil respiration. Kang et al. (2004) found that about 75 % of seasonal variation in soil respiration in such mesic ecosystems can be explained by variation in soil temperature. Because soil temperature is simulated to be the same between the two simulations, this greatly contributed to the similarity (defined as −2 % < RA difference < 2 %) of the spatial pattern

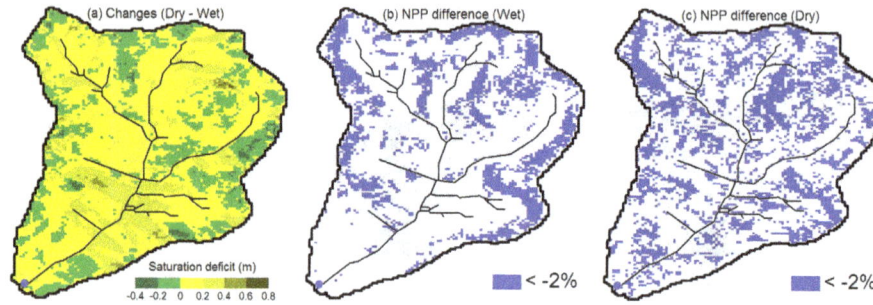

Fig. 10. **(a)** Deteriortation of soil moisture condition under the dry scenario compared to the wet scenario resulted in NPP decreases occurring in more areas of the watershed **(c)** under the dry scenario than that **(b)** under the wet scenario. The white areas show no significant differences.

of simulated soil RA in 79.9 % of areas in the watershed (Fig. 6). Indeed, the calculation of root RA in R-RHESS is mainly treated as a function of soil temperature, following Ryan (1991). Because saturation deficit was higher when water routing was considered, and because soil water deficit limits root production resulting from reduced NPP, the consequent soil RA is smaller under the simulation considering water routing (Fig. 6a–c). In fact, for cells where simulated NPP decreased by less than −2 % between the two contrasting simulations (Fig. 5c), 99.9 % experienced a decrease in RA, ranging from −1.4 to −100 % (Fig. 6c). Linear regression also suggested that soil RA is negatively and significantly correlated with saturation deficit (Fig. 9b).

Although soil temperature plays a dominant role in regulating soil RH, changes in soil water content due to lateral flow and connectivity affect litter production and soil microbial activity, which in turn affect soil RH. Riveros-Iregui et al. (2011) indicated that growing-season soil CO_2 efflux is known to vary laterally by as much as sevenfold within small subalpine watersheds in the northern Rocky Mountains, and the variability was strongly related to the landscape-imposed redistribution of soil water. Because soil RH in R-RHESSys is treated as a function of soil moisture following Parton et al. (1996), this explains that the simulated soil RH is spatially more variable (higher standard deviation) among cells when water routing is considered (Table 2). In our simulation, for cells where forest NPP decreased by less than −2 % (Fig. 5c), 97 % experienced a decrease in RH due to reduction of litter production (Fig. 6f). In a semiarid subalpine watershed, Riveros-Iregui and McGlynn (2009) observed that the highest soil CO_2 efflux rates often occur in areas with persistently high soil moisture, whereas lower soil CO_2 efflux rates are on forested uplands in subalpine watersheds. Such patterns were captured better under simulation considering water routing (Fig. 6d) than that ignoring water routing (Fig. 6e), partially because soil RH was generally simulated to be low in areas of steeper slopes (Figs. 1 and 6) and because forest NPP and litter production were low in these areas. Compared to soil RA, differences in simulated soil RH between the two contrasting simulations can be either negative or positive due to combined effects of soil temperature, moisture, and litter inputs

on RH. Overall, soil RH was negatively correlated to saturation deficit in our simulation, suggesting that neglect of water routing has the potential to cause the model to overestimate soil RH (Fig. 9c).

4.6 Effects of water routing on C and water dynamics under dry conditions

Consideration of water routing in model simulations had greater effects on simulated C and water dynamics under the dry scenario than under the wet scenario, largely because of deterioration of soil moisture condition under the dry scenario (Fig. 10a). For example, when averaged for the entire watershed, soil saturation deficit increased by 14 % under the dry scenario (0.72 m) compared to the wet scenario (0.63 m). The deterioration of soil moisture condition caused the number of cells where the difference in simulated monthly NPP was greater than 2 % to increase by 138 % under the dry scenario (6031 cells) (Fig. 10c) compared to the wet scenario (2531 cells, Fig. 10b). This explained why the absolute magnitudes of the simulated differences in monthly values of C and water fluxes were greater for those months, in which time series of daily precipitation were set to zero under the dry scenario. Our findings of the greater effects of consideration of water routing on simulated C and water dynamics under the dry scenario was consistent with Band (1993), who found that spatial variations in available soil water can have significant effects on areal averaged C and water fluxes rates, particularly under dry conditions.

5 Conclusions

Based on R-RHESSys and by keeping all model parameters and their parameterizations identical, this model-based comparison study indicated the following:

1. R-RHESSys is able to correctly simulate streamflow and baseflow for Biscuit Brook regardless of whether water routing is considered in the model simulation or not. When water routing is considered, however, R-RHESSys captures better our preconception of the

spatial patterns of water table depth and saturation deficit. In contrast, when water routing is neglected, the simulation has a tendency to underestimate water table depth and saturation deficit. Simulated patterns of water table depth and saturation deficit differ from our preconception of the two quantities across the landscape.

2. Differences in simulated water table depth and saturation deficit between simulations with and without water routing affect subsequent water fluxes from land to the atmosphere. At the individual cell level, simulated evaporation, transpiration and ET were spatially more heterogeneous across the landscape when water routing was considered. Although differences in simulated evaporation, plant transpiration, and ET are not significant (absolute difference < 2 %) in most areas of the watershed, when averaged for the entire watershed, evaporation, transpiration, and ET were simulated to be 4 to 18 % greater under simulation considering water routing than that ignoring water routing.

3. Forest productivity was generally simulated to be smaller and spatially more variable under simulation with water routing due to higher and more variable saturation deficit. Lower forest productivity and root production caused simulated soil RA to be lower when water routing was considered. In contrast, simulated soil RH with water routing can be either greater or smaller than without water routing due to the combined effects of soil moisture, temperature and litter inputs. When averaged for the entire watershed, forest productivity and soil respiration were modeled to be 8 to 11 % less under simulation considering water routing than that ignoring water routing.

Overall, this study indicated that lateral water flow exerts strong control on the spatial pattern and variability of water table depth and saturation deficit (e.g., Band et al., 1993), and such effects are more apparent in dry conditions (e.g., Band, 1993). When averaged for the entire watershed, simulated water fluxes from land to the atmosphere were higher, while forest productivity and soil respiration were less under simulation with water routing than those without water routing. Results of this study further demonstrated that the spatial pattern of soil moisture is fundamental to spatially distributed modeling of eco-hydrological processes (e.g., Band, 1993; Chamran et al., 2002; Hebrard et al., 2006) and suggested that ecosystem and C cycle models need to explicitly represent water routing because simulation with water routing better captures the patterns of water table depth and saturation deficit across landscapes.

Acknowledgements. The authors thank the New York City Department of Environmental Protection that provided funding for the research that led to this manuscript. We greatly appreciate Genevieve Ali and other two anonymous reviewers for their constructive comments on earlier version of this manuscript. We sincerely thank Lawrence Band, Elliot Schneiderman, Don Pierson, and Mark Zion for their valuable comments that greatly helped improve this manuscript. This study also benefited from the NSF EPSCoR grant (NSF 0814372) for Nevada.

Edited by: N. Basu

References

Asbjornsen, H., Goldsmith, G. R., Alvarado-Barrientos, M. S., Rebel, K., Van Osch, F. P., Rietkerk, M., Chen, J., Gotsch, S., Tobon, C., Geissert, D. R., Gomez-Tagle, A., Vache, K., and Dawson, T. E.: Ecohydrological advances and applications in plant-water relations research: a review, J. Plant. Ecol., 4, 3–22, 2011.

Band, L. E.: Effects of land surface representation on forest water and carbon budgets, J. Hydrol., 150, 749–772, 1993.

Band, L. E., Patterson, P., Nemani, R., and Running, S. W.: Forest ecosystem processes at the watershed scale: incorporating hillslope hydrology, Agr. Forest Meteorol., 63, 93–126, 1993.

Bertoldi, G., Rigon, R., and Over, T. H.: Impact of watershed geomorphic characteristics on the energy and water budgets, J. Hydrometeorol., 7, 389–403, 2006.

Beven, K. and Kirkby, M.: A physically-based variable contributing area model of basin hydrology, Hydrol. Sci. Bull., 24, 43–69, 1979.

Brooks, E. S., Boll, J., and McDanil, P. A.: Distributed and integrated response of a geographic information system-based hydrologic model in the eastern Palouse region, Idaho, Hydrol. Process., 21, 110–122, 2007.

Caylor, K. K., Manfreda, S., and Rodriguez-Iturbe, I.: On the coupled geomorphological and ecohydrological organization of river basins, Adv. Water Resour., 28, 69–86, 2005.

Chamran, F., Gessler, P. E., and Chadwick, O. A.: Spatially explicit treatment of soil-water dynamics along a semiarid catena, Soil Sci. Soc. Am. J., 66, 1571–1583, 2002.

Chen, J. M., Chen, X., Ju, W., and Geng, X.: Distributed hydrological model for mapping evapotranspiration using remote sensing inputs, J. Hydrol., 305, 15–39, 2005.

Christensen, L., Tague, C. L., and Baron, J. S.: Spatial patterns of simulated transpiration response to climate variability in a snow dominated mountain ecosystem, Hydrol. Process., 22, 3576–3588, 2008.

Crave, A. and Gascuel-Odoux, C.: The influence of topography on time and space distribution of soil surface water content, Hydrol. Process., 11, 203–210, 1997.

Doten, C. O., Bowling, L. C., Lanini, J. S., Maurer, E. P., and Lettenmaier, D. P.: A spatially distributed model for the dynamic prediction of sediment erosion and transport in mountainous forested watersheds, Water Resour. Res., 42, W04417, doi:10.1029/2004WR003829, 2006.

Gérard-Marchant, P., Hively, W. D., and Steenhuis, T. S.: Distributed hydrological modelling of total dissolved phosphorus transport in an agricultural landscape, part I: distributed runoff generation, Hydrol. Earth Syst. Sci., 10, 245–261, doi:10.5194/hess-10-245-2006, 2006.

Goovaerts, P.: Ordinary cokriging revisited, Math. Geol., 30, 21–42, 1998.

Guntner, A. and Bronstert, A.: Representation of landscape variability and lateral redistribution processes for large-scale hydrological modelling in semi-arid areas, J. Hydrol., 297, 136–161, 2004.

Hebrard, O., Voltz, M., Andrieux, P., and Moussa, R.: Spatiotemporal distribution of soil surface moisture in a heterogeneously farmed Mediterranean catchment, J. Hydrol., 329, 110–121, 2006.

Hopp, L., Harman, C., Desilets, S. L. E., Graham, C. B., McDonnell, J. J., and Troch, P. A.: Hillslope hydrology under glass: confronting fundamental questions of soil-water-biota co-evolution at Biosphere 2, Hydrol. Earth Syst. Sci., 13, 2105–2118, doi:10.5194/hess-13-2105-2009, 2009.

Hotta, N., Tanaka, N., Sawano, S., Kuraji, K., Shiraki, K., and Suzuki, M.: Changes in groundwater level dynamics after low-impact forest harvesting in steep, small watersheds, J. Hydrol., 385, 120–131, 2010.

Hwang, T., Band, L. E., Vose, J. M., and Tague, C.:. Ecosystem processes at the watershed scale: Hydrologic vegetation gradient as an indicator for lateral hydrologic connectivity of headwater catchments, Water Resour. Res., 48, W06514, doi:10.1029/2011WR011301, 2012.

Ivanov, V. Y., Bras, R. L., and Vivoni, E. R.: Vegetation-hydrology dynamics in complex terrain of semiarid areas: 2. Energy-water controls of vegetation spatiotemporal dynamics and topographic niches of favorability, Water Resour. Res., 44, W03430, doi:10.1029/2006WR005595, 2008.

Ju, W., Chen, J. M., Black, A. B., Barr, A. G., McCaughey, H., and Roulet, N. T.: Hydrological effects on carbon cycles of Canada's forests and wetlands, Tellus B, 58, 16–30, 2006.

Kang, S., Lee, D., and Kimball, J. S.: The effects of spatial aggregation of complex topography on hydro-ecological process simulations within a rugged forest landscape: development and application of a satellite-based topoclimatic model, Can. J. For. Res., 34, 519–530, 2004.

Kim, Y. and Eltahir, E. A. B.: Role of topography in facilitating coexistence of trees and grasses within savannas, Water Resour. Res., 40, W07505, doi:10.1029/2003WR002578, 2004.

Lane, S. N., Reaney, S. M., and Heathwaite, A. L.: Representation of landscape hydrological connectivity using a topographically driven surface flow index, Water Resour. Res., 45, W08423, doi:10.1029/2008WR007336, 2009.

Moore, R. D. and Thompson, J. C.: Are water table variations in a shallow forest soil consistent with the TOPMODEL concept?, Water Resour. Res., 32, 663–669, 1996.

Nash, J. E. and Sutcliffe, J. V.: River flow forecasting through conceptual models part I – A discussion of principles, J. Hydrol., 10, 282–290, 1970.

Pan, Y., Birdsey, R., Hom, J., McCullough, K., and Clark, K.: Improved estimates of net primary productivity from MODIS satellite data at regional and local scales, Ecol. Appl., 16, 125–132, 2006.

Parton, W. J., Mosier, A. R., Ojima, D. S., Valentine, D. W., Schimel, D. S., Weier, K., and Kulmala, A. E.: Generalized model for N_2 and N_2O production from nitrification and denitrification, Global Biogeochem. Cy., 10, 401–412, 1996.

Pockman, W. T. and Small, E. E.: The Influence of Spatial Patterns of Soil Moisture on the Grass and Shrub Responses to a Summer Rainstorm in a Chihuahuan Desert Ecotone, Ecosystems, 13, 511–525, 2010.

Price, M. F., Gratzer, G., Duguma, L. A., Kohler, T., Maselli, D., and Romeo, R. (Eds.): Mountain Forests in a Changing World - Realizing Values, addressing challenges, FAO/MPS and SDC, Rome, 2011.

Quinn, P., Beven, K., Chevallier, P., and Planchon, O.: The prediction of hillslope flow paths for distributed hydrological modeling using digital terrain models, Hydrol. Process., 5, 59–79, 1991.

Ridolfi, L., D'Odorico, P., Porporato, A., and Rodriguez-Iturbe, I.: Stochastic soil moisture dynamics along a hillslope, J. Hydrol., 272, 264–275, 2003.

Riveros-Iregui, D. A. and McGlynn, B. L.: Landscape structure control on soil CO_2 efflux variability in complex terrain: Scaling from point observations to watershed scale fluxes, J. Geophys. Res.-Biogeo., 114, G02010, doi:10.1029/2008JG000885, 2009.

Riveros-Iregui, D. A., McGlynn, B. L., Marshall, L. A., Welsch, D. L., Emanuel, R. E., and Epstein, H. E.: A watershed-scale assessment of a process soil CO_2 production and efflux model, Water Resour. Res., 47, W00J04, doi:10.1029/2010WR009941, 2011.

Running, S. W., Nemani, R. R., and Hungerford, R. D.: Extrapolation of Synoptic Meteorological Data in Mountainous Terrain and Its Use for Simulating Forest Evapotranspiration and Photosynthesis, Can. J. Forest Res., 17, 472–483, 1987.

Ryan, M. G.: Effects of climate change on plant respiration, Ecol. Appl., 1, 157–167, 1991.

Salvucci, G. D. and Entekhabi, D.: Hillslope and Climatic Controls on Hydrologic Fluxes, Water Resour. Res., 31, 1725–1739, 1995.

Scurlock, J. M. O., Asner, G. P., and Gower, S. T.: Global Leaf Area Index from Field Measurements, 1932–2000, Data source: Oak Ridge National Laboratory Distributed Active Archive Center, Oak Ridge, Tennessee, USA, http://www.daac.ornl.gov, 2001.

Singh, R., Jhorar, R. K., van Dam, J. C., and Feddes, R. A.: Distributed ecohydrological modelling to evaluate irrigation system performance in Sirsa district, India II: Impact of viable water management scenarios, J. Hydrol., 329, 714–723, 2006.

Smith, M. B., Koren, V., Reed, S., Zhang, Z., Zhang, Y., Moreda, F., Cui, Z., Naoki, M., Anderson, E. A., and Cosgrove, B. A.: The distributed model intercomparison project – Phase 2: Motivation and design of the Oklahoma experiments, J. Hydrol., 418–419, 3–16, 2012.

Smith, M. W., Bracken, L. J., and Cox, N. J.: Toward a dynamic representation of hydrological connectivity at the hillslope scale in semiarid areas, Water Resour. Res., 46, W12540, doi:10.1029/2009WR008496, 2010.

Sonnentag, O., Chen, J. M., Roulet, N. T., Ju, W., and Govind, A.: Spatially explicit simulation of peatland hydrology and carbon dioxide exchange: Influence of mesoscale topography, J. Geophys. Res.-Biogeo., 113, G02005, doi:10.1029/2007JG000605, 2008.

Sponseller, R. A. and Fisher, S. G.: The influence of drainage networks on patterns of soil respiration in a desert catchment, Ecology, 89, 1089–1100, 2008.

Svoray, T. and Karnieli, A.: Rainfall, topography and primary production relationships in a semiarid ecosystem, Ecohydrology, 4, 56–66, 2011.

Tague, C. L. and Band, L. E.: Evaluating explicit and implicit routing for watershed hydro-ecological models of forest hydrology at the small catchment scale, Hydrol. Process., 15, 1415–1439, 2001.

Tague, C. L. and Band, L. E.: RHESSys: regional hydro-ecologic simulation system – an objected-oriented approach to spatially distributed modeling of carbon, water and nutrient cycling, Earth Interact., 8, 1–42, 2004.

Tang, G. and Arnone III, J.: Trends in surface air temperature and temperature extremes in the Great Basin during the 20th century from ground-based observations, J. Geophys. Res.-Atmos., 118, 3579–3589, 2013.

Tang, G. and Beckage, B.: Projecting the distribution of forests in New England in response to climate change, Divers. Distrib., 16, 144–158, 2010.

Tang, G., Beckage, B., Smith, B., and Miller, P. A.: Estimating potential forest NPP, biomass and their climatic sensitivity in New England using a regional dynamic ecosystem model, Ecosphere, 1, 1–20, 2010.

Thompson, J. C. and Moore, R. D.: Relations between topography and water table depth in a shallow forest soil, Hydrol. Process., 10, 1513–1525, 1996.

Thompson, S., Katul, G., Konings, A., and Ridolfi, L.: Unsteady overland flow on flat surfaces induced by spatial permeability contrasts, Adv. Water Resour., 34, 1049–1058, 2011.

Urgeghe, A. M., Breshears, D. D., Martens, S. N., and Beeson, P. C.: Redistribution of Runoff Among Vegetation Patch Types: On Ecohydrological Optimality of Herbaceous Capture of Run-On, Rangeland Ecol. Manage., 63, 497–504, 2010.

Vogelmann, J. E., Sohl, T., Campbell, P. V., and Shaw, D. M.: Regional land cover characterization using landsat thematic mapper data and ancillary data sources, Environ. Monit. Assess., 51, 415–428, 1998a.

Vogelmann, J. E., Sohl, T., and Howard, S. M.: Regional characterization of land cover using multiple sources of data, Photogramm. Eng. Remote. Sens., 64, 45–47, 1998b.

Wang, J. H., Yang, H., Li, L., Gourley, J. J., Sadiq, I. K., Yilmaz, K. K., Adler, R. F., Policelli, F. S., Habib, S., Irwn, D., Limaye, A. S., Korme, T., and Okello, L.: The coupled routing and excess storage (CREST) distributed hydrological model, Hydrolog. Sci. J., 56, 84–98, 2011.

Wang, L., Koike, T., Yang, K., Jackson, T. J., Bindlish, R., and Yang, D.: Development of a distributed biosphere hydrological model and its evaluation with the Southern Great Plains Experiments (SGP97 and SGP99), J. Geophys. Res.-Atmos., 114, D08107, doi:10.1029/2008JD010800, 2009.

Wigmosta, M., Vail, L., and Lettenmaier, D.: Distributed hydrology–vegetation model for complex terrain. Water Resour. Res., 30, 1665–1679, 1994.

Dominant effect of increasing forest biomass on evapotranspiration: interpretations of movement in Budyko space

Fernando Jaramillo[1,2,3], Neil Cory[4], Berit Arheimer[5], Hjalmar Laudon[6], Ype van der Velde[7], Thomas B. Hasper[1], Claudia Teutschbein[8], and Johan Uddling[1]

[1]Department of Biological and Environmental Sciences, University of Gothenburg, 40530 Gothenburg, Sweden
[2]Department of Physical Geography, Stockholm University, 106 91, Stockholm, Sweden
[3]Stockholm Resilience Center, Stockholm University, 106 91, Stockholm, Sweden
[4]Department of Forest Resource Management; Division of Forest Resource Data, Swedish University of Agricultural Sciences, Umeå, Sweden
[5]Swedish Meteorological and Hydrological Institute, 601 76 Norrköping, Sweden
[6]Department of Forest Ecology and Management, Swedish University of Agricultural Sciences, 750 07 Umeå, Sweden
[7]Faculty of Earth and Life Sciences, University of Amsterdam, 1081 HV, Amsterdam, the Netherlands
[8]Department of Earth Sciences, Uppsala University, 75236, Uppsala, Sweden

Correspondence: Fernando Jaramillo (fernando.jaramillo@natgeo.su.se)

Abstract. During the last 6 decades, forest biomass has increased in Sweden mainly due to forest management, with a possible increasing effect on evapotranspiration. However, increasing global CO_2 concentrations may also trigger physiological water-saving responses in broadleaf tree species, and to a lesser degree in some needleleaf conifer species, inducing an opposite effect. Additionally, changes in other forest attributes may also affect evapotranspiration. In this study, we aimed to detect the dominating effect(s) of forest change on evapotranspiration by studying changes in the ratio of actual evapotranspiration to precipitation, known as the evaporative ratio, during the period 1961–2012. We first used the Budyko framework of water and energy availability at the basin scale to study the hydroclimatic movements in Budyko space of 65 temperate and boreal basins during this period. We found that movements in Budyko space could not be explained by climatic changes in precipitation and potential evapotranspiration in 60 % of these basins, suggesting the existence of other dominant drivers of hydroclimatic change. In both the temperate and boreal basin groups studied, a negative climatic effect on the evaporative ratio was counteracted by a positive residual effect. The positive residual effect occurred along with increasing standing forest biomass in the temperate and boreal basin groups, increasing forest cover in the temperate basin group and no apparent changes in forest species composition in any group. From the three forest attributes, standing forest biomass was the one that could explain most of the variance of the residual effect in both basin groups. These results further suggest that the water-saving response to increasing CO_2 in these forests is either negligible or overridden by the opposite effect of the increasing forest biomass. Thus, we conclude that increasing standing forest biomass is the dominant driver of long-term and large-scale evapotranspiration changes in Swedish forests.

1 Introduction

Boreal and temperate forests provide important ecosystem services at local, regional and global scales. These services include water regulation, soil stabilization, biodiversity preservation and the provisioning of timber, fibre, fuel, food and cultural values for humans (Chopra, 2005). Changes to the attributes of these forests, such as forest coverage, have had important implications for the functioning of the Earth's climate and water and carbon systems (Abbott et al., 2016; Sterling et al., 2013). The return flow of water vapour from the Earth's surface to the atmosphere, known as evapotranspiration, is a key hydroclimatic variable linking these systems. Major regulators of forest evapotranspiration include forest biomass (Feng et al., 2016), leaf stomatal conductance (Lin et al., 2015), canopy leaf area index (LAI) (Mu et al., 2013), tree hydraulic traits (Gao et al., 2014) and stand surface roughness (Donohue et al., 2007). In the context of temperate and boreal Sweden, changes in forest attributes are likely to play an important role because of the higher proportion of available energy partitioned into latent heat than in other ecosystems in these regions, such as grassland, wetlands and tundra (Baldocchi et al., 2000; Kasurinen et al., 2014; van der Velde et al., 2013).

However, the dominant effects of forest change on evapotranspiration are still a matter of debate. Zeng et al. (2016) state that the increase in total canopy leaf area linked to the recent "greening" of the Earth is responsible for more than 50 % of the increase in terrestrial evapotranspiration during the last 30 years. In contrast, Betts et al. (2007) argue for a dominant global decrease of evapotranspiration induced by a water-saving response to increasing carbon dioxide (CO_2) concentrations. Recent modelling studies have found that ignoring the existence of this water-saving response may overpredict future terrestrial evapotranspiration and drought (Milly and Dunne, 2016; Prudhomme et al., 2014; Swann et al., 2016). Piao et al. (2007) argue instead that this physiological effect is cancelled by simultaneous CO_2-induced increases in plant growth and total canopy leaf area, and that changes in climate and land use are the dominant drivers of evapotranspiration changes.

Tree or plot experiments in temperate and boreal forests focusing on the effect on evapotranspiration and water yield relate generally to forest management. These experiments are generally of short duration due to the difficulty and the costs of performing long-term controlled experiments in the field. As forest clear-cutting has been found to reduce evapotranspiration and increase runoff, reforestation or regrowth has resulted in an opposite effect (Andréassian, 2004; Sørensen et al., 2009). For long-term and large-scale studies of this effect, basin-scale hydrological assessments are otherwise required (Andréassian, 2004). Long-term availability of precipitation and runoff data can be used to estimate evapotranspiration changes by water balance over long periods and to study the combined effect of climatic and forest change on forest water use (Hasper et al., 2016). However, the Budyko framework (Budyko, 1974), which links water and energy availability to basin-scale hydrology, is required to further separate this combined effect (Zhang et al., 2001). The Budyko framework has already been applied to understand impacts on forest evapotranspiration in Germany by air quality (Renner et al., 2014) and in North America to differentiate the responses of forested basins to climate and human drivers (Creed et al., 2014; Jones et al., 2012; Wang and Hejazi, 2011; Williams et al., 2012). It has also been applied in China and at the global scale to study the hydrological effects of reforestation programmes and the forest controls on

water partitioning (Fang et al., 2001; Huang et al., 2003; Li et al., 2016; Xu et al., 2014; X. Zhang et al., 2008; Zhou et al., 2015).

Observation-based studies of tree water use responses to elevated CO_2 are also limited to tree- or plot-scale experiments of a few years (with the exception of a 17-year study by Tor-ngern et al., 2015). Although most of these short-term tree field experiments have found a stomatal water-saving response that reduces stomatal conductance and transpiration (Ainsworth and Rogers, 2007; Assmann, 1999; Medlyn et al., 2001), the response is often small or absent in northern tree species. In addition, stomatal conductance tends to be less sensitive to high CO_2 concentrations in gymnosperms (e.g. conifers) than in angiosperms (Hasper et al., 2017; Medlyn et al., 1999). Nordic experiments with Norway spruce showed no leaf water-saving response to elevated CO_2 (Hasper et al., 2016; Sigurdsson et al., 2013), while experiments with another conifer species, Scots pine, showed either a rather small reduction or no significant change in stomatal conductance under elevated CO_2 (Kellomäki and Wang, 1996; Sigurdsson et al., 2002, respectively). In the deciduous species silver birch, however, substantial CO_2-induced leaf water-saving responses were found (Rey and Jarvis, 1998). Stomatal water-saving responses may thus be expected to be small in conifer-dominated northern forests, but this remains to be tested at the basin level.

Our main objective was to determine in Swedish forests the dominant forest effect(s) driving changes in the ratio of actual evapotranspiration to precipitation, known as the evaporative ratio, from a basin-scale approach. Thus, since 1900, forests in Sweden have seen important changes in forest structure and composition (Antonson and Jansson, 2011). For this purpose, we used a former application of the Budyko framework (Jaramillo and Destouni, 2014, 2015; van der Velde et al., 2014) to study movement of basins in the space comprising the aridity index and the evaporative ratio, known as the Budyko space. We separated the climatic effect on evapotranspiration that relates to changes in the aridity index from a residual effect that relates to other drivers of change. We further explored the relationships between this residual effect and the forest attributes of standing forest biomass, forest cover area or species composition. We chose the period 1961–2012 for analysing the change due to the large availability of runoff and forest attribute data across Sweden during this period.

2 Materials and methods

2.1 Hydroclimatic data

We gathered data on the daily runoff (R) for 65 unregulated Swedish basins (Fig. 1) monitored by the Swedish Hydrologic and Meteorological Institute (SMHI) and compiled by Arheimer and Lindström (2015). The daily R data

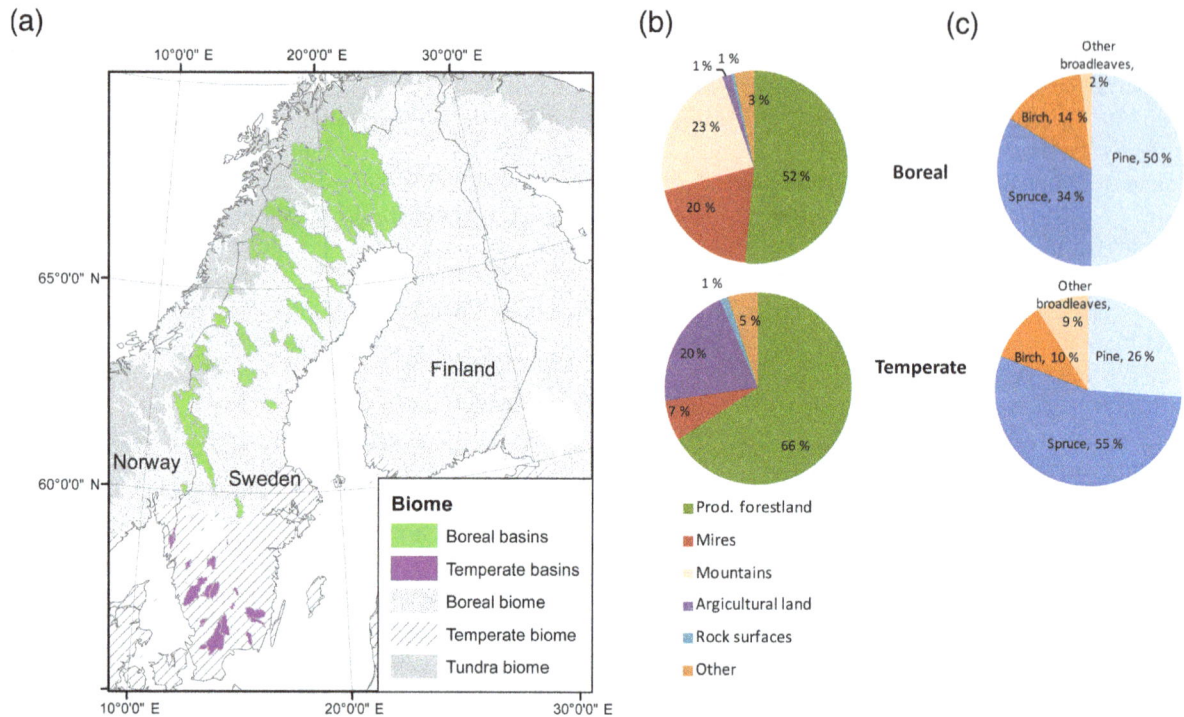

Figure 1. Location, land cover and forest change in the 65 basins used in this study. **(a)** The location of boreal (green) and temperate (purple) basins, **(b)** land cover in each basin group in terms of relative area and **(c)** composition of the forest by coniferous (blue) and deciduous (orange) species.

for the 65 basin outlets and corresponding basin boundaries were obtained from SMHI's hydrologic server (https://vattenwebb.smhi.se/station/) for the period 1961–2012. The daily R data were aggregated to annual values and we considered only complete years in the analysis, defined as years with at least 98 % data capture. We obtained the corresponding daily precipitation (P) data for the basins selected from 68 climatic stations from SMHI's online server. In order to deal with precipitation uncertainty, we also calculated daily P from the Climatic Research Unit gridded P product CRU TS3.23 (Harris et al., 2014) and an additional P product of the Luftwebb portal (http://luftwebb.smhi.se/) of SMHI. The latter is a bias-corrected gridded dataset of precipitation for Sweden during the period 1961–2014 with a 4 km × 4 km horizontal resolution and based in data collected from over 87 precipitation stations around the country (Johansson, 2000; Johansson and Chen, 2003). We made a spatial interpolation by Thiessen polygon to obtain three different estimates of mean P for each basin and used the annual P and R data to calculate actual evapotranspiration (E) as

$$E = P - R - \mathrm{d}S/\mathrm{d}t, \tag{1}$$

where $\mathrm{d}S/\mathrm{d}t$ is the change in water storage within the basin. We performed our assessments of hydroclimatic change during the period 1961–2012 by calculating the inter-annual means of the two 26-year subperiods 1961–1986 and 1987–2012 and defined all hydroclimatic and forest changes as the

difference between these means. Using such long-term periods becomes an advantage since when averaged over long periods $\mathrm{d}S/\mathrm{d}t$ should be considerably smaller than P, R and E, permitting the basin-scale assumption of essentially zero long-term water storage change. These assumptions are required in order to calculate E by Eq. (1).

We also obtained mean, minimum and maximum daily temperature data (T, $T_{\min}$, $T_{\max}$) from the climatic stations with P data for calculations of potential evapotranspiration (E_0). Estimates of annual E_0 in the 68 stations were obtained by three models: (1) the Langbein model in terms of T (Langbein, 1949), (2) the Hargreaves model in terms of $T_{\min}$ and $T_{\max}$ (Hargreaves et al., 1985) and (3) the gridded E_0 product from the Climatic Research Unit CRU TS3.23 (Harris et al., 2014). For the first two models, the E_0 station estimates were interpolated spatially to each basin area by a trivariate spline which allows a spatially varying relationship between E_0 and elevation (Tait and Woods, 2007). We obtained a mean E_0 from the three estimates to perform an uncertainty assessment and to minimize the potential problems of relying on a single model.

2.2 Budyko framework

We used the Budyko framework (Budyko, 1974) to characterize the observed partitioning of P into R and E during the period 1961–2012. The Budyko framework is based on the

(a)

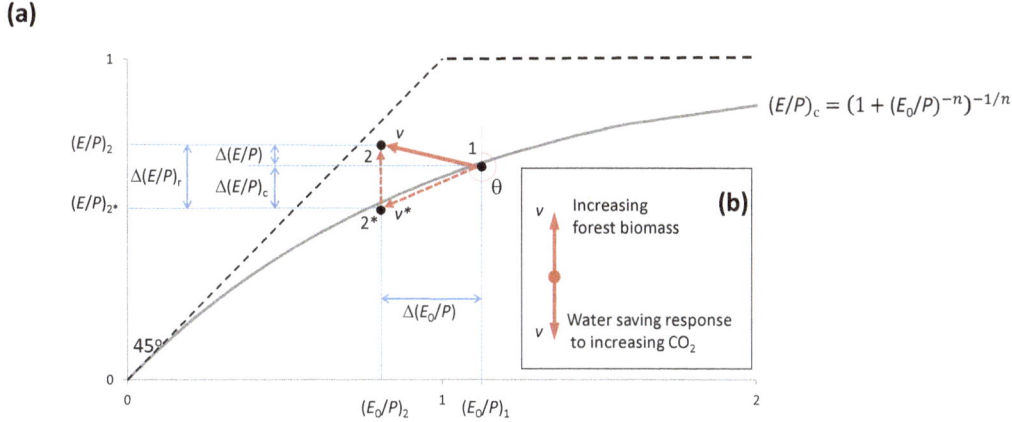

Figure 2. Schematic representation of movement in Budyko space. **(a)** Evaporative ratio (E/P) vs. aridity index (E_0/P) and the parameters describing movement in such space from period 1 (t_1) to period 2 (t_2) by a vector (v). The t_1 and t_2 represent mean conditions of E/P and E_0/P during the periods 1961–1986 and 1987–2012, respectively. The change in evaporative ratio ($\Delta E/P$) is divided into an effect from changes in the aridity index, $\Delta(E/P)_c$, termed the climatic effect and represented by the vector (v^*), and an effect from other drivers, termed the residual effect, $\Delta(E/P)_r$. The elliptic curve describes the Budyko-shaped curve of Choudhury's relationship between E/P and E_0/P. **(b)** The expected directions of movements in Budyko space with fixed E_0/P for increasing forest standing biomass and water-saving response to increasing atmospheric CO_2 concentrations under constant precipitation.

relationship of the partitioning of water and energy on land, and states that evapotranspiration is limited by the supply of water (i.e. P) and energy (i.e. E_0). This relationship (Fig. 2a) is often represented in the space (i.e. Budyko space) created between the ratio of actual evapotranspiration to precipitation (known as the evaporative ratio, E/P) and the ratio of potential evapotranspiration to precipitation (known as the aridity index, E_0/P).

The relationship between E/P and E_0/P is represented by Budyko-type curves expressing the first in terms of the latter: $E/P = F(E_0/P)$. The E/P is expected to increase linearly with E_0/P at low values of E_0/P, while gradually reducing its increasing rate at higher E_0/P values. Here, we use the "Budyko-type" formulation of Yang et al. (2008) which is a climatic model of the evaporative ratio, $(E/P)_c$, in terms of E_0/P and a parameter representing the effect of the contribution of catchment characteristics, such as vegetation, soils and topography for each basin (n). This formulation has the same functional form as that of Choudhury (1999) and has been synthesized by Zhang et al. (2015) as

$$(E/P)_c = \left(1 + (E_0/P)^{-n}\right)^{-1/n}. \tag{2}$$

For this study, a mean n value was calculated for each basin by solving Eq. (2) with the mean E_0/P and E/P values of the initial period 1961–1986, as done by Wang and Hejazi (2011). A basin that changes hydroclimatic conditions from a subperiod 1 (t_1) to a subperiod 2 (t_2) can be represented in Budyko space by a point moving from initial conditions (t_1: E_0/P_1, E/P_1) (Fig. 2). If the movement is only due to change in the aridity index ($\Delta(E_0/P)$), the movement will occur along the corresponding Budyko-type curve to a new point (t_{2*}: E_0/P_{2*}, E/P_{2*}), implying a change in

the evaporative ratio, termed henceforth as the climatic effect ($\Delta(E/P)_c$). However, a basin will most certainly move in time due to a combination of change in the aridity index and other drivers of change (Gudmundsson et al., 2016; Jaramillo and Destouni, 2014). Under such realistic circumstances, the basin will move to a new location not falling on the initial Budyko-type curve (t_2: E_0/P_2, E/P_2), implying a corresponding observed total change in the evaporative ratio ($\Delta(E/P)$).

Regarding some possible forest-related effects on evapotranspiration, increasing forest biomass should increase E due to an increase in root depth and total canopy leaf area, which, under constant conditions of precipitation, would imply an upward movement in Budyko space (Fig. 2b). In a similar way, a tree water-saving response to increasing atmospheric CO_2 concentrations reduces stomata conductance and should decrease E, implying a downward movement in Budyko space under constant P.

Following van der Velde et al. (2014), we represented movement due to changes in the aridity index as a vector (v^*) with direction of movement (θ) and magnitude (r) calculated as

$$\theta = b - \arctan\left(\frac{\Delta(E/P)_c}{\Delta(E_0/P)}\right), \tag{3}$$

$$r = \sqrt{(\Delta(E/P)_c)^2 + (\Delta(E_0/P))^2}, \tag{4}$$

where b is a constant and θ is in degrees ($0° < \theta < 360°$) starting clockwise from the upper vertical ($b = 90°$ when $\Delta(E_0/P) > 0$ and $b = 270°$ when $\Delta(E_0/P) < 0$). In a similar way, we represented the total movement vector (v)

based on R and P observations by replacing $\Delta(E/P)_c$ with $\Delta(E/P)$ in Eqs. (3) and (4).

We used the approach of Jaramillo and Destouni (2014) to synthesize climate-driven and total movements for the 65 basins as typical wind roses of direction and magnitude. The roses illustrate in the same plot the combination of changes in the aridity index and evaporative ratios for all basins. It is worth noting that since the magnitude and direction of change in Budyko space depends on the specific hydroclimatic conditions of each basin (Greve et al., 2016; Gudmundsson et al., 2016; Jaramillo and Destouni, 2014) and on the definition of the space in which such movement occurs, such wind roses oversimplify the variability of movements in Budyko space. However, using these wind roses is a simple way to synthesize general tendencies of movement in large sets of basins and enables a first-order identification of the importance of drivers of change different from the aridity index. For instance, directions of movement that go beyond the range of slopes of any Budyko-type curve (i.e. $45° < \theta < 90°$ and $225° < \theta < 270°$) imply that drivers different from long-term changes in the aridity index dominate the observed changes in the evaporative ratio. This when the n parameter in Eq. (2) is held constant in time.

2.3 Forest attributes

According to recent data from the Swedish National Inventory (NFI; www. slu.se/en/Collaborative-Centres-and-Projects/ the-swedish-national-forest-inventory/), productive forestland now accounts for 57 % of the Swedish land surface after constantly expanding throughout the 20th century (Nilsson et al., 2016). The productive forest standing volume has increased by 85 % since the first NFI took place in 1923 (KSLA, 2015). In Sweden, forest management is responsible for most changes in the abundance of coniferous and deciduous species (Elmhagen et al., 2015; Laudon et al., 2011). The forest data from the NFI here used quantifies the surface area of all types of land use (i.e. productive forestland, mires, mountainous regions, agricultural land, rock surfaces and urban areas); however, the most comprehensive data are collected for forestland. The NFI utilizes a stratified systematic sample based upon clustered sample plots designed to deliver statistics at the county level and as of today accounts for changes in methodology across time. The standing forest volume is differentiated into several categories (i.e. species, diameter and age composition, forest management stage). The species differentiated are Norway spruce (*Picea abies*), Scots pine (*Pinus sylvestris* L.), silver birch (*Betula pendula*) and other deciduous broadleaf species. The strata of the NFI are the Swedish counties; the sample plots have been distributed within each stratum. A single sample distribution is completed every 5 years – however, as each year (representing a fifth of the sample) is evenly distributed over the country, any consecutive 5-year period can be used. A detailed description of the inventory design and methods was provided by Fridman et al. (2014).

The main forest attributes studied here were forest cover area and standing volume of productive forest, extracted for the period 1961–2012 from the plots falling within the 65 basins. The sample plots are not restricted to provide only county-wise estimates – every sample plot has an upscaling factor that can be used to create estimates of forest attributes for a given area after grouping the sample plots falling in it. However, a larger area will result in more sample plots being used and a lower standard error. Many of the original 65 basins contained too few sample plots to provide meaningful estimates and therefore the basins were aggregated into larger groups. We thus aggregated the 65 basins into two main basin groups (i.e. temperate and boreal) according to a terrestrial ecoregion classification (Olson et al., 2001) to show the change in forest statistics within the 65 basins and reduce the sampling standard error.

For the analysis, forest cover and standing forest volume (i.e. biomass) were divided by the area of each of the two basin groups to obtain relative values of area (A in $km^2\,km^{-2}$) and volume (V in $km^3\,km^{-2}$), respectively. According to the NFI data, in both basin groups more than 50 % of the area is covered by forests (Fig. 1b), and these forests consist mainly of coniferous species (Fig. 1c). Although both the boreal and temperate basin groups have a similar proportion of coniferous and deciduous species – species distribution varies between them in terms of standing volume. In the boreal group, Scots pine accounts for 50 % of all trees whereas in the temperate group Norway spruce is the dominating species (55 % of all trees). Furthermore, as a proxy for species composition, we studied the ratio of the deciduous standing forest volume, V_d, to the total standing volume ($Q_v = V_d/V$). We assumed a conservative sampling standard error of A (2.7 %), V (5 %) following (Fridman et al., 2014) and a corresponding propagated error for Q_v (10 %) (Taylor, 1996).

2.4 Calculation of the residual effect on the evaporative ratio

In order to identify possible forest-related effects on the evaporative ratio, we separated the non-climatic residual effect from the climatic effect. The residual effect on the evaporative ratio from 1961–1986 to 1987–2012 ($\Delta(E/P)_r$) was calculated as

$$\Delta(E/P)_r = \Delta(E/P) - \Delta(E/P)_c. \tag{5}$$

The combination of three P products with the three E_0 products resulted in nine possible combinations of P and E_0 for each basin. This is important since the evaporative ratio estimates depend on the sources of P and E_0 and the models being used (Greve et al., 2015; Wang et al., 2015). For each basin, we built the corresponding uncertainty ranges for n, $(E/P)_c$, E/P, $\Delta(E/P)$ and the components $\Delta(E/P)_c$ and

$\Delta(E/P)_{\mathrm{r}}$. The arithmetic and the area-weighted (i.e. based on basin areas) means of $\Delta(E/P)$, $\Delta(E/P)_{\mathrm{c}}$ and $\Delta(E/P)_{\mathrm{r}}$ for each basin group, temperate and boreal, were calculated by averaging the values of all basins within that group that represent the largest area – that is, excluding any nested basins. In a similar way, the uncertainty ranges of each basin were also aggregated into basin group uncertainty ranges.

2.5 Regression analysis of the residual effect and forest attributes

In order to determine which of the forest attributes (A, V, Q_{v}) could better explain the changes in the residual effect on the evaporative ratio, we performed for each basin group linear regressions between all annual values (52 values) of the residual of the evaporative ratio, $(E/P)_{\mathrm{r}} = E/P - (E/P)_{\mathrm{c}}$, and each of the attributes A, V, Q_{v}. A bootstrapping procedure was performed to account for the sampling standard errors of the forest attribute data in the regression analysis. We constructed a normal distribution of 20 data points for each annual value of each forest attribute with the given mean and sampling standard error of each attribute data point. The coefficient of determination (R^2) and the statistical significance of the linear regression ($p < 0.05$) were calculated for each of 1000 random samples of the annual time series 1961–2012. We finally obtained the mean of the 1000 combinations of R^2 and p values to determine the attributes that could best explain the residual effect.

The assumption of steady-state conditions ($\mathrm{d}S/\mathrm{d}t \approx 0$) of the Budyko framework required to maintain the constant water supply limit may also become a potential source of uncertainty for calculations of $(E/P)_{\mathrm{r}}$ (Chen et al., 2013; Greve et al., 2016; Wang and Tang, 2014; L. Zhang et al., 2008). It is possible that a time interval of 1 year may not guarantee stationary conditions (i.e. $\mathrm{d}S/\mathrm{d}t \neq 0$) in these basins, affecting the calculations of E (Eq. 1) (Bring et al., 2015; Budyko, 1974; Moussa and Lhomme, 2016). We therefore applied the same bootstrapping and regression analysis to the 17 values and 10 mean values resulting from averaging the annual data of the period 1961–2012 into 3- and 5-year means, respectively. The 3- and 5-year means were also applied on the original moving 5-year consecutive average periods of the NFI data (see Sect. 2.3).

Additionally, recent findings have shown that changes in the ratio of precipitation in the form of snow to total precipitation (f_{s}) might affect E/P (Berghuijs et al., 2014). We therefore additionally calculated f_{s} per year and for each basin based on the mentioned collected daily P and T data; we assumed that in days with mean temperatures below $1\,^{\circ}\mathrm{C}$ precipitation fell as snow and above $1\,^{\circ}\mathrm{C}$ as rain, following Berghuijs et al. (2014). Finally, we spatially aggregated these values to the basin group level and incorporated f_{s} into the regression analysis previously described.

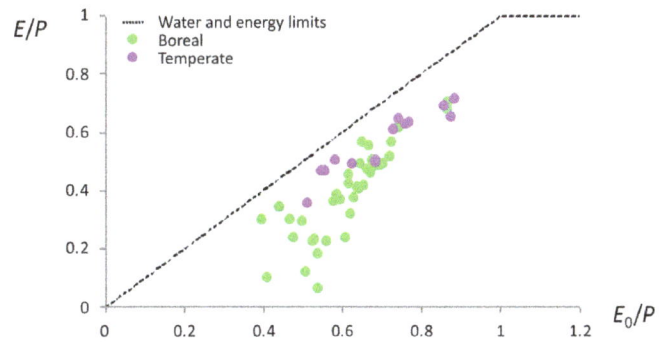

Figure 3. Hydroclimatic conditions in Budyko space. **(a)** Mean hydroclimatic conditions of the 65 basins during the period 1961–2012 illustrated in Budyko space, in terms of the aridity index (E_0/P; x axis) and evaporative ratio (E/P; y axis) for temperate (purple) and boreal (green) basins.

3 Results

3.1 Movement in Budyko space and separation of effects

Most of the 65 basins presented energy-limited conditions since their aridity index E_0/P fell below 1 (Fig. 3). This means that evapotranspiration in these basins was more limited by energy than by water availability. The mean E_0/P and E/P values of the boreal basins were generally lower than those of the temperate basins due to their northerly and cool location. Some basins plotted low in Budyko space, i.e. evaporative ratio E/P near zero, possibly due to underestimates of precipitation due to precipitation undercatch in rain gauges because of falling snow and/or wind, unrealistic measurements of runoff or significant groundwater flux across the basin boundary.

The roses of movements in Budyko space (Fig. 4) show the direction and magnitude of movement of each basin from the period 1961–1986 to the period 1987–2012, in the same way that a typical wind rose shows wind direction and wind speed. The roses show that from the period 1961–1986 to the period 1987–2012, all basins in temperate and boreal biomes experienced a decrease in the aridity index ($180° < \theta < 360°$; Fig. 4a, c). The roses also evidence an increase in the evaporative ratio in most basins in both the temperate (60 %) and boreal (61 %) groups, moving them upwards in Budyko space (Fig. 4a, c), with the direction of movement in both biomes being similar. Furthermore, the movements of highest magnitude occurred in upward directions, with increasing E/P. However, in the absence of other drivers of change, a decrease in the E_0/P can only result in a decrease in the E/P (Fig. 4b, d). Hence, the occurrence of the upward movements ($270° < \theta < 360°$) suggests the influence of other drivers of that movement different from long-term

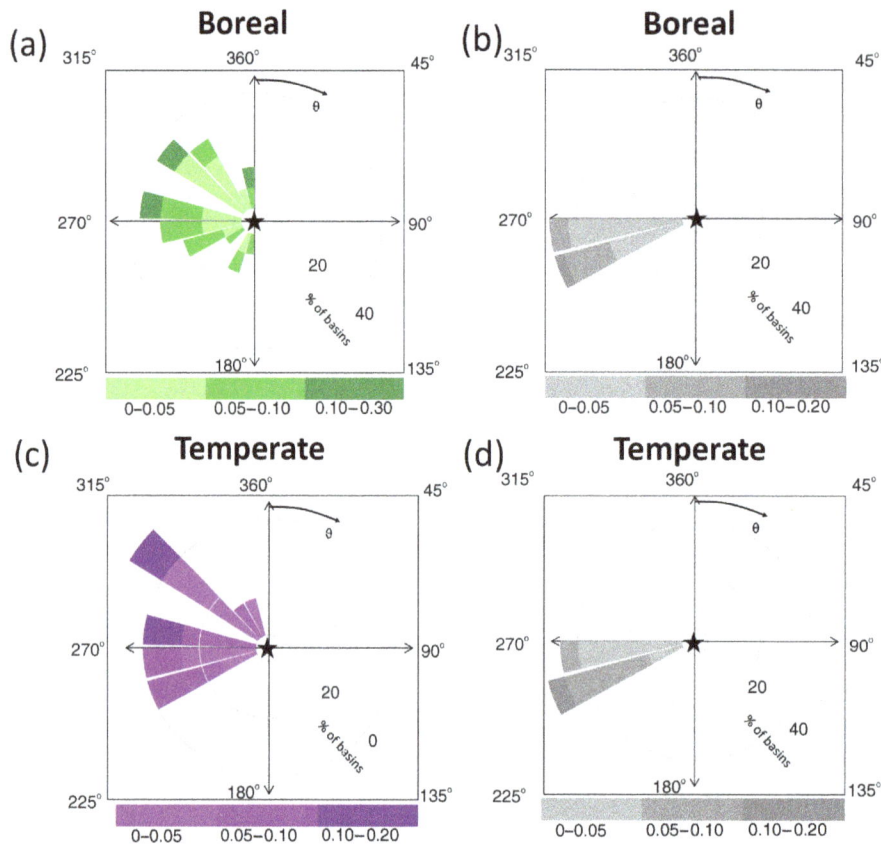

Figure 4. Hydroclimatic movement in Budyko space for the 65 basins due to changes in aridity index (E_0/P) and evaporative ratio (E/P) between the two comparative periods 1961–1986 and 1987–2012. Wind roses of individual basin movements of the boreal basin-group according to **(a)** the combined effect (green) of all drivers of change by calculating E/P from runoff observations (i.e. E/P by Eq. 2) and **(b)** the effect (grey) of only change in the aridity index by calculating E/P from modelled data (i.e. $(E/P)_c$ by Eq. 4), with fixed n and varying E_0/P). **(c–d)** Similar wind roses for temperate basins (purple and grey, respectively). The range of directions of movement ($0 < \theta < 360°$) is divided into 15° interval paddles that group all basins moving in each direction interval, with directions θ starting from the upper vertical and clockwise. We chose this 15° value arbitrarily, to provide sufficient detail of direction. Intensity of colour intervals represent the magnitude of the movement (r) in Budyko space in given direction θ. As an example, 37 % of all boreal basins (Fig. 4a) moved in the range of directions 270° $< \theta <$ 295°. Colour intensity describes the range of magnitude (r: –) of those movements in Budyko space. As an example, of those boreal basins moving in the range of direction previously described (37 % of the boreal basins), 14 % have moved with magnitudes between 0 and 0.05 (light green), 14 % with magnitudes between 0.05 and 0.10 (medium green) and 9 % with magnitudes between 0.10 and 0.30 (dark green). In contrast, no boreal basin moved in this range of direction (270° $< \theta <$ 295°) when using the Choudhury equation (Fig. 4b).

changes in E_0/P. As such, changes in the aridity index are not the only or the most important drivers in these basins.

The general decreases in the E_0/P appeared to be the result of an increase in P between the two time periods that was considerably larger than an increase in E_0 (Fig. 5). The increase in P occurred mostly around winter (January and February) and late spring–summer (May–August), and was larger in the temperate than in the boreal group of basins. For instance, the maximum increase of P in the boreal group occurred in June and reached 25 mm yr^{-1}. In contrast, although the increase in E_0 was larger in the temperate basin group, it was rather small with the highest value occurring in April in both biomes (7 mm yr^{-1} across biomes).

3.2 The residual effect on the evaporative ratio

The separation of the climatic and residual effects on the evaporative ratio at the basin group scale shows that the evaporative ratios in the temperate and boreal basin groups have experienced a decreasing climatic effect driven by less arid conditions, $\Delta(E/P)_c < 0$, between the periods 1961–1986 and 1987–2012 (Fig. 6). They have also experienced an increasing residual effect due to other drivers of change, $\Delta(E/P)_r > 0$. Note that the distribution of change for all basins shows that these counteracting effect applies to the median, arithmetic and area-weighted means and interquartile ranges of both basin groups. The uncertainty assessment of the arithmetic and area-weighted means of $\Delta(E/P)_c$ and

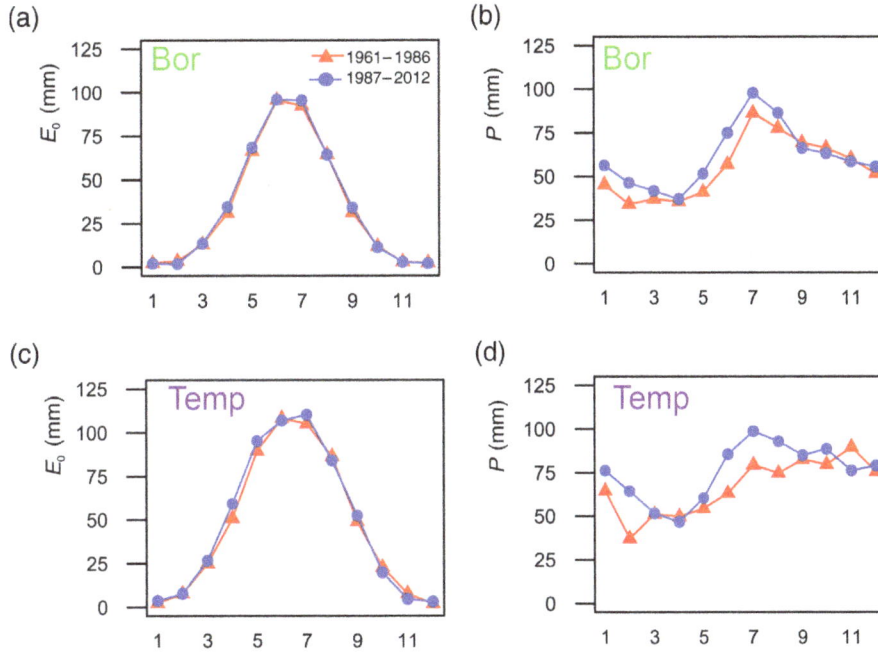

Figure 5. Intra-annual change in precipitation (P) and potential evapotranspiration (E_0). Changes in the mean of P and E_0 between the periods 1961–1986 and 1987–2012 in the boreal **(a–b)** and temperate **(c–d)** basin groups for each month – January (1) to December (12).

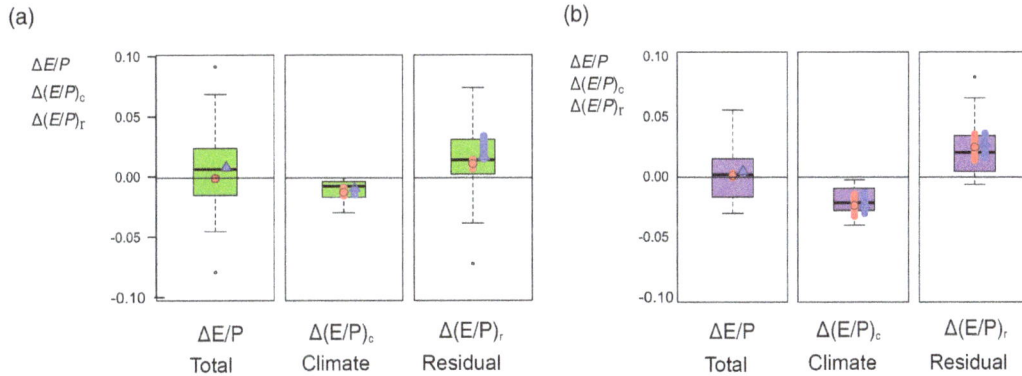

Figure 6. Climate and residual effects on the evaporative ratio. Distribution of changes from 1961–1986 to 1987–2012 in the evaporative ratio, $\Delta E/P$, and its climatic, $\Delta(E/P)_c$, and residual, $\Delta(E/P)_r$, effect components in the **(a)** boreal and **(b)** temperate groups of basins. Boxplot statistics include arithmetic mean (blue triangles), area-weighted mean (pink circles), median (thick horizontal black line), interquartile range (IQR) (boxes), whiskers (confidence interval of $\pm 1.58 \times \mathrm{IQR}\sqrt{N}$, where N is the number of basins in each biome group) and outliers (small black circles). Corresponding uncertainty ranges of the calculated arithmetic and area-weighted means of $\Delta E/P$, $\Delta(E/P)_c$ and $\Delta(E/P)_r$ for each basin group are shown as blue and pink vertical ranges of uncertainty, respectively. These uncertainty ranges arise from the use of three different precipitation and potential evapotranspiration data products.

$\Delta(E/P)_r$ demonstrates that, regardless of the data sources and models of P, T, E_0 used, $\Delta(E/P)_c$ and $\Delta(E/P)_r$ always show these counteracting effects in both basins groups.

3.3 Forest change and effects on the evaporative ratio $(E/P)_r$

The positive residual effect $\Delta(E/P)_r$ in both biomes might be the result of increasing E due to increasing forest volume or surface cover. Indeed, between the two 26-

year periods, the forest cover area A and the standing forest biomass V increased beyond the standard sampling error in the group of temperate basins (Fig. 7a–b). In the group of boreal basins, V also increased more than the standard sampling error but A remained stable. There was no statistically significant change in the overall forest composition in any of the basin groups; the change in Q_v was considerably smaller than the propagated sampling standard error (Fig. 7c).

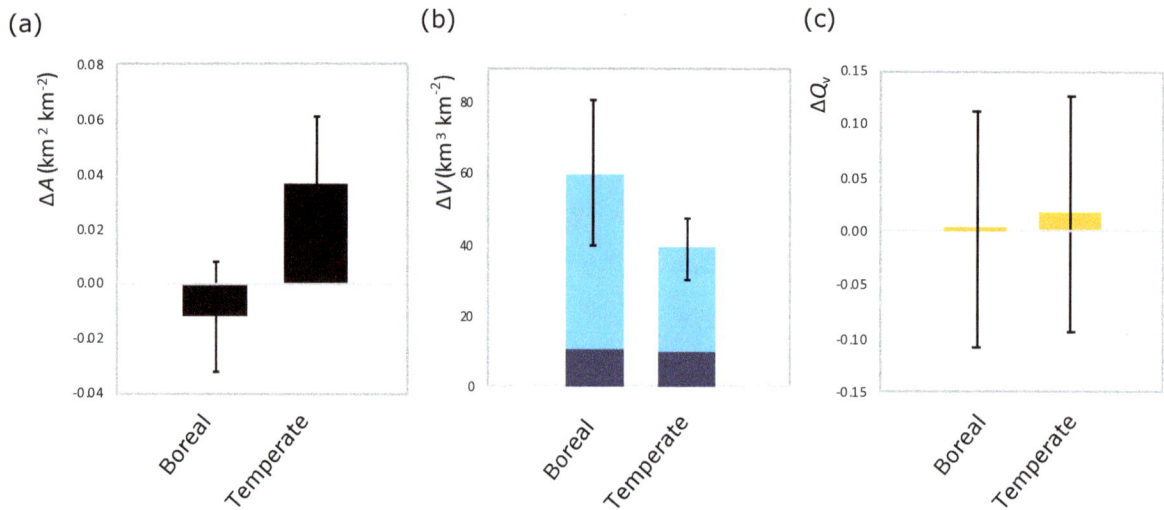

Figure 7. Changes in forest attributes per biome. Changes in both basin groups between the comparative 26-year periods 1961–1986 and 1987–2012, in **(a)** forest cover (ΔA), **(b)** standing forest volume (ΔV) and **(c)** forest composition represented as the ratio of the standing deciduous-forest volume to total standing forest volume (ΔQ_v). The whiskers represent the uncertainty range of ΔA, ΔV and ΔQ_v corresponding to the sampling standard errors for A (2.7 %), V (5 %) and propagated Q_v (10 %). For **(b)**, the dark blue represents ΔV of deciduous trees and light blue of coniferous trees.

The regression analysis between the annual $(E/P)_r$ and V in the period 1961–2012 presented a significant ($p < 0.05$) positive relationship in the boreal basin group when using any of the 1-year, 3-year and 5-year mean options used to divide the period 1961–2012 (Fig. 8 and Table S1 of the Supplement). Hence, V was the attribute that could best explain the residual effect on the evaporative ratio in this basin group. The A, Q_v and f_s could not explain significantly ($p > 0.05$) the variance of $(E/P)_r$ in any case, although f_s was close to be significant ($p = 0.06$) based on data with a 3-year time step. The V was also the only attribute that could significantly ($p < 0.05$) explain the variance of $(E/P)_r$ in the temperate basin group, but only when using annual data. The relationship of $(E/P)_r$ and V was almost significant ($p = 0.07$) for 3-year mean data and that of $(E/P)_r$ and A when using annual data ($p = 0.08$). Hence, among the four parameters studied, V was the attribute that could best explain the residual effect on the evaporative ratio also in the temperate group.

4   Discussion

4.1   Dominant forest-related effects on the evaporative ratio

Our results show that the increase in the residual effect on the evaporative ratio $\Delta(E/P)_r$ during the period 1961–2012 was best explained by an increase in standing forest biomass. Under a hypothetical situation with no changes in climate and stomatal conductance, increasing standing forest biomass should increase transpiration if the canopy LAI and

root biomass increased. At least in the case of LAI, an increase in Sweden has been indicated by remote sensing and modelled data (Zhu et al., 2016), and LAI and standing forest biomass are typically strongly correlated in Scandinavian forests (Heiskanen, 2006). Studies in many forest ecosystems across the world have found with reforestation a similar increase in evapotranspiration and/or decrease in runoff at the plot scale (Andréassian, 2004; Bosch and Hewlett, 1982; Sørensen et al., 2009) and basin scale (Fang et al., 2001; Huang et al., 2003; Li et al., 2016; Xu et al., 2014; X. Zhang et al., 2008).

4.2   No evidence of effect from the water-saving response or intra-annual climatic changes

If the water-saving response caused by reduced stomatal conductance under rising atmospheric CO_2 concentrations was a dominant driver of changes in evapotranspiration in the Swedish forests, a decrease in $(E/P)_r$ should have been observed in our study. However, our results show that in both basin groups $(E/P)_r$ rather increased. Hence, the increase in $(E/P)_r$ found in the two basin groups between the two periods implies that this water-saving response is either negligible or overridden by the positive effect from increasing standing forest biomass. The trees in both the temperate and boreal basin groups are dominated (> 80 %) by coniferous tree species and we found no statistically significant change in forest composition in the two basin groups towards more deciduous trees that could have affected $(E/P)_r$ by increasing the CO_2-induced water-saving response. This response has been found to be less pronounced in coniferous when com-

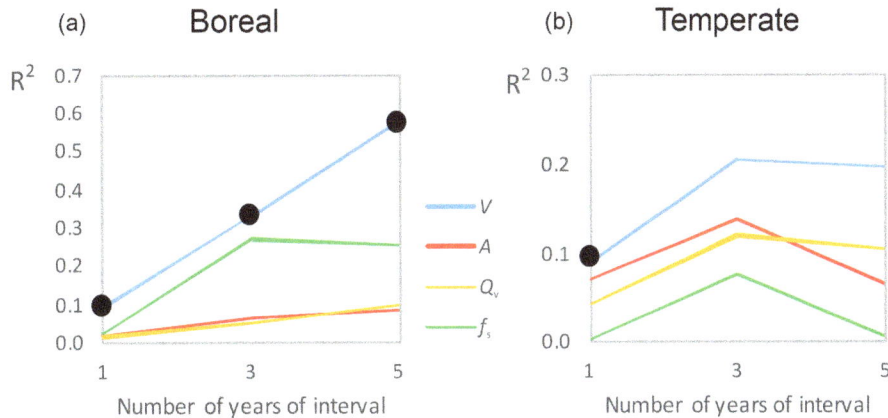

Figure 8. Performance of forest attributes as predictors of the residual of the evaporative ratio. Coefficients of determination (R^2) of the linear regressions between the residual of the evaporative ratio, $(E/P)_r$, and the forest attributes of forest biomass (V), forest cover (A), forest composition (Q_v) and fraction of precipitation falling as snow (f_s), in the boreal and temperate basin groups. The Pearson coefficient of determination (R^2) is shown for the regressions using annual ($N = 52$), 3-year ($N = 17$) and 5-year means ($N = 10$) during the period 1961–2012, with N being the number of values available for each regression with each interval selection. The significance ($p < 0.05$) of the R^2 of each linear regression is shown with a black dot. See Table S1 in the Supplement for values.

pared to deciduous tree species (Hasper et al., 2017; Medlyn et al., 1999).

Intra-annual or seasonal changes of precipitation and temperature with no corresponding change in the annual absolute values of these variables may also affect $(E/P)_r$ (Chen et al., 2013; Milly, 1993; Zanardo et al., 2012). However, the general seasonality pattern of potential evapotranspiration and precipitation in northern latitudes is persistent in time (Fig. 5). Furthermore, although the fraction of precipitation falling as snow f_s decreased from the period 1961–1986 to the period 1987–2012 from 0.20 to 0.14 in the temperate basin group and from 0.45 to 0.43 in the boreal basin group, there was no consistent or significant relationship between f_s and $(E/P)_r$ (Fig. 8 and Table S1).

4.3 Uncertainty on other possible drivers of change on the evaporative ratio

Apart from the inherent uncertainty in the hydroclimatic and forest-attribute data, are there any other factors that may explain the increase in $(E/P)_r$ in both basin groups? The areal extent of forest fires and wind throws in Sweden is small, and such effects should thus have minor influences on large-scale evapotranspiration. Drainage practices used in forestry and agriculture may have played a role in the partitioning of water in Swedish landscapes, although the magnitude of this effect is uncertain and differs between different drainage techniques (Wesström et al., 2003). Farmers implemented open ditches in forests throughout the 20th century, mainly during the 1930s, before the present study period, and the extent of functioning drainage in Sweden today is about 5 % of the total land area (Berglund et al., 2009).

Flow regulation and hydropower development have been shown to increase the evaporative ratio at local to global

scales (Destouni et al., 2013; Jaramillo and Destouni, 2015; Levi et al., 2015). However, all the basins used in this study are unregulated (Arheimer and Lindström, 2015). Technological improvements over time in rain gauges to reduce losses of water due to wind, evaporation and snow undercatch may also lead to a false increase in precipitation and corresponding increase evapotranspiration. Nevertheless, in Sweden, these improvements occurred before the period of the present study; windshields were introduced in Sweden in 1853, and the last installations were reported in 1935. Even the replacement of the old cans made out of galvanized iron for new aluminium ones occurred around 1958 (Sofokleous, 2016).

5 Conclusion

We used the Budyko framework to study movements in Budyko space of 65 unregulated Swedish basins, aiming to detect the dominating effect(s) of forest. In all basins, the aridity index decreased due to an increase in precipitation larger than a corresponding increase in potential evapotranspiration. In 60 % of the basins, this decrease was accompanied by an increase in the evaporative ratio, which, as the Budyko framework of water and energy availability implies, cannot be explained by changes in precipitation or potential evapotranspiration. In both the temperate and boreal basin groups studied, a positive residual effect on the evaporative ratio counteracted the negative climatic effect, maintaining the evaporative ratio relatively stable over time. The positive residual effect occurred along with increasing standing forest biomass in both the temperate and boreal basin groups, increasing forest cover in the temperate basin group, and with no apparent changes in forest species composition.

In general, standing forest biomass was the forest attribute that best explained the variance of the residual effect on the evaporative ratio across basin groups. These results further suggest that the water-saving response to increasing CO_2 in these forests or any other negative effect is either negligible or overridden by the opposite effect of the increasing forest biomass. We conclude that forest expansion is the dominant driver of long-term and large-scale evapotranspiration changes in Swedish forests.

Competing interests. The authors declare that they have no conflict of interest.

Acknowledgements. The strategic research area Biodiversity and Ecosystem Services in a Changing Climate BECC of Lund University and the University of Gothenburg (http://www.becc.lu.se/), the Swedish Research Council (VR, project 2015-06503) and the Swedish Research Council for Environment, Agricultural Sciences and Spatial Planning (942-2015-740) funded this study. We thank the four anonymous reviewers for their valuable critique, contributions and suggestions, which greatly improved the paper.

Edited by: Sally Thompson

References

Abbott, B. W., Jones, J. B., Schuur, E. A. G., III, F. S. C., Bowden, W. B., Bret-Harte, M. S., Epstein, H. E., Flannigan, M. D., Harms, T. K., Hollingsworth, T. N., Mack, M. C., McGuire, A. D., Natali, S. M., Rocha, A. V., Tank, S. E., Turetsky, M. R., Vonk, J. E., Wickland, K. P., Aiken, G. R., Alexander, H. D., Amon, R. M. W., Benscoter, B. W., Yves Bergeron, Bishop, K., Blarquez, O., Bond-Lamberty, B., Breen, A. L., Buffam, I., Yihua Cai, Carcaillet, C., Carey, S. K., Chen, J. M., Chen, H. Y. H., Christensen, T. R., Cooper, L. W., Cornelissen, J. H. C., de Groot, W. J., DeLuca, T. H., Dorrepaal, E., Fetcher, N., Finlay, J. C., Forbes, B. C., French, N. H. F., Gauthier, S., Girardin, M. P., Goetz, S. J., Goldammer, J. G., Gough, L., Grogan, P., Guo, L., Higuera, P. E., Hinzman, L., Hu, F. S., Gustaf Hugelius, Jafarov, E. E., Jandt, R., Johnstone, J. F., Karlsson, J., Kasischke, E. S., Gerhard Kattner, Kelly, R., Keuper, F., Kling, G. W., Kortelainen, P., Kouki, J., Kuhry, P., Hjalmar, L., Laurion, I., Macdonald, R. W., Mann, P. J., Martikainen, P. J., McClelland, J. W., Ulf Molau, Oberbauer, S. F., Olefeldt, D., Paré, D., Parisien, M.-A., Payette, S., Changhui Peng, Pokrovsky, O. S., Rastetter, E. B., Raymond, P. A., Raynolds, M. K., Rein, G., Reynolds, J. F., Robards, M., Rogers, B. M., Schädel, C., Schaefer, K., Schmidt, I. K., Anatoly Shvidenko, Sky, J., Spencer, R. G. M., Starr, G., Striegl, R. G., Teisserenc, R., Tranvik, L. J., Virtanen, T., Welker, J. M., and Zimov, S.: Biomass offsets little or none of permafrost carbon release from soils, streams, and wildfire: an expert assessment, Environ. Res. Lett., 11, 034014, https://doi.org/10.1088/1748-9326/11/3/034014, 2016.

Ainsworth, E. A. and Rogers, A.: The response of photosynthesis and stomatal conductance to rising [CO_2]: mechanisms and environmental interactions, Plant Cell Environ., 30, 258–270, https://doi.org/10.1111/j.1365-3040.2007.01641.x, 2007.

Andréassian, V.: Waters and forests: from historical controversy to scientific debate, J. Hydrol., 291, 1–27, https://doi.org/10.1016/j.jhydrol.2003.12.015, 2004.

Antonson, H. and Jansson, U. (Eds.): Agriculture and forestry in Sweden since 1900: geographical and historical studies, The Royal Swedish Academy of Agriculture and Forestry, Stockholm, 2011.

Arheimer, B. and Lindström, G.: Climate impact on floods: changes in high flows in Sweden in the past and the future (1911–2100), Hydrol. Earth Syst. Sci., 19, 771–784, https://doi.org/10.5194/hess-19-771-2015, 2015.

Assmann, S. M.: The cellular basis of guard cell sensing of rising CO_2, Plant Cell Environ., 22, 629–637, https://doi.org/10.1046/j.1365-3040.1999.00408.x, 1999.

Baldocchi, D., Kelliher, F. M., Black, T. A., and Jarvis, P.: Climate and vegetation controls on boreal zone energy exchange, Glob. Change Biol., 6, 69–83, https://doi.org/10.1046/j.1365-2486.2000.06014.x, 2000.

Berghuijs, W. R., Woods, R. A., and Hrachowitz, M.: A precipitation shift from snow towards rain leads to a decrease in streamflow, Nat. Clim. Change, 4, 583–586, https://doi.org/10.1038/nclimate2246, 2014.

Berglund, Ö., Berglund, K., and Sohlenius, G.: Organogen jordbruksmark i Sverige 1999–2008, Rapport 12, Inst. för markvetenskap, Avd. för hydroteknik, Sveriges Lantbruksuniversitet, Uppsala, 2009.

Betts, R. A., Boucher, O., Collins, M., Cox, P. M., Falloon, P. D., Gedney, N., Hemming, D. L., Huntingford, C., Jones, C. D., Sexton, D. M. H., and Webb, M. J.: Projected increase in continental runoff due to plant responses to increasing carbon dioxide, Nature, 448, 1037–1041, https://doi.org/10.1038/nature06045, 2007.

Bosch, J. and Hewlett, J.: A review of catchment experiments to determine the effect of vegetation changes on water yield and evapotranspiration, J. Hydrol., 55, 3–23, https://doi.org/10.1016/0022-1694(82)90117-2, 1982.

Bring, A., Asokan, S. M., Jaramillo, F., Jarsjö, J., Levi, L., Pietroń, J., Prieto, C., Rogberg, P., and Destouni, G.: Implications of freshwater flux data from the CMIP5 multimodel output across a set of Northern Hemisphere drainage basins, Earths Future, 3, 206–217, https://doi.org/10.1002/2014EF000296, 2015.

Budyko, M. I.: Climate and life, Academic Press, vol. 18, 507 pp., 1974.

Chen, X., Alimohammadi, N., and Wang, D.: Modeling interannual variability of seasonal evaporation and storage change based on the extended Budyko framework, Water Resour. Res., 49, 6067–6078, https://doi.org/10.1002/wrcr.20493, 2013.

Chopra, K. R.: Ecosystems and human well-being?: policy responses: findings of the Responses Working Group of the Millennium Ecosystem Assessment, Island Press, Washington, DC, 2005.

Choudhury, B.: Evaluation of an empirical equation for annual evaporation using field observations and results from a biophysical model, J. Hydrol., 216, 99–110, https://doi.org/10.1016/S0022-1694(98)00293-5, 1999.

Creed, I. F., Spargo, A. T., Jones, J. A., Buttle, J. M., Adams, M. B., Beall, F. D., Booth, E. G., Campbell, J. L., Clow, D., Elder,

K., Green, M. B., Grimm, N. B., Miniat, C., Ramlal, P., Saha, A., Sebestyen, S., Spittlehouse, D., Sterling, S., Williams, M. W., Winkler, R., and Yao, H.: Changing forest water yields in response to climate warming: results from long-term experimental watershed sites across North America, Glob. Change Biol., 20, 3191–3208, https://doi.org/10.1111/gcb.12615, 2014.

Destouni, G., Jaramillo, F., and Prieto, C.: Hydroclimatic shifts driven by human water use for food and energy production, Nat. Clim. Change, 3, 213–217, https://doi.org/10.1038/nclimate1719, 2013.

Donohue, R. J., Roderick, M. L., and McVicar, T. R.: On the importance of including vegetation dynamics in Budyko's hydrological model, Hydrol. Earth Syst. Sci., 11, 983–995, https://doi.org/10.5194/hess-11-983-2007, 2007.

Elmhagen, B., Destouni, G., Angerbjörn, A., Borgström, S., Boyd, E., Cousins, S., Dalén, L., Ehrlén, J., Ermold, M., Hambäck, P., Hedlund, J., Hylander, K., Jaramillo, F., Lagerholm, V., Lyon, S., Moor, H., Nykvist, B., Pasanen-Mortensen, M., Plue, J., Prieto, C., van der Velde, Y., and Lindborg, R.: Interacting effects of change in climate, human population, land use, and water use on biodiversity and ecosystem services, Ecol. Soc., 20, 1–10, https://doi.org/10.5751/ES-07145-200123, 2015.

Fang, J., Chen, A., Peng, C., Zhao, S., and Ci, L.: Changes in Forest Biomass Carbon Storage in China Between 1949 and 1998, Science, 292, 2320–2322, https://doi.org/10.1126/science.1058629, 2001.

Feng, X., Fu, B., Piao, S., Wang, S., Ciais, P., Zeng, Z., Lü, Y., Zeng, Y., Li, Y., Jiang, X., and Wu, B.: Revegetation in China's Loess Plateau is approaching sustainable water resource limits, Nat. Clim. Change, 6, 1019–1022, https://doi.org/10.1038/nclimate3092, 2016.

Fridman, J., Holm, S., Nilsson, M., Nilsson, P., Ringvall, A., and Ståhl, G.: Adapting National Forest Inventories to changing requirements – the case of the Swedish National Forest Inventory at the turn of the 20th century, Silva Fenn., 48, 1095, https://doi.org/10.14214/sf.1095, 2014.

Gao, H., Hrachowitz, M., Schymanski, S. J., Fenicia, F., Sriwongsitanon, N., and Savenije, H. H. G.: Climate controls how ecosystems size the root zone storage capacity at catchment scale, Geophys. Res. Lett., 41, 7916–7923, https://doi.org/10.1002/2014GL061668, 2014.

Greve, P., Gudmundsson, L., Orlowsky, B., and Seneviratne, S. I.: Introducing a probabilistic Budyko framework, Geophys. Res. Lett., 42, 2261–2269, https://doi.org/10.1002/2015GL063449, 2015.

Greve, P., Gudmundsson, L., Orlowsky, B., and Seneviratne, S. I.: A two-parameter Budyko function to represent conditions under which evapotranspiration exceeds precipitation, Hydrol. Earth Syst. Sci., 20, 2195–2205, https://doi.org/10.5194/hess-20-2195-2016, 2016.

Gudmundsson, L., Greve, P., and Seneviratne, S. I.: The sensitivity of water availability to changes in the aridity index and other factors – A probabilistic analysis in the Budyko space, Geophys. Res. Lett., 43, 6985–6994, https://doi.org/10.1002/2016GL069763, 2016.

Hargreaves, G., Hargreaves, G., and Riley, J.: Irrigation Water Requirements for Senegal River Basin, J. Irrig. Drain. Eng., 111, 265–275, https://doi.org/10.1061/(ASCE)0733-9437(1985)111:3(265), 1985.

Harris, I., Jones, P. D., Osborn, T. J. and Lister, D. H.: Updated high-resolution grids of monthly climatic observations – the CRU TS3.10 Dataset, Int. J. Climatol., 34, 623–642, https://doi.org/10.1002/joc.3711, 2014.

Hasper, T. B., Wallin, G., Lamba, S., Hall, M., Jaramillo, F., Laudon, H., Linder, S., Medhurst, J. L., Räntfors, M., Sigurdsson, B. D., and Uddling, J.: Water use by Swedish boreal forests in a changing climate, Funct. Ecol., 30, 690–699, https://doi.org/10.1111/1365-2435.12546, 2016.

Hasper, T. B., Dusenge, M. E., Breuer, F., Uwizeye, F. K., Wallin, G., and Uddling, J.: Stomatal CO_2 responsiveness and photosynthetic capacity of tropical woody species in relation to taxonomy and functional traits, Oecologia, 184, 43–57, https://doi.org/10.1007/s00442-017-3829-0, 2017.

Heiskanen, J.: Estimating aboveground tree biomass and leaf area index in a mountain birch forest using ASTER satellite data, Int. J. Remote Sens., 27, 1135–1158, https://doi.org/10.1080/01431160500353858, 2006.

Huang, M., Zhang, L., and Gallichand, J.: Runoff responses to afforestation in a watershed of the Loess Plateau, China, Hydrol. Process., 17, 2599–2609, https://doi.org/10.1002/hyp.1281, 2003.

Jaramillo, F. and Destouni, G.: Developing water change spectra and distinguishing change drivers worldwide, Geophys. Res. Lett., 41, 8377–8386, https://doi.org/10.1002/2014GL061848, 2014.

Jaramillo, F. and Destouni, G.: Local flow regulation and irrigation raise global human water consumption and footprint, Science, 350, 1248–1251, https://doi.org/10.1126/science.aad1010, 2015.

Johansson, B.: Areal Precipitation and Temperature in the Swedish Mountains, Hydrol. Res., 31, 207–228, 2000.

Johansson, B. and Chen, D.: The influence of wind and topography on precipitation distribution in Sweden: statistical analysis and modelling, Int. J. Climatol., 23, 1523–1535, https://doi.org/10.1002/joc.951, 2003.

Jones, J. A., Creed, I. F., Hatcher, K. L., Warren, R. J., Adams, M. B., Benson, M. H., Boose, E., Brown, W. A., Campbell, J. L., Covich, A., Clow, D. W., Dahm, C. N., Elder, K., Ford, C. R., Grimm, N. B., Henshaw, D. L., Larson, K. L., Miles, E. S., Miles, K. M., Sebestyen, S. D., Spargo, A. T., Stone, A. B., Vose, J. M., and Williams, M. W.: Ecosystem Processes and Human Influences Regulate Streamflow Response to Climate Change at Long-Term Ecological Research Sites, BioScience, 62, 390–404, https://doi.org/10.1525/bio.2012.62.4.10, 2012.

Kasurinen, V., Alfredsen, K., Kolari, P., Mammarella, I., Alekseychik, P., Rinne, J., Vesala, T., Bernier, P., Boike, J., Langer, M., Belelli Marchesini, L., van Huissteden, K., Dolman, H., Sachs, T., Ohta, T., Varlagin, A., Rocha, A., Arain, A., Oechel, W., Lund, M., Grelle, A., Lindroth, A., Black, A., Aurela, M., Laurila, T., Lohila, A., and Berninger, F.: Latent heat exchange in the boreal and arctic biomes, Glob. Change Biol., 20, 3439–3456, https://doi.org/10.1111/gcb.12640, 2014.

Kellomäki, S. and Wang, K.-Y.: Photosynthetic responses to needle water potentials in Scots pine after a four-year exposure to elevated CO_2 and temperature, Tree Physiol., 16, 765–772, https://doi.org/10.1093/treephys/16.9.765, 1996.

KSLA: Forests and Forestry in Sweden, The Royal Swedish Academy of Agriculture and Forestry (KSLA), available at: http://www.ksla.se/wp-content/uploads/2015/08/

Forests-and-Forestry-in-Sweden_2015.pdf (last access: 1 September 2017), 2015.

Langbein, W. B.: Annual runoff in the United States, United States Department of the Interior, Geological Survey Circular 52, 1949.

Laudon, H., Sponseller, R. A., Lucas, R. W., Futter, M. N., Egnell, G., Bishop, K., Ågren, A., Ring, E., and Högberg, P.: Consequences of More Intensive Forestry for the Sustainable Management of Forest Soils and Waters, Forests, 2, 243–260, https://doi.org/10.3390/f2010243, 2011.

Levi, L., Jaramillo, F., Andričević, R., and Destouni, G.: Hydroclimatic changes and drivers in the Sava River Catchment and comparison with Swedish catchments, Ambio, 44, 624–634, https://doi.org/10.1007/s13280-015-0641-0, 2015.

Li, Y., Liu, C., Zhang, D., Liang, K., Li, X., and Dong, G.: Reduced Runoff Due to Anthropogenic Intervention in the Loess Plateau, China, Water, 8, 1–16, https://doi.org/10.3390/w8100458, 2016.

Lin, Y.-S., Medlyn, B. E., Duursma, R. A., Prentice, I. C., Wang, H., Baig, S., Eamus, D., de Dios, V. R., Mitchell, P., Ellsworth, D. S., de Beeck, M. O., Wallin, G., Uddling, J., Tarvainen, L., Linderson, M.-L., Cernusak, L. A., Nippert, J. B., Ocheltree, T. W., Tissue, D. T., Martin-StPaul, N. K., Rogers, A., Warren, J. M., De Angelis, P., Hikosaka, K., Han, Q., Onoda, Y., Gimeno, T. E., Barton, C. V. M., Bennie, J., Bonal, D., Bosc, A., Löw, M., Macinins-Ng, C., Rey, A., Rowland, L., Setterfield, S. A., Tausz-Posch, S., Zaragoza-Castells, J., Broadmeadow, M. S. J., Drake, J. E., Freeman, M., Ghannoum, O., Hutley, L. B., Kelly, J. W., Kikuzawa, K., Kolari, P., Koyama, K., Limousin, J.-M., Meir, P., Lola da Costa, A. C., Mikkelsen, T. N., Salinas, N., Sun, W., and Wingate, L.: Optimal stomatal behaviour around the world, Nat. Clim. Change, 5, 459–464, https://doi.org/10.1038/nclimate2550, 2015.

Medlyn, B. E., Badeck, F.-W., De Pury, D. G. G., Barton, C. V. M., Broadmeadow, M., Ceulemans, R., De Angelis, P., Forstreuter, M., Jach, M. E., Kellomäki, S., Laitat, E., Marek, M., Philippot, S., Rey, A., Strassemeyer, J., Laitinen, K., Liozon, R., Portier, B., Roberntz, P., Wang, K., and Jstbid, P. G.: Effects of elevated [CO_2] on photosynthesis in European forest species: a meta-analysis of model parameters, Plant Cell Environ., 22, 1475–1495, https://doi.org/10.1046/j.1365-3040.1999.00523.x, 1999.

Medlyn, B. E., Barton, C. V. M., Broadmeadow, M. S. J., Ceulemans, R., De Angelis, P., Forstreuter, M., Freeman, M., Jackson, S. B., Kellomäki, S., Laitat, E., Rey, A., Roberntz, P., Sigurdsson, B. D., Strassemeyer, J., Wang, K., Curtis, P. S., and Jarvis, P. G.: Stomatal conductance of forest species after long-term exposure to elevated CO_2 concentration: a synthesis, New Phytol., 149, 247–264, https://doi.org/10.1046/j.1469-8137.2001.00028.x, 2001.

Milly, P. C. D.: An analytic solution of the stochastic storage problem applicable to soil water, Water Resour. Res., 29, 3755–3758, https://doi.org/10.1029/93WR01934, 1993.

Milly, P. C. D. and Dunne, K. A.: Potential evapotranspiration and continental drying, Nat. Clim. Change, 6, 946–949, https://doi.org/10.1038/nclimate3046, 2016.

Moussa, R. and Lhomme, J.-P.: The Budyko functions under non-steady-state conditions, Hydrol. Earth Syst. Sci., 20, 4867–4879, https://doi.org/10.5194/hess-20-4867-2016, 2016.

Mu, Q., Zhao, M., and Running, S. W.: MODIS Global Terrestrial Evapotranspiration (ET) Product (NASA MOD16A2/A3), Algorithm Theoretical Basis Document, Collection 5, available at: http://www.ntsg.umt.edu/sites/ntsg.umt.edu/files/MOD16_ATBD.pdf (last access: 16 November 2016), 2013.

Nilsson, P., Cory, N., and Wikberg, P.-E.: Skogsdata 2016: aktuella uppgifter om de svenska skogarna från Riksskogstaxeringen, Tema: Skogen då, nu och i framtiden, Institutionen för skoglig resurshållning, Sveriges lantbruksuniversitet, SLU, Umeå, available at: http://urn.kb.se/resolve?urn=urn:nbn:se:slu:epsilon-e-3484, last access: 10 September 2016.

Olson, D. M., Dinerstein, E., Wikramanayake, E. D., Burgess, N. D., Powell, G. V. N., Underwood, E. C., D'amico, J. A., Itoua, I., Strand, H. E., Morrison, J. C., Loucks, C. J., Allnutt, T. F., Ricketts, T. H., Kura, Y., Lamoreux, J. F., Wettengel, W. W., Hedao, P., and Kassem, K. R.: Terrestrial Ecoregions of the World: A New Map of Life on Earth A new global map of terrestrial ecoregions provides an innovative tool for conserving biodiversity, BioScience, 51, 933–938, https://doi.org/10.1641/0006-3568(2001)051[0933:TEOTWA]2.0.CO;2, 2001.

Piao, S., Friedlingstein, P., Ciais, P., de Noblet-Ducoudre, N., Labat, D., and Zaehle, S.: Changes in climate and land use have a larger direct impact than rising CO_2 on global river runoff trends, P. Natl. Acad. Sci. USA, 104, 15242–15247, https://doi.org/10.1073/pnas.0707213104, 2007.

Prudhomme, C., Giuntoli, I., Robinson, E. L., Clark, D. B., Arnell, N. W., Dankers, R., Fekete, B. M., Franssen, W., Gerten, D., Gosling, S. N., Hagemann, S., Hannah, D. M., Kim, H., Masaki, Y., Satoh, Y., Stacke, T., Wada, Y., and Wisser, D.: Hydrological droughts in the 21st century, hotspots and uncertainties from a global multimodel ensemble experiment, P. Natl. Acad. Sci. USA, 111, 3262–3267, https://doi.org/10.1073/pnas.1222473110, 2014.

Renner, M., Brust, K., Schwärzel, K., Volk, M., and Bernhofer, C.: Separating the effects of changes in land cover and climate: a hydro-meteorological analysis of the past 60 yr in Saxony, Germany, Hydrol. Earth Syst. Sci., 18, 389–405, https://doi.org/10.5194/hess-18-389-2014, 2014.

Rey, A. and Jarvis, P. G.: Long-term photosynthetic acclimation to increased atmospheric CO_2 concentration in young birch (Betula pendula) trees, Tree Physiol., 18, 441–450, https://doi.org/10.1093/treephys/18.7.441, 1998.

Sigurdsson, B. D., Roberntz, P., Freeman, M., Naess, M., Saxe, H., Thorgeirsson, H., and Linder, S.: Impact studies on Nordic forests: effects of elevated CO_2 and fertilization on gas exchange, Can. J. Forest Res., available at: http://agris.fao.org/agris-search/search.do?recordID=US201302929023 (last access: 25 April 2016), 2002.

Sigurdsson, B. D., Medhurst, J. L., Wallin, G., Eggertsson, O., and Linder, S.: Growth of mature boreal Norway spruce was not affected by elevated [CO_2] and/or air temperature unless nutrient availability was improved, Tree Physiol., 33, 1192–1205, https://doi.org/10.1093/treephys/tpt043, 2013.

Sofokleous, I.: Correction of Inhomogeneous Data in the Precipitation Time Series of Sweden Due to the Wind Shield Introduction, available at: http://uu.diva-portal.org/smash/record.jsf?pid=diva2:916093 (last access: 6 January 2017), 2016.

Sørensen, R., Ring, E., Meili, M., Högbom, L., Seibert, J., Grabs, T., Laudon, H., and Bishop, K.: Forest harvest increases runoff most during low flows in two boreal streams, Ambio, 38, 357–363, 2009.

Sterling, S. M., Ducharne, A., and Polcher, J.: The impact of global land-cover change on the terrestrial water cycle, Nat. Clim. Change, 3, 385–390, https://doi.org/10.1038/nclimate1690, 2013.

Swann, A. L. S., Hoffman, F. M., Koven, C. D., and Randerson, J. T.: Plant responses to increasing CO2 reduce estimates of climate impacts on drought severity, P. Natl. Acad. Sci. USA, 113, 10019–10024, https://doi.org/10.1073/pnas.1604581113, 2016.

Tait, A. and Woods, R.: Spatial Interpolation of Daily Potential Evapotranspiration for New Zealand Using a Spline Model, J. Hydrometeorol., 8, 430–438, https://doi.org/10.1175/JHM572.1, 2007.

Taylor, J. R.: An Introduction to Error Analysis: The Study of Uncertainties in Physical Measurements, 2nd edn., University Science Books, Sausalito, California, 1996.

Tor-ngern, P., Oren, R., Ward, E. J., Palmroth, S., McCarthy, H. R., and Domec, J.-C.: Increases in atmospheric CO_2 have little influence on transpiration of a temperate forest canopy, New Phytol., 205, 518–525, https://doi.org/10.1111/nph.13148, 2015.

van der Velde, Y., Lyon, S. W., and Destouni, G.: Data-driven regionalization of river discharges and emergent land cover–evapotranspiration relationships across Sweden, J. Geophys. Res.-Atmos., 118, 2576–2587, https://doi.org/10.1002/jgrd.50224, 2013.

van der Velde, Y., Vercauteren, N., Jaramillo, F., Dekker, S. C., Destouni, G., and Lyon, S. W.: Exploring hydroclimatic change disparity via the Budyko framework, Hydrol. Process., 28, 4110–4118, https://doi.org/10.1002/hyp.9949, 2014.

Wang, D. and Hejazi, M.: Quantifying the relative contribution of the climate and direct human impacts on mean annual streamflow in the contiguous United States, Water Resour. Res., 47, W00J12, https://doi.org/10.1029/2010WR010283, 2011.

Wang, D. and Tang, Y.: A one-parameter Budyko model for water balance captures emergent behavior in darwinian hydrologic models, Geophys. Res. Lett., 41, 4569–4577, https://doi.org/10.1002/2014GL060509, 2014.

Wang, W., Xing, W., and Shao, Q.: How large are uncertainties in future projection of reference evapotranspiration through different approaches?, J. Hydrol., 524, 696–700, https://doi.org/10.1016/j.jhydrol.2015.03.033, 2015.

Wesström, I., Ekbohm, G., Linnér, H., and Messing, I.: The effects of controlled drainage on subsurface outflow from level agricultural fields, Hydrol. Process., 17, 1525–1538, https://doi.org/10.1002/hyp.1197, 2003.

Williams, C. A., Reichstein, M., Buchmann, N., Baldocchi, D., Beer, C., Schwalm, C., Wohlfahrt, G., Hasler, N., Bernhofer, C., Foken, T., Papale, D., Schymanski, S., and Schaefer, K.: Climate and vegetation controls on the surface water balance: Synthesis of evapotranspiration measured across a global network of flux towers, Water Resour. Res., 48, W06523, https://doi.org/10.1029/2011WR011586, 2012.

Xu, X., Yang, D., Yang, H., and Lei, H.: Attribution analysis based on the Budyko hypothesis for detecting the dominant cause of runoff decline in Haihe basin, J. Hydrol., 510, 530–540, https://doi.org/10.1016/j.jhydrol.2013.12.052, 2014.

Yang, H., Yang, D., Lei, Z., and Sun, F.: New analytical derivation of the mean annual water-energy balance equation, Water Resour. Res., 44, W03410, https://doi.org/10.1029/2007WR006135, 2008.

Zanardo, S., Harman, C. J., Troch, P. A., Rao, P. S. C., and Sivapalan, M.: Intra-annual rainfall variability control on interannual variability of catchment water balance: A stochastic analysis, Water Resour. Res., 48, W00J16, https://doi.org/10.1029/2010WR009869, 2012.

Zeng, Z., Zhu, Z., Lian, X., Li, L. Z. X., Chen, A., He, X., and Piao, S.: Responses of land evapotranspiration to Earth's greening in CMIP5 Earth System Models, Environ. Res. Lett., 11, 104006, https://doi.org/10.1088/1748-9326/11/10/104006, 2016.

Zhang, D., Cong, Z., Ni, G., Yang, D., and Hu, S.: Effects of snow ratio on annual runoff within the Budyko framework, Hydrol. Earth Syst. Sci., 19, 1977–1992, https://doi.org/10.5194/hess-19-1977-2015, 2015.

Zhang, L., Dawes, W. R., and Walker, G. R.: Response of mean annual evapotranspiration to vegetation changes at catchment scale, Water Resour. Res., 37, 701–708, https://doi.org/10.1029/2000WR900325, 2001.

Zhang, L., Potter, N., Hickel, K., Zhang, Y., and Shao, Q.: Water balance modeling over variable time scales based on the Budyko framework – Model development and testing, J. Hydrol., 360, 117–131, https://doi.org/10.1016/j.jhydrol.2008.07.021, 2008.

Zhang, X., Zhang, L., Zhao, J., Rustomji, P., and Hairsine, P.: Responses of streamflow to changes in climate and land use/cover in the Loess Plateau, China, Water Resour. Res., 44, W00A07, https://doi.org/10.1029/2007WR006711, 2008.

Zhou, G., Wei, X., Chen, X., Zhou, P., Liu, X., Xiao, Y., Sun, G., Scott, D. F., Zhou, S., Han, L., and Su, Y.: Global pattern for the effect of climate and land cover on water yield, Nat. Commun., 6, 5918, https://doi.org/10.1038/ncomms6918, 2015.

Zhu, Z., Piao, S., Myneni, R. B., Huang, M., Zeng, Z., Canadell, J. G., Ciais, P., Sitch, S., Friedlingstein, P., Arneth, A., Cao, C., Cheng, L., Kato, E., Koven, C., Li, Y., Lian, X., Liu, Y., Liu, R., Mao, J., Pan, Y., Peng, S., Peñuelas, J., Poulter, B., Pugh, T. A. M., Stocker, B. D., Viovy, N., Wang, X., Wang, Y., Xiao, Z., Yang, H., Zaehle, S., and Zeng, N.: Greening of the Earth and its drivers, Nat. Clim. Change, 6, 791–795, https://doi.org/10.1038/nclimate3004, 2016.

3

Hydrological connectivity inferred from diatom transport through the riparian-stream system

N. Martínez-Carreras[1], C. E. Wetzel[1], J. Frentress[1], L. Ector[1], J. J. McDonnell[2,3], L. Hoffmann[1], and L. Pfister[1]

[1]Luxembourg Institute of Science and Technology, Department Environmental Research and Innovation, Belvaux, Luxembourg
[2]Global Institute for Water Security, University of Saskatchewan, Saskatoon, Canada
[3]School of Geosciences, University of Aberdeen, Aberdeen, Scotland, UK

Correspondence to: N. Martínez-Carreras (nuria.martinez@list.lu)

Abstract. Diatoms (*Bacillariophyta*) are one of the most common and diverse algal groups (ca. 200 000 species, ≈ 10–$200\,\mu m$, unicellular, eukaryotic). Here we investigate the potential of aerial diatoms (i.e. diatoms nearly exclusively occurring outside water bodies, in wet, moist or temporarily dry places) to infer surface hydrological connectivity between hillslope-riparian-stream (HRS) landscape units during storm runoff events. We present data from the Weierbach catchment (0.45 km^2, northwestern Luxembourg) that quantify the relative abundance of aerial diatom species on hillslopes and in riparian zones (i.e. surface soils, litter, bryophytes and vegetation) and within streams (i.e. stream water, epilithon and epipelon). We tested the hypothesis that different diatom species assemblages inhabit specific moisture domains of the catchment (i.e. HRS units) and, consequently, the presence of certain species assemblages in the stream during runoff events offers the potential for recording whether there was hydrological connectivity between these domains or not. We found that a higher percentage of aerial diatom species was present in samples collected from the riparian and hillslope zones than inside the stream. However, diatoms were absent on hillslopes covered by dry litter and the quantities of diatoms (in absolute numbers) were small in the rest of hillslope samples. This limits their use for inferring hillslope-riparian zone connectivity. Our results also showed that aerial diatom abundance in the stream increased systematically during all sampled events ($n = 11$, 2011–2012) in response to incident precipitation and increasing discharge. This transport of aerial diatoms during events suggested a rapid connectivity between the soil surface and the stream. Diatom transport data were compared to two-component hydrograph separation, and end-member mixing analysis (EMMA) using stream water chemistry and stable isotope data. Hillslope overland flow was insignificant during most sampled events. This research suggests that diatoms were likely sourced exclusively from the riparian zone, since it was not only the largest aerial diatom reservoir, but also since soil water from the riparian zone was a major streamflow source during rainfall events under both wet and dry antecedent conditions. In comparison to other tracer methods, diatoms require taxonomy knowledge and a rather large processing time. However, they can provide unequivocal evidence of hydrological connectivity and potentially be used at larger catchment scales.

1 Introduction

The generation of storm runoff is strongly linked to hydrological connectivity – surface and subsurface – that controls threshold changes in flow and concomitant flushing of solutes and labile nutrients (McDonnell, 2013). To date, various approaches to quantifying hydrological connectivity have been presented, including hydrometric mapping at hillslope (Tromp-van Meerveld and McDonnell, 2006) and catchment scales (Spence, 2010), connectivity metrics (Ali and Roy, 2010) and high-frequency water table monitoring (Jencso et al., 2009). Perhaps the most popular tool has been the use of environmental tracers for characterizing and understanding complex water flow connections within catchments, between soils, channels, overland surfaces, and hill-slopes (Buttle, 1998). Chemical tracers and stable isotopes of the water molecule have been widely used for quantifying the temporal sources of storm flow (i.e. event and pre-event water) using mass balance equations (see Klaus and McDonnell, 2013, for a review). These tracers have also been used together to quantify the geographic sources of runoff using end-member mixing models (EMMA) (see Hooper, 2001, for a review).

Despite their usefulness, chemical and isotope tracer-based hydrograph separations do not provide unequivocal evidence of hillslope-riparian-stream (HRS) connectivity. This has been identified as perhaps the key feature for improving our understanding of water origin and the processes that sustain stream flow (Jencso et al., 2010). Consequently, new techniques are desperately needed to gain a process-based understanding of hydrological connectivity (Bracken et al., 2013).

Here we build on recent work by Pfister et al. (2009, 2015) and Wetzel et al. (2013) to examine the use of aerial diatoms (i.e. diatoms nearly exclusively occurring outside water bodies, and in wet, moist or temporarily dry places; Van Dam et al., 1994), as natural tracers to infer connectivity in the HRS system. Diatoms are one of the most common and diverse algal groups (ca. 200 000 species; Round et al., 1990). Due to their small size (~ 10–$200\,\mu$m; Mann, 2002), they can be easily transported by flowing water within or between elements of the hydrological cycle (Pfister et al., 2009). Diatoms are present in most terrestrial habitats and their diversified species distributions are largely controlled by physio-geographical factors (e.g. light, temperature, pH and moisture) and anthropogenic pollution (Dixit et al., 2002; Ector and Rimet, 2005).

Our work tests the hypothesis that different diatom species assemblages inhabit specific moisture domains of the HRS system and, consequently, the presence of certain species assemblages in the stream during runoff events has the ability to record periods of hydrological connectivity between these watershed components. We compare diatom results with traditional two-component hydrograph separation, and end-member mixing analysis (EMMA) using stream water chemistry and stable isotope data. We also present soil water content and groundwater level data within the HRS system to facilitate a somewhat holistic understanding of catchment runoff processes (as advocated by Bonell, 1998; Burns, 2002; Lischeid, 2008). Specifically, we addressed the following questions.

1. Can aerial diatom transport reveal hydrological connectivity within the HRS system?

2. How do diatom results compare to traditional tracer-based and hydrometric methods to infer hydrological connectivity?

3. Can aerial diatoms be established as a new hydrological tracer?

2 Study area

Our study site is the Weierbach catchment (0.45 km^2; 49°49′ N, 5°47′ E), a sub-catchment of the Attert River and located in the northwestern part of the Grand Duchy of Luxembourg (Fig. 1). The region is known as the Oesling, an elevated sub-horizontal plateau cut by deep V-shaped valleys and with average altitudes ranging between 450 and 500 m.

Weierbach has a temperate, semi-oceanic climate regime. Annual precipitation in the Attert River basin ranges from 950 mm on the western border to 750 mm on the eastern border (average from 1971 to 2000; Pfister et al., 2005). Precipitation is relatively uniform throughout the year, although strong seasonality in low flow exists due to higher evapotranspiration from July to September. The annual runoff ratio is high (~ 55 % based on 2005 to 2011 streamflow data) and flow sometimes ceases during summer months.

The geology of the catchment is dominated by Devonian schists, phyllades and quartzite. The schist bedrock is covered by Pleistocene periglacial slope deposits (Juilleret et al., 2011). Soil depths are shallow (< 1 m) and dominated by cambisoils, rankers, lithosoils and colluvisoils. Soil texture is dominated by silt mixed with gravels. The schist bedrock is relatively impermeable, while the soil surface and the Pleistocene periglacial slope deposits exhibit high infiltration rates and high storage capacity (Wrede et al., 2014).

Vegetation in the study catchment is mainly mixed oak–beech hardwood deciduous forest (76 % of the land cover, *Fagus sylvatica* L. and *Quercus petraea* (Matt.) Liebl.) where the soil surface is covered with fallen leaves. Conifers cover a smaller part (24 % land cover) of the catchment (*Pseudotsuga menziessii* (Mirb.) Franco and *Picea abies* (L.) H. Karst), and the soil surface beneath conifers is covered mainly by bryophytes. A well-defined riparian zone extends up to 3 m away from the stream channel. Vegetation in the riparian zone includes *Dryopteris carthusiana* (Vill.) H. P. Fuchs, *Impatiens noli-tangere* L., *Chrysosplenium oppositifolium* L. and *Oxalis acetosella* L.

3 Methodology

3.1 Hydrometric monitoring

Table 1 shows a summary of collection methods, sampling resolution and locations in the Weierbach catchment. Stream water depth at the catchment outlet was measured using a differential pressure transducer at a 15 min interval (ISCO 4120 Flow Logger) (Fig. 1). Stream electrical conductivity at the outlet was also measured at 15 min intervals using a conductivity meter (WTW). Rainfall was measured with a tipping bucket rain gauge (52203 model, manufactured by

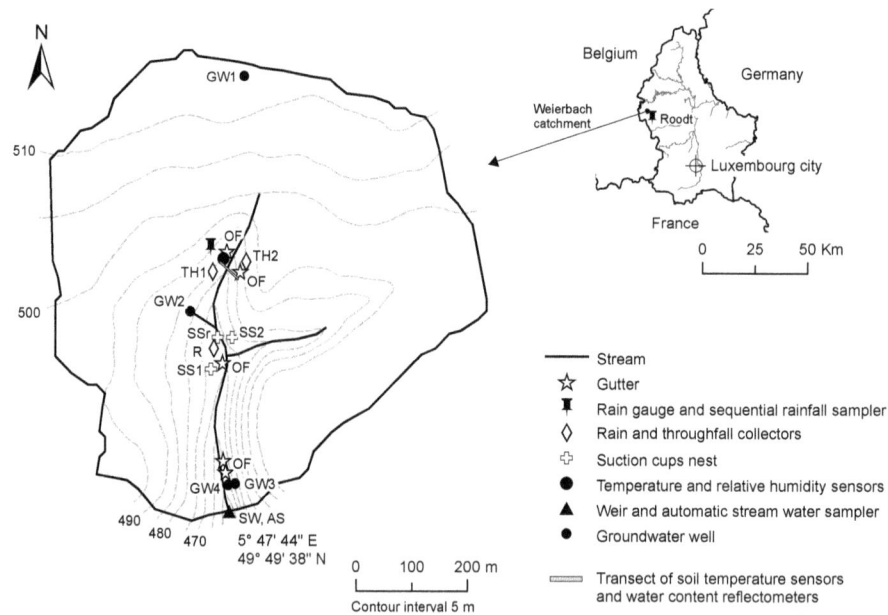

Figure 1. Detailed map of topography and instrumentation locations in the Weierbach catchment (northwest of Luxembourg City).

Table 1. Summary of collection methods, sampling resolution and locations in the Weierbach catchment.

	Component	Resolution	Method	No. of locations
Hydrology	Discharge	15 min	Stage-discharge rating curve	1 (outlet)
	Precipitation	15 min	Tipping bucket	2
	Water table depth	15 min	TD driver	4
	Soil moisture	30 min	Water content reflectometer	4
	Stream conductivity	15 min	Conductivity meter	1 (outlet)
	Groundwater conductivity	30 min	Conductivity meter	2
Geochemistry and isotopes	Groundwater	Fortnightly	Manual	4
	Overland flow (hillslope)	Accum. events	Gutters	5
	Precipitation	Accum. fortnightly	Rain gauge	1
	Precipitation	~2.5 mm increments	Sequential rainfall sampler	1
	Snow	Sporadic	Manual	Spots
	Soil water	Accum. fortnightly	Suction cups	3
	Stream water	1–6 h (events)	ISCO automatic sampler	1 (outlet)
	Stream water	Fortnightly	Manual	3
	Throughfall	Accum. fortnightly	Rain gauge	2
Diatoms	Epilithon	Once per season	Manual	3
	Epipelon	Once per season	Manual	3
	Overland flow (hillslope)	Accum. events	Gutters	5
	Stream water	1–6 h (events)	ISCO automatic sampler	1 (outlet)
	Stream water	Monthly	Manual	1 (outlet)
	Substrates	Once per season	Manual	16

Young, Campbell Scientific Ltd.). One rain gauge was installed within a small clearing of the study catchment (see Fig. 1), and another one installed in an open area at the Roodt meteorological station, located ≈ 3.5 km distant from the Weierbach one (49°48'22.2" N, 5°49'52.7" E). Data gaps

due to instrument failure were filled with rainfall data from a nearby weather station (49°47'39.2" N, 5°49'13.2" E).

Four groundwater wells were instrumented with real-time TD-Divers data loggers (Schlumberger Water Services) and WTW conductivity meters – each recording at 15 min intervals. GW1 was located in a plateau, and GW2, GW3 and GW4 in the transition zone between riparian and hillslope settings (Fig. 1). Wells were around 2 m deep and were screened at least for the lowest 50 cm up to a metre.

The volumetric water content (VWC) of soils was measured using water content reflectometers (CS616-L model, Campbell Scientific), which use the time-domain reflectometry method. Four probes were installed at 10 cm depth, parallel to the surface and along a 5 m transect perpendicular to the stream (Fig. 1): riparian zone, foot of the hillslope, mid-hillslope and plateau positions.

3.2 Water sampling and laboratory methods

Fortnightly, cumulative rainfall (R) and throughfall samples under deciduous trees (TH1) and coniferous trees (TH2) were collected using conical, volumetric rain gauges. A ten-bottle sequential rainfall sampler was installed at the rain gauge located within the Weierbach (modified from Kennedy et al., 1979). Three automatic water samplers (ISCO 3700 FS and 6712 FS) were installed immediately upstream of the weir to collect stream water samples (AS) frequently (0.5 to 4 h) during storm events. Sampling was triggered by flow conditions. Events were considered separately if they were separated by a period of at least 24 h without rainfall. Stream water at the catchment outlet (SW) and wells (GW1 to GW4) were sampled fortnightly, as well as prior to, during, and following precipitation events. Soil water was sampled fortnightly using Teflon suction lysimeters, installed at three locations: deciduous hillslope (SS1), coniferous hillslope (SS2), and riparian zone (SSr). Three soil depths for each location: 10 cm for the organic layer (Ah horizon), 20 and 60 cm for the mineral layers (B and C horizons). Overland flow (OF) that occurred on lower hillslope was sampled using 1 and 2 m long gutters sealed to the soil surface, which diverted surface runoff to 1 or 2 L plastic, blackened (to prevent light penetration which causes diatom growth) water bottles. Note that what we refer to as OF might in fact originate within the forest litter layer (Buttle and Turcotte, 1999; Sidle et al., 2007). All gutters were covered to avoid the influence of precipitation. Gutters were regularly cleaned with Milli-Q water to avoid diatom growth on their surfaces.

All water samples were analysed for electrical conductivity (EC), anion and cation concentrations (Cl^-, NO_3^-, SO_4^{2-}, Na^+, K^+, Mg^{2+}, Ca^{2+}), silica (SiO_2) and UV absorbance at 254 nm (Abs 254 nm). UV absorbance at 254 nm can be considered as a proxy of DOC (Edzwald et al., 1985). Samples were analysed at the Luxembourg Institute of Science and Technology chemistry laboratory after filtration through WHATMAN GF / C glass fibre filters ($< 0.45 \mu m$). Prior to analysis, samples were stored at 4 °C. Dissolved anions and cations were analysed by ion chromatography (Dionex HPLC), SiO_2 by spectrophotometry (ammonium molybdate method), and UV absorbance was measured by a Beckmann Coulter spectrophotometer. Isotopic analyses of $^{18}O / ^{16}O$ and $^2H / H$ were conducted using a LGR Liquid-Water Iso-

tope Analyser (LWIA) at the Luxembourg Institute of Science and Technology (model DLT-100, version 908-0008) (Penna et al., 2010). The analyser was connected to a LC PAL liquid auto-injector for the automatic and simultaneous measurement of $^2H / H$ and $^{18}O / ^{16}O$ ratios in water samples. According to the manufacturer's specifications (Los Gatos Research Inc., 2008), the DLT-100 908-0008 LWIA provides isotopic measurements with a precision below 0.6‰ for $^2H / H$ and 0.2‰ for $^{18}O / ^{16}O$. Data were transformed into δ notation relative to Vienna Standard Mean Ocean Water (VSMOW) standards (δ^2H and $\delta^{18}O$ in ‰).

3.3 Diatom sampling, sample preparation and analysis

Diatom analysis was conducted for multiple sample types: stream water, overland flow, epilithon, epipelon, and diatoms attached to different substrates outside the streambed (i.e. litter, bryophytes, vegetation and soils).

A small set of stream water and overland flow samples was set aside for geochemical and isotopic analysis (≈ 70 mL); the rest of the sample was centrifuged (1250 rpm, 8 min) to concentrate the diatoms.

In addition to high-frequency sampling during rainfall events, seasonal sampling campaigns were carried out throughout the Weierbach catchment to assess the geographic and intra-annual variability of diatom communities. The following substrates were sampled in the catchment: (i) litter, bryophytes from the two hillslope classifications (hardwood and coniferous) and surface soil samples; and (ii) litter, bryophytes, and vegetation in the riparian zone. Each sample was comprised of five sub-samples collected on a 5 m transect parallel to the stream (a subsample collected every metre). Only material from the top surface, where there was greatest incident sunlight, was collected into 1 L plastic bottles. Sample bottles containing different substrata were filled with carbonated water (1 L), carefully shaken and left to settle overnight at 0 °C. The next day, the diatom-filled, carbonated water was recovered by passing it through a 1 mm screen. Sample substrate was then rinsed with additional carbonated water to remove as many diatoms from the sampled substrate as possible. This procedure was repeated several times until a 2 L sample volume was achieved. The recovered sample, now with substrate removed, was stored at 0 °C for a minimum of 8 h to allow diatoms to settle, and the supernatant removed by aspiration.

During the same catchment-wide campaigns, epilithic (in-stream stone substrata) and epipelic (in-stream sediment or soil substrata) samples were also collected, treated and counted following European standards CEN 13946 and CEN 14407 (European Committee for Standardization, 2003, 2004). For epilithic samples a minimum of five stones from the main flow and well-lit stream reaches were brushed to collect the diatom biofilm, while epipelic samples were collected by disturbing small pools with sediment bottoms and then pipetting a superficial layer of 5–10 mm of sediment from reach pools.

All samples were preserved with 4% formaldehyde and treated with hot hydrogen peroxide to obtain clean frustule suspensions. After eliminating the organic matter from the diatom suspensions, diluted HCl was added to remove the calcium carbonate and avoid its precipitation later, which would make diatom frustule observation difficult. Finally, oxidized samples were rinsed with deionized water by decantation of the suspension several times, and permanent slides were mounted with Naphrax®.

Diatom valves were identified and counted (≈ 400 valves) on microscopic slides with a light microscope (Leica DMRX®). For the autecological assignment of the diatom species we relied on (1) the Denys (1991) diatom ecological classification system refined by Van Dam et al. (1994), which is, as far as we know, the only formal classification of the occurrence of freshwater diatoms in relation to moisture; and (2) the associated hydrological units assigned by Pfister et al. (2009) to the five diatom occurrence classes defined by Van Dam et al. (1994). We express these results as relative abundance (percentage) of aerial valves, i.e. categories 4 and 5 of Van Dam's et al. (1994) classification.

3.4 Hydrograph separation

Two-component hydrograph separation was performed using $\delta^{18}O$ isotopic composition and the mass balance approach (Pinder and Jones, 1969; Sklash and Farvolden, 1982; Pearce et al., 1986; Sklash et al., 1986). The incremental mean method proposed by McDonnell et al. (1990) was used to adjust $\delta^{18}O$ rainfall isotopic composition, so that the bulk isotopic composition of rainfall from the beginning of the event to the time of stream sampling was calculated (i.e. rain that had not yet fallen was excluded from the estimate).

Spatial end-member contributions to stream water were explored using EMMA (Christophersen and Hooper, 1992), which assumes that (i) the stream water is a mixture of end-member solutions with a fixed composition, (ii) the mixing model is linear and relies on hydrodynamic mixing, (iii) the solutes used as tracers are conservative, and (iv) the end-member solutions are distinguishable from one another. Catchment end-members included shallow groundwater (GW1-4), soil water ($SS1_{20}$, $SS1_{60}$, $SS2_{60}$), soil water from the riparian zone (SSr), rainfall (R), throughfall (TH1-2), snow (SN) and overland flow (OF). We applied the diagnostic tools of Hooper (2003), which have been recently applied in the literature (James and Roulet, 2006; Ali et al., 2010; Barthold et al., 2011; Neill et al., 2011; Inamdar et al., 2013). Our approach followed three main steps.

1. We identified tracers that exhibit conservative linear mixing assuming that stream water chemistry is controlled by physical mixing of different sources of water and not by equilibrium mixing (Christophersen and Hooper, 1992; Hooper, 2003; Liu et al., 2008). The latest would imply equilibrium reactions among solutes of different charge, which may be approximated by high-order polynomials. Hooper (2003) suggested that conservative and linear mixing of tracers can be evaluated using bivariate scatter plots. In this study, stream water concentrations and isotopic compositions (of all samples collected during storm events and low flows at the catchment outlet) were considered conservative when they exhibited at least one linear trend with one other tracer (i.e. $r^2 > 0.5$, p value < 0.01) (James and Roulet, 2006; Ali et al., 2010; Barthold et al., 2011).

2. We performed a principal component analysis (PCA) on the stream water data. The PCA was applied on the correlation matrix of the standardized values of tracers selected in step (i) (i.e. by subtracting the mean concentration or isotopic composition of each solute and dividing by its standard deviation) (Christophersen and Hooper, 1992). For each water tracer, residuals were defined by subtracting the original value from its orthogonal projection. A "good" mixing subspace was indicated by a random pattern of residuals plotted against the concentration or isotopic composition of the original values. On the contrary, structure or curvature in the subspace indicates violation against one of the assumptions of the EMMA approach (i.e. solutes do not mix conservatively) (Hooper, 2003). Eigenvectors were retained until there was no structure to the residuals. Standardized data were multiplied by the eigenvectors and projected into the new U space.

3. Finally, potential end-members were standardized using the mean and standard deviation of the stream water data. Their inter-quartile values (i.e. 25 and 75 %) were then multiplied by the eigenvectors and projected into the U space of the stream water samples. Those end-members that best met the constraints of the mixing model theory as described by Christophersen and Hooper (1992) and Hooper (2003) were identified. Similar to previous studies, rather than calculating precise end-member contributions, we investigated the arrangement and relative positioning of all potential end-members with respect to stream flow in the U space (Inamdar et al., 2013). In order to account for end-member temporal variability, end-member concentrations and isotopic compositions for specific storm events were determined by considering the samples collected during the event, as well as the preceding and following months (Inamdar et al., 2013).

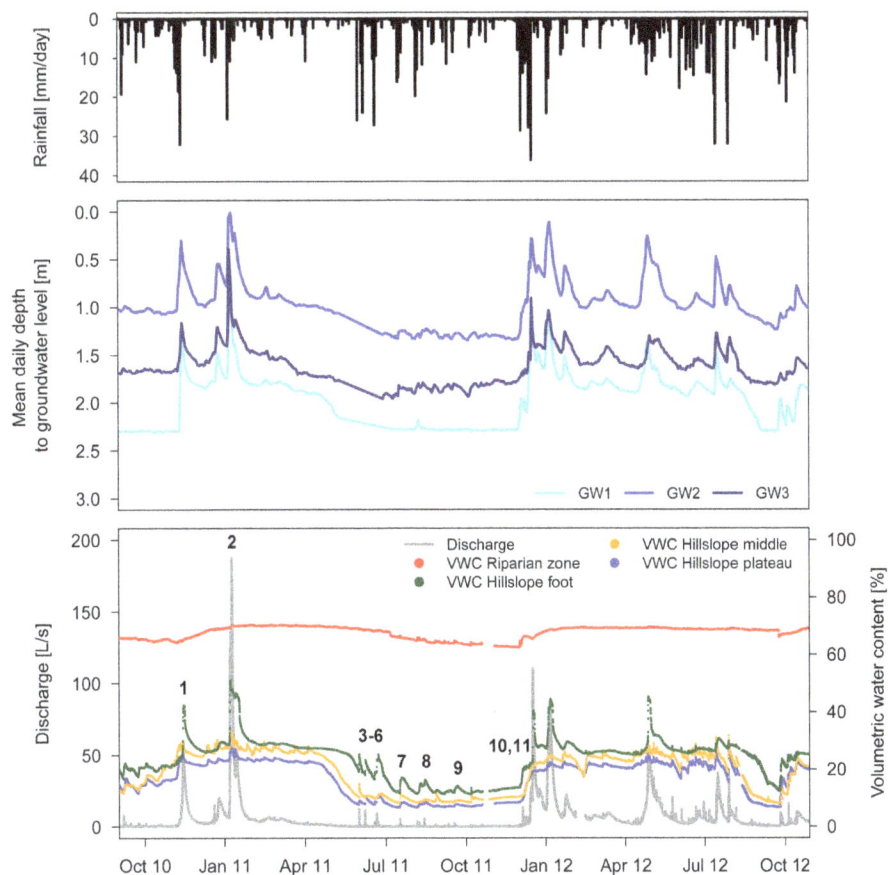

Figure 2. Time series of daily rainfall measured at the Roodt meteorological station (≈ 3.5 km distant from the Weierbach) (upper plot), mean daily groundwater depth at three different locations (GW1: plateau; GW2: close to a spring; and GW3: hillslope foot) (middle plot) and soil volumetric water content measured in a transect from the hillslope plateau to the riparian zone along with corresponding water discharge (lower plot). Numbers in the lower plot identify sampled storm events.

4 Results

4.1 Hydrometric response

The hydrometric response for water years 2011–2012 is shown in Fig. 2. Diatom sampling commenced in November 2010 when the catchment started to progressively wet up (see groundwater depths and soil volumetric water content in Fig. 2). Annual precipitation for the water year 2011 was 671 mm, a $\sim 20\%$ decrease compared to the average of the preceding 4 years (873 mm, as measured by the nearby meteorological station, Roodt), and 838 mm for the water year 2012. In January 2011, a 10-year return period rain-on-snow event produced a peak flow of $1.5\,\mathrm{mm\,h^{-1}}$. The high winter discharge levels decreased progressively from February to June 2011 due to reduced precipitation during this period. Afterwards, a dry period extended from July to November 2011. A longer wet period was measured the following year (from December 2011 to July 2012).

During wet antecedent conditions, streamflow response of the basin was double peaked, with a first peak timing coincident with the rainfall input and the second, delayed peak coming a few hours later. On the contrary, when the catchment was dry, the hydrological response was shorter and only a single sharp peak occurred.

We determined hydrological connectivity along a HRS transect via hydrometric observations. Water tables in the saprolite and fractured schist bedrock responded significantly to rainfall events. The magnitude of water level change was well correlated with the precipitation amount. Soil volumetric water content (VWC) decreased with distance upslope (VWC hillslope foot > VWC hillslope middle > VWC hillslope plateau (Fig. 2)). The riparian zone showed unchanging values close to saturation during wet periods ($\approx 70\,\%$), which decreased slightly when the catchment was dry ($\approx 65\,\%$). For all monitored events, VWC at 10 cm depth responded quickly to incident rainfall at all transect locations (i.e. hillslope foot, middle and plateau), suggesting a vertically infiltrating, wetting front.

During dry antecedent conditions (summer and spring), threshold-like behaviour between soil moisture and discharge was observed at the hillslope foot (Fig. 3a). Only

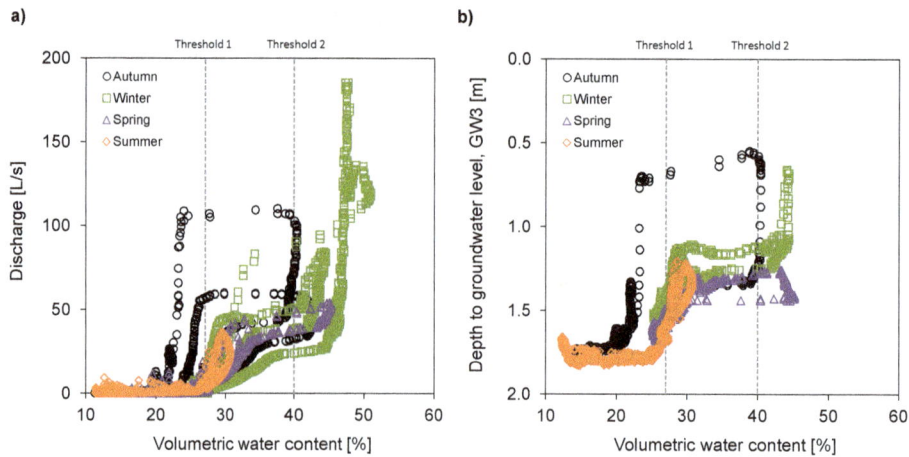

Figure 3. Relationship between (**a**) volumetric water content (hillslope foot) and discharge, and (**b**) between volumetric water content and depth to groundwater level for the period plotted in Fig. 2. Vertical dashed lines represent two threshold values (see details in the text).

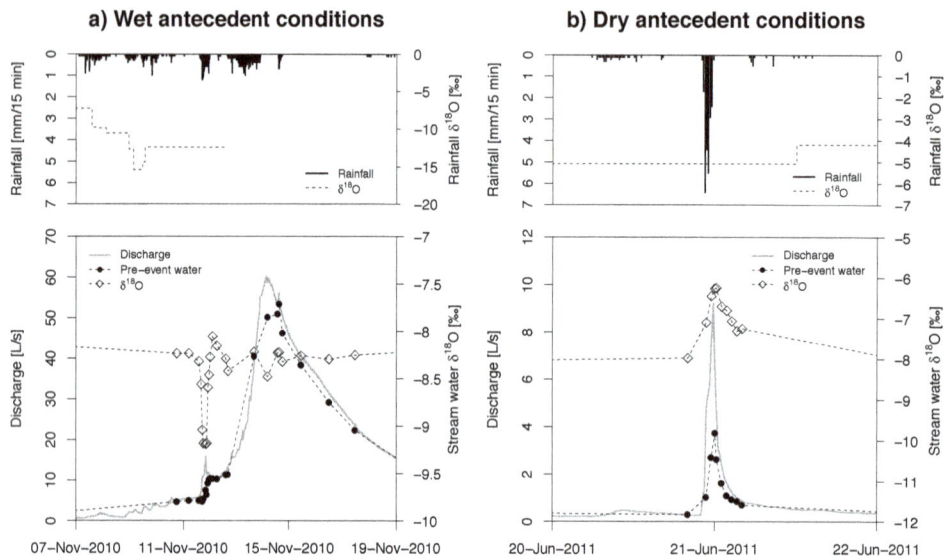

Figure 4. Two-component hydrograph separation for (**a**) the 7 November 2010 event (wet antecedent conditions) and (**b**) 20 June 2011 event (dry antecedent conditions) using $\delta^{18}O$ isotopic composition.

when the VWC was higher than ≈ 27–30% did discharge increase significantly (threshold 1 in Fig. 3a). A second threshold appeared when the catchment was wet (autumn and winter); stream discharge increased significantly when VWC was above 40% (threshold 2 in Fig. 3a). This likely indicated connectivity between the hillslope and riparian compartments and the stream channel. A similar relationship was observed between VWC and depth to groundwater levels (i.e. GW1, GW2 and GW3; Fig. 3b).

4.2 Hydrograph separation

Two-component hydrograph separation results using $\delta^{18}O$ isotopic composition (i.e. pre-event water vs. event water) showed that, in winter, when the catchment was wet and

flow response was double-peaked, the first peak had a larger contribution of event water than the delayed peak. For instance, the first peak of the November 2010 event showed a maximum of 50% event water contribution. This contrasted with the delayed peak that exhibited only a maximum of 16% event water contribution (Fig. 4a). When the catchment was dry, the response consisted of one sharp peak composed largely of event water. A maximum event-water contribution of 60% was estimated for a storm event that occurred in June 2011 (Fig. 4b).

Twelve different tracers measured in the different water compartments of the catchment were used to assess end-member contributions to stream water (Fig. 5). Ten out of the twelve tracers presented linear trends in the solute–solute plots of stream water samples with at least one other tracer

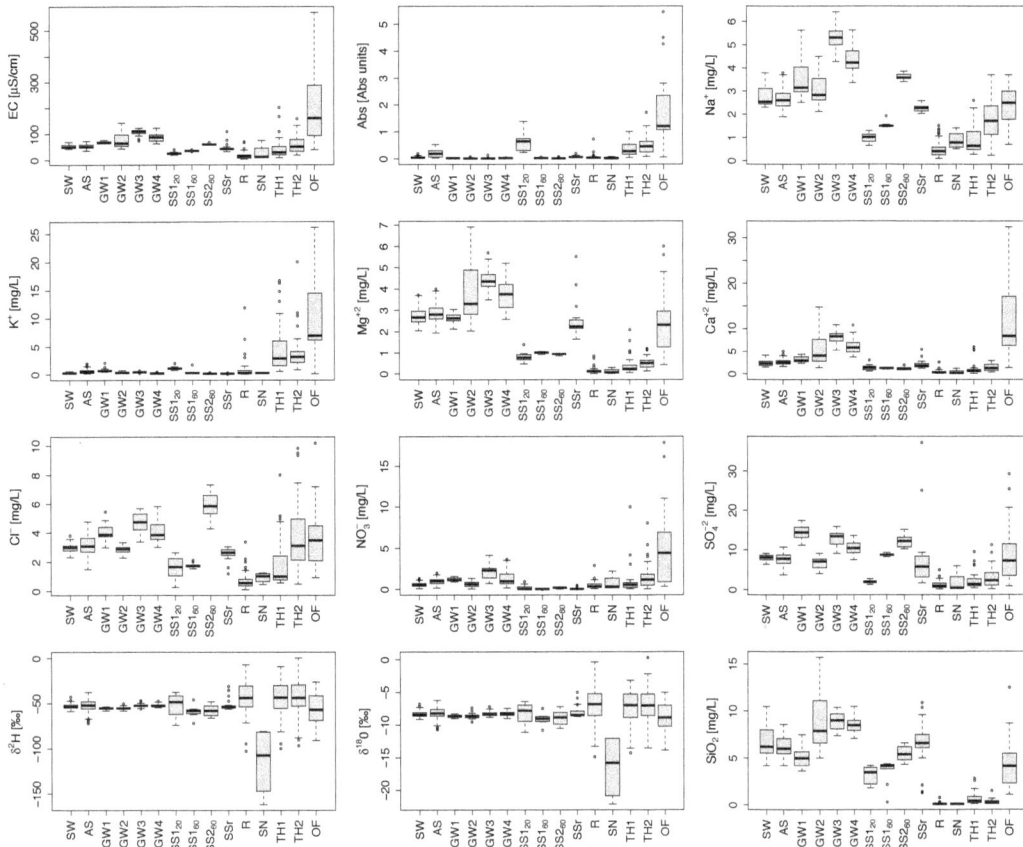

Figure 5. Boxplots of tracers measured for stream water sampled fortnightly (SW, $n = 47$) and using automatic samplers (AS, $n = 179$), groundwater (GW1, $n = 24$; GW2, $n = 49$; GW3, $n = 49$; GW4, $n = 47$), soil water (SS1$_{20}$, $n = 22$; SS1$_{60}$, $n = 10$; SS2$_{60}$, $n = 9$), soil water from the riparian zone (SSr, $n = 21$), rainfall (R, $n = 44$), snow (SN, $n = 4$), throughfall (TH1, $n = 35$; TH2, $n = 38$) and overland flow (OF, $n = 21$). Outliers were discarded.

(EC, Cl^-, Na^+, K^+, Mg^{2+}, Ca^{2+}, SiO_2, Abs, δ^2H and $\delta^{18}O$; $r^2 > 0.5$, p value < 0.01, Fig. 6). These tracers were retained for the PCA analysis. Weaker linear trends were found between NO_3^- and the other tracers ($r^2 < 0.13$) and between SO_4^{2-} and the other tracers ($r^2 < 0.43$). NO_3^- and SO_4^{2-} did not reach the pre-defined threshold of collinearity ($r^2 > 0.5$), and were therefore not retained.

A PCA was performed on the correlation matrix of stream concentrations and isotopic compositions for the ten selected tracers. The first three principal components explained 91.3 % of the variance in stream concentrations and isotopic compositions and were selected to generate a three-dimensional mixing space (U space, Table 2). Plots of residuals of each solute plotted against observed concentrations and isotopic compositions suggested that three components were needed to obtain a well-defined mixing subspace. End-member tracer concentrations and isotopic compositions were then projected into the mixing space (Fig. 7). All stream water samples are plotted inside the mixing domain defined by the end-members. Rainfall, throughfall, soil water and soil water from the riparian zone end-members are plotted in the

upper right quadrant of the U1–U2 mixing space (Fig. 7a). Shallow groundwater samples were located in the lower left quadrant and snow in the lower right quadrant. Overland flow is plotted in the upper left quadrant and was located furthest away from stream water samples and with the largest interquartile ranges. Most of the stream water samples were clustered in the immediate vicinity of the soil water from the riparian zone samples, half-way between the throughfall and the groundwater samples. Snow seems to contribute to some stream water samples that are placed slightly more toward the lower right quadrant (Fig. 7a). The large distance between stream water and overland flow samples suggests a minor role of the latter in total runoff generation. Event peak-flow samples are highlighted in Fig. 7b. In general, results show that when the catchment was wet, there was a higher contribution of groundwater to streamflow (events 1–2 and 10–11) than when the catchment antecedent condition was dry (events 3–9). However, compared to winter (events 1–2), a much higher contribution of throughfall was estimated during summer (events 5–8), when the pre-storm catchment state was dry.

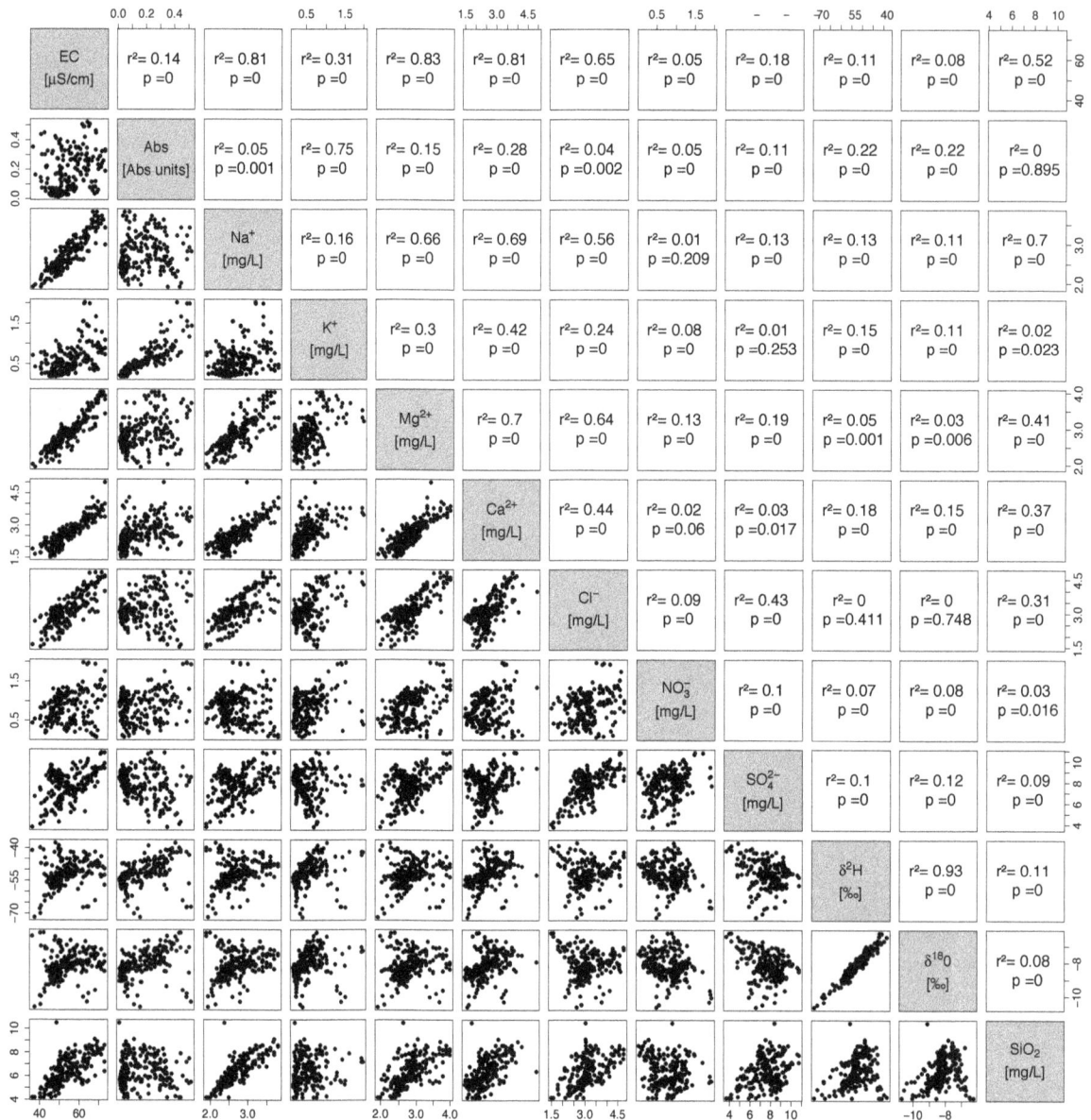

Figure 6. Bivariate plots of stream water chemistry and water stable isotope data collected at the outlet of the Weierbach catchment ($n = 226$; SW and AS displayed in Fig. 5). The upper part of the diagonal shows the Pearson correlation coefficient and its significance at the 0.95 confidence level.

In order to better understand water pathways during each event separately, we plotted stream water samples collected for each event and end-member tracer signatures in the previously determined two-dimensional mixing space (Figs. 8 and 9). We accounted for end-member temporal variability by plotting not only end-member samples collected the same month as the event occurred, but also the preceding and following months. Groundwater and rainfall signals remained relatively constant throughout the year, whereas throughfall, riparian and soil water presented higher temporal variability. Results showed that runoff mixing patterns changed between events. During autumn and winter, when the catchment was wet (events 1–2, and 10–11), stream water signal composi-

tion was most similar to riparian, soil water and groundwater. Only samples collected during the rain-on-snow event (event 2) might have a small contribution of not only overland flow but also snow. Mixing patterns changed during spring and summer when the catchment was drier (i.e. events 3 to 9). As previously seen in Fig. 7b, groundwater seems to have a much lower contribution to stream water, since stream water samples are now plotted in an intermediate position between throughfall and soil water from the riparian zone (with the exception of event 3, which still has a significant groundwater contribution). Note that overland flow did not occur and the soils were dry during these spring and summer events.

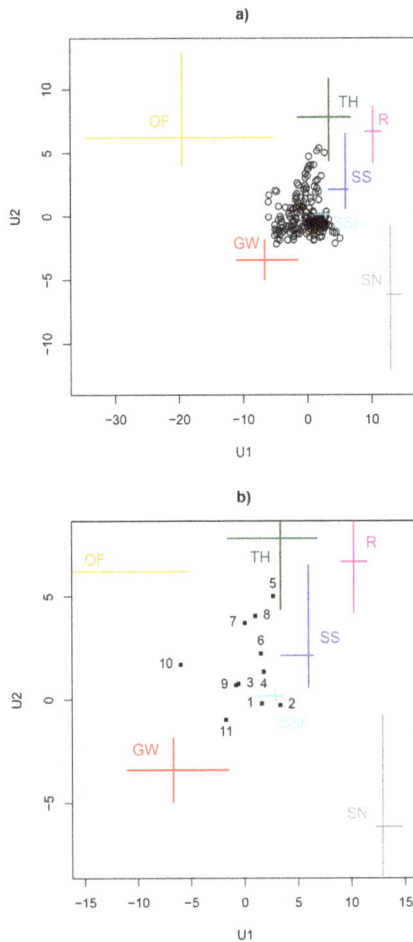

Table 2. Variance explained by each eigenvector ($n = 210$).

Eigenvectors	Proportion of variance explained, %	Accumulated variance explained, %
1	57.6	57.6
2	20.5	78.1
3	13.2	91.3
4	2.8	94.0
5	2.3	96.4
6	1.4	97.8
7	0.8	98.6
8	0.6	99.2
9	0.5	99.7
10	0.3	100

Figure 7. (a) U1–U2 mixing diagram of stream water tracers (black circles; AS + SW in Fig. 5) and **(b)** zoom into the U1–U2 mixing diagram showing event peakflow stream water samples (black squares; numbers identify storm events in Fig. 2). Sampling points data plotted in Fig. 5 were grouped into seven end-members and the interquartile ranges of each end-member were projected into the new mixing space (U space; GW: groundwater; SN: snow; SS: soil water; SSr: soil water from the riparian zone; OF: overland flow; R: rainfall; TH: throughfall). Because **(b)** is a zoom into the U1–U2 mixing diagram, the interquartile ranges of some end-members are not fully represented.

4.3 Seasonal and geographic variability in aerial diatom communities in the hillslope-riparian-stream system

The qualitative and semi-quantitative analysis of diatom microflora revealed 230 taxa in the Weierbach catchment. Diatom communities from samples collected during the seasonal campaigns in the streambed (i.e. epilithon, epipelon and stream water samples) during low flow were usually composed of species from oligotrophic environments, mainly occurring in water bodies, but also rather regularly on wet and moist surfaces (i.e. the riparian zone hydrological func-

tional unit of Pfister et al., 2009), such as *Achnanthes saxonica* Krasske, *Achnanthidium kranzii* (Lange-Bertalot) Round & Bukthiyarova, *Fragilariforma virescens* (Ralfs) D. M. Williams & Round, *Eunotia botuliformis* F. Wild, Nörpel & Lange-Bertalot, and *Planothidium lanceolatum* (Brébisson) Lange-Bertalot. Important seasonal changes in the relative abundance of aerial diatoms amongst the sampled habitats were not observed (Table 3). The null hypothesis of equal distributions was tested with the Mann–Whitney U test for the samples from the riparian zone and the hillslope (too small an amount of stream water at low flow and streambed samples). P values were high (0.21 and 0.73 for the riparian zone and the hillslope samples, respectively) and the null hypothesis was accepted. No diatom valves were found in groundwater or rainfall samples.

The riparian zone was characterized by several species that prefer aerial habitats, mainly living on exposed soils or epiphytically on bryophytes. Such species occur mainly in wet and moist or temporarily dry places or live nearly exclusively outside water bodies (categories 4 and 5 of Pfister et al., 2009), such as *Chamaepinnularia evanida* (Hustedt) Lange-Bertalot, *C. parsura* (Hustedt) C. E. Wetzel & Ector, *Eunotia minor* (Kützing) Grunow, *Hantzschia abundans* Lange-Bertalot, *Nitzschia harderi* Hustedt, *Orthoseira dendroteres* (Ehrenberg) Round, R. M. Crawford & D. G. Mann, *Pinnularia borealis* Ehrenberg, *P. perirrorata* Krammer, *Stauroneis parathermicola* Lange-Bertalot and *S. thermicola* (J. B. Petersen) J. W. G. Lund.

Diatoms were completely absent in samples from dry litter on the hillslope and only occurred on bryophytes. Almost no diatoms were found in overland flow samples. The relative abundance of aerial valves was higher in hillslopes and riparian samples compared to streambed samples (Table 3). However, we found a higher number of aerial diatoms (in absolute numbers) in the riparian zone. This emphasizes the importance of the riparian zones as the main terrestrial diatom source during rainfall, when diatoms are mobilized

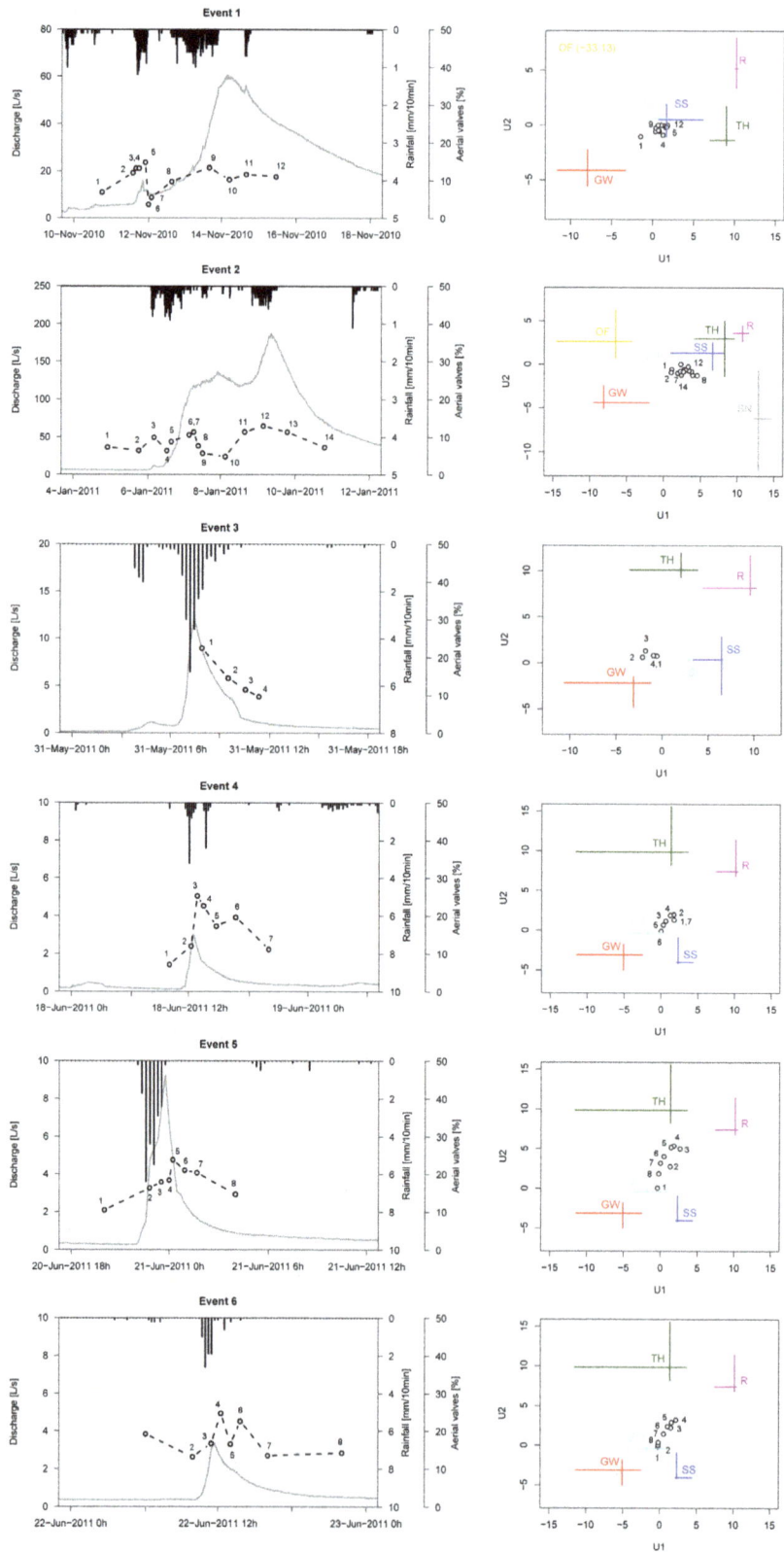

Figure 8. Hydrograph, hyetograph and percentage of aerial valves in the stream water for events 1–6 in the Weierbach catchment (left), and U1–U2 mixing diagrams for each event. End-members are rainfall (R), throughfall (TH), snow (SN), soil water (SS), soil water from the riparian zone (SSr) and groundwater (GW). Bars represent end-member values' interquartile ranges of samples collected during the month when the event occurred, as well as the previous and following months.

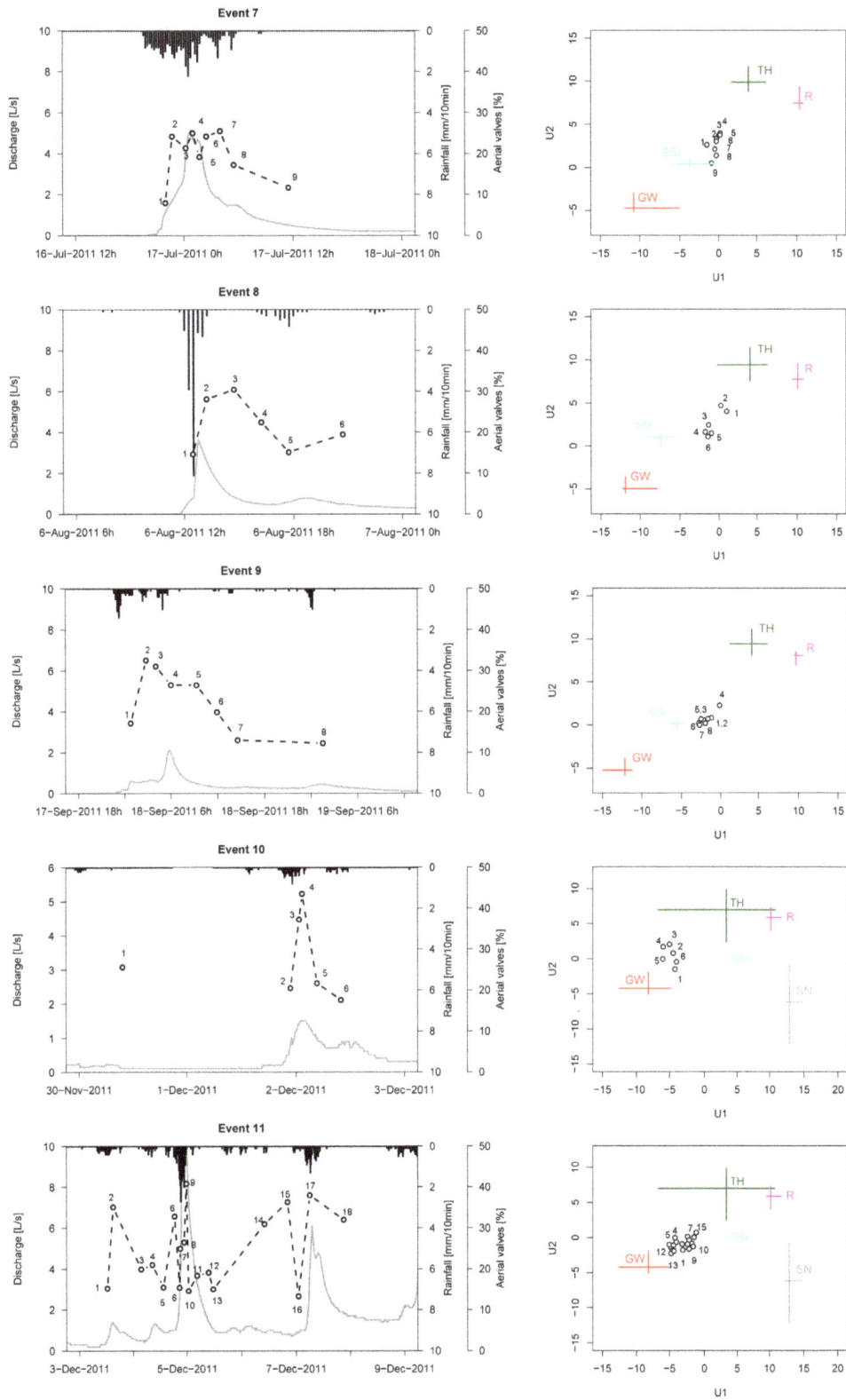

Figure 9. Hydrograph, hyetograph and percentage of aerial valves in the stream water for events 7–11 in the Weierbach catchment (left), and U1–U2 mixing diagrams for each event. End-members are rainfall (R), throughfall (TH), snow (SN), soil water (SS), soil water from the riparian zone (SSr) and groundwater (GW). Bars represent end-member values' interquartile ranges of samples collected during the month when the event occurred, as well as the previous and following months.

Table 3. Relative percentage of aerial valves quantified in distinct zones of the Weierbach catchment. Streambed samples refer to epilithon samples. Riparian zone samples include litter, bryophytes and vegetation. Hillslope samples include litter, bryophytes and surface soil samples. Diatoms were absent on hillslopes covered by dry litter and samples were discarded.

	Sample	n	Min (%)	Max (%)	Mean (%)	SD (%)
Summer 2010	Stream water at low flow	3	10.1	19.4	14.9	4.6
	Streambed	6	14.8	21.7	19.0	2.7
	Riparian zone	25	8.5	61.5	22.9	16.9
	Hillslope	12	11.6	96.6	36.5	27.0
Winter 2011	Stream water at low flow	8	5.9	16.1	9.8	3.3
	Streambed	2	5.0	8.8	6.9	2.7
	Riparian zone	39	12.4	67.2	21.9	12.0
	Hillslope	16	11.3	100.0	40.4	26.4

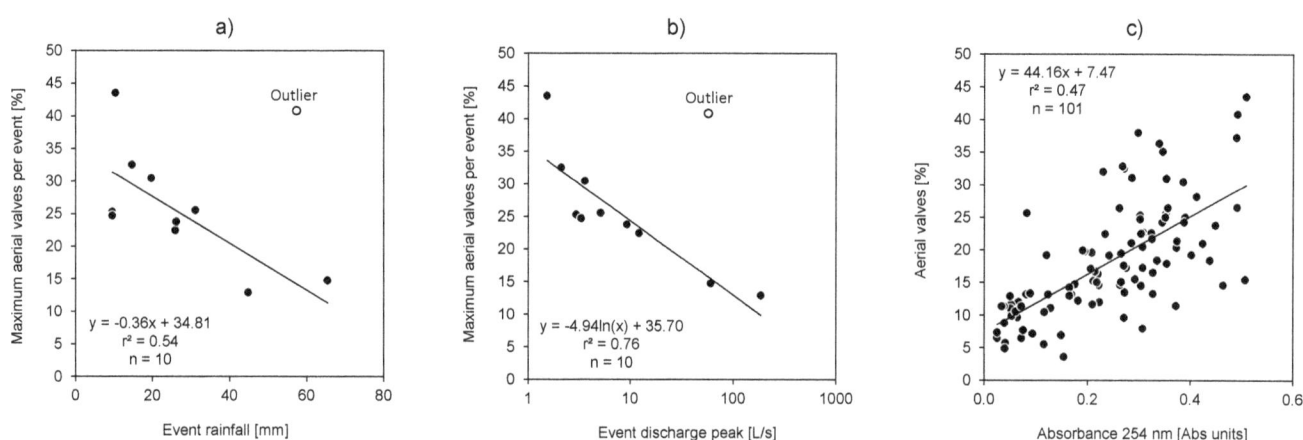

Figure 10. Correlations between (**a**) maximum percentage of aerial valves in the stream water per event and event rainfall, (**b**) maximum percentage of aerial valves in the stream water per event and maximum event discharge, and (**c**) percentage of aerial valves in the stream water and UV absorbance at 254 nm.

from moist or temporarily dry habitats into the stream channel (Table 3).

4.4 Aerial diatom transport during rainfall events

A series of 11 rainfall events were sampled from November 2010 to December 2011 during both wet and dry catchment conditions (Table 4 and Fig. 2). The main aerial species found in stream water during storm events were as follows: *Chamaepinnularia evanida*, *C. obsoleta* (Hustedt) C. E. Wetzel & Ector, *C. parsura*, *Humidophila brekkaensis* (J. B. Petersen) R. L. Lowe et al., *H. perpusilla* (Grunow) R. L. Lowe et al., *Eolimna tantula* (Hustedt) Lange-Bertalot, *Eunotia minor*, *Pinnularia obscura* Krasske, *P. perirrorata*, *Stauroneis parathermicola*, and *S. thermicola*.

Stream water samples taken throughout storm hydrographs showed a systematic increase in aerial diatoms as a response to incident precipitation and increasing discharge (Figs. 8 and 9). During events, the minimum increment of aerial valves' relative abundance was 8.1 % (event 2),

whereas the maximum increment was 27 % (event 11). The maximum percentage of aerial valves was 43.5 % (event 10).

No significant relationship was found between the percentage of aerial diatoms and instantaneous discharge ($r^2 = 0.13$, $n = 101$; discharge on the x axis), most probably due to different diatom abundances on the rising limb of the hydrograph than on the recession limb (i.e. hysteretic effects). Two events showed clockwise hysteretic loops (events 1 and 2); five events showed counter-clockwise hysteretic loops (events 4, 5, 6, 8, and 10) and three showed figure-eight shaped hysteretic loops (events 7, 9 and 11). Although a clear pattern was not observed, results suggest that clockwise hysteretic loops predominated during wet conditions (the greater percentages of aerial diatoms in streamflow were immediately before peakflow), and counter-clockwise hysteretic loops during dry conditions (the greater percentages were immediately after peakflow).

Aerial valves comprised less than 15 % of the total diatoms in low flow samples for all events except 6, 9 and 10 (which had 19.2, 17.1, and 25.6 %, respectively). Due to technical problems, no low-flow sample was collected for event 3. No

Table 4. General hydrological characteristics of the sampled rainfall-runoff events that occurred from October 2010 to December 2011 in the Weierbach catchment.

	Beginning of precipitation	Duration (h)	Total P (mm)	Maximum intensity (mm 15 min^{-1})	Antecedent P, 10 days (mm)	Antecedent P, 20 days (mm)	Pre-event discharge (L s^{-1})	Maximum discharge (L s^{-1})
Event 1	11 Nov 2010	154	65	1.2	42	49	5.4	60.4
Event 2	06 Jan 2011	142	45	0.9	–	–	6.1	187.5
Event 3	31 May 2011	14	26	5.4	1	4	0.1	12.2
Event 4	18 Jun 2011	10	10	3.2	8	71	0.1	3.0
Event 5	20 Jun 2011	14	26	6.4	25	62	0.3	9.2
Event 6	22 Jun 2011	13	10	2.6	51	89	0.4	3.4
Event 7	16 Jul 2011	29	31	2.2	6	8	0	5.2
Event 8	06 Aug 2011	12	20	8.1	7	21	0	3.6
Event 9	17 Sep 2011	49	15	1.4	12	22	0	2.1
Event 10	01 Dec 2011	46	10	0.8	2	3	0.1	1.5
Event 11	03 Dec 2011	124	57	2.7	13	14	0.2	13.1

relationship was observed between antecedent event rainfall and the percentage of aerial valves observed during low flow ($n = 10$, $r^2 = 0.08$ and 0.09 for 10 and 20 days of antecedent rainfall, respectively).

At event scale, there were significant correlations between maximum percentage of aerial diatoms and event rainfall and maximum event discharge ($r^2 = 0.54$, $p < 0.05$, $n = 10$, Fig. 10a; $r^2 = 0.76$, $p < 0.05$, $n = 10$, Fig. 10b, respectively; the multi-peak event sampled in December 2011 was considered as an outlier). High percentages ($> 35\%$) of aerial diatom relative abundance were measured during dry catchment conditions, compared to when the catchment was wet, where maximum relative abundances were low ($< 15\%$). Alternatively, higher maximum percentages of aerial diatom proportions ($> 35\%$) were measured during dry catchment conditions, when events were shorter and more intense.

A significant correlation between percentage of aerial diatoms with UV absorbance at 254 nm was found ($r^2 = 0.55$, $p < 0.05$, $n = 76$, Fig. 10c). During rainfall events in the Weierbach catchment, the relative abundance of aerial diatoms was associated with increased organic matter concentrations in the stream. A similar trend was observed with K$^+$ ($r^2 = 0.25$, $p < 0.05$, $n = 76$), which is also associated with organic matter content. The relative abundance of aerial diatoms was not correlated with any other tracers.

5 Discussion

5.1 Can aerial diatoms transport reveal hydrological connectivity within the hillslope-riparian-stream system?

Our central hypothesis for this study was that aerial diatoms could indicate connectivity within the HRS system. In order to test this hypothesis, we sampled from potential upland catchment sources (i.e. hillslope and riparian zones), and within the streambed (i.e. epilithon, epipelon and stream water samples).

Before testing our central hypothesis, we tested for the existence of distinguishable diatom species assemblages on the hillslope, the riparian zone and the stream. Only if diatom assemblages are distinguishable between these zones can their presence in the channel during rainfall events serve as a proxy for HRS connectivity. Results showed clear differences in diatom species assemblages between the hillslopes, riparian zone and streams, with higher relative abundance of aerial diatoms on the hillslopes and in the riparian zones compared to the stream (Table 3). Diatoms are usually abundant in moist environments (Van de Vijver and Beyens, 1999; Nováková and Poulíčková, 2004; Chen et al., 2012; Vacht et al., 2014), but in spite of the presence of diatoms in bryophyte-covered areas of the hillslopes, we did not find any diatom valves in hillslopes covered by dry litter. Moreover, the quantities of aerial diatoms found on the hillslopes covered by bryophytes and in the overland flow gutter samples were small and sometimes not sufficient to fully characterize the zone (due to the rarity of some species but also linked to sampling difficulties). This constrained the use of aerial diatoms to infer hillslope-riparian zone connectivity in some parts of the Weierbach catchment because of a limited diatom reservoir on hillslopes.

Despite the highest relative abundance of aerial valves on the hillslope compared to the riparian zone, the riparian zone was still the largest aerial diatom reservoir (in absolute numbers) with the highest probability of connecting to the stream (Table 3). We did not observe significant seasonal differences in diatom species assemblages among the different sampled habitats.

We examined the aerial diatoms transported in the stream water during runoff events. We observed an increase in the relative abundance of aerial diatoms with discharge for all sampled events regardless of antecedent wetness conditions.

Hence, during storm events there was an increase in the relative proportion of diatoms in categories 4 and 5 of Van Dam's et al. (1994) classification. Similar results were reported by Pfister et al. (2009). These observations imply hydrological connectivity between the riparian soil surface and the stream for all events. The use of aerial diatoms to infer hydrological connectivity in the Weierbach catchment thus remains limited to the riparian-stream system as no diatoms were found on the hillslopes covered by dry litter.

Even though aerial diatoms do not live in microhabitats with flowing water, they were found in stream water samples during low flow conditions preceding storm events (Table 3). This indicated that the "stock" of aerial diatoms in the catchment before the sampled events was not completely exhausted during previous events. Similar conclusions were drawn by Coles et al. (2015), who examined diatom population depletion effects during rainfall and found that while aerial diatom populations in the riparian zone were depleted in response to rainfall disturbance, rainfall was unlikely to completely exhaust the diatom reservoir.

We hypothesize that the transport of diatoms from the riparian zone to the stream might take place either through (i) a network of macropores in the shallow soils of the riparian zone or (ii) overland flow in the riparian zone. The potential for diatoms to be transported through the subsurface matrix was investigated using fluorescent diatoms and soil columns by Tauro et al. (2015). Results demonstrated that sub-surface transport of diatoms through the sub-surface matrix was unlikely. However, the potential for transport of diatoms through heterogeneous macropore networks remains unexplored. The increased relative abundance of aerial diatoms in the stream event water could also be explained by as yet undocumented surface or near-surface pathways.

5.2 How do diatom results compare to the other methods to infer hydrological connectivity?

Two-component hydrograph separation and EMMA provide valuable information on water sources and flowpaths. Using these methods we learned that in the Weierbach catchment, during spring and summer, the hydrological response was largely composed of event water (see an example of dry antecedent catchment conditions in Fig. 4b). Similar conclusions were drawn by Wrede et al. (2014) using dissolved silica. Accordingly, EMMA results suggest canopy throughfall, rainfall and riparian soil water were the main water sources (Figs. 8 and 9). As observed in other headwater catchments (e.g. Penna et al., 2011), discharge likely increased due to channel interception and riparian runoff leading to clear and singular hydrograph peaks (Fig. 4b). During fall and winter, when the catchment was at its wettest state, double peaked hydrographs characterized the event hydrological response. Hydrograph separation indicated that the first peak was mainly event water and the delayed, second peak was mostly pre-event water (Fig. 4a; Wrede et al., 2014). During

these events, soil water, groundwater, and throughfall contributed substantially to total discharge (Figs. 8 and 9). Hillslope overland flow was insignificant during most sampled events. Only for event 2 – the largest storm on record – was overland flow a significant contributor to stream discharge, likely due to rapid snowmelt onto a surface-saturated area (Figs. 8 and 9).

During all sampled events the relative abundance of aerial diatoms increased with discharge indicating hydrological connectivity between the riparian zone and the stream. These findings are consistent with the hydrograph separation results. Aerial diatoms could reach the stream as saturated areas expand during rainfall events. Accordingly, we found a significant correlation between percentage of aerial diatoms with UV absorbance (proxy of DOC). DOC concentrations associated with runoff storm often come mainly from the near-stream riparian zones (Boyer et al., 1997). Controls on surface saturated and subsurface mixing processes are currently being investigated in the Weierbach riparian zone using infrared imagery and groundwater metrics (Pfister et al., 2010).

Hydrological connectivity between hillslopes and the stream has also been previously defined by water table connections between the hillslope and the riparian zone (Vidon and Hill, 2004; Ocampo et al., 2006; Jencso et al., 2010; McGuire and McDonnell, 2010). While our results showed that overland flow did not occur on hillslopes during most sampled events, the VWC measurements and timing of the hydrograph response suggest that subsurface hydrological connectivity along the HRS system occurs during wet catchment conditions (Fig. 3). Hence, if aerial diatoms found on the hillslopes, might reach the stream through sub-surface flowpaths remains unknown. Others have demonstrated that tracer transport can occur on larger timescales that extend beyond individual events (McGuire and McDonnell, 2010). Whether this may also be true for diatoms remains to be explored.

5.3 Can aerial diatoms be established as a new hydrological tracer?

Storm hydrograph separation using stable isotope tracers has resulted in major advances in catchment hydrology. However, despite their usefulness, these methods do not provide unequivocal evidence of hydrological connectivity in the HRS system. In comparison, diatoms can provide evidence of riparian-stream connectivity. Further research is needed to better understand diatom transport processes (and associated water flowpaths) in headwater catchments. Future studies should focus on expanding our understanding of terrestrial diatom taxonomy and ecology, which are scarce or lacking for a large number of taxa (Wetzel et al., 2013, 2014). Even though this new data source will have its own individual measurement uncertainty (McMillan et al., 2012), diatoms offer the possibility to tackle open questions in hydrology and eco-hydrology.

A key issue with the concept of hydrological connectivity is how it can be applied across and between environments. Uncertainties increase when applying two-component hydrograph separation at large scales. For instance, Klaus and McDonnell (2013) note that quantifying the spatial variability in the isotope signal of rainfall and snowmelt can be difficult in large catchments and in catchments with complex topography. Similarly, some studies showed that, for mesoscale catchments, only qualitative results of the contribution of a runoff component can be obtained by the hydrograph separation techniques (Uhlenbrook and Hoeg, 2003). For aerial diatoms to be useful and a way forward to increase our understanding of hydrological pathways at a range of scales, they must be also relevant across environments and scales (Bracken et al., 2013). The current concepts related to HRS connectivity are best suited to humid, temperate settings (Beven, 1997; Bracken and Croke, 2007) and represent only very specific settings (Bracken et al., 2013). Previous investigations in Luxembourg have shown that freshwater diatom assemblages in headwater streams have regional distributions strongly affected by geology, as well as anthropogenic factors (e.g. organic pollution sources and eutrophication) (Rimet et al., 2004). Hence, we speculated that diatoms have potential in headwater systems, and at larger catchment scales to determine connectivity between contrasting geological zones.

The need to account for the temporal variability in end-member chemistry and to collect high-frequency data on both – stream water as well as potential runoff end-members – has been well recognized (Inamdar et al., 2013). As noted by Tetzlaff et al. (2010), seasonality should also be considered when using living organisms to trace water flowpaths. Diatom end-members must be sampled seasonally in order to ensure that populations have not undergone demographic changes. Indeed, this increases the sampling needs and the overall laboratory procedures of an already time-consuming approach (i.e. sampling, pre-treating the samples, mounting permanent slides and diatom identification). A potential alternative to reduce processing time is to develop new techniques such as to dye diatom valves and use them to trace water flowpaths (see Tauro et al., 2015). The use of dyed diatoms under field conditions for experimental hydrology remains unexplored.

6 Conclusions

We investigated the potential for aerial diatoms, i.e. diatoms nearly exclusively occurring outside water bodies and in wet and moist or temporarily dry places (Van Dam et al., 1994), to serve as natural tracers capable of detecting connectivity within the HRS system. We found that the relative abundance of aerial diatoms in stream water samples collected during storm events increased with runoff during all seasons. Sampling of the potential catchment sources of diatoms in the HRS system and inside the stream channel (i.e. epilithon, epipelon and stream water samples) indicated that riparian zones appear to be the largest aerial diatom reservoir. Few diatom valves were found in overland flow samples and diatoms were completely absent on leaf-covered hillslopes, occurring only in hillslope samples with bryophytes and limiting the use of aerial diatoms to infer hillslope-riparian zone connectivity. Nonetheless, we have shown the use of diatoms to quantify riparian-stream connectivity as the relative abundance of aerial diatoms increased with discharge during all sampled events. Although further research is needed to determine the exact pathways that aerial diatoms use to reach the stream, diatoms offer the possibility of address open questions in hydrology at small and large catchment scales.

Acknowledgements. Funding for this research was provided by the Luxembourg National Research Fund (FNR) in the framework of the BIGSTREAM (C09/SR/14), ECSTREAM (C12/SR/40/8854) and CAOS (INTER/DFG/11/01) projects. We are most grateful to the Administration des Services Techniques de l'Agriculture (ASTA) for providing meteorological data. We also acknowledge Delphine Collard for technical assistance in diatom sample treatment and preparation, François Barnich for the water chemistry analyses, and Jean-François Iffly, Christophe Hissler, Jérôme Juilleret, Laurent Gourdol and Julian Klaus for their constructive comments on the project and technical assistance in the field.

Edited by: A. Butturini

References

Ali, G. A. and Roy, A. G.: Shopping for hydrologically representative connectivity metrics in a humid temperate forested catchment, Water Resour. Res., 46, W12544, doi:10.1029/2010WR009442, 2010.

Ali, G. A., Roy, A. G., Turmel, M. C., and Courchesne, F.: Source-to-stream connectivity assessment through end-member mixing analysis, J. Hydrol., 392, 119–135, doi:10.1016/j.jhydrol.2010.07.049, 2010.

Barthold, F. K., Tyralla, C., Schneider, K., Vaché, K. B., Frede, H. G., and Breuer, L.: How many tracers do we need for end member mixing analysis (EMMA)? A sensitivity analysis, Water Resour. Res., 47, W08519, doi:10.1029/2011WR010604, 2011.

Beven, K.: TOPMODEL: a critique, Hydrol. Process., 11, 1069–1085, 1997.

Bonell, M.: Selected challenges in runoff generation research in forests from the hillslope to headwater drainage basin scale, J. Am. Water Resour. As., 34, 765–785, doi:10.1111/j.1752-1688.1998.tb01514.x, 1998.

Boyer, E. W., Hornberger, G. M., Bencala, K. E., and Mcknight, D. M.: Response characteristics of DOC flushing in an alpine catchment, Hydrol. Process., 11, 1635–1647, doi:10.1002/(SICI)1099-1085(19971015)11:12<1635::AID-HYP494>3.0.CO;2-H, 1997.

Bracken, L. J. and Croke, J.: The concept of hydrological connectivity and its contribution to understanding runoff-dominated geomorphic systems, Hydrol. Process., 21, 1749–1763, doi:10.1002/hyp.6313, 2007.

Bracken, L. J., Wainwright, J., Ali, G. A., Tetzlaff, D., Smith, M. W., Reaney, S. M., and Roy, A. G.: Concepts of hydrological connectivity: Research approaches, pathways and future agendas, Earth-Sci. Rev., 119, 17–34, doi:10.1016/j.earscirev.2013.02.001, 2013.

Burns, D. A.: Stormflow-hydrograph separation based on isotopes: the thrill is gone – what's next?, Hydrol. Process., 16, 1515–1517, doi:10.1002/hyp.5008, 2002.

Buttle, J. M.: Fundamentals of small catchment hydrology, edited by: Kendall, C. and McDonnel, J. J.: Isotope tracers in catchment hydrology, Elsevier, Amsterdam, 1–49, 1998.

Buttle, J. M. and Turcotte, D. S.: Runoff processes on a forested slope on the Canadian Shield, Nord. Hydrol., 30, 1–20, doi:10.2166/nh.1999.001, 1999.

Chen, X., Bu, Z., Yang, X., and Wang, S.: Epiphytic diatoms and their relation to moisture and moss composition in two montane mires, Northeast China, Fund. Appl. Limnol., 181, 197–206, doi:10.1127/1863-9135/2012/0369, 2012.

Christophersen, N. and Hooper, R. P.: Multivariate analysis of stream water chemical data: The use of principal components analysis for the end-member mixing problem, Water Resour. Res., 28, 99–107, doi:10.1029/91WR02518, 1992.

Coles, A. E., Wetzel, C. E., Martínez-Carreras, N., Ector, L., McDonnell, J. J., Frentress, J., Klaus, J., Hoffmann, L., and Pfister, L.: Diatoms as a tracer of hydrological connectivity: are they supply limited?, Ecohydrology, accepted, doi:10.1002/eco.1662, 2015.

Denys, L.: A check-list of the diatoms in the Holocene deposits of the Western Belgian coastal plain with a survey of their apparent ecological requirements. I. Introduction, ecological code and complete list, Service Géologique de Belgique-Belgische Geologische Dienst, Professional Paper 1991/2-246, 41 pp., 1991.

Dixit, S. S., Dixit, A. S., and Smol, J. P.: Diatom and chrysophyte transfer functions and inferences of post-industrial acidification and recent recovery trends in Killarney lakes (Ontario, Canada), J. Paleolimnol., 27, 79–96, doi:10.1023/A:1013571821476, 2002.

Ector, L. and Rimet, F.: Using bioindicators to assess rivers in Europe: an overview, edited by: Lek, S., Scardi, M., Verdonschot, P. F. M., Descy, J. P., and Park, Y. S.: Modelling Community Structure in Freshwater Ecosystems, Springer-Verlag, Berlin, Heidelberg, 7–19, 2005.

Edzwald, J. K., Becker, W. C., and Wattier, K. L.: Surrogate parameters for monitoring organic matter and THM precursors, J. Am. Water Works Assess., 77, 122–132, 1985.

European Committee for Standardization: Water quality – guidance standard for the routine sampling and pretreatment of benthic diatoms from rivers, [EN 13946:2003], European Committee for Standardization, Brussels, 2003.

European Committee for Standardization: Water quality – guidance standard for the identification, enumeration and interpretation of benthic diatom samples from running waters, [EN 14407:2004]. European Committee for Standardization, Brussels, 2004.

Hooper, R. P.: Applying the scientific method to small catchment studies: a review of the Panola Mountain experience, Hydrol. Process., 15, 2039–2050, doi:10.1002/hyp.255, 2001.

Hooper, R. P.: Diagnostic tools for mixing models of stream water chemistry, Water Resour. Res., 39, 1055, doi:10.1029/2002WR001528, 2003.

Inamdar, S., Dhillon, G., Singh, S., Dutta, S., Levia, D., Scott, D., Mitchell, M., Van Stan, J., and McHale, P.: Temporal variation in end-member chemistry and its influence on runoff mixing patterns in a forested, Piedmont catchment, Water Resour. Res., 49, 1828–1844, doi:10.1002/wrcr.20158, 2013.

James, A. L. and Roulet, N. T.: Investigating the applicability of end-member mixing analysis (EMMA) across scale: A study of eight small, nested catchments in a temperate forested watershed, Water Resour. Res., 42, W08434, doi:10.1029/2005WR004419, 2006.

Jencso, K. G., McGlynn, B. L., Gooseff, M. N., Wondzell, S. M., Bencala, K. E., and Marshall, L. A.: Hydrologic connectivity between landscapes and streams: Transferring reach- and plot-scale understanding to the catchment scale, Water Resour. Res., 45, W04428, doi:10.1029/2008WR007225, 2009.

Jencso, K. G., McGlynn, B. L., Gooseff, M. N., Bencala, K. E., and Wondzell, S. M.: Hillslope hydrologic connectivity controls riparian groundwater turnover: Implications of catchment structure for riparian buffering and stream water sources, Water Resour. Res., 46, W10524, doi:10.1029/2009WR008818, 2010.

Juilleret, J., Iffly, J. F., Pfister, L., and Hissler, C.: Remarkable Pleistocene periglacial slope deposits in Luxembourg (Oesling): pedological implication and geosite potential, B. Soc. Naturalistes Luxemb., 112, 125–130, 2011.

Kennedy, V. C., Zellweger, G. W., and Avanzino, R. J.: Variation of rain chemistry during storms at two sites in northern California, Water Resour. Res., 15, 687–702, doi:10.1029/WR015i003p00687, 1979.

Klaus, J. and McDonnell, J. J.: Hydrograph separation using stable isotopes: Review and evaluation, J. Hydrol., 505, 47–64, doi:10.1016/j.jhydrol.2013.09.006, 2013.

Lischeid, G.: Combining hydrometric and hydrochemical data sets for investigating runoff generation processes: Tautologies, inconsistencies, and possible explanations, Geogr. Compass, 2, 255–280, doi:10.1111/j.1749-8198.2007.00082.x, 2008.

Liu, F., Bales, R. C., Conklin, M. H., and Conrad, M. E.: Streamflow generation from snowmelt in semi-arid, seasonally snow-covered, forested catchments, Valles Caldera, New Mexico, Water Resour. Res., 44, W12443, doi:10.1029/2007WR006728, 2008.

Los Gatos Research, Inc.: Liquid-Water Isotope Analyser, Automated Injection, 2008.

Mann, D. G.: Diatoms: organism and image, edited by: du Buf, H. and Bayer, N. M.: Automatic Diatom Identification, World Scientific Publishing, Singapore, 9–40, 2002.

McDonnell, J. J.: Are all runoff processes the same?, Hydrol. Process., 27, 4103–4111, doi:10.1002/hyp.10076, 2013.

McDonnell, J. J., Bonell, M., Stewart, M. K., and Pearce, A. J.: Deuterium variations in storm rainfall: Implications for stream hydrograph separation, Water Resour. Res., 26, 455–458, doi:10.1029/WR026i003p00455, 1990.

McGuire, K. J. and McDonnell, J. J.: Hydrological connectivity of hillslopes and streams: Characteristic time scales and nonlinearities, Water Resour. Res., 46, W10543, doi:10.1029/2010WR009341, 2010.

McMillan, H., Krueger, T., and Freer, J.: Benchmarking observational uncertainties for hydrology: rainfall, river discharge and water quality, Hydrol. Process., 26, 4078–4111, doi:10.1002/hyp.9384, 2012.

Neill, C., Chaves, J. E., Biggs, T., Deegan, L. A., Elsenbeer, H., Figueiredo, R. O., Germer, S., Johnson, M. S., Lehmann, J., Markewitz, D., and Piccolo, M. C.: Runoff sources and land cover change in the Amazon: an end-member mixing analysis from small watersheds, Biogeochemistry, 105, 7–18, doi:10.1007/s10533-011-9597-8, 2011.

Nováková, J. and Poulíčková, A.: Moss diatom (Bacillariophyceae) flora of the Nature Reserve Adršpašsko-Teplické Rocks (Czech Republic), Czech Phycol., 4, 75–86, 2004.

Ocampo, C. J., Sivapalan, M., and Oldham, C.: Hydrological connectivity of upland-riparian zones in agricultural catchments: Implications for runoff generation and nitrate transport, J. Hydrol., 331, 643–658, doi:10.1016/j.jhydrol.2006.06.010, 2006.

Pearce, A. J., Stewart, M. K., and Sklash, M. G.: Storm runoff generation in humid headwater catchments: 1. Where does the water come from?, Water Resour. Res., 22, 1263–1272, doi:10.1029/WR022i008p01263, 1986.

Penna, D., Stenni, B., Šanda, M., Wrede, S., Bogaard, T. A., Gobbi, A., Borga, M., Fisher, B. M. C., and Bonazza, M.: On the reproducibility and repeatability of laser absorption spectroscopy measurements for δ^2H and δ^{18}O isotopic analysis, Hydrol. Earth Syst. Sci., 14, 1551–1566, doi:10.5194/hess-14-1551-2010, 2010.

Penna, D., Tromp-Van Meerveld, H. J., Gobbi, A., Borga, M., and Dalla Fontana, G.: The influence of soil moisture on threshold runoff generation processes in an Alpine headwater catchment, Hydrol. Earth Syst. Sci., 15, 689–702, doi:10.5194/hess-15-689-2011, 2011.

Pfister, L., Wagner, C., Vansuypeene, E., Drogue, G., and Hoffmann, L.: Atlas climatique du Grand-Duché de Luxembourg, Musée National d'Histoire Naturelle, Société des Naturalistes Luxembourgeois, Centre de Recherche Public – Gabriel Lippmann, Administration des Services Techniques de l'Agriculture, Luxembourg, 80 pp., 2005.

Pfister, L., McDonnell, J. J., Wrede, S., Hlúbiková, D., Matgen, P., Fenicia, F., Ector, L., and Hoffmann, L.: The rivers are alive: on the potential for diatoms as a tracer of water source and hydrological connectivity, Hydrol. Process., 23, 2841–2845, doi:10.1002/hyp.7426, 2009.

Pfister, L., McDonnell, J. J., Hissler, C., and Hoffmann, L.: Ground-based thermal imagery as a simple, practical tool for mapping saturated area connectivity and dynamics, Hydrol. Process., 24, 3123–3132, doi:10.1002/hyp.7840, 2010.

Pfister, L., Wetzel, C. E., Martínez-Carreras, N., Iffly, J. F., Klaus, J., Holko, L., and McDonnell, J. J.: Examination of aerial diatom flushing across watersheds in Luxembourg, Oregon and Slovakia for tracing episodic hydrological connectivity, J. Hydrol. Hydromech., 63, 239–249, doi:10.1515/johh-2015-0031, 2015.

Pinder, G. F. and Jones, J. F.: Determination of the groundwater component of peak discharge from the chemistry of total runoff, Water Resour. Res., 5, 438–445, doi:10.1029/WR005i002p00438, 1969.

Rimet, F., Ector, L., Cauchie, H. M., and Hoffmann, L.: Regional distribution of diatom assemblages in the headwater streams of Luxembourg, Hydrobiologia, 520, 105–117, doi:10.1023/B:HYDR.0000027730.12964.8c, 2004.

Round, F. E., Crawford, R. M., and Mann, D. G.: The diatoms. Biology and morphology of the genera, Cambridge University Press, Cambridge, 1990.

Sidle, R. C., Hirano, T., Gomi, T., and Terajima, T.: Hortonian overland flow from Japanese forest plantations – an aberration, the real thing, or something in between?, Hydrol. Process., 21, 3237–3247, doi:10.1002/hyp.6876, 2007.

Sklash, M. G. and Farvolden, R. N.: The use of environmental isotopes in the study of high-runoff episodes in streams, in: Isotope Studies of Hydrologic Processes, edited by: Perry, E. C. J. and Montgomery, C. W., Northern Illinois University Press, Illinois, 65–73, 1982.

Sklash, M. G., Stewart, M. K., and Pearce, A. J.: Storm runoff generation in humid headwater catchments: 2. A case study of hillslope and low-order stream response, Water Resour. Res., 22, 1273–1282, doi:10.1029/WR022i008p01273, 1986.

Spence, C.: A paradigm shift in hydrology: Storage thresholds across scales influence catchment runoff generation, Geogr. Compass, 4, 819–833, doi:10.1111/j.1749-8198.2010.00341.x, 2010.

Tauro, F., Martínez-Carreras, N., Barnich, F., Juilleret, J., Wetzel, C. E., Ector, L., Hissler, C., and Pfister, L.: Diatom percolation through soils: a proof of concept laboratory experiment, Ecohydrology, under review, 2015.

Tetzlaff, D., Soulsby, C., and Birkel, C.: Hydrological connectivity and microbiological fluxes in montane catchments: the role of seasonality and climatic variability, Hydrol. Process., 24, 1231–1235, doi:10.1002/hyp.7680, 2010.

Tromp-van Meerveld, H. J. and McDonnell, J. J.: Threshold relations in subsurface stormflow: 1. A 147-storm analysis of the Panola hillslope, Water Resour. Res., 42, W02410, doi:10.1029/2004WR003778, 2006.

Uhlenbrook, S. and Hoeg, S.: Quantifying uncertainties in tracer-based hydrograph separations: a case study for two-, three- and five-component hydrograph separations in a mountainous catchment, Hydrol. Process., 17, 431–453, doi:10.1002/hyp.1134, 2003.

Vacht, P., Puusepp, L., Koff, T., and Reitalu, T.: Variability of riparian soil diatom communities and their potential as indicators of anthropogenic disturbances, Est. J. Ecol., 63, 168–184, doi:10.3176/eco.2014.3.04, 2014.

Van Dam, H., Mertens, A., and Sinkeldam, J.: A coded checklist and ecological indicator values of freshwater diatoms from The Netherlands, Neth. J. Aquat. Ecol., 28, 117–133, doi:10.1007/BF02334251, 1994.

Van de Vijver, B. and Beyens, L.: Moss diatom communities from Ile de la Possession (Crozet, Subantarctica) and their relationship with moisture, Polar Biol., 22, 219–231, doi:10.1007/s003000050414, 1999.

Vidon, P. G. F. and Hill, A. R.: Landscape controls on the hydrology of stream riparian zones, J. Hydrol., 292, 210–228, doi:10.1623/hysj.51.6.1021, 2004.

Wetzel, C. E., Martínez-Carreras, N., Hlúbiková, D., Hoffmann, L., Pfister, L., and Ector, L.: New combinations and type analysis of *Chamaepinnularia* species (Bacillariophyceae) from aerial habitats, Cryptogamie Algol., 34, 149–168, doi:10.782/crya.v34.iss2.2013.149, 2013.

Wetzel, C. E., Van de Vijver, B., Kopalová, K., Hoffmann, L., Pfister, L., and Ector, L.: Type analysis of the South American diatom *Achnanthes haynaldii* (Bacillariophyta) and description of *Planothidium amphibium* sp. nov., from aerial and aquatic environments in Oregon (USA), Plant Ecol. Evol., 147, 439–454.

Wrede, S., Fenicia, F., Martínez-Carreras, N., Juilleret, J., Hissler, C., Krein, A., Savenije, H. H. G., Uhlenbrook, S., Kavetski, D., and Pfister, L.: Towards more systematic perceptual model development: a case study using 3 Luxembourgish catchments, Hydrol. Process., doi:10.1002/hyp.10393, 2014.

4

Changes in dissolved organic matter quality in a peatland and forest headwater stream as a function of seasonality and hydrologic conditions

Tanja Broder[1,2], **Klaus-Holger Knorr**[2], **and Harald Biester**[1]

[1]IGÖ, Umweltgeochemie, TU Braunschweig, Langer Kamp 19c, 38106 Braunschweig, Germany
[2]ILÖK, Hydrologie, WWU Münster, Heisenbergstr. 2, 48149 Münster, Germany

Correspondence to: Tanja Broder (broder@uni-muenster.de)

Abstract. Peatlands and peaty riparian zones are major sources of dissolved organic matter (DOM), but are poorly understood in terms of export dynamics and controls thereof. Thereby quality of DOM affects function and behavior of DOM in aquatic ecosystems, but DOM quality can also help to track DOM sources and their export dynamics under specific hydrologic preconditions. The objective of this study was to elucidate controls on temporal variability in DOM concentration and quality in stream water draining a bog and a forested peaty riparian zone, particularly considering drought and storm flow events. DOM quality was monitored using spectrofluorometric indices for aromaticity ($SUVA_{254}$), apparent molecular size (S_R) and precursor organic material (FI), as well as PARAFAC modeling of excitation emission matrices (EEMs).

Indices for DOM quality exhibited major changes due to different hydrologic conditions, but patterns were also dependent on season. Stream water at the forested site with mineral, peaty soils generally exhibited higher variability in DOM concentrations and quality compared to the outflow of an ombrotrophic bog, where DOM was less susceptible to changes in hydrologic conditions. During snowmelt and spring events, near-surface protein-like DOM pools were exported. A microbial DOM fraction originating from groundwater and deep peat layers was increasing during drought, while a strongly microbially altered DOM fraction was also exported by discharge events with dry preconditions at the forested site. This might be due to accelerated microbial activity in the peaty riparian zone of the forested site under these preconditions. Our study demonstrated that DOM export dynamics are not only a passive mixing of different hydrological sources, but monitoring studies have to consider that DOM quality depends on hydrologic preconditions and season. Moreover, the forested peaty riparian zone generated the most variability in headwater DOM quantity and quality, as could be tracked by the used spectrofluorometric indices.

1 Introduction

Dissolved organic matter (DOM) is ubiquitous in soils and aqueous ecosystems. It plays a fundamental role in surface water chemistry, e.g., in metal bioavailability and mobility (Tipping et al., 2002), nutrient cycling (Jansson et al., 2012), pH buffering and ionic balance (Hruška et al., 2003). It affects light penetration (Karlsson et al., 2009), the aquatic food web structure (Jansson et al., 2007), is an energy source for microbial metabolism (Cole et al., 2007; Amon and Benner, 1996) and is part of the carbon cycle (Cole et al., 2007). But not only DOM quantity is of great interest, as the DOM quality strongly affects function and behavior of DOM in aquatic ecosystems.

Most DOM input to aquatic systems is of terrestrial origin (see Mulholland, 2003). Concentrations and characteristics of DOM vary strongly among surface waters depending on catchment, climate and hydrology (Ågren et al., 2014; Laudon et al., 2004; Frost et al., 2006; Winterdahl et al., 2014). However, DOM concentrations and characteristics can also vary largely over time due to seasonal changes in production, consumption and transport of DOM (e.g., Fellman et al., 2009; Perdrial et al., 2014; Wallin et al., 2015). Peatlands, which store large amounts of carbon, have thereby received attention as a major source of DOM to surface water (Worrall et al., 2002; Aitkenhead et al., 1999). But wet riparian zones with organic-rich layers are also recognized as a DOM source (e.g., Bishop et al., 2004; Seibert et al., 2009; Laudon et al., 2004, Ledesma et al., 2015). The annual dissolved organic carbon (DOC) concentration dynamics and long-term DOC concentration increase, observed for many catchments (Monteith et al., 2007; Worrall et al., 2004), points out the importance to understand DOM origin and factors controlling DOM export. Storm events have been shown to be quantitatively important for DOM exports to streams in peatland catchments (Clark et al., 2007) and carbon-rich riparian zones, as they generate high DOM concentration peaks. DOM quality, as well as DOM quantity, is especially important for drinking-water production. Aromatic structures of DOM could cause disinfection by-product (DBP) generation during drinking water treatment (Korshin et al., 1997). More aromatic, humic DOM also decreases light penetration and DOM photo-degradation potential in surface waters (Cory et al., 2007; Ward and Cory, 2016). The easily biodegradable DOM (BDOM) fraction – mainly fresh, protein-like DOM, derived from root or leaf exudates, litter decay, or leachates – can be readily utilized and serves as important nitrogen and phosphorus source in aquatic systems (Fellman et al., 2009). Microbially processed DOM are residual and recalcitrant substances. Assessing such variability in DOM quality can be a valuable tool to track DOM sources and transport mechanisms (Singh et al., 2014), which is crucial for predicting DOM exports and quality.

Comprehensive studies have mainly focused on total DOM concentration and much is known about the DOM export from peatland and forested catchments (e.g., Laudon et al., 2011; Grabs et al., 2012; Clark et al., 2009). Trends of DOM quality during storm flow events or differences depending on catchment type are scarce. Inamdar et al. (2011) and Hood et al. (2006) characterized DOM during storm events in a temperate forest catchment dominated by mineral soils. A high contribution of aromatic structures during storm flow was ascribed to flushing of humic-rich near-surface soil layers and lower contribution of shallow groundwater. Organic soil layer DOM can be highly aromatic or humic, reflecting decomposition of complex plant and soil organic matter. As DOM percolates through the soil, sorption to mineral phases preferentially removes larger, aromatic components (e.g., Meier et al., 2004; Kaiser and Zech, 2000), and longer residence times enhance alteration of DOM by microbial processes. Thus groundwater DOM is mostly of microbial origin and of apparently smaller molecular size (e.g., Inamdar et al., 2012; Singh et al., 2014). While Singh et al. (2014) described a strong pulse of protein-like DOM during fall leaf fall, Perdrial et al. (2014) perceived only modest shifts in DOM quality over seasons in a forested catchment in New Mexico. Fellman et al. (2009) focused on the bioavail-

able fraction of DOM from wetland and forest soils, and found a strong biotic control on BDOM interacting with abiotic processes and hydrologic flow paths. The BDOM fraction was highest during spring due to a low biotic demand and shallow flow paths. Ågren et al. (2008) reported higher aromaticity and apparent molecular size during snowmelt at a wetland catchment compared to a forest catchment. However, a comparison to other high discharge events during the growing season is lacking.

A limitation in DOM quality studies is that determination of DOM structures is elaborate and expensive, while large datasets and high temporal resolution would be desirable (Strohmeier et al., 2013). UV-Vis and fluorescence spectroscopy are limited in data interpretation in terms of specific chemical structures, but due to low cost and rapid analysis they enable us to generate a comprehensive dataset covering a wide range of hydrologic and seasonal conditions. It allows for distinction between different DOM constituents and the disentanglement of their specific export behavior, and might also be used to trace different DOM sources (Hood et al., 2006). Several optical indices describe the nature of DOM: specific ultra-violet absorbance at 254 nm ($SUVA_{254}$) is commonly used as indicator for the proportion of aromatic structures (Weishaar et al., 2003). The spectral slope ratio (S_R) (Helms et al., 2008) is used as a proxy for apparent DOM molecular size. The humification index (HIX) (Ohno, 2002) and fluorescence index (FI) (Cory and McKnight, 2005) are derived from fluorescence-based excitation–emission matrices (EEMs). While HIX describes the degree of humification, FI differentiates between plant-derived and microbial- or planktonic-derived DOM. The fluorescence EEMs can be further analyzed using parallel factor (PARAFAC) analysis, decomposing the EEMs into hypothetic fluorophores related to differences in composition of DOM (Stedmon and Bro, 2008; Murphy et al., 2013).

Following up on a previous study describing DOC fluxes and concentration dynamics from a bog catchment (Broder and Biester, 2015), the present study intends to elucidate different spatiotemporal dynamics in DOM quality over a year, comparing a bog and a forested peaty riparian zone, as those landscape types are considered as the main sources of stream DOM. We hypothesized that DOM quality is highly variable in a headwater stream depending on hydrologic conditions and season. In addition, we expected that DOM quality at the forested site is more affected by changes in hydrologic conditions than at the bog site. Furthermore, we tested if spectrofluorometric indices can be used to track DOM sources and their dynamics under specific hydrologic (pre)conditions. We expected short-term DOM quality changes due to high discharge events, which cause changes in hydrologic flow paths in the catchment, such as a development of surface flow networks or a connection of organic-rich surface layers to discharging waters. Changes in DOM composition in stream water might further reflect DOM sources of shallow groundwater or deeper peat layers versus organic-rich upper soil layers or near-surface peat layers. This short-term pattern was expected to be overlain by seasonal DOM changes due to changes of DOM production and consumption over the year. General differences in DOM quality between the bog and forested riparian zone catchments are caused by differences in vegetation, water level fluctuations and an existence of mineral soil layers.

To test our hypotheses, we chose a headwater stream catchment to compare DOM export from discrete landscape units. The Oder catchment in the Harz Mountains (Germany) is particularly suitable for our study as the stream originates within a bog and enables us to retrieve an exclusively bog-derived DOM signal within this headwater stream. Short residence times make in-stream processes negligible and allow landscape-type-specific studies. The effects of storm events, hypothesized to induce major DOM dynamics, were particularly considered in our sampling design. For DOM characterization and source identification we applied spectrofluorometric indices like $SUVA_{254}$, S_R, HIX and FI, as well as PARAFAC modeling of excitation–emission matrices. Seasonality effects were assessed considering mean daily air temperatures.

2 Materials and methods

2.1 Study site

The study site is located within the nature protection area of the Harz Mountains. The Odersprung bog exhibits an erosion rill, draining the peatland (Fig. 1). The catchment responds quickly to rainfall events and discharge is mainly fed by near-surface waters. A more detailed hydrologic description is given in Broder and Biester (2015). The bog vegetation is dominated by *Sphagnum magellanicum* and *S. rubellum*, associated with *Eriophorum angustifolium* and *Molinia caerulae* (Baumann, 2009). The peatland is surrounded by spruce forest growing on a cambic podzol soil at the hillslopes, and peaty soils with deep organic topsoil layer in the riparian zone. One discharge sampling was conducted directly at the rill outflow, where all water originates exclusively from the domed bog. Another sampling spot was established about 20 m further downstream where the small headwater stream increasingly receives water from the surrounding forested, organic-rich mineral soils and peaty riparian zone (Fig. 1). The catchment is underlain by granitic bedrock. The mean peat thickness of the bog is about 3 m, while the mineral soils are shallow at the hillslope (30 cm) and deeper in depressions (100 cm). Organic content of the soil varies between 30 and 97 % in the organic-rich surface layers (Broder and Biester, unpublished data).

2.2 Sampling and field measurements

Stream water sampling at each sampling spot was conducted from snowmelt to the beginning of snowfall in 2013. Water

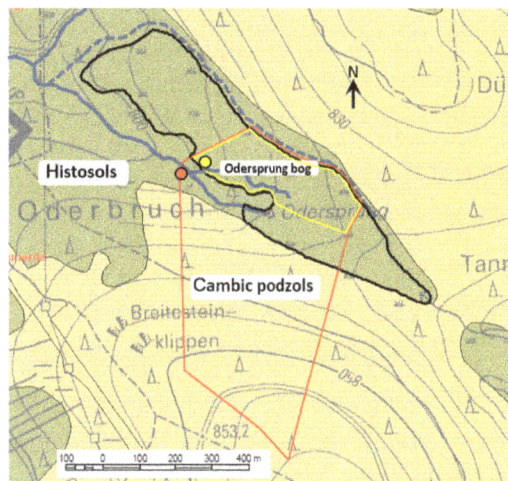

Figure 1. Location of the study area in the Harz Mountains, Germany. Red and yellow lines indicate each catchment boundary, circles represent the discharge monitoring spots (yellow – bog catchment; red – forest catchment with peaty riparian zone). Green areas indicate peaty soils, beige-colored areas outline mineral cambic podzol soils. The bog area is confined by the bold black line. Map source: NIBIS mapserver, Lower Saxony authority for mining, energy and geology (LBEG).

samples of 500 mL volume were taken by an automated water sampler (Teledyne ISCO, USA) in 6-day intervals summing up to 44 samples. Additional grab samples were taken every 2 to 3 weeks (30 samples in total) in polyethylene (PE) tubes, which were previously rinsed twice with sample water. High-frequency storm event sampling was conducted on several occasions in 3 h intervals resulting in 191 samples. A V-notch weir was installed at the bog outlet for discharge quantification. Water stage at the weir as well as at the bog site was recorded at 10 min time resolution by a water level logger (Odyssey dataflow systems, New Zealand) installed in a slotted PVC piezometer tube of 4 cm diameter. Temperature, humidity and precipitation were monitored on-site at the same resolution as the water level (using a tipping-bucket rain-gauge and tinytag tgp 4500 and 4810, Gemini, Belgium).

2.3 Laboratory analysis, indices and PARAFAC modeling

Water samples were vacuum filtered with a 0.45 μm nylon filter (Merck Millipore, Germany) and stored in the dark at 4 °C. All water samples were analyzed for DOC by thermo-catalytic oxidation using the NPOC method (non-purgeable organic carbon; multi N/C 2100S, Analytik Jena, Germany). UV-VIS spectra of all samples were recorded with a Lambda 25 (Perkin Elmer, USA) in the range of 200–800 nm at 0.5 nm resolution. A possible iron interference was excluded as the maximum iron concentration of 500 μg L^{-1} (and an iron to carbon ratio of about 0.01) was well below published critical concentration levels (see Weishaar et al., 2003; Xiao et al., 2013; Poulin et al., 2014). For subsequent fluorescence spectroscopy, samples were diluted to absorption < 0.3 at 254 nm to reduce inner-filter effects. Absorbance at 254 nm wavelength ($abs_{254\,nm}$, m^{-1}) was used as an indicator for the absolute aromaticity of DOM samples as conjugated systems like aromatic molecules have the greatest absorption in the UV range of 200–380 nm (Weishaar et al., 2003). $SUVA_{254}$ was calculated by dividing absorbance at 254 nm (m^{-1}) by the DOC concentration ($mg\,L^{-1}$) according to Weishaar et al. (2003), with increasing $SUVA_{254}$ values indicating a higher aromaticity. The spectral slope ratio (S_R), a proxy inversely related to molecular weight, was calculated after Helms et al. (2008) by dividing the slope in the interval of 275–295 nm by the slope at 350–400 nm. Slopes were determined using linear regression of log-transformed absorption spectra.

Fluorescence spectroscopy was conducted through measurement campaigns at 3-month intervals. During this time filtered samples were stored at 4 °C in the dark. Fluorescence EEMs were collected with a Cary eclipse fluorescence spectrometer (Agilent, USA) in 5 nm steps over an excitation range of 240–450 and 2 nm steps over an emission range of 300–600 nm. Inner filter correction, blank subtraction and Raman normalization was performed using the drEEM 0.2.0 toolbox from Murphy et al. (2013) and MATLAB (Version 2013a, MathWorks, USA). Reshaped EEMs were subjected to PARAFAC analysis to obtain hypothetical fluorophores for DOM fingerprinting. In total, 435 samples were included in the PARAFAC model, with both discharge and pore water samples originating from different sites. Samples examined in this study accounted for 242 samples within this model. A model with five fluorescence components could be obtained and split-half validated following the drEEM and N-way toolbox (Murphy et al., 2013; Stedmon and Bro, 2008). The sum of fluorescence intensities of the modeled components thereby represents the total fluorescence of a sample. The contribution of fluorescent DOM (fDOM) to total DOC was evaluated by normalizing total fluorescence by DOC concentrations (fDOM / DOC ratio).

The FI was calculated by the ratio of fluorescence emission intensities at 470 and 520 nm at an excitation wavelength of 370 nm (Cory and McKnight, 2005). The FI differentiates between plant-derived (FI: 1.3–1.4) and microbial- or planktonic-derived DOM (FI: 1.7–2.0) (McKnight et al., 2001) as the ratio represents the greater decrease in emission with increasing wavelengths of microbial-derived DOM. As our study site is a headwater catchment we assume that all DOM is of terrestrial origin and therefore, we interpret an FI > 1.7 as microbially derived or microbially processed DOM. The HIX was calculated after the modified equation of Ohno (2002) whereby higher values in a range of 0 to 1 indicate a red shift of spectral emission and a higher degree of DOM humification.

2.4 Statistical analyses

Statistics were performed using IBM SPSS 24. The dataset was split regarding sampling sites (bog or forest) and was further divided into a seasonal (6-day interval) and an event record (high-resolution sampling campaigns). On each subdataset descriptive statistics of mean, median, minimum and maximum value, and standard deviation (SD) were performed. As all datasets were neither normally distributed (after the Shapiro–Wilk test), nor have a homogeneity of variance (Levene's test), Spearman's rank correlation was used to test correlations of specific parameters. Accordingly, the Mann–Whitney or Kruskall–Wallis test with Bonferroni correction was applied to test significant differences between non-parametric datasets (0.05 level of significance).

3 Results

3.1 Seasonal trends

3.1.1 Hydrologic conditions and DOC concentrations

The DOC concentration record and hydrologic characteristic at the bog site has been described previously in Broder and Biester (2015). In short, bog discharge exhibited a flashy regime with an instantaneous response to rain events. The rain event with the highest recorded discharge peak occurred in spring, while in summer a longer drought period resulted in very low discharge and little response to rainfall due to recovery of water storage within the bog. More frequent rain events in fall at wetter antecedent moisture conditions caused again more flashy discharge and concentration responses.

The variability of DOC concentrations over the year ranged between 5.0 and 45.8 $mg\,L^{-1}$ (SD of 7.0 $mg\,L^{-1}$ at the forested site and 8.1 $mg\,L^{-1}$ at the bog site) and was larger than during single rain events at both sites, where standard deviations ranged between 1.3 and 1.9 $mg\,L^{-1}$ at the bog site and from 0.6 to 4.2 $mg\,L^{-1}$ at the forested site (Figs. 2 and 3 and Table 1). However, the rain event in fall was responsible for the highest recorded DOC concentrations of 37.3 and 45.8 $mg\,L^{-1}$ during the entire study period at both sites. The concentration trend generally followed the vegetation period with highest concentrations in late summer and fall. Spearman's correlation of DOC concentrations with mean daily air temperature was significant and positive at the bog site (coefficient of 0.591; $p < 0.01$, two-tailed), but not at the forested site. The lowest concentrations, of 5.0 and 10.2 $mg\,L^{-1}$, were measured during snowmelt at both sites. Concentrations of DOC were significantly higher at the forested site (median of 32.1 $mg\,L^{-1}$) than at the bog outlet (median of 25.7 $mg\,L^{-1}$) over the whole sampling period (see Table 1).

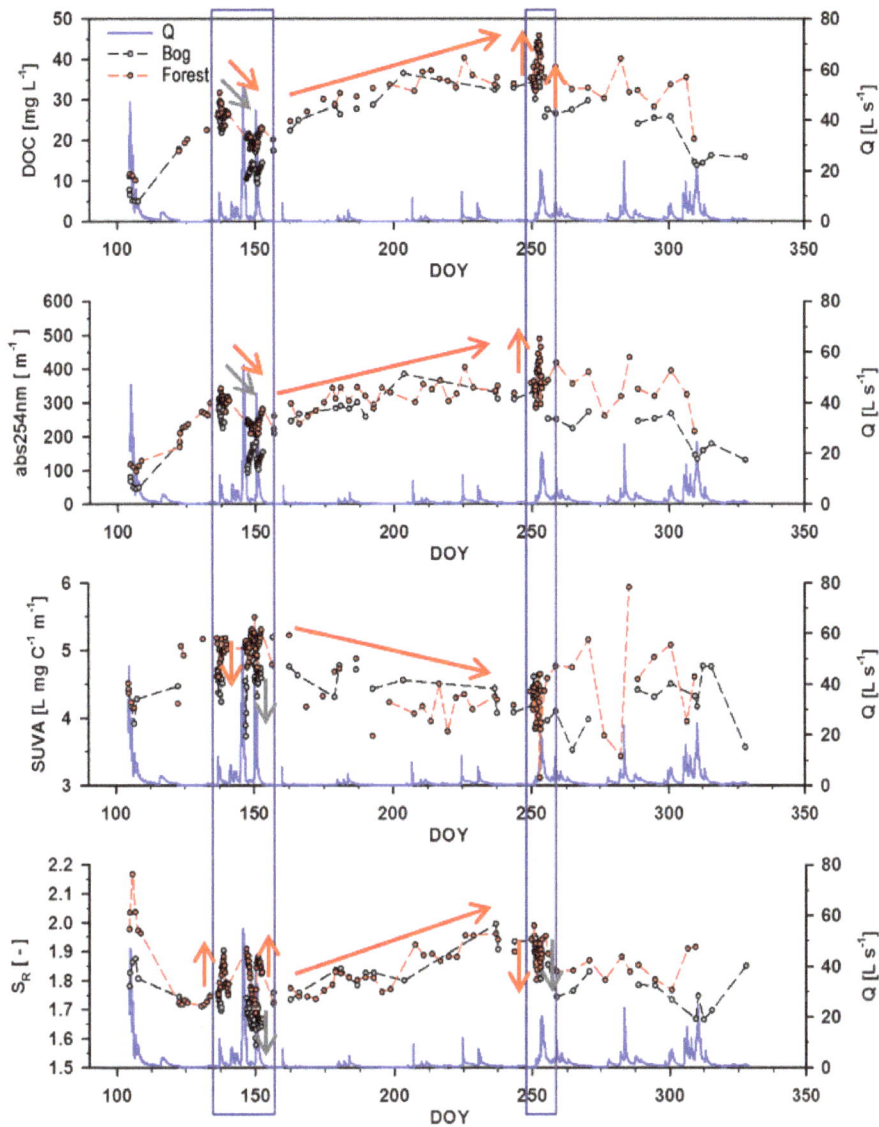

Figure 2. Annual records of DOC concentrations, $abs_{254\,nm}$, $SUVA_{254}$ and S_R from top to bottom (DOY – day of the year) in 2013. The blue line represents the bog discharge (Q). Grey circles represent the bog site, red circles the forested site, while arrows indicate concentration or index trends during rain events and summer drought at the different sites. Sampled rain events in spring and fall are highlighted by blue boxes.

3.1.2 DOM quality using spectrofluorometric indices

The $abs_{254\,nm}$ as index for total aromaticity of the DOM exhibited a similar trend to the DOC concentrations at both sites over the year (Fig. 2). Nonetheless, $SUVA_{254}$ values as index for proportional aromaticity of DOM varied between 3.5 and 5.2 at the bog site and between 3.4 and 5.9 at the forested site, but with no seasonal trend as observed for DOC concentrations. According to Weishaar et al. (2003), calculated $SUVA_{254}$ values corresponded to a DOM aromaticity of 27–38 % for the bog site and 29–42 % for the forested site. Variations were mainly induced by hydrologic conditions with high values during rain events in spring (up to 5.5). Mean

$SUVA_{254}$ values were higher at the forested site ($SUVA_{254}$ of 4.6, SD of 0.5), but showed a larger variability than at the bog site (mean $SUVA_{254}$ of 4.4, SD of 0.4, see also Table 1). During the summer drought period $SUVA_{254}$ values decreased at both sites, but this decrease was stronger at the forested site (Fig. 2).

As expected, the S_R as reciprocally proportional index for molecular weight of DOM exhibited an opposite trend to $SUVA_{254}$ (Fig. 2), expressed in a negative correlation on a 0.01 level of significance (Spearman correlation coefficient of -0.616 at the bog site and -0.598 at the forest site). The annual dynamic was similar at both sites, but with higher S_R during snowmelt (S_R up to 2.2) and sampled rain events

Table 1. General descriptive statistics (mean, median, standard deviation (SD), minimum and maximum values) for DOC concentrations, and DOM quality parameters ($abs_{254\,nm}$, SUVA, S_R, FI, HIX, C1 %, C2 %, C3 %, C4 % and C5 %) over the whole sampling period. High-resolution rain event data are excluded to project seasonal variability.

	DOC		Abs254		SUVA$_{254}$		S_R		FI		HIX	
	Bog*	Forest*	Bog*	Forest*	Bog	Forest	Bog*	Forest*	Bog*	Forest*	Bog	Forest
N	32	37	32	42	32	36	32	42	22	20	22	20
Mean	23.0	30.4	233	304	4.4	4.6	1.80	1.84	1.58	1.63	0.91	0.92
Median	25.7	32.1	250	313	4.4	4.6	1.78	1.83	1.58	1.64	0.92	0.92
SD	8.1	7.0	79	79	0.4	0.5	0.09	0.08	0.04	0.05	0.04	0.02
Min	5.0	11.2	49	113	3.5	3.4	1.67	1.71	1.52	1.55	0.76	0.86
Max	36.6	45.8	384	489	5.2	5.9	2.00	1.98	1.68	1.75	0.95	0.95

	C1 %		C2 %		C3 %		C4 %		C5 %	
	Bog	Forest	Bog*	Forest*	Bog*	Forest*	Bog*	Forest*	Bog	Forest
N	22	20	22	20	22	20	22	20	22	20
Mean	44	44	26	28	14	17	10	6	6	5
Median	44	44	26	28	14	17	10	7	5	5
SD	1	1	2	1	1	2	2	4	3	2
Min	40	42	23	25	12	13	4	0	4	3
Max	47	46	29	30	15	21	13	12	14	10

* Significant differences between bog and forested site (Mann–Whitney; two-tailed, $\rho < 0.05$).

(S_R 1.7–2.0) at the forested site (Fig. 3), indicating a lower molecular weight than at the bog site. During the summer drought S_R steadily increased from 1.7 to 2.0, indicating decreasing molecular weight. With the onset of fall rain events, molecular weight increased again, indicated by lower S_R values.

The HIX, as well as the FI, also exhibited no annual trend (Fig. 4). At the bog and the forested site HIX only varied during snowmelt and spring events with lower values down to 0.76 and 0.81, respectively, compared to HIX in summer and fall, where values remained between 0.90 and 0.94 and between 0.91 and 0.95, respectively. FI exhibited values between 1.5 and 1.75, with significantly higher values at the forested site. At the forested site FI increased during summer drought (from 1.5 to 1.7).

3.1.3 Hypothetical fluorophores modeled by PARAFAC

To facilitate a description of the PARAFAC results, the identified components are briefly described here and compared to hypothetical fluorophores typically observed in other studies. PARAFAC modeling resulted in a five-component model with four humic-like and one protein-like hypothetic fluorophores. Excitation–emission regions of each component can be found in Fig. 5. The modeled PARAFAC component C1 can be compared to a terrestrial, humic-like fluorophore originating from forest and wetland soils, as described by Perdrial et al. (2014), and C2 can again be described as humic-like (see, e.g., C3 of Singh et al., 2014), but the excitation–emission region is shifted to higher excitation and emission wavelengths compared to C1, indicating more

conjugated and more aromatic fluorescent molecules. A component similar to C3 has previously been described as humic, but also of terrestrial origin, small molecular size, recalcitrant and reduced (Cory and McKnight, 2005; Singh et al., 2014; Fellman et al., 2008; Perdrial et al., 2014). C4 is only slightly shifted compared to the excitation–emission region of C3 and compares to C2 from Fellman et al. (2008) and Ohno and Bro (2006), another humic-like fluorophore. C5 could be described as tryptophan-like, of microbial origin, labile and of recent biological production (described in e.g., Fellman et al., 2008 as C8). It can be used as proxy for BDOM (Fellman et al., 2008). In C1 and C3 fulvic-like fluorophores might also be included, which are more hydrophilic and therefore more mobile than the humic-like DOM (in Fellman et al., 2008: C3, C4), but could not be clearly separated into individual components.

3.1.4 DOM quality using PARAFAC

The fDOM, as the sum of all fluorescent components modeled by PARAFAC, showed changes with discharge events with both minimum and maximum intensities during sampled discharge events (Fig. 4). Fluorescence was elevated during summer and lower values occurred in spring and fall at both sites. Normalizing fDOM to DOC concentrations, a decrease in the fluorophore fraction in DOM from spring to fall could be observed (Fig. 4). This fluorophore fraction also decreased during individual sampled rain events.

The fDOM at the bog site showed few seasonal changes in the contribution of the four components over the year. The greatest changes were perceived during snowmelt with the

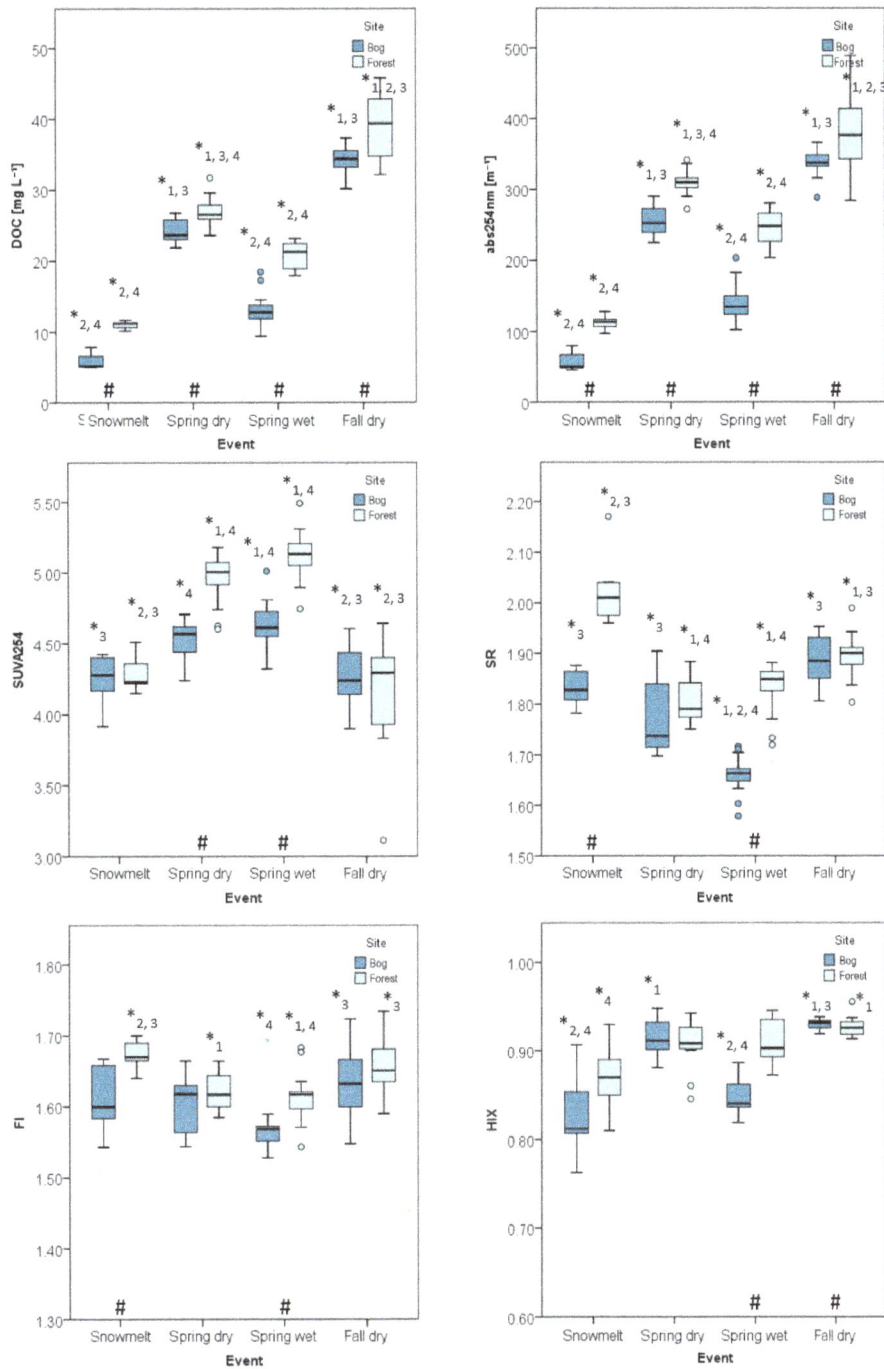

Figure 3. Box plots of SUVA, S_R, HIX, DOC concentrations, $abs_{254\,nm}$ and FI during events. Dark blue bars indicate the bog site, lighter blue bars the forested site. Asterisks indicate a significant difference between events (Kruskall–Wallis; $p < 0.05$), which are indicated by added numbers: 1 – snowmelt, 2 – spring dry, 3 – spring wet, 4 – fall dry. Significant differences between the bog and the forested site at an event (Mann–Whitney; $p < 0.05$) are indicated by a hash. A table of descriptive statistics for each event can be found in the Supplement.

highest protein-like C5 % ($\sim$ 10–17 %) and variable humic-like C1 %, C2 % and C4 % contributions at the bog site (Fig. 7). The protein-like C5 % contributed least to fDOM with about 5 % during most of the record. The components C1 % and C4 % increased during summer drought at the bog

site (Fig. 6). The protein-like C5 % contributed largely to fDOM during snowmelt and a wet spring event (Fig. 6). The humic like C1 % increased during summer drought, while C2 % decreased at the same time. The components C2 % and C3 % were significantly elevated at the forested site, while

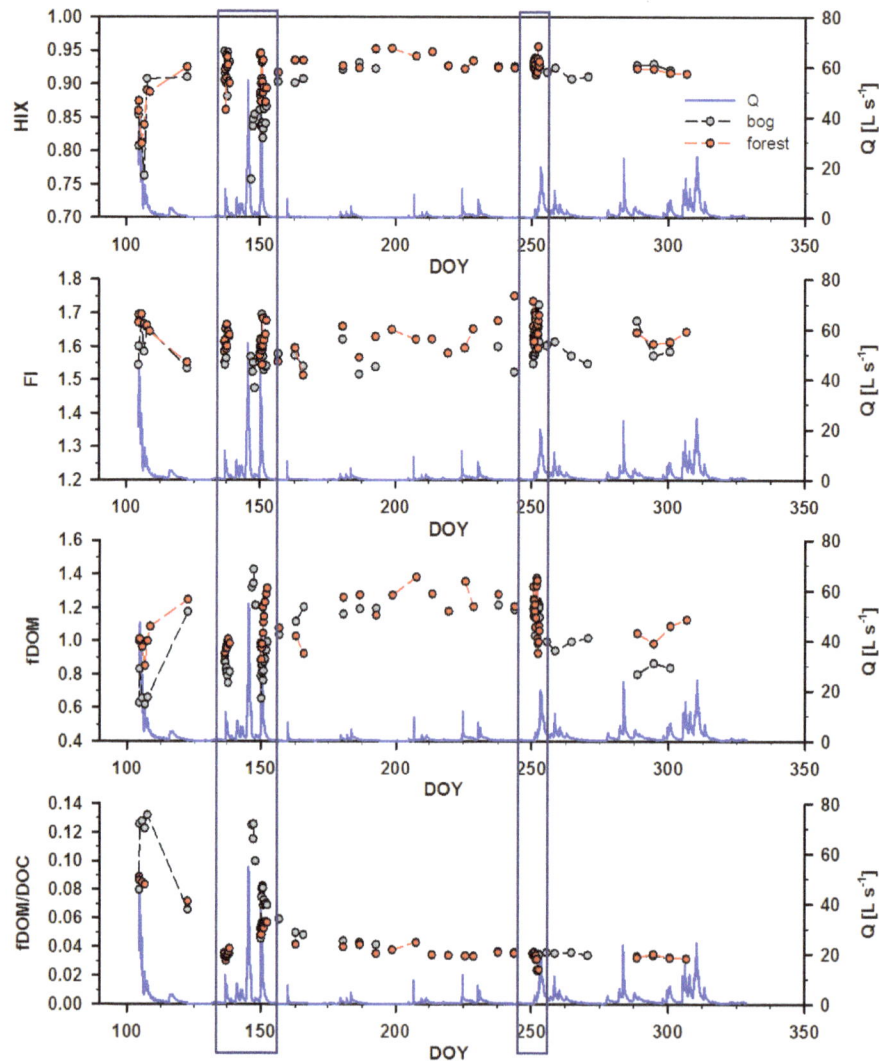

Figure 4. Annual record of the humification index (HIX), fluorescence index (FL), sum of all Fmax PARAFAC components values (fDOM) and fDOM to DOC concentration (mg L^{-1}) ratio against day of the year (DOY) in 2013. The blue line represents the bog discharge (Q). Grey circles represent the bog site, red circles the forested site. Sampled rain events in spring and fall are highlighted.

significantly higher contributions of C4 % were attributed to the bog site (Figs. 6 and 7).

3.2   Snowmelt, drought and rain events

3.2.1   Snowmelt

Snowmelt samples exhibited a distinct pattern compared to discharge events in other seasons. DOC concentrations (5.0–11.7 mg L^{-1}), abs$_{254\,nm}$ (45–128 m^{-1}), HIX (0.76–0.93) and fDOM (0.6–1.1) were lowest and normalized fDOM / DOC (0.06–0.14) was highest during snowmelt at both sites (Figs. 3 and 4). Differences to other events were significant, excluding the wet spring event, though. Values of S_R during snowmelt were high at the forested site with about 1.96–2.17, while S_R at the bog site exhibited mean values of 1.78–

1.88 (Fig. 3), which was still higher than following spring samples, though. SUVA$_{254}$ was significantly lower (4.2–4.5) than in following spring events at the forested site, while at the bog site SUVA$_{254}$ of 3.9–4.4 were similar to values of the rest of the record (Table 1, Fig. 2). The protein-like PARAFAC component C5 % was elevated during snowmelt (7–16 %) and was significantly higher than at the fall event at both sites (Fig. 7). Values of the humic-like C2 % (median of 23 %) and C4 % (median of 7 %) were low at the bog site compared to median annual values of 26 and 10 %, respectively. At the forested site the bog-derived C4 % was almost absent and the humic-like forest-derived C3 % was strongly elevated (contributing about 20 % to total fluorescence, Fig. 7). It should be noted, though, that the snowmelt event was not entirely covered by our sampling period as the upstream water was still covered with snow.

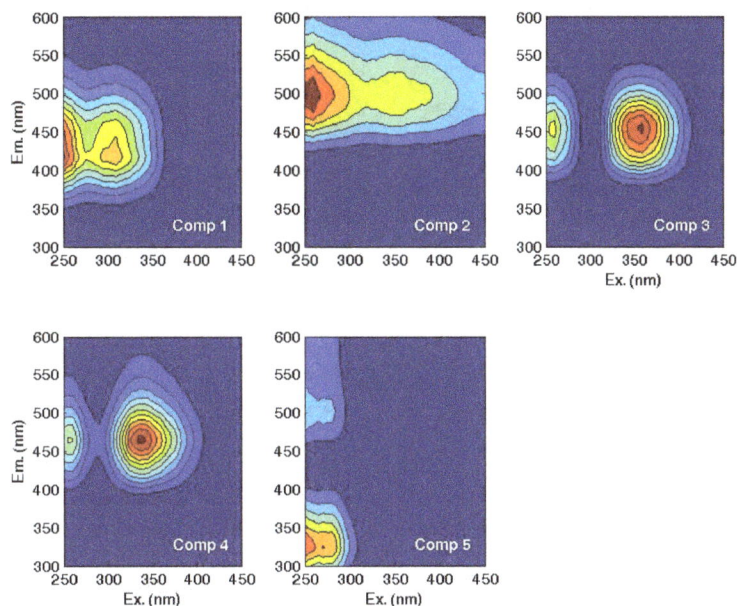

Figure 5. Characteristic EEMs of all five modeled PARAFAC components. For further component description see Sect. 3.1.3.

3.2.2 Spring

The sampled spring rain events could be differentiated by hydrologic preconditions as the first event occurred under dry preconditions, while the second described spring event followed after preceding rain events under wet hydrologic preconditions. DOC concentrations and $abs_{254\,nm}$ were significantly higher after dry preconditions (21.9–31.7 mg L^{-1} and 225–341 m^{-1}) than after wet preconditions (9.4–23.1 mg L^{-1} and 102–281 m^{-1}) at both sites (Fig. 3). SUVA$_{254}$ values at the forested site were persistently high during the whole spring time (4.5–5.5), but decreased during each rain event. At the bog site, SUVA$_{254}$ values were significantly lower than at the forested site during spring rain events, but were still significantly elevated compared to the fall event at the bog site. High-resolution sampling before the wet-precondition spring event during low-flow conditions showed a steady decrease in S_R until the onset of the rain event where values momentarily peaked at the forested site. However, at the first spring event with dry preconditions, the S_R values peaked later at the declining limb of the hydrograph. At the bog site S_R values during spring events were low, but again with increasing values at the declining limb of the hydrograph, which indicates a decrease in apparent molecular size.

Contribution of protein-like component C5 % exhibited elevated values during the second spring event with wet preconditions at the bog site (Figs. 6 and 7). The bog-derived humic-like C4 % decreased during spring events at both sites (Fig. 6). This trend was more distinct at the forested site, with a quick drop to zero at the second event. The humic-like C1 % dropped from the first sampled spring event to the

second one at the bog site, while the more aromatic C2 % increased at the forested site from the first to second event. The predominantly forest-derived humic-like C3 % also increased at all storm events and was significantly higher at the forested site.

3.2.3 Drought

The year 2013 was characterized by a strong summer drought, which caused low discharge over the summer months. During the prevailing drought period DOC concentrations, $abs_{254\,nm}$ and S_R increased, while SUVA$_{254}$ decreased, especially at the forested site (from 5.4 to 3.7, Fig. 2). The humic-like C1 % increased and the humic-like, but more aromatic C2 % decreased at the forested site, while at the bog outlet an increase in the humic-like bog-derived C4 % was perceived (Fig. 6).

3.2.4 Fall

The fall event following the summer drought generated the highest DOC concentrations of the annual record with 45.8 mg L^{-1} DOC at the forested site and 37.3 mg L^{-1} DOC at the bog site. Even though $abs_{254\,nm}$ was high and even increasing during the event at the forested site, SUVA$_{254}$ values were significantly lower than during spring events, indicating a lower aromaticity of DOM in fall (Fig. 3). Congruently, S_R values were higher at both sites, indicating smaller DOM during the fall event. The humic-like PARAFAC components C1 %, C4 % and the protein-like C5 % decreased during the fall rain event at the forested site, while the forest-derived humic-like C2 % and C3 % increased (Fig. 6). The bog-derived C4 % contributions were significantly higher than

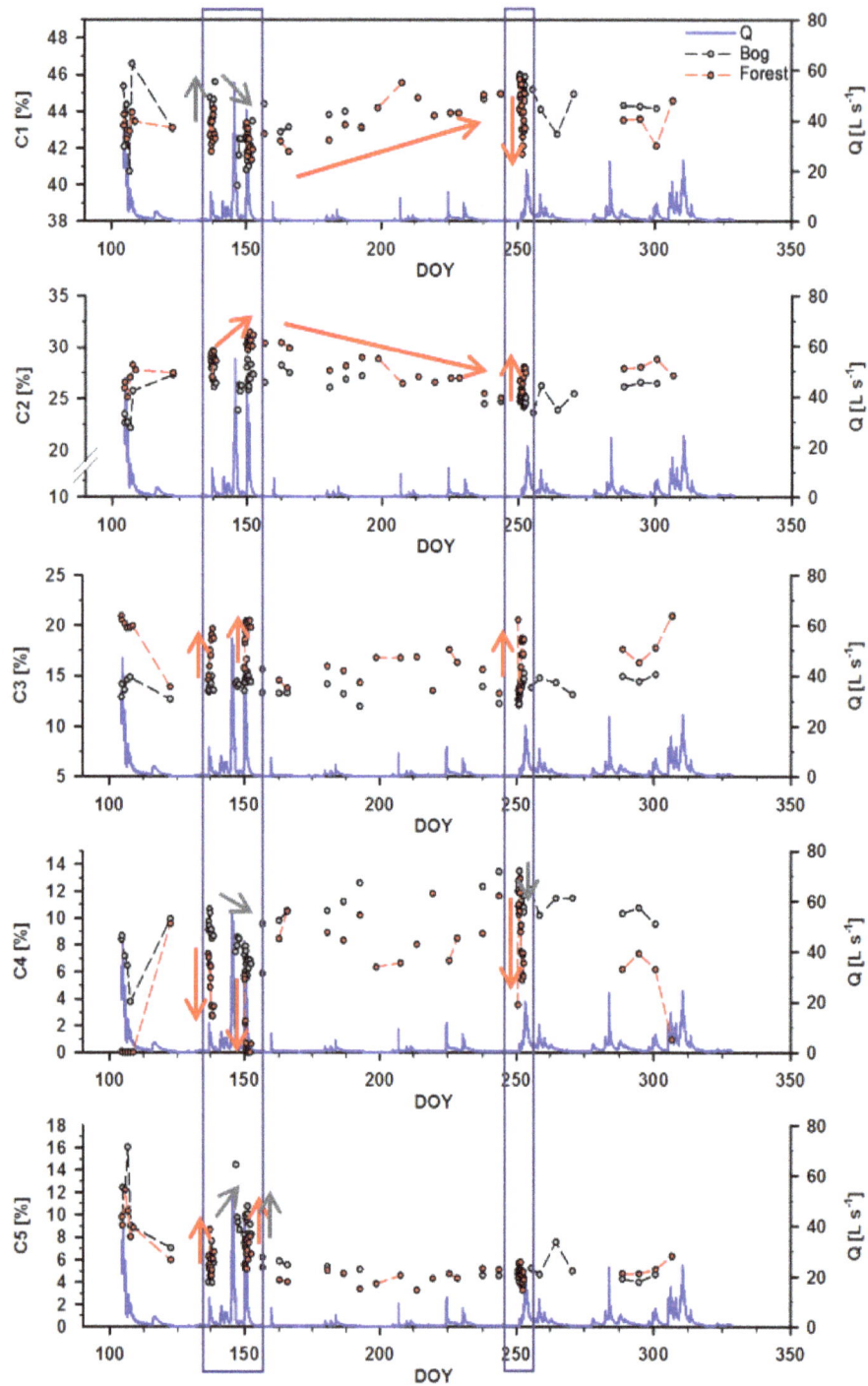

Figure 6. Percentage of PARAFAC components C1–C5. Blue line represents the bog discharge (Q). Grey dots indicate bog site values, red dots forest site values, while arrows indicate trends during rain events and summer drought at the different sites. Sampled rain events in spring and fall are highlighted.

during snowmelt and in spring at the bog site (Fig. 7), but exhibited the same decreasing trend with the ongoing rain event at both sites. The protein-like C5 % exhibited lower values during the fall event than at the spring rain.

4   Discussion

As expected, the spectroscopic indices for aromaticity ($SUVA_{254}$) and apparent molecular size (S_R) were inversely correlated, which implies that an increase in aro-

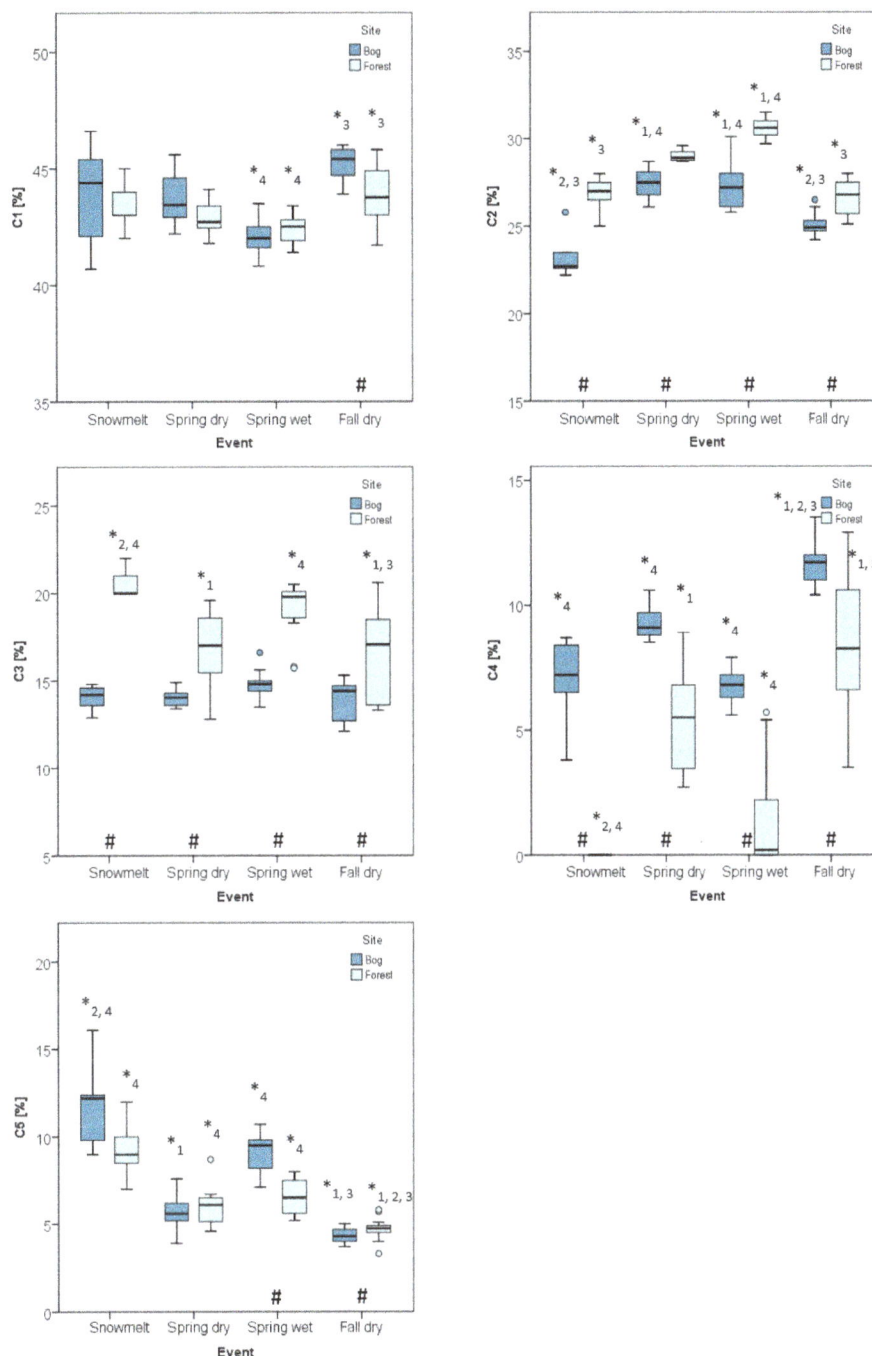

Figure 7. Box plots of PARAFAC components during events. Dark blue bars indicate the bog site, lighter blue bars the forested site. A table of descriptive statistics for each event can be found in the Supplement. Asterisks indicate a significant difference between events (Kruskall–Wallis; $\rho < 0.05$), which are indicated by added numbers: 1 – snowmelt, 2 – spring dry, 3 – spring wet, 4 – fall dry. Significant differences between the bog and the forested site at an event (Mann–Whitney; $\rho < 0.05$) are indicated by a hash. A table of descriptive statistics for each event can be found in the Supplement.

maticity is accompanied by an increase in apparent molecular size. This suggests that aromatic structures dominate the DOM fraction of apparently high molecular size in this study. Fluorometric indices exhibited different trends than UV-VIS indices even though they should reflect similar changes in DOM quality. HIX as fluorometric index for humic components showed less variability than $SUVA_{254}$. A difference between those two indices was also reported by Inamdar et al. (2011). As $SUVA_{254}$ is a proxy exclusively for aromatic DOM, HIX might include other humic or hydrophobic

DOM components derived from plant decomposition. Trends of HIX in this study rather indicate that HIX represents all fluorescent humic-like components, which resulted in less variability due to a domination of humic components in all catchment compartments.

The FI values of both sites varied between 1.5 and 1.75. This indicates a domination of microbial or microbially processed DOM over vascular-plant-derived DOM in the catchment (Cory and McKnight, 2005). As we assume that all DOM is of terrestrial origin, it implies that the majority of measured stream DOM is not fresh organic material as would be indicated by high contributions of C5 %, nor highly aromatic DOM derived from plant decomposition, but has been strongly modified by microbial processes within the soils. It also indicates long residence times of DOM in the soils, as produced DOM is mainly not immediately transported to surface waters, but gets altered or consumed within the soils of the catchment. This may be supported by the high SUVA$_{254}$ values, which indicate a strong aromatic fraction contributing 27 to 42 % to DOM, presumably due to residual enrichment.

In contrast to other studies, the forested site exhibited higher SUVA$_{254}$ values for aromaticity most of the time. For example, Ågren et al. (2008) and Wallin et al. (2015) found similar SUVA$_{254}$ values or even higher values at a peatland site, respectively, compared to mineral soils at a forested site. On the one hand this might be due to the dominating *Sphagnum* vegetation and peat at the bog site here, as *Sphagnum* is known to produce less aromatic organic matter than vascular plants, which also occur in peatlands, due to the lack of lignin in *Sphagnum* mosses (Spencer et al., 2008). On the other hand, the domination of peaty soils in the riparian zone and the domination of shallow sub-surface flow over groundwater contribution at the forested site might enhance DOM aromaticity due to the release of highly decomposed and modified organic matter. In contrast to the bog, the peaty riparian zone is subjected to great water level changes and accelerated dry–wet cycles, which result in repeated aeration and enhanced decomposition (Singh et al., 2014).

4.1 Seasonal trends in DOM concentrations and quality

Although major changes in DOC quality occurred due to hydrological changes as demonstrated by individual discharge events, seasonal patterns also occurred. DOC concentrations at the bog site were significantly correlated with daily mean air temperatures and generally followed the vegetation period with low concentrations during snowmelt and spring and highest DOC concentrations in early fall. This strong seasonal effect has been observed frequently and was ascribed to DOC production and solubility, but also enhanced litter decay by leaf fall in early fall (Christ and David, 1996; Singh et al., 2014; Wallin et al., 2015). Nonetheless, at the forested site especially, this seasonal trend was overprinted by hydrological events, which generated high DOC concentration

peaks up to 45.8 mg L^{-1} in fall due to rapid mobilization from hydrologically connected source areas.

All spectrofluorometric indices and PARAFAC components exhibited major changes during high discharge. Thus they were mainly controlled by hydrologic (pre)conditions, coinciding with few changes in DOM quality over the year as observed by Perdrial et al. (2014). Due to predominance of Norway spruce, pulses of protein-like DOM from leaf fall in the fall (Singh et al., 2014) would not be expected in our catchment. Between the two sampling sites, specific differences in PARAFAC component contributions were identified. The C4 % could be denoted as a bog-derived humic-like component, while C2 % and C3 % were predominantly forest-derived humic-like components. The humic-like C1 % and the protein-like C5 % could not be specifically attributed to one of the sites.

4.2 Event DOM characteristics and DOM sources

4.2.1 DOC concentrations

Organic-rich riparian zones are known to generate high DOC concentrations (Grabs et al., 2012). As a large part of the organic-rich upper soil layer is hydrologically connected and contributes to discharge only during events, high DOC concentrations at the forested site are due to the repeated flushing of peaty soils in the riparian zone. Additionally, upper organic layers of shallow hillslope soils may get connected via surface or near-surface flow networks during such events. Therefore, dry preconditions within the catchment facilitated high DOC concentrations during events of a certain magnitude when the upper soil layer gets hydrologically connected. The DOC concentrations at the bog site were less sensitive to rain events and not as elevated during those events than at the forested site. Here, partly decreasing concentrations were observed. Due to the usually high water level, rain events here do not connect additional DOM pools, but lead to dilution by surface flow or an exhaustion effect (Broder and Biester, 2015). We conclude that DOC concentration peaks during rain events were mainly induced by peaty forest soils and not by bogs. Although the latter are strong C sources to the aquatic, they are less susceptible to rain events. DOC concentration trends of the bog site were further disentangled in Broder and Biester (2015).

4.2.2 Snowmelt

During snowmelt, DOC concentration at both sites were lowest. This has been reported elsewhere (e.g., Laudon et al., 2004; Clark et al., 2008) and can be attributed not only to dilution by snow packs, but also to low microbial and plant activity. The absolute values of chromophore and fluorophore DOM were also lowest. Overall, spectrofluorometric indices point to rather small molecular size, which were less aromatic, especially at the forested site. Additionally, PARAFAC components indicate a strong flush of labile, protein-like DOM. This characteristic was also more pronounced at the forested site than at the bog site. The elevated export of a protein-like fraction, i.e., easily biodegradable DOM (Fellman et al., 2009), can be explained by less biotic demand and a domination of shallow flow paths during snowmelt bypassing large, strongly modified and aromatic DOM pools in the subsurface (Fellman et al., 2009). Also, near-surface freeze–thaw cycles during winter provide fresh DOM from microbial cell lysis and root mortality, which is not utilized due to the low productivity (Haei et al., 2012; Fellman et al., 2009). Differences between the two sites were more evident in other PARAFAC components. At the forested site a large fraction of humic-like fluorescence was attributed to C3 %, while C4 % was absent. C3 % only increased strongly at the forested site during rain events, while C3 % remained constant at the bog site. This suggests that component C3 % is predominantly sourced in the upper, organic layer of the forest soil and is only mobilized during rain events, when these near-surface layers get hydrologically connected, even during snowmelt. As this component increases the most out of all identified components, higher DOC concentrations during rain events should be mainly caused by a connection of near-surface organic-rich layers to the streams, irrespective of the season. In our study an increase of C3 % coincides with lower SUVA$_{254}$ and higher S_R and FI values. This points to a microbially modified, recalcitrant, but less aromatic DOM fraction of smaller apparent molecular size, confirming previous descriptions of a largely similar fluorophore (e.g., Fellman et al., 2008; Singh et al., 2014). Overall, the snowmelt DOM can be described as smaller, less chromophoric, less aromatic and more biodegradable due to a higher protein-like fraction, which is more pronounced at the forested site and has its source in the upper soil layer.

4.2.3 Drought

A drought period during summer 2013 caused a strong decrease in the bog water level down to 35 cm depth, and exceptionally low discharge representing pronounced baseflow conditions. This dry period induced concomitant changes in DOM quality in stream discharge. While DOC concentrations at both sites continuously increased, aromaticity and apparent molecular size decreased. While FI at the bog site remained constant (around 1.55), FI at the forested site increased during drought up to 1.75, indicating a greater fraction of microbially derived, strongly modified DOM. Under these drought conditions, the indices congruently illustrate a less aromatic, smaller, and more microbial DOM at the peaty forest soils compared to the bog site and to the rest of the record. This can be explained by a higher contribution of shallow groundwater and decreasing discharge through the peaty surface layers of the riparian zone. Due to the adsorption of larger, more aromatic compounds to mineral phases,

groundwater DOM is typically of smaller molecular size and less aromatic (Meier et al., 2004; Inamdar et al., 2012). For the bog site, increasing C1 % and bog-derived humic-like C4 %, as well as decreasing molecular size, also hints at a DOM source change toward deeper peat layers. Summarizing the summer drought fingerprint, DOM during this period approaches characteristics observed for shallow groundwater at the forested site and resembles DOM from deeper peat layers at the bog site. This caused a change in DOM quality to smaller, more microbial and less aromatic components.

4.2.4 Rain events

The three sampled rain events distributed over spring and fall clearly differed with respect to observed changes in DOM quality. The spring events can be differentiated between dry (first event) and wet (second event) preconditions, while the fall event occurred again under dry preconditions. The main difference between these two preconditions was a significantly higher DOC concentration under dry preconditions at both sites. While this difference in DOM concentration could be attributed to a dilution effect under wet preconditions at the bog site, notable differences in DOM quality also occurred: the non-fluorescent DOM fraction was clearly elevated following dry preconditions. This indicates that under dry preconditions a specific DOM fraction, which cannot be separated by fluorometric indices, is exported compared to wet preconditions. As non-fluorescent DOM is probably more easily degradable DOM like organic acids or products of biotic activity; this might be a flushing effect, when the upper soil layer gets hydrologically connected after prolonged aeration, decomposition and concomitant enrichment of potentially mobile DOM, while under wet preconditions this DOM fraction gets exhausted. As this flushing effect also occurred at the dry fall event, the export of this DOM fraction may be mainly attributed to the hydrologic preconditions and not to the low demand of labile DOM in early spring, as has been suggested elsewhere (Fellman et al., 2009). This would further imply that focusing on the protein-like C5 % as proxy for BDOM (Fellman et al., 2008) neglects a further labile DOM fraction, which might serve as a nutrient source downstream and does not correlate with protein-like fluorescence.

A major seasonal difference between the spring and fall events is that there were no elevated protein-like DOM exports in fall. However, elevated contributions of protein-like fluorescence as described before (Singh et al., 2014) were ascribed to leaf fall at that time of the year. As in our catchment only coniferous trees occurred, this may not be observed in this study. However, the contribution of the non-fluorescent fraction in fall was even higher than in spring. This might be indicative of higher biotic activity, generating small, non-fluorescent molecules, and demonstrates inherent limitations of the spectrofluorometric approach of DOM characterization.

At the forested site, trends of $SUVA_{254}$, HIX and S_R indicated a decreasing aromaticity and apparent molecular size during all rain events. This is confirmed by an increasing FI up to 1.74, meaning a shift to more microbial DOM. The DOM was of even smaller apparent molecular size and less aromatic during the fall event even though trends of the indices were partly reversed. This reversal was due to the extreme dry preconditions and prevailing baseflow with a high microbially derived DOM fraction, primarily originating from groundwater and deep peat. With the onset of a rain event the organic soil layers get hydrologically connected, leading to a decrease of microbial- and shallow-groundwater-derived compounds and an increase in more aromatic DOM originating from the strongly humified organic matter of the peaty layers. Taking together the trend of the humic-like C1 % at the forested site during events and drought periods reveals that this component probably represents a rather microbially processed DOM fraction from groundwater, while humic-like C2 % and C3 % largely contribute to DOM under high discharge conditions and therefore represent a DOM sourced in the upper soil layers. In contrast to other studies (Hood et al., 2006; Inamdar et al., 2011), a general increase of aromatic DOM during rain events at the forested site was not recorded. However, there is an increase in humic and microbial DOM which indicates that the humic-like C3 % is also microbially derived DOM, but from another source. However, under wet preconditions the decrease in $SUVA_{254}$ was less distinct and C2 %, indicative of a shift to rather aromatic structures, contributed more to fDOM than under dry preconditions. This indicates a stronger aromatic contribution to DOM export from upper soil layer under wet preconditions at the forested site, while under dry preconditions longer aeration may yield more modified, less aromatic DOM that has pooled up during drought.

Under wet preconditions, DOM quantity and quality differed significantly between sites, as expressed by significant differences of all indices except of the PARAFAC component C1 %. At the bog site much less DOC was exported than at the forested site, indicating an exhaustion, as well as a dilution effect of the surficial DOM pool here (Broder and Biester, 2015). Hydraulic conductivities at greater depths, where high DOM concentrations prevail, are presumably too low for rapid mobilization. Compared to changes at the forested site, aromaticity only moderately shifted during rain events and over seasons. However, even though aromaticity was not elevated at the bog site, S_R indicated DOM of a rather large molecular size being mobilized under wet preconditions. Also the protein-like component was highest at the wet spring event, which is explained by flushing of fresh biotic material in the surface layer. Moreover, during this event overland flow is very likely (Broder and Biester, 2015), which might further leach larger polymers of proteins, cellulose or polysaccharides from the living biomass. Also, Fellman et al. (2009) reported an increase in protein-like DOM export and related this to lower residence times and low

biotic demand in a wetland catchment. Unfortunately, S_R was not monitored in that study. The bog-specific humic-like C4 % component only moderately decreased during rain events and increased over summer drought at the bog site. This dynamic in C4 % presumably describes a component from a deeper peat layer, which is constantly exported over the year and gets diluted by upper surface or near-surface export during high discharge events. Therefore, this component may be used as tracer for deep bog porewater, and the observed dilution effect clearly points out that bogs do not primarily drive variations in DOM loads of streams.

Summarizing dynamics during events, DOM quality changes reflect different contributions of DOM pools depending on hydrologic preconditions and season. Fresh and labile DOM was exported during spring events at both sites, especially under wet preconditions at the bog site. Even though aromaticity of DOM in the studied catchment was high, events showed an increase in microbial or strongly microbially altered DOM. However, PARAFAC components show that this assumed microbial component is not only sourced in shallow groundwater, but that there is an additional microbial DOM pool in the upper soil layer that is especially mobilized under dry preconditions. Comparing the two sites, our results demonstrate that not only major dynamics in DOM quantities, but also variability in DOM quality was mainly driven by the forested site, i.e., by shallow peaty soils with stronger variations in water tables and thus hydraulic connectivity of the different layers.

5 Conclusion

Variability in stream water DOM quantity and quality was primarily generated at the forested site with peaty riparian zones and not at the bog site. Thereby, changes in headwater DOM quality were mainly induced by hydrologic conditions, which points out the importance of high-resolution studies and consideration of high-discharge events, which not only generate the highest DOC concentrations, but export different DOM pools with different chemical properties and fate in aquatic systems. Especially under wet preconditions, DOM quantity and quality differed significantly between the bog and the forested site. There was no clear seasonal trend in DOM quality, but DOM concentrations at the bog site generally followed the temperature trend. Nevertheless, the response of DOM quality to changes in hydrologic conditions also differed, probably depending on season. The export of protein-like DOM components was specific for snowmelt and for spring events after wet preconditions at this study site. Those DOM compounds might serve as an important nutrient source in the aquatic system. Nevertheless, not only during spring events, but also in fall a non-fluorescent DOM fraction of small apparent molecular size was exported, especially during events with dry preconditions that may be of similarly high bioavailability. During drought periods DOM export was limited to a deeper peat layer at the bog site and shallow groundwater at the forested site as could be tracked by indices displaying a microbial DOM signature originating from long DOM residence times in the soil and peat. At events with wet preconditions, additional near-surface DOM pools were connected due to increasing water levels in the catchment. Next to aromatic DOM compounds, a near-surface microbial DOM fraction was exported during those events, as could be tracked by specific PARAFAC components. While at the bog site a dilution effect of DOM concentration sets in under wet preconditions, the forested site generated the highest DOM concentration peaks under wet preconditions.

The different spectrofluorometric indices were generally suitable to track the origin and dynamics of DOM. However, it must be considered that this method could not display apparent dynamics in a non-fluorescent fraction, triggered by hydrologic conditions and season. The PARAFAC modeling of different DOM components proved a useful tool to track export dynamics of different DOM pools under different seasonal and hydrologic conditions, which could not have been resolved by the spectrofluorometric indices alone. Moreover, our study demonstrates the need for approaches tracking DOM sources to understand DOM export dynamics, while approaches based solely on hypothetic hydrological compartments, such as surface flow, soil water and groundwater, may be too simplistic. This understanding of how different DOM pools get exported might become even more important in view of future changes in the hydrologic regime due to climate change.

Competing interests. The authors declare that they have no conflict of interest.

Acknowledgements. This work was funded by the NTH graduate school "GeoFluxes" of the Federal State of Lower Saxony, Germany. We are grateful to the Nationalpark Harz for giving access to the site. The authors thank Adelina Calean, Petra Schmidt and Julian Fricke for help with lab and field work. UV-VIS and fluorescence spectroscopy were carried out in the laboratory of the Institute of Landscape Ecology at the University of Münster. We thank Johan Rydberg for helpful contributions and Christian Blodau for scientific and financial support.

Edited by: T. Blume

References

Ågren, A., Buffam, I., Berggren, M., Bishop, K., Jansson, M., and Laudon, H.: Dissolved organic carbon characteristics in boreal streams in a forest-wetland gradient during the transition between winter and summer, J. Geophys. Res., 113, G03031, doi:10.1029/2007JG000674, 2008.

Ågren, A. M., Buffam, I., Cooper, D. M., Tiwari, T., Evans, C. D., and Laudon, H.: Can the heterogeneity in stream dissolved organic carbon be explained by contributing landscape elements?, Biogeosciences, 11, 1199–1213, doi:10.5194/bg-11-1199-2014, 2014.

Aitkenhead, J. A., Hope, D., and Billett, M. F.: The relationship between dissolved organic carbon in stream water and soil organic carbon pools at different spatial scales, Hydrol. Process., 13, 1289–1302, doi:10.1002/(SICI)1099-1085(19990615)13:8<1289::AID-HYP766>3.0.CO;2-M, 1999.

Amon, R. and Benner, R.: Photochemical and microbial consumption of dissolved organic carbon and dissolved oxygen in the Amazon River system, Geochim. Cosmochim. Ac., 60, 1783–1792, doi:10.1016/0016-7037(96)00055-5, 1996.

Baumann, K.: Entwicklung der Moorvegetation im Nationalpark Harz, 1st Edn., Schriftenreihe aus dem Nationalpark Harz, 4, Nationalparkverwaltung Harz, Wernigerode, Germany, 244 pp., 2009.

Bishop, K., Seibert, J., Köhler, S., and Laudon, H.: Resolving the Double Paradox of rapidly mobilized old water with highly variable responses in runoff chemistry, Hydrol. Process., 18, 185–189, doi:10.1002/hyp.5209, 2004.

Broder, T. and Biester, H.: Hydrologic controls on DOC, As and Pb export from a polluted peatland – the importance of heavy rain events, antecedent moisture conditions and hydrological connectivity, Biogeosciences, 12, 4651–4664, doi:10.5194/bg-12-4651-2015, 2015.

Christ, M. and David, M.: Temperature and moisture effects on the production of dissolved organic carbon in a Spodosol, Soil Biol. Biochem., 28, 1191–1199, doi:10.1016/0038-0717(96)00120-4, 1996.

Clark, J. M., Lane, S. N., Chapman, P. J., and Adamson, J. K.: Export of dissolved organic carbon from an upland peatland during storm events: Implications for flux estimates, J. Hydrol., 347, 438–447, doi:10.1016/j.jhydrol.2007.09.030, 2007.

Clark, J. M., Lane, S. N., Chapman, P. J., and Adamson, J. K.: Link between DOC in near surface peat and stream water in an upland catchment: Biogeochemistry of forested ecosystem – Selected papers from BIOGEOMON, the 5th International Symposium on Ecosystem Behaviour, held at the University of California, Santa Cruz, on June 25–30, 2006, Sci. Total Environ., 404, 308–315, doi:10.1016/j.scitotenv.2007.11.002, 2008.

Clark, J. M., Ashley, D., Wagner, M., Chapman, P. J., Lane, S. N., Evans, C. D., and Heathwaite, A. L.: Increased temperature sensitivity of net DOC production from ombrotrophic peat due to water table draw-down, Global Change Biol., 15, 794–807, doi:10.1111/j.1365-2486.2008.01683.x, 2009.

Cole, J. J., Prairie, Y. T., Caraco, N. F., McDowell, W. H., Tranvik, L. J., Striegl, R. G., Duarte, C. M., Kortelainen, P., Downing, J. A., Middelburg, J. J., and Melack, J.: Plumbing the Global Carbon Cycle: Integrating Inland Waters into the Terrestrial Carbon Budget, Ecosystems, 10, 172–185, doi:10.1007/s10021-006-9013-8, 2007.

Cory, R. M. and McKnight, D. M.: Fluorescence Spectroscopy Reveals Ubiquitous Presence of Oxidized and Reduced Quinones in Dissolved Organic Matter, Environ. Sci. Technol., 39, 8142–8149, doi:10.1021/es0506962, 2005.

Cory, R. M., McKnight, D. M., Chin, Y.-P., Miller, P., and Jaros, C. L.: Chemical characteristics of fulvic acids from Arctic surface waters: Microbial contributions and photochemical transformations, J. Geophys. Res.-Biogeo., 112, G04S51, doi:10.1029/2006JG000343, 2007.

Fellman, J. B., D'Amore, D. V., Hood, E., and Boone, R. D.: Fluorescence characteristics and biodegradability of dissolved organic matter in forest and wetland soils from coastal temperate watersheds in southeast Alaska, Biogeochemistry, 88, 169–184, doi:10.1007/s10533-008-9203-x, 2008.

Fellman, J. B., Hood, E., D'Amore, D. V., Edwards, R. T., and White, D.: Seasonal changes in the chemical quality and biodegradability of dissolved organic matter exported from soils to streams in coastal temperate rainforest watersheds, Biogeochemistry, 95, 277–293, doi:10.1007/s10533-009-9336-6, 2009.

Frost, P. C., Larson, J. H., Johnston, C. A., Young, K. C., Maurice, P. A., Lamberti, G. A., and Bridgham, S. D.: Landscape predictors of stream dissolved organic matter concentration and physicochemistry in a Lake Superior river watershed, Aquat. Sci., 68, 40–51, doi:10.1007/s00027-005-0802-5, 2006.

Grabs, T., Bishop, K., Laudon, H., Lyon, S. W., and Seibert, J.: Riparian zone hydrology and soil water total organic carbon (TOC): implications for spatial variability and upscaling of lateral riparian TOC exports, Biogeosciences, 9, 3901–3916, doi:10.5194/bg-9-3901-2012, 2012.

Haei, M., Öquist, M. G., Ilstedt, U., and Laudon, H.: The influence of soil frost on the quality of dissolved organic carbon in a boreal forest soil: combining field and laboratory experiments, Biogeochemistry, 107, 95–106, doi:10.1007/s10533-010-9534-2, 2012.

Helms, J. R., Stubbins, A., Ritchie, J. D., Minor, E. C., Kieber, D. J., and Mopper, K.: Absorption spectral slopes and slope ratios as indicators of molecular weight, source, and photobleaching of chromophoric dissolved organic matter, Limnol. Oceanogr., 53, 955–969, doi:10.4319/lo.2008.53.3.0955, 2008.

Hood, E., Gooseff, M. N., and Johnson, S. L.: Changes in the character of stream water dissolved organic carbon during flushing in three small watersheds, Oregon, J. Geophys. Res., 111, G01007, doi:10.1029/2005JG000082, 2006.

Hruška, J., Köhler, S., Laudon, H., and Bishop, K.: Is a Universal Model of Organic Acidity Possible: Comparison of the Acid/Base Properties of Dissolved Organic Carbon in the Boreal and Temperate Zones, Environ. Sci. Technol., 37, 1726–1730, doi:10.1021/es0201552, 2003.

Inamdar, S., Singh, S., Dutta, S., Levia, D., Mitchell, M., Scott, D., Bais, H., and McHale, P.: Fluorescence characteristics and sources of dissolved organic matter for stream water during storm events in a forested mid-Atlantic watershed, J. Geophys. Res., 116, G03043, doi:10.1029/2011JG001735, 2011.

Inamdar, S., Finger, N., Singh, S., Mitchell, M., Levia, D., Bais, H., Scott, D., and McHale, P.: Dissolved organic matter (DOM) concentration and quality in a forested mid-Atlantic watershed, USA, Biogeochemistry, 108, 55–76, doi:10.1007/s10533-011-9572-4, 2012.

Jansson, M., Persson, L., de Roos, A. M., Jones, R. I., and Tranvik, L. J.: Terrestrial carbon and intraspecific size-variation shape lake ecosystems, Trends Ecol. Evol., 22, 316–322, doi:10.1016/j.tree.2007.02.015, 2007.

Jansson, M., Berggren, M., Laudon, H., and Jonsson, A.: Bioavailable phosphorus in humic headwater streams in boreal Sweden, Limnol. Oceanogr., 57, 1161–1170, doi:10.4319/lo.2012.57.4.1161, 2012.

Kaiser, K. and Zech, W.: Dissolved organic matter sorption by mineral constituents of subsoil clay fractions, J. Plant Nutr. Soil Sci., 163, 531–535, doi:10.1002/1522-2624(200010)163:5<531:AID-JPLN531>3.0.CO;2-N, 2000.

Karlsson, J., Byström, P., Ask, J., Ask, P., Persson, L., and Jansson, M.: Light limitation of nutrient-poor lake ecosystems, Nature, 460, 506–509, doi:10.1038/nature08179, 2009.

Korshin, G. V., Li, C.-W., and Benjamin, M. M.: Monitoring the properties of natural organic matter through UV spectroscopy: A consistent theory, Water Res., 31, 1787–1795, doi:10.1016/S0043-1354(97)00006-7, 1997.

Laudon, H., Köhler, S., and Buffam, I.: Seasonal TOC export from seven boreal catchments in northern Sweden, Aquat. Sci., 66, 223–230, doi:10.1007/s00027-004-0700-2, 2004.

Laudon, H., Berggren, M., Ågren, A., Buffam, I., Bishop, K., Grabs, T., Jansson, M., and Köhler, S.: Patterns and Dynamics of Dissolved Organic Carbon (DOC) in Boreal Streams: The Role of Processes, Connectivity, and Scaling, Ecosystems, 14, 880–893, doi:10.1007/s10021-011-9452-8, 2011.

Ledesma, J. L. J., Grabs, T., Bishop, K. H., Schiff, S. L., and Kohler, S. J.: Potential for long-term transfer of dissolved organic carbon from riparian zones to streams in boreal catchments, Global Change Biol., 21, 2963–2979, doi:10.1111/gcb.12872, 2015.

McKnight, D. M., Boyer, E. W., Westerhoff, P. K., Doran, P. T., Kulbe, T., and Andersen, D. T.: Spectrofluorometric characterization of dissolved organic matter for indication of precursor organic material and aromaticity, Limnol. Oceanogr., 46, 38–48, doi:10.4319/lo.2001.46.1.0038, 2001.

Meier, M., Chin, Y.-P., and Maurice, P.: Variations in the composition and adsorption behavior of dissolved organic matter at a small, forested watershed, Biogeochemistry, 67, 39–56, doi:10.1023/B:BIOG.0000015278.23470.f7, 2004.

Monteith, D. T., Stoddard, J. L., Evans, C. D., de Wit, H. A., Forsius, M., Høgåsen, T., Wilander, A., Skjelkvåle, B. L., Jeffries, D. S., Vuorenmaa, J., Keller, B., Kopácek, J., and Vesely, J.: Dissolved organic carbon trends resulting from changes in atmospheric deposition chemistry, Nature, 450, 537–540, doi:10.1038/nature06316, 2007.

Mulholland, P. J.: Large-Scale Patterns in Dissolved Organic Carbon Concentration, Flux and Sources, in: Aquatic Ecosystems: Interactivity of Dissolved Organic Matter, edited by: Findlay, S. E. G. and Sinsabaugh, R. L., Academic Press/Elsevier, San Diego, London, 139–160, 2003.

Murphy, K. R., Stedmon, C. A., Graeber, D., and Bro, R.: Fluorescence spectroscopy and multi-way techniques, PARAFAC, Analyt. Meth., 5, 6557–6566, doi:10.1039/c3ay41160e, 2013.

Ohno, T.: Fluorescence Inner-Filtering Correction for Determining the Humification Index of Dissolved Organic Matter, Environ. Sci. Technol., 36, 742–746, doi:10.1021/es0155276, 2002.

Ohno, T. and Bro, R.: Dissolved organic matter characterization using multiway spectral decomposition of fluorescence landscapes, Soil Sci. Soc. Am. J., 70, 2028–2037, doi:10.2136/sssaj2006.0005, 2006.

Perdrial, J. N., McIntosh, J., Harpold, A., Brooks, P. D., Zapata-Rios, X., Ray, J., Meixner, T., Kanduc, T., Litvak, M., Troch, P. A., and Chorover, J.: Stream water carbon controls in seasonally snow-covered mountain catchments: impact of inter-annual variability of water fluxes, catchment aspect and seasonal processes, Biogeochemistry, 118, 273–290, doi:10.1007/s10533-013-9929-y, 2014.

Poulin, B. A., Ryan, J. N., and Aiken, G. R.: Effects of Iron on Optical Properties of Dissolved Organic Matter, Environ. Sci. Technol., 48, 10098–10106, doi:10.1021/es502670r, 2014.

Seibert, J., Grabs, T., Köhler, S., Laudon, H., Winterdahl, M., and Bishop, K.: Linking soil- and stream-water chemistry based on a Riparian Flow-Concentration Integration Model, Hydrol. Earth Syst. Sci., 13, 2287–2297, doi:10.5194/hess-13-2287-2009, 2009.

Singh, S., Inamdar, S., Mitchell, M., and McHale, P.: Seasonal pattern of dissolved organic matter (DOM) in watershed sources: influence of hydrologic flow paths and autumn leaf fall, Biogeochemistry, 118, 321–337, doi:10.1007/s10533-013-9934-1, 2014.

Spencer, R. G. M., Aiken, G. R., Wickland, K. P., Striegl, R. G., and Hernes, P. J.: Seasonal and spatial variability in dissolved organic matter quantity and composition from the Yukon River basin, Alaska, Global Biogeochem. Cy., 22, GB4002, doi:10.1029/2008GB003231, 2008.

Stedmon, C. A. and Bro, R.: Characterizing dissolved organic matter fluorescence with parallel factor analysis: A tutorial, Limnol. Oceanogr.-Meth., 6, 572–579, doi:10.4319/lom.2008.6.572, 2008.

Strohmeier, S., Knorr, K.-H., Reichert, M., Frei, S., Fleckenstein, J. H., Peiffer, S., and Matzner, E.: Concentrations and fluxes of dissolved organic carbon in runoff from a forested catchment: insights from high frequency measurements, Biogeosciences, 10, 905–916, doi:10.5194/bg-10-905-2013, 2013.

Tipping, E., Rey-Castro, C., Bryan, S. E., and Hamilton-Taylor, J.: Al(III) and Fe(III) binding by humic substances in freshwaters, and implications for trace metal speciation, Geochim. Cosmochim. Ac., 66, 3211–3224, doi:10.1016/S0016-7037(02)00930-4, 2002.

Wallin, M. B., Weyhenmeyer, G. A., Bastviken, D., Chmiel, H. E., Peter, S., Sobek, S., and Klemedtsson, L.: Temporal control on concentration, character, and export of dissolved organic carbon in two hemiboreal headwater streams draining contrasting catchments, J. Geophys. Res.-Biogeo., 120, 832–846, doi:10.1002/2014JG002814, 2015.

Ward, C. P. and Cory, R. M.: Complete and Partial Photo-oxidation of Dissolved Organic Matter Draining Permafrost Soils, Environ. Sci. Technol., 50, 3545–3553, doi:10.1021/acs.est.5b05354, 2016.

Weishaar, J. L., Aiken, G. R., Bergamaschi, B. A., Fram, M. S., Fujii, R., and Mopper, K.: Evaluation of Specific Ultraviolet Absorbance as an Indicator of the Chemical Composition and Reactivity of Dissolved Organic Carbon, Environ. Sci. Technol., 37, 4702–4708, doi:10.1021/es030360x, 2003.

Winterdahl, M., Erlandsson, M., Futter, M. N., Weyhenmeyer, G. A., and Bishop, K.: Intra-annual variability of organic carbon concentrations in running waters: Drivers along a climatic gradient, Global Biogeochem. Cy., 28, 451–464, doi:10.1002/2013GB004770, 2014.

Worrall, F., Burt, T. P., Jaeban, R. Y., Warburton, J., and Shedden, R.: Release of dissolved organic carbon from upland peat, Hydrol. Process., 16, 3487–3504, doi:10.1002/hyp.1111, 2002.

Worrall, F., Burt, T., and Adamson, J.: Can climate change explain increases in DOC flux from upland peat catchments?, Sci. Total Environ., 326, 95–112, doi:10.1016/j.scitotenv.2003.11.022, 2004.

Xiao, Y.-H., Sara-Aho, T., Hartikainen, H., and Vähätalo, A. V.: Contribution of ferric iron to light absorption by chromophoric dissolved organic matter, Limnol. Oceanogr.-Meth., 58, 653–662, 2013.

5

Importance of considering riparian vegetation requirements for the long-term efficiency of environmental flows in aquatic microhabitats

Rui Rivaes[1], Isabel Boavida[2], José M. Santos[1], António N. Pinheiro[2], and Teresa Ferreira[1]

[1]Forest Research Centre, Instituto Superior de Agronomia, Universidade de Lisboa, Tapada da Ajuda 1349-017 Lisbon, Portugal

[2]CERIS, Civil Engineering Research Innovation and Sustainability Centre, Instituto Superior Técnico, Universidade de Lisboa, Av. Rovisco Pais, 1049-001 Lisbon, Portugal

Correspondence to: Rui Rivaes (ruirivaes@isa.ulisboa.pt)

Abstract. Environmental flows remain biased toward the traditional biological group of fish species. Consequently, these flows ignore the inter-annual flow variability that rules species with longer lifecycles and therefore disregard the long-term perspective of the riverine ecosystem. We analyzed the importance of considering riparian requirements for the long-term efficiency of environmental flows. For that analysis, we modeled the riparian vegetation development for a decade facing different environmental flows in two case studies. Next, we assessed the corresponding fish habitat availability of three common fish species in each of the resulting riparian landscape scenarios. Modeling results demonstrated that the environmental flows disregarding riparian vegetation requirements promoted riparian degradation, particularly vegetation encroachment. Such circumstance altered the hydraulic characteristics of the river channel where flow depths and velocities underwent local changes of up to 10 cm and 40 cm s^{-1}, respectively. Accordingly, after a decade of this flow regime, the available habitat area for the considered fish species experienced modifications of up to 110 % when compared to the natural habitat. In turn, environmental flows regarding riparian vegetation requirements were able to maintain riparian vegetation near natural standards, thereby preserving the hydraulic characteristics of the river channel and sustaining the fish habitat close to the natural condition. As a result, fish habitat availability never changed more than 17 % from the natural habitat.

1 Introduction

Freshwater ecosystems provide vital services for human existence but are on top of the world's most threatened ecosystems (Dudgeon et al., 2006; Revenga et al., 2000), primarily due to river damming (Allan and Castillo, 2007). The ability to provide sufficient water to ensure the functioning of freshwater ecosystems is an important concern as its capacity to provide goods and services is sustained by water-dependent ecological processes (Acreman, 2001). The relevance of this subject compelled the scientific community to appeal to all governments and water-related institutions across the globe to engage in environmental flow restoration and maintenance in every river (Brisbane Declaration, 2007). Actually, this issue is a global research topic, as all dams, weirs, and levees change the magnitudes of peak flood flows of rivers to a certain extent (e.g., FitzHugh and Vogel, 2010; Maheshwari et al., 1995; Miller et al., 2013; Nilsson and Berggren, 2000; Uddin et al., 2014a, b). As a result of this, there are still opportunities for the implementation of environmental flow restoration at hundreds of thousands of these structures worldwide (Richter and Thomas, 2007).

Environmental flows can be defined as "the quantity, timing and quality of water flows required to sustain freshwater and estuarine ecosystems, and the human livelihoods and wellbeing that depend upon these ecosystems" (Brisbane Declaration, 2007) and play an essential role in the conservation of freshwater ecosystems (Arthington et al., 2006; Hughes and Rood, 2003). It is now agreed that environmental flows must ideally be based on the ecological requirements of different biological communities (e.g., Acreman et al., 2009, 2014; Acreman and Ferguson, 2010; Arthington et al., 2010; Arthington, 2012; Arthington and Zalucki, 1998; Davis and Hirji, 2003; Dyson et al., 2003; Poff et al., 1997) and should present a dynamic and variable hydrological regime to maintain the native biodiversity and the ecological processes that represent every river (Bunn and Arthington, 2002; Lytle and Poff, 2004; Postel and Richter, 2003). In this sense, holistic methodologies meant to address river systems as a whole (Arthington et al., 1992; King and Tharme, 1994; King and Louw, 1998) are clearly being increasingly applied out of Australia and South Africa (Hirji and Davis, 2009), the origin countries of this holistic concept. However, the most commonly applied methods throughout the world are still hydrologically based methods (Dyson et al., 2003; Linnansaari et al., 2012; Tharme, 2003). Conversely, environmental flows ascertained through habitat simulation methods still persist generally based on the requirements of a single biological group, mostly fish (Acreman et al., 2009; Arthington, 2012; Tharme, 2003), and require an input from less typically monitored taxa (Gillespie et al., 2014). Accordingly, these approaches still disregard the inter-annual flow variability that rules species with longer lifecycles, like riparian vegetation, therefore lacking the long-term perspective of the riverine ecosystem (Stromberg et al., 2010). The feedbacks of these shortcomings on the riparian and aquatic communities were seldom estimated before and so, the efficiency of such approaches along with its long-term after-effects remains practically unknown.

Riparian vegetation is a suitable environmental change indicator (Benjankar et al., 2012; Nilsson and Berggren, 2000) that responds directly to a flow regime in an inter-annual time frame (Capon and Dowe, 2007; Naiman et al., 2005; Poff et al., 1997) and has a clear significance in the habitat improvement of aquatic systems (e.g., Broadmeadow and Nisbet, 2004; Chase et al., 2016; Dosskey et al., 2010; Gregory et al., 1991; Pusey and Arthington, 2003; Rood et al., 2015; Ryan et al., 2013; Salemi et al., 2012; Statzner, 2012; Tabacchi et al., 2000; Van Looy et al., 2013; Wootton, 2012). In fact, riparian vegetation and aquatic species interact biologically, physically, and chemically (Gregory et al., 1991). Riparian vegetation is capable of influencing aquatic species in several ways. It affects food webs by providing an important input of nutrients that are a major food source for invertebrates, which are in turn eaten by fishes (Wootton, 2012). It influences hydrological processes (Salemi et al., 2012; Tabacchi et al., 2000) and protects aquatic habitats by means of river bank stability (Rood et al., 2015) and providence of large woody debris (Fetherston et al., 1995). It provides thermal regulation of rivers by overshadowing (Ryan et al., 2013) and protects water quality both by trapping sediments and contaminants (Chase et al., 2016) as by chemi-

cal uptake and cycling (Dosskey et al., 2010). On the other hand, aquatic species also appear to be able to influence riparian zones, although at a much smaller magnitude, acting as ecosystem engineers (Statzner, 2012). For instance, fishes can dig in sand and gravel for food or reproductive purposes and therefore influence sediment surface characteristics and critical shear stress (e.g., Hassan et al., 2008; Statzner et al., 2003).

Accordingly, riparian restoration is an indispensable implementation measure to recover the natural river processes and is the most promising restoration action in many degraded rivers (Palmer et al., 2014). Hence, incorporating riparian vegetation requirements (the need for specific flows to preserve the naturalness of recruitment and meta-stability facing fluvial processes) into environmental flows could be an important contribution to fill in these gaps.

We have already noticed how environmental flow regimes disregarding riparian vegetation requirements allow for the degradation of riparian woodlands in the subsequent years following such river regulation (e.g., Rivaes et al., 2015). However, we are not aware of studies assessing the return effect of this degradation again on the efficiency of those environmental flow regimes. The purpose of this study is to evaluate the effect of disregarding riparian vegetation requirements in the efficiency of environmental flow regimes regarding fish habitat availability in the long-term perspective of the fluvial ecosystem. We used an approach from an ecohydraulic point of view to evaluate the effects of riparian landscape degradation on fish species. By riparian landscape we mean the specific spatial patterns of riparian vegetation that result from ecological, geomorphological, and hydrological processes and are depicted by the existing patch mosaic with different vegetation types and succession phases. We were particularly interested in answering the following questions: (i) are environmental flows exclusively addressing fish requirements capable of preserving the habitat availability of these aquatic species in the long term? (ii) If not, to what extent can the disregard for riparian vegetation requirements derail the goals of environmental flows addressing only aquatic species as a result of the riparian landscape degradation? (iii) Are environmental flows regarding riparian requirements able to maintain the habitat availability of fish species?

To approach these questions, we first modeled the structural response of riparian vegetation (please see Naiman et al., 2005, and NRC, 2002, for a better understanding about riparian vegetation structure) facing a decade of different environmental flows in two different case studies. Next, we performed an assessment of habitat availability for fish species in each of the resulting riparian landscape scenarios. We are not aware of such a modeling approach ever being used in the appraisal of the long-term efficiency of environmental flow regimes, which can provide an extremely valuable insight into the expected long-term effects of environmental flows in river ecosystems in advance.

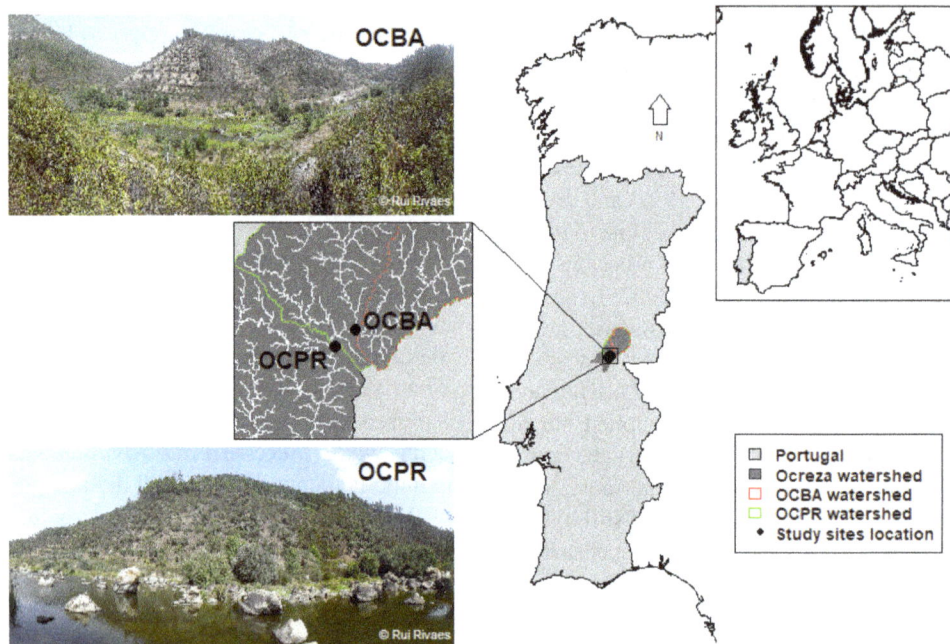

Figure 1. Location and characterization of the study sites OCBA and OCPR.

2 Methods

2.1 Study sites

The two study sites were selected in the Ocreza River, eastern Portugal (Fig. 1). This is a medium-sized stream that runs on schistose rocks for 94 km and drains a 1429 km^2 watershed with a mean annual flow of 16.5 m^3 s^{-1}. The flow regime is typically Mediterranean (Gasith and Resh, 1999), with a low flow period interrupted by flash floods in winter (the median of mean daily discharges in the winter months is 8.8 m^3 s^{-1} and maximum annual discharges with return periods of 2, 5, 10, and 100 years are, respectively, 323, 549, 718, and 1314 m^3 s^{-1}) and a very low flow, even null at times, during summer (the first quartile and the median of mean daily discharges in the summer months are, respectively, 0 and 0.1 m^3 s^{-1}). Two study sites were considered (OCBA and OCPR) to provide a broader analysis of the aquatic habitat modifications in different hydrogeomorphological contexts. The OCBA study site (39°44′07.05″ N, 7°44′16.51″ W) is located 30 km upstream from the river mouth and OCPR (39°43′16.88″ N, 7°46′01.05″ W) is approximately 5 km downstream of OCBA. Despite the relatively small distance between them, several characteristics differentiate the two study sites. While in OCBA, the river flows freely on a boulder substrate and is confined to steep valley hillsides, in OCPR, the river flows on a coarser boulder substrate with sparse bedrock presence and is located in a relatively wider valley section. OCBA and OCPR also differ in watershed areas, representing 54 and 72 % of the entire river basin, respectively. This feature further differ-

entiates the two case studies, as the intermediate watershed of OCPR collects water from a much rainier zone, thereby conferring an increased flow regime in this study site. The surveyed areas in the OCBA and OCPR study sites encompass a river length of approximately 500 and 300 m, respectively, laterally limited by the 100-year flooded zone, thus totaling approximately 4 and 3 ha for the OCBA and OCPR study sites, respectively. In both cases, the fish community is characterized by native cyprinid species, mainly *Luciobarbus bocagei* (Iberian barbel, hereafter barbel), *Pseudochondrostoma polylepis* (Iberian straight-mouth nase, hereafter nase), and *Squalius alburnoides* (calandino), whereas the local riparian vegetation is composed mostly of willows (*Salix salviifolia* Brot. and *Salix atrocinerea* Brot.) and ashes (*Fraxinus angustifolia* Vahl).

2.2 Data collection

2.2.1 Hydraulic data

The riverbed topography was surveyed in 2013 using a combination of a Nikon DTM330 total station and a Global Positioning System (GPS) (Ashtech, model Pro Mark2). Altogether, 7707 points were surveyed at OCBA and 25 132 at OCPR. Trees, boulders, and large objects emerging from the water were defined by marking the object intersection with the riverbed and by surveying the points necessary to approximately define its shape.

Hydraulic data, i.e., water velocities and depths, were measured as a series of points along several cross sections in the study sites. Depths were measured with a ruler and water velocities with a flow probe (model 002, Valeport) positioned at 60 % of the local depth below the surface (Bovee and Milhous, 1978). Additionally, the substrate composition was visually assessed and mapped to determine posteriorly the effective roughness heights of the riverbed. These data were used to calculate river discharge in each study site and to calibrate the model. Additional information about hydraulic data and channel bed characteristics is provided as the Supplement (Sect. S1 – Tables S1, S2, S3, and S4).

2.2.2 Riparian vegetation data

The riparian vegetation was assessed in 2013 to support the calibration and validation of the riparian vegetation model. This task consisted in recording the location and shape of all homogeneous vegetation patches with a sub-meter precision handheld GPS (Ashtech, Mobile Mapper 100), while dendrochronological methods were used to determine the approximate age of the patches. Two or three of the largest individuals in each patch were cored with a standard 5 mm increment borer, taking two perpendicular cores at breast height in adult trees (Mäkinen and Vanninen, 1999). For individuals with a diameter smaller than 5 cm at breast height, discs were obtained for age calculation purposes, and on multistemmed trees, the cores/discs were taken from the largest stem. The patches were later classified by succession phase according to its corresponding development stage. Patch georeferencing, patch aging, and succession phase classification followed the methodology used by Rivaes et al. (2013).

Five succession phases were identified in the study sites: Initial phase (IP), Pioneer phase (PP), Early Successional Woodland phase (ES), Established Forest phase (EF), and Mature Forest phase (MF). Initial phase was attributed to all patches dominated by gravel bars, sometimes covered by herbaceous vegetation but without woody arboreal species. The patches dominated by the recruitment of woody arboreal species were considered to be the Pioneer phase. The Early Successional Woodland phase classification was attributed to all patches with a high standing biomass and well-established individuals, dominated by pioneer watertable-dependent species, such as willows and alders (*Alnus glutinosa*). Older patches dominated by macrophanerophytes, such as ash trees, were considered to be in the Established Forest phase. The Mature Forest phase was considered at patches where terrestrial vegetation was also present, determining the transition phase to the upland vegetation communities. Further information on the characterization of succession phases is provided as the Supplement (Sect. S2 – Table S6 and Figs. S1 and S2).

2.2.3 Fish data

Fish populations were sampled during 2012 and 2013 at undisturbed or minimally disturbed sites in the Ocreza basin, an essential requisite when studying habitat preferences of stream fishes in order to reflect their optimal habitat (Gorman and Karr, 1978). Sampling occurred in autumn (November 2012), spring (May 2013) and early summer (June 2013) when there is full connectivity among instream habitats. Overall, four native species (cyprinids) were found – barbel, nase, calandino and the Southern Iberian chub (*Squalius pyrenaicus*). The latter was however excluded from the present study, as an insufficient number of individuals were collected to draw unbiased conclusions. Non-native fish (the gudgeon *Gobio lozanoi*) occurred in the study area, but in very low density. Field procedures followed those by Boavida et al. (2011, 2015). Fish sampling was performed during daylight using pulsed DC electrofishing (SAREL model WFC7-HV; Electracatch International, Wolverhampton, UK), with low voltage (250 V) and a 30 cm diameter anode to reduce the effect of positive galvanotaxis. A 200 m long reach at each site was surveyed by wading upstream in a zigzag pattern to ensure full coverage of available habitats. To avoid displacements of individuals from their original positions, a modified point electrofishing procedure was employed (Copp, 1989). Sampling points were approached discreetly, and the activated anode was swiftly immersed in the water for five seconds. Upon sighting a fish or a shoal of fishes, a numbered location marker was anchored to the streambed for subsequent microhabitat use measurements. Fish were immediately collected by means of a separate dip net held by another operator, quickly measured for total length (TL), and then placed in buckets with portable ELITE aerators to avoid continuous shocking and repeated counting, before being returned alive to the river. Ensuing fish sampling, microhabitat measurements of flow depth (cm), mean water velocity (cm s^{-1}) and dominant substrate composition were taken in 0.8 by 0.8 m quadrats at the location where each fish was captured. Microhabitat availability measurements were made using the same variables by quantifying randomly selected points along 15–25 m equidistant transects perpendicular to the flow at each sampling site. To develop habitat suitability curves (HSCs) for target fish size classes, microhabitat variables (flow depth, water velocity, dominant substrate and cover) were divided into classes, and histograms of frequencies of use and availability were constructed (Boavida et al., 2011). A summary of collected fish data, as well as data analysis to determine habitat use, availability, and preference of fish species regarding the considered variables, is provided as the Supplement (Sect. S2 – Table S7 and Figs. S3 to S12).

2.3 Flow regime definition

Three flow regimes were considered for the modeling of riparian vegetation: (i) the natural flow regime (hereafter named natural flow regime), (ii) an environmental flow regime considering only fish requirements (hereafter named the Eflow regime), and (iii) an environmental flow regime considering both fish and riparian requirements (hereafter named the Eflow&Flush regime). The natural flow regime data was obtained from the Portuguese Water Resources National Information System (SNIRH, 2010). The environmental flow regimes used in this study are an adaptation from the environmental flow regime created by Ferreira et al. (2014) for the location of the study sites (Fig. 2). These authors determined an environmental flow regime presented in a multiannual fashion considering a decadal time frame and accounting for two different flow regime components: a monthly flow regime addressing fish requirements and a multiannual flow regime composed by floods with different recurrence intervals addressing riparian vegetation requirements. The first component, i.e., the flow regime addressing fish requirements (Eflow), was determined according to the instream flow incremental methodology (Bovee, 1982) and was built on a monthly basis to embody the intra-annual variability ruling the main lifecycle events of this biological group (Encina et al., 2006; Gasith and Resh, 1999). These mean monthly discharges addressing fish requirements that compose the Eflow aimed for the following goals: (i) maximize the habitat of the target species while attributing the same weight for each species; (ii) privilege the spawning months (spring; Santos et al., 2005) and promote the younger life stages during summer; (iii) maintain the characteristic intra-annual variability of the river flow; and (iv) preserve the natural regime whenever the environmental flows suggest higher discharges. The second component of the environmental flow regime (floods with a certain recurrence interval) proposed by Ferreira et al. (2014) was determined according to Rivaes et al. (2015) and intends to characterize the inter-annual flow variability to which the arrangement of riparian vegetation communities respond (Hughes, 1997). The flushing flows addressing riparian requirements in the Eflow&Flush regime were defined based on the need of riparian communities for the minimum necessary flushing flow regime to maintain the viability and sustainability of riparian vegetation, particularly, avoiding vegetation encroachment and conserving the ecological succession equilibrium of the riparian ecosystem (Rivaes et al., 2015). Therefore, the environmental flow regimes used in this study are considered an adaptation from Ferreira et al. (2014) as we used just the fish-addressing component (only mean monthly discharges) as the standard procedure of an environmental flow regime considering only fish requirements (Eflow) and both components (mean monthly discharges and flushing flows) for the environmental flow regime addressing fish and riparian requirements (Eflow&Flush).

2.4 Riparian vegetation modeling

The riparian vegetation modeling was performed using the *CASiMiR-vegetation* model (Benjankar et al., 2009). This tool simulates the succession dynamics of riparian vegetation, based on the existing relationships of the ecological relevant hydrological elements (Poff et al., 1997) and the veg-

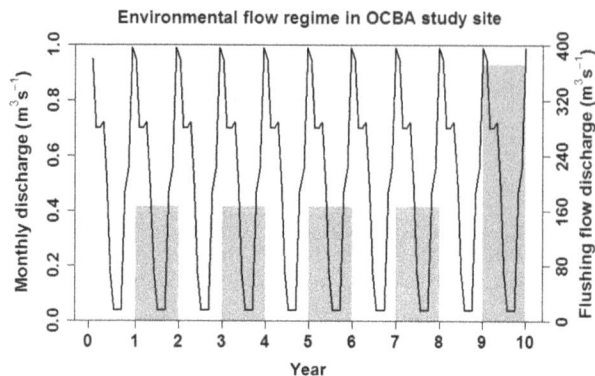

Figure 2. Environmental flow regime addressing fish (black line, left axis) and riparian (grey bars, right axis) requirements considered for the habitat modeling in the OCBA study site. Fish requirements are addressed by a constant monthly discharge and riparian requirements by a flushing flow in the years in which they are planned (the duration of the flushing flow is similar to a natural flood with an equal recurrence interval). The hydrograph for the Eflow&Flush flow regime is similar in the OCPR study site.

etation metrics that reflect riparian communities to such hydrological alterations (Merritt et al., 2010). The strengths of this model are the capacity to incorporate the past patch dynamics into every model run, the ability to work at a response guild level by using succession phases as modeling units, and the ability to provide the outputs in a spatially explicit way. In turn, the main disadvantages of this model can be attributed to the non-existence of a plant competition module or the lack of an incorporated hydrodynamic model.

The rationale of this model is based on the fact that riparian communities respond to the hydrological and habitat variations on a timescale between the year and the decade (Frissell et al., 1986; Thorp et al., 2008), being that the flood pulse is the predominant factor in these population dynamics (Thoms and Parsons, 2002). For these reasons, the hydrological regime is inputted into the model in terms of maximum annual discharges as these discharges are considered to be the annual threshold for riparian morphodynamic disturbance that determine the succession or retrogression of vegetation. Notwithstanding this, the model also predicts the annual riparian adjustments according to its vital rates in relation to groundwater depth, as well as the annual recruitment areas, based on the annual minimum mean daily discharges. The groundwater depth corresponding to the mean annual discharge of the river is also a model input used as a reference for the general habitat conditions that determine the expected riparian landscape according to the calibrated thresholds of the riparian succession phases. Thus, the magnitude and duration of extreme low flows are accounted for by the CASiMiR-vegetation model. A complete detailing of model rationale and parameterization can be found in Politti and Egger (2011) and Benjankar et al. (2011). Model calibration was carried out in accordance with the methodology

Table 1. Maximum annual discharges ($m^3 s^{-1}$) considered in the CASiMiR-vegetation model for each study site.

Year	OCBA			OCPR		
	Natural	Eflow	Eflow&Flush	Natural	Eflow	Eflow&Flush
1	671	0.99	0.99	951	5.51	5.51
2	203	0.99	167	287	5.51	237
3	327	0.99	0.99	464	5.51	5.51
4	217	0.99	167	308	5.51	237
5	316	0.99	0.99	449	5.51	5.51
6	371	0.99	167	526	5.51	237
7	702	0.99	0.99	995	5.51	5.51
8	202	0.99	167	286	5.51	237
9	195	0.99	0.99	276	5.51	5.51
10	440	0.99	371	624	5.51	527

described in previous studies (García-Arias et al., 2013; Rivaes et al., 2013). In particular, calibration was performed by running the CASiMiR-vegetation model for a decade to simulate the effect of the local historic flow regime on riparian vegetation. The result of the model was then compared with an observed vegetation map that was surveyed in the same year as the one corresponding to the result of the model. This is an iterative process of trial and error where the parameter of the shear stress resistance threshold of each succession phase is tuned to obtain the best calibration outcome (see Wainwright and Mulligan, 2004, for a better understanding). All the other parameters, namely, patch age and height above water table ranges, were determined based on the data collected in the field. This information is provided as the Supplement (Sect. S1 – Table S5). During calibration, the riparian vegetation model achieved an agreement evaluation of 0.61 by the quadratic weighted kappa (Cohen, 1960), which is considered to be in good agreement with the observed riparian landscape (Altman, 1991; Viera and Garrett, 2005). This agreement evaluation can be understood as a classification 61 % better than what would be expected by a random assignment of classes. The riparian vegetation model was further validated in this specific watershed (Ferreira et al., 2014), with even better results (quadratic weighted kappa of 0.68). After calibration and validation (calibrated parameters provided as the Supplement; Sect. S1 – Table S5), the riparian vegetation was modeled for periods of 10 years according to the corresponding flow regimes (Table 1). Such a modeling period was considered to be long enough to avoid the influence of the initial vegetation conditions, while river morphological changes still do not assume importance in vegetation development (Politti et al., 2014). Furthermore, during modeling, riverbed topography was considered fixed for several reasons: the study sites are located in a fairly steep valley in which the river is not allowed to meander considerably during such a short timescale; the typical substrate of both study sites is armored and very coarse (boulders, large boulders, and bedrock); in these conditions the small monthly

discharges intended to maintain aquatic fauna requirements are not able to create water depths and flow velocities capable of moving or eroding particles of the size of those found as substrate in the considered study sites (for a better understanding, please see Alexander and Cooker, 2016; Clarke and Hansen, 1996; Hjulström, 1939); no significant differences were found during the substrate analysis of the different succession phases; prior knowledge of the authors shows that the considered floods do not bring noteworthy changes to river geomorphology during this period (Rivaes et al., 2015); the model calibration and validation results exhibited a good agreement with the observed riparian landscape while using the same methodology; and by using a fixed topography it is possible to analyze the exclusive effect of riparian landscape degradation on the river hydraulics.

The resulting riparian vegetation maps were then used as the respective riparian landscapes (hereafter named the natural, Eflow, and Eflow&Flush landscapes) in the hydrodynamic modeling of the fish habitat in each study site.

2.5 Hydrodynamic modeling of fish habitats

The hydrodynamic modeling was performed using a calibrated version of the River2D model (Steffler et al., 2002). This is a finite element model widely used in fluvial modeling studies for the assessment of habitat availability (Boavida et al., 2011; Jalón and Gortázar, 2007) that brings together a 2-D hydrodynamic model and a habitat model to simulate the flow conditions of the river stretch and estimate its potential habitat value according to the fish habitat preferences. The strengths of this model are the fact of being public domain software and being technically robust throughout a wide range of modeling circumstances. On the other hand, some limitations of this model are the non-incorporation of a morphodynamic module or the ability to embody fuzzy logic rules during the computation of species habitat availability.

The calibration procedure followed the methodology proposed by Boavida et al. (2013, 2015). Calibration was performed by iteratively adjusting the bed channel roughness to attain a good agreement of the simulated versus surveyed water surface elevations and velocity profiles in the surveyed cross sections. Boundary conditions were set according to the water surface elevations measured at the upstream and downstream cross sections. Calibrated parameters are provided in the Supplement (Sect. S1 – Tables S1, S2, S3, and S4).

The hydrodynamic modeling comprised the Eflow discharge ranges in the study sites (0–2 and 0–5.5 $\mathrm{m}^3\,\mathrm{s}^{-1}$ for OCBA and OCPR, respectively) and was accomplished for each riparian landscape scenario. The different riparian landscapes were represented in the hydrodynamic model by changing the channel roughness according to the spatial extent of the riparian succession phases; i.e., the channel roughnesses inputted to the model are the riparian landscape maps converted into channel roughness maps. Roughness is a critical feature influencing the physical variables of flow hydraulics (Chow, 1959; Curran and Hession, 2013), whose distinct combinations typify diverse functional habitats, which are selected by fish according to its preference. The roughness classification of riparian vegetation succession phases was determined based on the roughness measurement literature on similar vegetation types (Chow, 1959; Wu and Mao, 2007) and expert judgment during model calibration.

After modeling the Eflow discharges in each of the riparian landscape scenarios of the two study sites, the hydraulic characteristics of each riparian landscape (roughness, flow depth and velocity) were compared using a t-test (confidence level of 99 %) in R environment (R Development Core Team, 2011) in order to determine the existence of mean significant differences between riparian landscapes. Habitat simulation was achieved by the combination of the hydraulic modeling (flow depth and velocity) with preference curves information for the considered target species. The riverbed characteristics of substrate and cover were kept unchanged during the hydrodynamic modeling. Changing the substrate according to the modifications in succession phase disposal seemed to be an incorrect practice in this case because during data treatment, no significant differences were detected in riverbed substrate between succession phases. Cover modification was also disregarded because the CASiMiR-vegetation model only reproduces the riparian area, not the aquatic zone (note that this *aquatic zone* is a definition *sensu* CASiMiR-vegetation model, designating the area of the river channel that is permanently submerged throughout the hydrologic year and where riparian vegetation is unable to establish and develop. It corresponds to only a fraction of the wetted area by river flow during the discharges considered in the subsequent hydrodynamic modeling.) and therefore, this feature cannot be correctly modeled by the riparian vegetation model. Notwithstanding, the most important variables determining fish habitat availability influenced by riparian vegetation degradation were considered, namely, depth, velocity and substrate (Parasiewicz, 2007).

The Habitat Suitability Index (HSI) was determined for each species and life stage regarding the product of the velocity (Velocity Suitability Index – VSI), depth (Depth Suitability Index – DSI) and substrate (Substrate Suitability Index – SSI) variables, according with Eq. (1):

$$HSI = VSI \times DSI \times SSI. \tag{1}$$

The product of the HSI by the influencing area (A) of the corresponding model ith node defines the weighted usable area (WUA) of that node. The sum of the WUA result in the total amount of habitat suitability for the study site, as described by Eq. (2), is

$$WUA = \sum_{n=1}^{i} A_i \times HSI_i = f(Q). \tag{2}$$

Considering that the BACI approach (before-after control-impact) is generally the best way of detecting impacts or beneficial outcomes in river systems (Downes et al., 2002), the resulting WUAs were then compared to the natural habitat in a census-based benchmark. The equality of proportions between habitat availabilities was tested using the χ^2 test for proportions in the R environment, while deviations were measured using the most commonly used measures of forecast accuracy, namely, root mean square deviation (RMSD), mean absolute deviation (MAD), and mean absolute percentage deviation (MAPD). In all cases, smaller values of these measures indicate better performance in parameter estimation.

2.6   Workflow of the modeling procedure

The workflow of the modeling procedure is presented in Fig. 3. Firstly, the calibrated version of the riparian vegetation model is used to produce the riparian landscape scenarios according to each of the considered flow regimes. In each modeling run, this model uses as inputs one of the specific flow regimes mentioned and models the effects of a decade of such a flow regime in the local riparian vegetation. The output of the model is an expected riparian vegetation landscape map (detailed by succession phases) resulting from the inputted flow regime. This map is converted into a channel roughness map by attributing to each riparian succession phase a specific effective roughness height based on the expert knowledge of the authors, on the literature (e.g., Barnes, 1967; Chow, 1959; Fisher and Dawson, 2003), and on the calibration results of the models. The considered roughness values of each succession phase are provided as the Supplement (Sect. S1 – Tables S3 and S4). These roughness maps are one of the inputs of the River2D model.

Secondly, the River2D hydrodynamic model is used to determine the water depths and flow velocities at the microhabitat scale (already considering each of the roughness maps coming from the conversion of the CASiMiR-vegetation output vegetation maps) and to compute the weighted usable

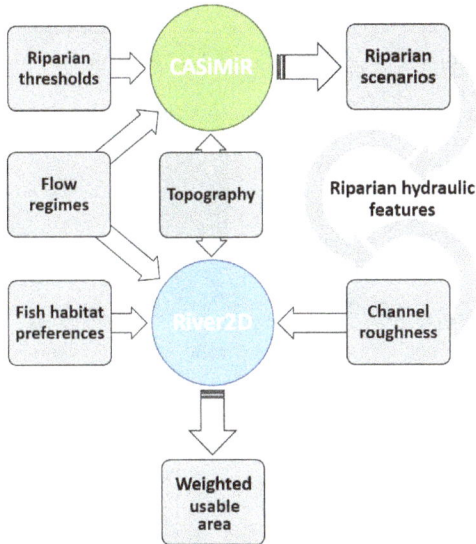

Figure 3. Methodological scheme representing the workflow of the modeling procedure. White arrows stand for direct inputs, striped white arrows for model outputs, and grey arrows for variable conversion processes.

areas of the considered fish species using the previous calculated variables and the inputted information regarding the observed fish species habitat preferences for water depth and flow velocity. This is done similarly using each of the riparian landscape scenarios. For each scenario run, the outcome of this model is therefore the weighted usable area of each of the considered species and life stages for each of the discharges considered in the Eflow regime.

3 Results

3.1 Riparian vegetation modeling

Different riparian landscapes resulted from the riparian vegetation modeling according to the considered flow regimes in both case studies (Fig. 4). Nonetheless, the modeled response of riparian vegetation to each flow regime is similar in the two study sites. The riparian landscape, driven by the natural flow regime, presents a river channel that is largely devegetated, where the Initial (IP) and Pioneer (PP) phases together represent approximately 43 and 35 % of the study site areas in OCBA and OCPR, respectively. In this riparian landscape, the Early Succession Woodland phase (ES) can only settle in approximately 8 % of OCBA and 1 % of OCPR areas. The floodplain succession phases, namely, the Established Forest phase (EF) and Mature Forest phase (MF), represent nearly 40 and 10 % of the study area for OCBA, and close to 42 and 23 % for OCPR, respectively.

In contrast, the riparian landscape created by the Eflow regime is where the riparian vegetation encroachment is more prominent. Herein, riparian vegetation settles in the

channel and evolves toward mature phases due to the lack of the river flood disturbance. IP is now reduced to approximately 3 % in OCBA and 6 % in OCPR, while PP is nonexistent in both cases. ES covers up to approximately 48 and 26 % of the corresponding study areas, whereas EF and MF maintain about the same area in both case studies.

The riparian landscape driven by the Eflow&Flush regime shows the capacity of this flow regime to hold back vegetation encroachment in both cases. In this riparian landscape scenario, IP and PP are maintained at approximately 30 % of the study site area in both case studies, whereas ES is kept under 21 % in OCBA and only 2 % in OCPR. Once again, EF and MF preserve their areas in both case studies.

Summing up, the results of the riparian vegetation modeling show a riparian landscape degradation by vegetation encroachment in the Eflow landscape scenario when compared with the natural riparian landscape. Instead, the Eflow&Flush landscape scenario keeps approximately the same patch disposal and succession phase's proportion as the natural landscape and therefore does not present evidence of riparian landscape degradation.

3.2 Hydrodynamic modeling

The changes undertaken by the riparian vegetation facing different flow regimes are able to modify the hydraulic characteristics of the river stretches (Fig. 5). Channel effective roughness heights (k_s) change dramatically according to the considered riparian landscapes, increasing proportionally to the encroachment level of vegetation in the study sites. In both case studies, the k_s values of the Eflow landscape are clearly distinct and higher compared to the other two riparian landscapes (Fig. 5). The k_s values in the Eflow&Flush landscape were found to be between the values of Eflow and natural landscapes in the case of OCBA, and were very similar to the natural landscape in the case of OCPR (Fig. 5). Notwithstanding this, in both case studies, the k_s mean values are statistical significantly different between all three riparian landscapes (test results in the Supplement; Sect. S3 – Table S8). The mean k_s of the Eflow, Eflow&Flush, and natural landscapes are 0.999, 0.709, and 0.462 m, respectively, in OCBA, and 1.034, 0.742, and 0.7178 m, respectively, in OCPR.

Changes also occur in flow depth and flow velocity for the considered discharge range of the proposed environmental flows (Fig. 5). Although not so noticeable due to the great amount of data, differences are statistically significant. In OCBA, the Eflow landscape creates a circumstance with statistically significant higher depths (mean depth is 0.402 m) and lower flow velocities (mean flow velocity is 0.128 m s^{-1}) than the natural and Eflow&Flush landscapes. The t-tests on water depths (H0: the true difference in the means is equal to 0) revealed highly significant p-values (< 0.001), respectively, for the comparisons between Eflow and natural flow regimes, and Eflow and Eflow&Flush flow regimes.

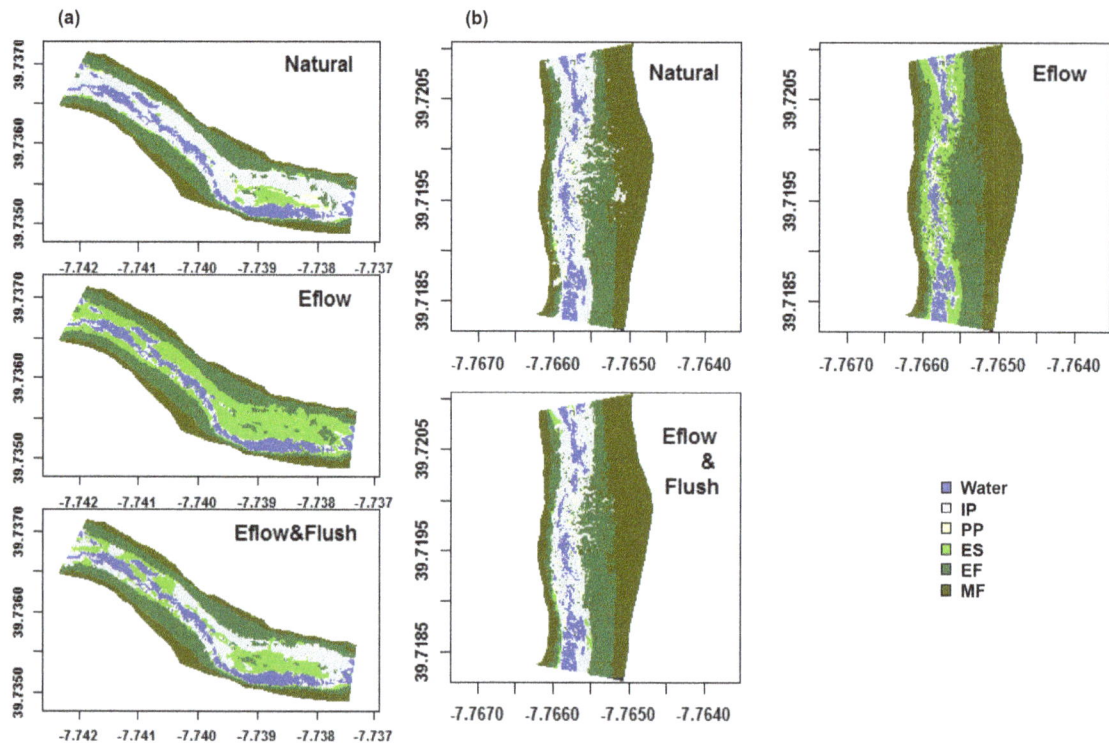

Figure 4. Expected patch mosaic of the riparian vegetation habitats shaped by the natural, Eflow, and Eflow&Flush flow regimes (detailed by succession phase, namely, initial phase – IP, pioneer phase – PP, early succession woodland phase – ES, established forest phase – EF, and mature forest phase – MF) in the OCBA study site (**a**) and in the OCPR study site (**b**).

The t-tests on flow velocities also derived a highly significant p-value (< 0.001) in both the comparisons of natural versus Eflow regimes and Eflow versus Eflow&Flush flow regimes (test results in the Supplement; Sect. S3 – Tables S9 and S10). In contrast, depth and flow velocity are not significantly distinguishable between the natural and Eflow&Flush landscapes, where mean depth and flow velocity are 0.397 m and 0.136 m s^{-1}, respectively, in the former, and 0.399 m and 0.135 m s^{-1}, respectively, in the latter.

For the OCPR study site, flow depths are not significantly different (t-tests obtained p-values of 0.122 for natural versus Eflow regimes and 0.098 for Eflow versus Eflow&Flush flow regimes). Mean values of flow depth for Eflow, Eflow&Flush, and natural landscapes are 0.420, 0.417, and 0.418, respectively. Nonetheless, flow velocities are different with statistical significance as the p-values of the t-tests for natural versus Eflow and for Eflow versus Eflow&Flush were highly significant (< 0.001). The Eflow landscape creates statistically significantly lower flow velocities (0.271 m s^{-1}) when compared to the statistically significantly indistinct Eflow&Flush (0.277 m s^{-1}) and natural (0.278 m s^{-1}) landscapes (test results in the Supplement; Sect. S3 – Tables S9 and S10).

Furthermore, when comparing water depths and flow velocities point by point, one can find differences between scenarios of up to 10 cm in water depth and more than 40 cm s^{-1}

in flow velocity. Accordingly, there are locations where the considered hydraulic parameters change considerably, shifting the habitat preference of fishes in one or two classes of the corresponding habitat preference curves.

In general, the Eflow landscapes present an increased channel roughness interfering with river flow and creating increased water depths and slower flow velocities when compared with the natural landscape. By contrast, despite the increased channel roughness of the Eflow&Flush landscape, the water depths and flow velocities are very similar to the ones in the natural landscape. These results demonstrate that an environmental flow addressing exclusively fish requirements is not capable of preserving the habitat availability of the aquatic species for which it was proposed in the long term.

3.3 Analysis of the aquatic habitat suitability for fish species

During a hydrological year, each riparian landscape provides different WUAs for the target fish species, with the same environmental flow regime addressing fish species (Fig. 6). Differences from the natural habitat suitability are greater in the Eflow landscape for both case studies. In OCBA, major differences in the WUA can be found almost all year round for the barbel juveniles, throughout autumn and winter months

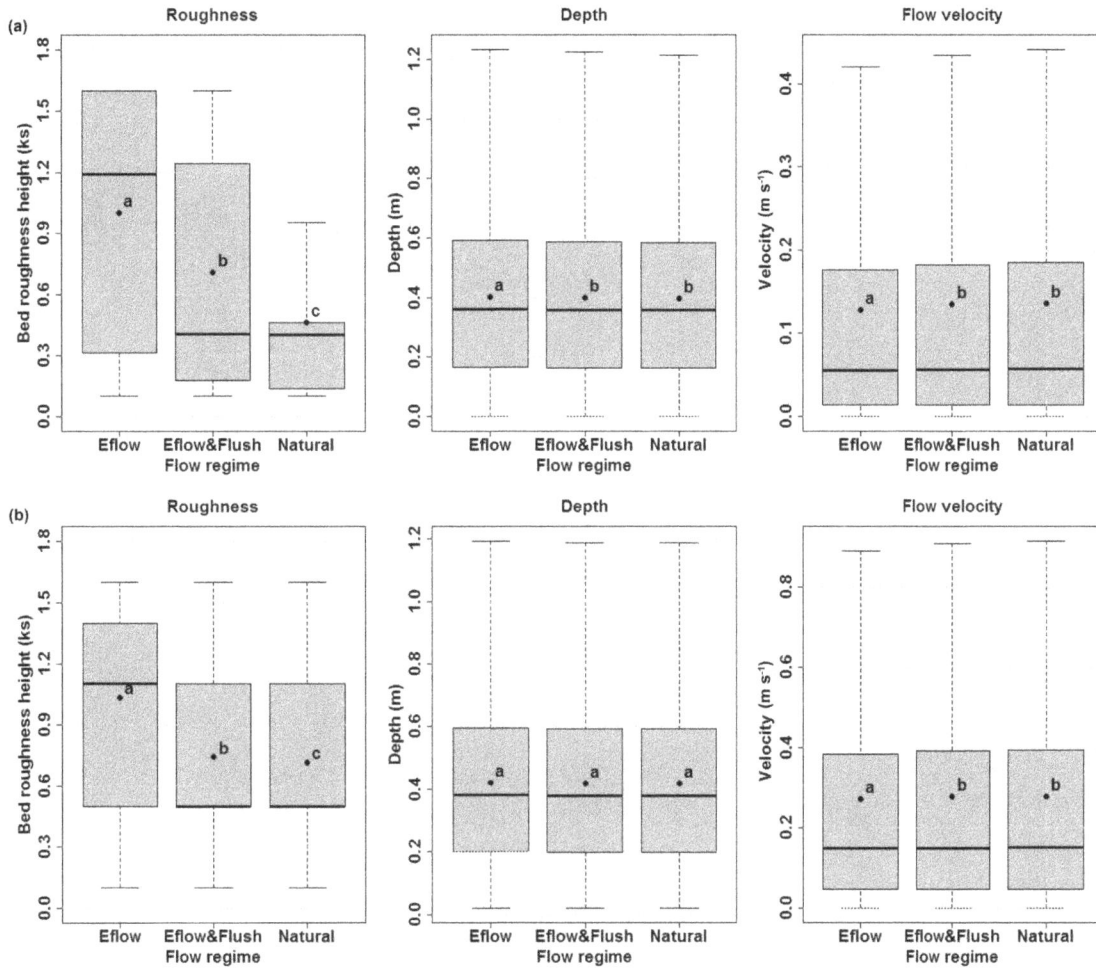

Figure 5. Hydraulic characterization of OCBA **(a)** and OCPR **(b)** according to the different expected riparian vegetation habitats driven by the Eflow, Eflow&Flush, and natural flow regimes (data obtained from 2-D hydrodynamic modeling). Different letters stand for statistically significant differences between groups (*t*-test). Boxplots portray the non-outlier value range, thick black lines the median value, and black dots the mean values.

for the nase juveniles and during spring months for the calandino. Compared to the natural landscape, the WUA modifications instilled by the Eflow landscape are on average approximately 12 %, and are higher than 17 % in a quarter of the cases reaching 80 % in an extreme situation. Particularly, the Eflow landscape provides less habitat suitability during autumn and winter months for the barbel and nase juveniles, c. 17 and 14 %, respectively. Likewise, in this riparian landscape, the habitat suitability during spring months increases approximately 23 % for the barbel juveniles and approximately 20 and 27 % for the calandino juveniles and adults, respectively. On the other hand, throughout the year, the Eflow&Flush landscape provides a WUA very similar to the natural landscape. The habitat changes created by the Eflow&Flush landscape are on average approximately 2 % and never reach 8 % for all species and life stages.

As for OCPR, major differences in WUA are seen almost all year round for calandino and nase, and exist particularly

in spring months for barbel. WUA modifications due to the Eflow landscape are on average near 29 %, a quarter being more than 50 % and reaching up to more than 100 % different in the most extreme case. The Eflow landscape consistently provides less habitat suitability during the autumn and winter months for the nase juveniles and adults, ca. 50 and 38 %, respectively, while the habitat suitability increases by approximately 46 % in calandino. Moreover, the Eflow landscape provides an increased WUA during spring months in approximately 18 % of the barbel adults and 71 % of the calandino adults, while it decreases the habitat on average for approximately 7 % of the remaining species and life stages. Also in this case study, the Eflow&Flush landscape provides a WUA very similar to the natural landscape throughout the year. The habitat changes created by the Eflow&Flush landscape are on average near 3 % and always less than 17 % for all species and life stages. Accordingly, in both case studies, the WUA differences evidenced in the Eflow landscape

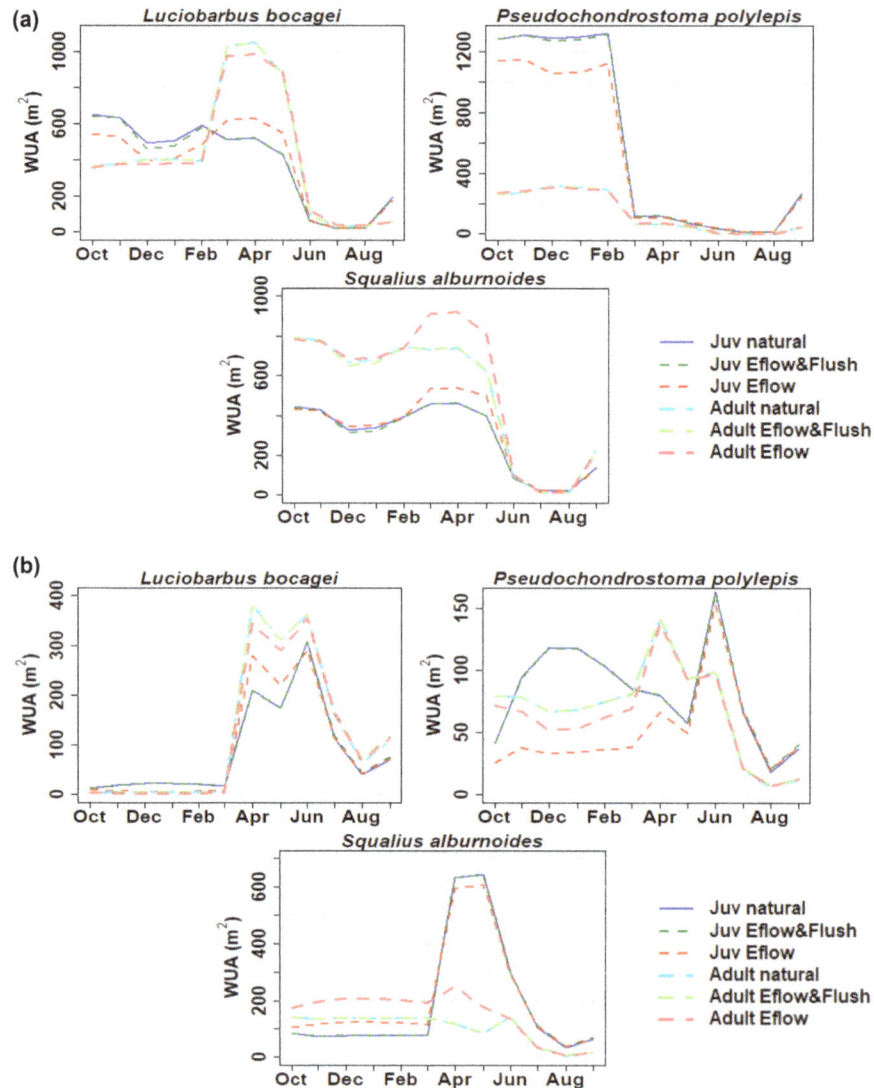

Figure 6. Fish-weighted usable areas provided by the fish-addressed environmental flow regime (Eflow) flowing through the different riparian landscape scenarios that originated from a decade of three different flow regimes (natural, Eflow&Flush, and Eflow) at the OCBA **(a)** and OCPR **(b)** study sites.

proved to be significant in several months by the χ^2 test, whereas this was never the case for the Eflow&Flush landscapes (test results provided in the Supplement; Sect. S3 – Tables S11, S12, S13, and S14).

The riparian-induced modifications on the WUAs are also confirmed by all the employed deviation measures (Table 2). According to RMSD, MAD, and MAPD, the habitat provided by the Eflow landscape is always farther apart from the natural habitat for all species and life stages. In OCBA, the larger deviations occur for the barbel juveniles and nase adults, whereas in OCPR, the calandino adults and the barbel juveniles are the ones enduring greater habitat deviations from the natural circumstances. All together, these results reveal that the disregard of riparian requirements into environmental flows can derail the goals of environmental flows ad-

dressing only aquatic species by an extent of approximately an average of 12 to 29 % of the fish WUAs in the considered study sites as a result of the riparian landscape degradation. On the other hand, results reveal that environmental flows regarding riparian requirements are able to maintain the habitat availability of fish species as the WUAs in the study sites never change on average more the 3 % in a decade.

4 Discussion

This study evaluated the benefits of incorporating riparian requirements into environmental flows by estimating the expected repercussions of riparian changes driven by regulated flow regimes on the fish long-term habitat suitability. To this

Table 2. Deviation analysis of the weighted usable areas for the considered regulated flow regimes benchmarked by the natural flow regime (RMSD – root mean square deviation, MAD – mean absolute deviation, MAPD – mean absolute percentage deviation). Values stand for the habitat availability deviation, in area and percentage, of the environmental flow regimes compared to the natural habitat availability of each species and life stage.

| | OCBA study site | | | | | | OCPR study site | | | | | |
| | Eflow | | | Eflow&Flush | | | Eflow | | | Eflow&Flush | | |
	RMSD (m²)	MAD (m²)	MAPD (%)	RMSD (m²)	MAD (m²)	MAPD (%)	RMSD (m²)	MAD (m²)	MAPD (%)	RMSD (m²)	MAD (m²)	MAPD (%)
Laciobarbus bocagei (juv.)	86.00	72.10	15.40	12.17	7.24	2.52	26.23	17.37	35.55	2.51	1.50	0.63
Laciobarbus bocagei (adult)	29.46	20.55	5.83	2.87	2.12	1.55	12.94	7.73	23.15	3.44	1.79	3.01
Pseudochondrostoma polypepis (juv.)	128.21	86.14	11.58	9.42	5.72	2.26	45.42	32.71	34.43	1.55	0.92	2.51
Pseudochondrostoma polypepis (adult)	7.32	5.85	18.70	2.17	1.37	2.10	9.90	7.00	10.34	0.51	0.35	2.42
Squalius alburnoides (juv.)	44.05	28.16	8.46	6.20	4.06	2.10	33.10	27.78	28.37	2.44	1.35	2.18
Squalius alburnoides (adult)	92.41	52.47	10.23	7.49	5.31	2.37	61.76	47.83	40.54	0.96	0.63	2.90

end, the riparian vegetation was modeled for 10-year periods according to three different flow regimes and results were inputted as the habitat basis for the hydrodynamic modeling and subsequent assessment of the fish habitat suitability in those riparian landscapes. Such ecological modeling approach, where a joint analysis is performed while embracing a suitable time response for the ecosystems involved, enables a realistic biological-response modeling and substantiates the long-term research that is required in environmental flow science (Arthington, 2015; Petts, 2009). Furthermore, this approach allows one to foresee and assess the outcome of recommended flow regimes, which is an essential topic but has been poorly considered in environmental flow science (Davies et al., 2013; Gippel, 2001). This research provides an insight of the expected long-term effects of environmental flows in river ecosystems, therefore unveiling the potential remarkable role of riparian vegetation on the support of environmental flows efficiency, which can transform the actual paradigm in environmental flow science.

During modeling, geomorphology was considered immutable and sediment transport that originated from the environmental flow regimes was disregarded. River morphodynamics and their interactions with riparian vegetation constitute an important river process in many rivers, particularly in fine sediment rivers (e.g., Corenblit et al., 2009, 2011; Gurnell et al., 2012; Gurnell, 2014). However, the research on the temporal scales of geomorphic and ecological processes is still scarce in coarse-bed rivers (Corenblit et al., 2011), and simultaneously more complex and uncertain (Yasi et al., 2013). The error predictions from the best hydraulic predictors in this type of river can range from 50 to 200 % (Van Rijn, 1993; Yasi et al., 2013). Disregarding such processes in these study sites was carefully considered. Given the above and the arguments mentioned in the methods section, we are confident that this option in this case will not bring tangible shortcomings to this research. Furthermore, the possible riverbed degradation effects due to the release of sediment-starving floods by the dam were not tested because according to our expert knowledge this will not pose a problem in this case. Such floods with similar recurrence intervals were already tested by Rivaes et al. (2015) in two river stretches of much smaller grain size (pebbles and sand) and results showed in both cases that such flood discharges were not relevant for riverbed degradation. The influence of fish species on geomorphology and riparian vegetation by ecosystem engineering, as was mentioned in the introduction, was not considered either during this study as it seemed fairly unrealistic in these case studies due to the general dimension of riverbed particles.

The results of the vegetation modeling illustrate how the natural flow regime generates morphodynamic disturbances, without which the riparian vegetation is able to settle and age in the river channel. This is an important outcome that is essential to remember when providing environmental flow instructions. Subsequently, microhabitat analysis demonstrated that changes in the riparian landscape induce modifications in the hydraulic characteristics of the river stretches. The differences in the mean values of these parameters are subtle between riparian landscapes, but are statistically significant. Furthermore, a detailed analysis using a pairwise comparison of flow depths and velocities between scenarios shows that modifications can reach 10 cm in water depth and more than $40 \, \text{cm s}^{-1}$ in flow velocity in some places. The hydrodynamic modeling results show that the water flowing near the margins is more affected than the water flowing in deeper areas of the river channel. One reason for these results is certainly because this study is about the effects of riparian vegetation encroachment on the physical habitat due to the colonization of the river margins by woody riparian vegetation.

Accordingly, there are locations where the considered hydraulic parameters change considerably, shifting the habitat preference of fishes in one or two classes of the corresponding habitat preference curves. These changes are particularly important considering that an alteration of one class regarding these parameters is sufficient to change fish preferences from near null to maximum and vice versa in many cases, as can be seen in the preference curves provided in the Supplement (Sect. S3 – Figs. S10, S11, and S12).

The hydrodynamic modeling also indicated changes directly affecting the habitat suitability of the existing fish species according to the riparian landscape. Through time, the riparian landscape shaped by the Eflow regime diverged in habitat suitability from the natural and Eflow&Flush landscapes, and there were cases where the habitat suitability was modified by more than double. The relationship between fish assemblages and habitat has long been acknowledged (e.g., Clark et al., 2008; Matthews, 1998; Pusey et al., 1993) and can have a significant impact on the ecological status and function of the existing fish communities (Freeman et al., 2001; Jones et al., 1996; Randall and Minns, 2000). Effectively, habitat loss is the major threat concerning fish population dynamics and biodiversity (Bunn and Arthington, 2002), thereby promoting population changes with a proportional response to the enforced habitat change (Cowley, 2008). This is particularly true for the fish species considered in this study (Cabral et al., 2006). The habitat decrease for barbel and nase during the autumn and winter months jeopardizes this species survival by refuge loss, which is particularly important in flashy rivers (Hershkovitz and Gasith, 2013), such as the Ocreza River and Mediterranean rivers in general. On the other hand, the habitat change during the spring months undermines the spawning activity and consequently the sustainability of future population stocks (Lobón-Cerviá and Fernandez-Delgado, 1984). The habitat increase in calandino during this period can be ecologically tricky due to the habitat plasticity of this species (Doadrio, 2011; Gomes-Ferreira et al., 2005), as well as its characteristic adoption for an r-selection strategy as an evolutionary response to frequently disturbed environments (Bernardo et al., 2003). Above all, one should not ignore the fact that the relationships between

fish assemblages and habitat are extremely complex (e.g., Diana et al., 2006; Hubert and Rahel, 1989; Santos et al., 2011), a consequence of the actual natural conditions (Poff and Allan, 1995; Poff et al., 1997) that when disrupted may allow the expansion of more generalist and opportunistic fauna (Poff and Ward, 1989).

Our results indicate that environmental flows taking into account riparian vegetation requirements are able to preserve the naturalness of the riparian landscape and, consequently, the maintenance of the fish habitat suitability. Accordingly, the implementation of such measures in place of using environmental flows addressing only fish requirements can provide significant positive ecological effects in downstream reaches (Lorenz et al., 2013; Pusey and Arthington, 2003) and additional ecosystem services like stream bank stability, flood risk reduction, or wildlife habitat (Berges, 2009; Blackwell and Maltby, 2006) while imposing minor revenue losses on dam managers (Rivaes et al., 2015).

The implementation of such environmental flows could provide an additional way to attain the "good ecological status" required by the Water Framework Directive (WFD). In addition, taking up a procedure such as this one can act both as "win–win" and "no-regret" adaptation measures during the second phase of the WFD, because it potentiates the improvement of other ecological indicators and mitigates the impacts of flow regulation, while being robust enough to account for different scenarios of climate change (EEA, 2005).

Water science still lacks strong links between flow restoration and its ecological benefits (Miller et al., 2012), particularly regarding long-term monitoring of environmental flow performance (King et al., 2015, and citations therein). Nevertheless, the outcomes of this study are a product of long-term simulations by models that were calibrated and validated for the corresponding watershed with local data in natural river flow conditions. This standard procedure in modeling strengthens confidence in our predictions as the models proved to correctly replicate the response of the riparian and fish communities when paralleled with simultaneous observational data. In addition, model uncertainty due to estimation uncertainty in input parameters was previously assessed by means of sensitivity analyses of both models. In either case the models proved to be quite robust to the uncertainty of estimated parameter inputs (see Rivaes et al., 2013, and Boavida et al., 2013), which reveals a relatively small uncertainty in the model outputs and provides additional confidence in the results.

In conclusion, we predict a change in fish habitat suitability according to the long-term structural adjustments that riparian landscapes endure following river regulation. These changes can be attributed to the effects that altered riparian landscapes have on the hydraulic characteristics of the river stretches. In our view, environmental flow regimes considering only the aquatic biota are expected to become obsolete in a few years due to the alteration of the habitat premises on which they were based. This situation points to the unsustainability of these environmental flows in the long term, failing to achieve the desired effects on aquatic communities for which those were proposed in the first place. An environmental flow regime that simultaneously considers riparian vegetation requirements contributes to the preservation of the hydraulic characteristics of the river channel at the natural riverine habitat standards, thereby maintaining the habitat assumptions that support the environmental flow regimes regarding aquatic communities. Consequently, accounting for riparian vegetation requirements poses an essential measure to ensure the effectiveness of environmental flow regimes in the long-term perspective of the fluvial ecosystem.

Competing interests. The authors declare that they have no conflict of interest.

Acknowledgements. This research was financially supported by the Fundação para a Ciência e a Tecnologia (FCT) under project UID/AGR/00239/2013. Rui Rivaes benefited from a PhD grant sponsored by the FCT (SFRH/BD/52515/2014). Isabel Boavida was supported by a postdoctoral grant (SFRH/BPD/90832/2012) also sponsored by the FCT. José Maria Santos was supported by a postdoctoral grant from the MARS project (http://www.mars-project.eu). The Portuguese Institute for Nature Conservation and Forests (ICNF) provided the necessary fishing and handling permits.

Edited by: Stan Schymanski

References

Acreman, M.: Ethical aspects of water and ecosystems, Water Policy, 3, 257–265, https://doi.org/10.1016/S1366-7017(01)00009-5, 2001.

Acreman, M. C. and Ferguson, J. D.: Environmental flows and the European Water Framework Directive, Freshwater Biol., 55, 32–48, https://doi.org/10.1111/j.1365-2427.2009.02181.x, 2010.

Acreman, M. C., Aldrick, J., Binnie, C., Black, A., Cowx, I., Dawson, H., Dunbar, M., Extence, C., Hannaford, J., Harby, A., Holmes, N., Jarritt, N., Old, G., Peirson, G., Webb, J., and Wood, P.: Environmental flows from dams: the water framework directive, P. I. Civil Eng.-Eng. Su., 162, 13–22, https://doi.org/10.1680/ensu.2009.162.1.13, 2009.

Acreman, M., Arthington, A. H., Colloff, M. J., Couch, C., Crossman, N. D., Dyer, F., Overton, I., Pollino, C. A., Stewardson, M. J., and Young, W.: Environmental flows for natural, hybrid, and novel riverine ecosystems in a changing world, Front. Ecol. Environ., 12, 466–473, https://doi.org/10.1890/130134, 2014.

Alexander, J. and Cooker, M. J.: Moving boulders in flash floods and estimating flow conditions using boulders in ancient deposits, Sedimentology, 63, 1582–1595, https://doi.org/10.1111/sed.12274, 2016.

Allan, J. D. and Castillo, M. M.: Stream Ecology: Structure and function of running waters, Second edition ed., Springer, Dordrecht, NL, 436 pp., 2007.

Altman, D. G.: Practical Statistics for Medical Research, Chapman & Hall, London, UK, 613 pp., 1991.

Arthington, A. H.: Environmental flows: saving rivers in the third millennium, Freshwater Ecology Series, 4, Univ. of California Press, 406 pp., 2012.

Arthington, A. H.: Environmental flows: a scientific resource and policy framework for river conservation and restoration, Aquat. Conserv., 25, 155–161, https://doi.org/10.1002/aqc.2560, 2015.

Arthington, A. H. and Zalucki, J. M.: Comparative Evaluation of Environmental Flow Assessment Techniques: Review of Methods, in: Land and Water Resources Research and Development Corporation, edited by: Arthington, A. H. and Zalucki, J. M., Canberra, AUSTRALIA, 141 pp., 1998.

Arthington, A. H., King, J. M., O'Keeffe, J. H., Bunn, S. E., Day, J. A., Pusey, B. J., Blüdhorn, D. R., and Tharme, R. E.: Development of an holistic approach for assessing environmental water requirements of riverine ecosystems, Proceedings of an International Seminar and Workshop on Water Allocation for the Environment, Armidale, Australia, 1992, 69–76, 1992.

Arthington, A. H., Bunn, S. E., Poff, L. N., and Naiman, R. J.: The challenge of providing environmental flow rules to sustain river ecosystems, Ecol. Appl., 16, 1311–1318, 2006.

Arthington, A. H., Naiman, R. J., McClain, M. E., and Nilsson, C.: Preserving the biodiversity and ecological services of rivers: new challenges and research opportunities, Freshwater Biol., 55, 1–16, https://doi.org/10.1111/j.1365-2427.2009.02340.x, 2010.

Barnes, H. H.: Roughness Characteristics of Natural Channels, Washington, USA, 1967.

Benjankar, R., Egger, G., and Jorde, K.: Development of a dynamic floodplain vegetation model for the Kootenai river, USA: concept and methodology, 7th ISE and 8th HIC, 2009.

Benjankar, R., Egger, G., Jorde, K., Goodwin, P., and Glenn, N. F.: Dynamic floodplain vegetation model development for the Kootenai River, USA, J. Environ. Manage., 92, 3058–3070, https://doi.org/10.1016/j.jenvman.2011.07.017, 2011.

Benjankar, R., Jorde, K., Yager, E. M., Egger, G., Goodwin, P., and Glenn, N. F.: The impact of river modification and dam operation on floodplain vegetation succession trends in the Kootenai River, USA, Ecol. Eng., 46, 88–97, https://doi.org/10.1016/j.ecoleng.2012.05.002, 2012.

Berges, S. A.: Ecosystem services of riparian areas: stream bank stability and avian habitat, Master of Science, Iowa State University, Ames, Iowa, USA, 106 pp., 2009.

Bernardo, J. M., Ilhéu, M., Matono, P., and Costa, A. M.: Interannual variation of fish assemblage structure in a Mediterranean river: implications of streamflow on the dominance of native or exotic species, River Res. Appl., 19, 521–532, https://doi.org/10.1002/rra.726, 2003.

Blackwell, M. S. A. and Maltby, E.: How to use floodplains for flood risk reduction, European Communities, Luxembourg, Belgium, 144 pp., 2006.

Boavida, I., Santos, J., Cortes, R., Pinheiro, A., and Ferreira, M.: Assessment of instream structures for habitat improvement for two critically endangered fish species, Aquat. Ecol., 45, 113–124, https://doi.org/10.1007/s10452-010-9340-x, 2011.

Boavida, I., Santos, J. M., Katopodis, C., Ferreira, M. T., and Pinheiro, A.: Uncertainty in predicting the fish-response to two-dimensional habitat modeling using field data, River Res. Appl., 29, 1164–1174, https://doi.org/10.1002/rra.2603, 2013.

Boavida, I., Santos, J. M., Ferreira, M. T., and Pinheiro, A. N.: Barbel habitat alterations due to hydropeaking, J. Hydro-Environ.

Res., 9, 237–247, https://doi.org/10.1016/j.jher.2014.07.009, 2015.

Bovee, K. D.: A guide to stream habitat analysis using the Instream Flow Incremental Methodology, U.S.D.I. Fish and Wildlife Service, Office of Biological Services, Washington, 131 pp., 1982.

Bovee, K. D. and Milhous, R. T.: Hydraulic simulation in instream flow studies: Theory and techniques. Instream Flow Information Paper: No. 5. FWS/OBS-78/33, Fish and Wildlife Service, Fort Collins, Colorado, USA, 1978.

Brisbane Declaration: The Brisbane Declaration. Environmental flows are essential for freshwater ecosystem health and human well-being, Declaration of the 10th International River*symposium* and International Environmental Flows Conference, Brisbane, AUS, 2007, 1–7, 2007.

Broadmeadow, S. and Nisbet, T. R.: The effects of riparian forest management on the freshwater environment: a literature review of best management practice, Hydrol. Earth Syst. Sci., 8, 286–305, https://doi.org/10.5194/hess-8-286-2004, 2004.

Bunn, S. E. and Arthington, A. H.: Basic Principles and Ecological Consequences of Altered Flow Regimes for Aquatic Biodiversity, Environ. Manage., 30, 492–507, https://doi.org/10.1007/s00267-002-2737-0, 2002.

Cabral, M. J., Almeida, J., Almeida, P. R., Dellinger, T., Ferrand de Almeida, N., Oliveira, M. E., Palmeirim, J. M., Queiroz, A. I., Rogado, L., and Santos-Reis, M.: Livro vermelho dos vertebrados de Portugal, Instituto da Conservação da Natureza/Assírio & Alvim, Lisboa, 660 pp., 2006.

Capon, S. J. and Dowe, J. L.: Diversity and dynamics of riparian vegetation, in: Principles for riparian lands management, edited by: Lovett, S. and Price, P., Land & Water Australia, Canberra, AUS, 3–33, 2007.

Chase, J. W., Benoy, G. A., Hann, S. W. R., and Culp, J. M.: Small differences in riparian vegetation significantly reduce land use impacts on stream flow and water quality in small agricultural watersheds, J. Soil Water Conserv., 71, 194–205, https://doi.org/10.2489/jswc.71.3.194, 2016.

Chow, V. T.: Open channel hydraulics, McGraw-Hill, New York, USA, 680 pp., 1959.

Clark, J. S., Rizzo, D. M., Watzin, M. C., and Hession, W. C.: Spatial distribution and geomorphic condition of fish habitat in streams: an analysis using hydraulic modelling and geostatistics, River Res. Appl., 24, 885–899, https://doi.org/10.1002/rra.1085, 2008.

Clarke, A. O. and Hansen, C. L.: The Recurrence of Large Boulder Movement in Small Watersheds of the Anza Borrego Desert, California, Yearbook of the Association of Pacific Coast Geographers, 58, 28–61, 1996.

Cohen, J.: A coefficient of agreement for nominal scales, Educ. Psychol. Meas., XX, 37–46, https://doi.org/10.1177/001316446002000104, 1960.

Copp, G. H.: The habitat diversity and fish reproductive function of floodplain ecosystems, Environ. Biol. Fish., 26, 1–27, https://doi.org/10.1007/bf00002472, 1989.

Corenblit, D., Steiger, J., Gurnell, A. M., and Naiman, R. J.: Plants intertwine fluvial landform dynamics with ecological succession and natural selection: a niche construction perspective for riparian systems, Global Ecol. Biogeogr., 18, 507–520, https://doi.org/10.1111/j.1466-8238.2009.00461.x, 2009.

Corenblit, D., Baas, A. C. W., Bornette, G., Darrozes, J., Delmotte, S., Francis, R. A., Gurnell, A. M., Julien, F., Naiman, R. J., and Steiger, J.: Feedbacks between geomorphology and biota controlling Earth surface processes and landforms: A review of foundation concepts and current understandings, Earth-Sci. Rev., 106, 307–331, https://doi.org/10.1016/j.earscirev.2011.03.002, 2011.

Cowley, D. E.: Estimating required habitat size for fish conservation in streams, Aquat. Conserv., 18, 418–431, https://doi.org/10.1002/aqc.845, 2008.

Curran, J. C. and Hession, W. C.: Vegetative impacts on hydraulics and sediment processes across the fluvial system, J. Hydrol., 505, 364–376, https://doi.org/10.1016/j.jhydrol.2013.10.013, 2013.

Davis, R. and Hirji, R.: Environmental flows: concepts and methods. Water Resources and Environment Technical Note no C1, Environmental Flow Assessment series, World Bank, Washington, DC, 27 pp., 2003.

Davies, P. M., Naiman, R. J., Warfe, D. M., Pettit, N. E., Arthington, A. H., and Bunn, S. E.: Flow-ecology relationships: closing the loop on effective environmental flows, Mar. Freshwater Res., 65, 133–141, https://doi.org/10.1071/MF13110, 2013.

Diana, M., Allan, J. D., and Infante, D.: The influence of physical habitat and land use on stream fish assemblages in southeastern Michigan, Am. Fish. S. S., 48, 359–374, 2006.

Doadrio, I.: Ictiofauna continental española: bases para su seguimiento, Ministerio de Medio Ambiente y Medio Rural y Marino, Centro de Publicaciones, Madrid, Spain, 2011.

Dosskey, M. G., Vidon, P., Gurwick, N. P., Allan, C. J., Duval, T. P., and Lowrance, R.: The Role of Riparian Vegetation in Protecting and Improving Chemical Water Quality in Streams1, J. Am. Water Resour. As., 46, 261–277, https://doi.org/10.1111/j.1752-1688.2010.00419.x, 2010.

Downes, B. J., Barmuta, L. A., Fairweather, P. G., Faith, D. P., Keough, M. J., Lake, P., Mapstone, B. D., and Quinn, G. P.: Monitoring ecological impacts: concepts and practice in flowing waters, Cambridge University Press, 434 pp., 2002.

Dudgeon, D., Arthington, A. H., Gessner, M. O., Kawabata, Z. I., Knowler, D. J., Lévêque, C., Naiman, R. J., Prieur-Richard, A. H., Soto, D., Stiassny, M. L. J., and Sullivan, C. A.: Freshwater biodiversity: importance, threats, status and conservation challenges, Biol. Rev., 81, 163–182, https://doi.org/10.1017/S1464793105006950, 2006.

Dyson, M., Bergkamp, G., and Scanion, J.: Flow. The Essentials of Environmental Flows, in: IUCN, edited by: Dyson, M., Bergkamp, G., and Scanion, J., Gland, Switzerland and Cambridge, UK, 118 pp., 2003.

EEA: Vulnerability and adaptation to climate change in Europe. Technical report No. 7/2005, European Environment Agency, Copenhagen, DNK, 79 pp., 2005.

Encina, L., Granado-Lorencio, C. A., and Rodríguez Ruiz, A.: The Iberian ichthyofauna: ecological contributions, Limnetica, 25, 349–368, 2006.

Ferreira, M. T., Pinheiro, A. N., Santos, J. M., Boavida, I., Rivaes, R., and Branco, P.: Determinação de um regime de caudais ecológicos a jusante do empreendimento de Alvito, Instituto Superior de Agronomia, Universidade de Lisboa, Lisboa, 136 pp., 2014.

Fetherston, K. L., Naiman, R. J., and Bilby, R. E.: Large woody debris, physical process, and riparian forest development in montane river networks of the Pacific Northwest, Geomorphology, 13, 133–144, https://doi.org/10.1016/0169-555X(95)00033-2, 1995.

Fisher, K. and Dawson, H.: Reducing uncertainty in river flood conveyance – roughness review, Department for Environment, Food & Rural Affairs, Environment Agency, Lincoln, UK, 209 pp., 2003.

FitzHugh, T. W. and Vogel, R. M.: The impact of dams on flood flows in the United States, River Res. Appl., 27, 1192–1215, https://doi.org/10.1002/rra.1417, 2010.

Freeman, M. C., Bowen, Z. H., Bovee, K. D., and Irwin, E. R.: Flow and Habitat Effects on Juvenile Fish Abundance in Natural and Altered Flow Regimes, Ecol. Appl., 11, 179–190, https://doi.org/10.2307/3061065, 2001.

Frissell, C., Liss, W., Warren, C., and Hurley, M.: A hierarchical framework for stream habitat classification: Viewing streams in a watershed context, Environ. Manage., 10, 199–214, 1986.

García-Arias, A., Francés, F., Ferreira, T., Egger, G., Martínez-Capel, F., Garófano-Gómez, V., Andrés-Doménech, I., Politti, E., Rivaes, R., and Rodríguez-González, P. M.: Implementing a dynamic riparian vegetation model in three European river systems, Ecohydrology, 6, 635–651, https://doi.org/10.1002/eco.1331, 2013.

Gasith, A. and Resh, V. H.: Streams in Mediterranean Climate Regions: abiotic influences and biotic responses to predictable seasonal events, Annu. Rev. Ecol. Syst., 30, 51–81, 1999.

Gillespie, B. R., Desmet, S., Kay, P., Tillotson, M. R., and Brown, L. E.: A critical analysis of regulated river ecosystem responses to managed environmental flows from reservoirs, Freshwater Biol., 60, 410–425, https://doi.org/10.1111/fwb.12506, 2014.

Gippel, C.: Australia's environmental flow initiative: Filling some knowledge gaps and exposing others, Water Sci. Technol., 43, 73–88, 2001.

Gomes-Ferreira, A., Ribeiro, F., Moreira da Costa, L., Cowx, I. G., and Collares-Pereira, M. J.: Variability in diet and foraging behaviour between sexes and ploidy forms of the hybridogenetic Squalius alburnoides complex (Cyprinidae) in the Guadiana River basin, Portugal, J. Fish Biol., 66, 454–467, https://doi.org/10.1111/j.0022-1112.2005.00611.x, 2005.

Gorman, O. T. and Karr, J. R.: Habitat Structure and Stream Fish Communities, Ecology, 59, 507–515, https://doi.org/10.2307/1936581, 1978.

Gregory, S. V., Swanson, F. J., McKee, W. A., and Cummins, K. W.: An Ecosystem Perspective of Riparian Zones: Focus on links between land and water, Bioscience, 41, 540–551, https://doi.org/10.2307/1311607, 1991.

Gurnell, A.: Plants as river system engineers, Earth Surf. Proc. Land., 39, 4–25, https://doi.org/10.1002/esp.3397, 2014.

Gurnell, A. M., Bertoldi, W., and Corenblit, D.: Changing river channels: The roles of hydrological processes, plants and pioneer fluvial landforms in humid temperate, mixed load, gravel bed rivers, Earth-Sci. Rev., 111, 129–141, https://doi.org/10.1016/j.earscirev.2011.11.005, 2012.

Hassan, M. A., Gottesfeld, A. S., Montgomery, D. R., Tunnicliffe, J. F., Clarke, G. K. C., Wynn, G., Jones-Cox, H., Poirier, R., MacIsaac, E., Herunter, H., and Macdonald, S. J.: Salmon-driven bed load transport and bed morphology in mountain streams, Geophys. Res. Lett., 35, L04405, https://doi.org/10.1029/2007GL032997, 2008.

Hershkovitz, Y. and Gasith, A.: Resistance, resilience, and community dynamics in mediterranean-climate streams, Hydrobiologia, 719, 59–75, https://doi.org/10.1007/s10750-012-1387-3, 2013.

Hirji, R. and Davis, R.: Environmental Flows in Water Resources Policies, Plans, and Projects, Environment and Development, The World Bank C1 – Findings and Recommendations, 189 pp., 2009.

Hjulström, F. H.: Transportation of detritus by moving water: Part 1. Transportation, in: Sp 10: Recent Marine Sediments, 5–31, 1939.

Hubert, W. A. and Rahel, F. J.: Relations of Physical Habitat to Abundance of Four Nongame Fishes in High-Plains Streams: A Test of Habitat Suitability Index Models, N. Am. J. Fish. Manage., 9, 332–340, https://doi.org/10.1577/1548-8675(1989)009<0332:rophta>2.3.co;2, 1989.

Hughes, F. M. R.: Floodplain biogeomorphology, Prog. Phys. Geog., 21, 501–529, 1997.

Hughes, F. M. R. and Rood, S. B.: Allocation of River Flows for Restoration of Floodplain Forest Ecosystems: A Review of Approaches and Their Applicability in Europe, Environ. Manage., 32, 12–33, https://doi.org/10.1007/s00267-003-2834-8, 2003.

Jalón, D. G. D. and Gortázar, J.: Evaluation of instream habitat enhancement options using fish habitat simulations: case-studies in the river Pas (Spain), Aquat. Ecol., 41, 461–474, https://doi.org/10.1007/s10452-006-9030-x, 2007.

Jones, M. L., Randall, R. G., Hayes, D., Dunlop, W., Imhof, J., Lacroix, G., and Ward, N. J. R.: Assessing the ecological effects of habitat change: moving beyond productive capacity, Can. J. Fish. Aquat. Sci., 53, 446–457, 1996.

King, J. M. and Tharme, R. E.: Assessment of the Instream Flow Incremental Methodology and Initial Development of Alternative Instream Flow Methodologies for South Africa, South African Water Research Commission, 604 pp., 1994.

King, J. and Louw, D.: Instream flow assessments for regulated rivers in South Africa using the Building Block Methodology, Aquat. Ecosyst. Health, 1, 109–124, https://doi.org/10.1016/S1463-4988(98)00018-9, 1998.

King, A., Gawne, B., Beesley, L., Koehn, J., Nielsen, D., and Price, A.: Improving Ecological Response Monitoring of Environmental Flows, Environ. Manage., 55, 991–1005, https://doi.org/10.1007/s00267-015-0456-6, 2015.

Linnansaari, T., Monk, W. A., Baird, D. J., and Curry, R. A.: Review of approaches and methods to assess Environmental Flows across Canada and internationally. Research Document 2012/039, Canadian Science Advirosy Secretariat (CSAS), 75 pp., 2012.

Lobón-Cerviá, J. and Fernandez-Delgado, C.: On the biology of the barbel (Barbus barbus bocagei) in the Jarama River, Folia zoologica, 33, 371–384, 1984.

Lorenz, A. W., Stoll, S., Sundermann, A., and Haase, P.: Do adult and YOY fish benefit from river restoration measures?, Ecol. Eng., 61, 174–181, https://doi.org/10.1016/j.ecoleng.2013.09.027, 2013.

Lytle, D. A. and Poff, N. L.: Adaptation to natural flow regimes, Trends Ecol. Evol., 19, 94–100, https://doi.org/10.1016/j.tree.2003.10.002, 2004.

Maheshwari, B. L., Walker, K. F., and McMahon, T. A.: Effects of regulation on the flow regime of the river Murray, Australia, Regul. River., 10, 15–38, https://doi.org/10.1002/rrr.3450100103, 1995.

Mäkinen, H. and Vanninen, P.: Effect of sample selection on the environmental signal derived from tree-ring series, Forest Ecol. Manag., 113, 83–89, 1999.

Matthews, W. J.: Patterns in freshwater fish ecology, Springer Science & Business Media, Norman, Oklahoma, USA, 756 pp., 1998.

Merritt, D. M., Scott, M. L., Poff, L. N., Auble, G. T., and Lytle, D. A.: Theory, methods and tools for determining environmental flows for riparian vegetation: riparian vegetation-flow response guilds, Freshwater Biol., 55, 206–225, https://doi.org/10.1111/j.1365-2427.2009.02206.x, 2010.

Miller, K. A., Webb, J. A., de Little, S. C., and Stewardson, M.: Will environmental flows increase the abundance of native riparian vegetation on lowland rivers? A systematic review protocol, Environmental Evidence, 1, 1–9, https://doi.org/10.1186/2047-2382-1-14, 2012.

Miller, K. A., Webb, J. A., de Little, S. C., and Stewardson, M. J.: Environmental Flows Can Reduce the Encroachment of Terrestrial Vegetation into River Channels: A Systematic Literature Review, Environ. Manage., 52, 1202–1212, https://doi.org/10.1007/s00267-013-0147-0, 2013.

Naiman, R. J., Décamps, H., and McClain, M. E.: Riparia – Ecology, conservation and management of streamside communities, Elsevier academic press, London, UK, 430 pp., 2005.

Nilsson, C. and Berggren, K.: Alterations of Riparian Ecosystems Caused by River Regulation, Bioscience, 50, 783–792, https://doi.org/10.1641/0006-3568(2000)050[0783:aorecb]2.0.co;2, 2000.

NRC, N. R. C.: Riparian Areas: Functions and Strategies for Management, The National Academies Press, Washington, DC, USA, 444 pp., 2002.

Palmer, M. A., Hondula, K. L., and Koch, B. J.: Ecological Restoration of Streams and Rivers: Shifting Strategies and Shifting Goals, Annu. Rev. Ecol. Evol. S., 45, 247–269, https://doi.org/10.1146/annurev-ecolsys-120213-091935, 2014.

Parasiewicz, P.: Using MesoHABSIM to develop reference habitat template and ecological management scenarios, River Res. Appl., 23, 924–932, https://doi.org/10.1002/rra.1044, 2007.

Petts, G.: Instream flow science for sustainable river management, J. Am. Water Resour. As., 45, 1071–1086, https://doi.org/10.1111/j.1752-1688.2009.00360.x, 2009.

Poff, N. L. and Allan, J. D.: Functional Organization of Stream Fish Assemblages in Relation to Hydrological Variability, Ecology, 76, 606–627, https://doi.org/10.2307/1941217, 1995.

Poff, N. L. and Ward, J. V.: Implications of streamflow variability and predictability for lotic community structure: a regional analysis of streamflow patterns, Can. J. Fish. Aquat. Sci., 46, 1805–1818, 1989.

Poff, L. N., Allan, J. D., Bain, M. B., Karr, J. R., Prestegaard, K. L., Richter, B. D., Sparks, R. E., and Stromberg, J. C.: The natural flow regime, Bioscience, 47, 769–784, 1997.

Politti, E. and Egger, G.: Casimir Vegetation Manual, Environmental consulting Ltd, Klagenfurt, AT, 76 pp., 2011.

Politti, E., Egger, G., Angermann, K., Rivaes, R., Blamauer, B., Klösch, M., Tritthart, M., and Habersack, H.: Evaluating climate change impacts on Alpine floodplain vegetation, Hydrobiologia, 737, 225–243, https://doi.org/10.1007/s10750-013-1801-5, 2014.

Postel, S. and Richter, B.: Rivers for life: managing water for people and nature, Island Press, Washington DC, USA, 243 pp., 2003.

Pusey, B. J. and Arthington, A. H.: Importance of the riparian zone to the conservation and management of freshwater fish: a review, Mar. Freshwater Res., 54, 1–16, https://doi.org/10.1071/MF02041, 2003.

Pusey, B. J., Arthington, A. H., and Read, M. G.: Spatial and temporal variation in fish assemblage structure in the Mary River, south-eastern Queensland: the influence of habitat structure, Environ. Biol. Fish., 37, 355–380, https://doi.org/10.1007/bf00005204, 1993.

R Development Core Team: R: A language and environment for statistical computing, R Foundation for Statistical Computing, Vienna, AT, 2011.

Randall, R. G. and Minns, C. K.: Use of fish production per unit biomass ratios for measuring the productive capacity of fish habitats, Can. J. Fish. Aquat. Sci., 57, 1657–1667, https://doi.org/10.1139/cjfas-57-8-1657, 2000.

Revenga, C., Brunner, J., Henninger, N., Kassem, K., and Payne, R.: Pilot Analysis of Global Ecosystems: Freshwater Systems, World Resources Institute, Washington, DC, 80 pp., 2000.

Richter, B. D. and Thomas, G. A.: Restoring environmental flows by modifying dam operations, Ecol. Soc., 12, 1–12, 2007.

Rivaes, R., Rodríguez-González, P. M., Albuquerque, A., Pinheiro, A. N., Egger, G., and Ferreira, M. T.: Riparian vegetation responses to altered flow regimes driven by climate change in Mediterranean rivers, Ecohydrology, 6, 413–424, https://doi.org/10.1002/eco.1287, 2013.

Rivaes, R., Rodríguez-González, P. M., Albuquerque, A., Pinheiro, A. N., Egger, G., and Ferreira, M. T.: Reducing river regulation effects on riparian vegetation using flushing flow regimes, Ecol. Eng., 81, 428–438, https://doi.org/10.1016/j.ecoleng.2015.04.059, 2015.

Rivaes, R., Boavida, I., Santos, J. M., Pinheiro, A. N., and Ferreira, M. T.: Data availability to ensure the reproducibility of the results of Rivaes et al. (2017) in the journal HESSD, https://doi.org/10.5281/zenodo.839531, last access: 6 August 2017.

Rood, S. B., Bigelow, S. G., Polzin, M. L., Gill, K. M., and Coburn, C. A.: Biological bank protection: trees are more effective than grasses at resisting erosion from major river floods, Ecohydrology, 8, 772–779, https://doi.org/10.1002/eco.1544, 2015.

Ryan, D. K., Yearsley, J. M., and Kelly-Quinn, M.: Quantifying the effect of semi-natural riparian cover on stream temperatures: implications for salmonid habitat management, Fisheries Manag. Ecol., 20, 494–507, https://doi.org/10.1111/fme.12038, 2013.

Salemi, L. F., Groppo, J. D., Trevisan, R., Marcos de Moraes, J., de Paula Lima, W., and Martinelli, L. A.: Riparian vegetation and water yield: A synthesis, J. Hydrol., 454, 195–202, https://doi.org/10.1016/j.jhydrol.2012.05.061, 2012.

Santos, J. M., Ferreira, M. T., Godinho, F. N., and Bochechas, J.: Efficacy of a nature-like bypass channel in a Portuguese lowland river, J. Appl. Ichthyol., 21, 381–388, https://doi.org/10.1111/j.1439-0426.2005.00616.x, 2005.

Santos, J. M., Reino, L., Porto, M., Oliveira, J. O., Pinheiro, P., Almeida, P., Cortes, R., and Ferreira, M.: Complex size-dependent habitat associations in potamodromous fish species, Aquat. Sci., 73, 233–245, https://doi.org/10.1007/s00027-010-0172-5, 2011.

SNIRH: National Water Resources Information System, Instituto da Água, I. P. (INAG), 2010.

Statzner, B.: Geomorphological implications of engineering bed sediments by lotic animals, Geomorphology, 157–158, 49–65, https://doi.org/10.1016/j.geomorph.2011.03.022, 2012.

Statzner, B., Sagnes, P., Champagne, J.-Y., and Viboud, S.: Contribution of benthic fish to the patch dynamics of gravel and sand transport in streams, Water Resour. Res., 39, 1309, https://doi.org/10.1029/2003WR002270, 2003.

Steffler, P., Ghanem, A., Blackburn, J., and Yang, Z.: River2D, University of Alberta, Alberta, CANADA, 2002.

Stromberg, J. C., Tluczek, M. G. F., Hazelton, A. F., and Ajami, H.: A century of riparian forest expansion following extreme disturbance: Spatio-temporal change in Populus/Salix/Tamarix forests along the Upper San Pedro River, Arizona, USA, Forest Ecol. Manag., 259, 1181–1198, https://doi.org/10.1016/j.foreco.2010.01.005, 2010.

Tabacchi, E., Lambs, L., Guilloy, H., Planty-Tabacchi, A.-M., Muller, E., and Décamps, H.: Impacts of riparian vegetation on hydrological processes, Hydrol. Process., 14, 2959–2976, 2000.

Tharme, R. E.: A global perspective on environmental flow assessment: emerging trends in the development and application of environmental flow methodologies for rivers, River Res. Appl., 19, 397–441, https://doi.org/10.1002/rra.736, 2003.

Thoms, M. C. and Parsons, M.: Eco-geomorphology: an interdisciplinary approach to river science, The Structure and Management Implications of Fluvial Sedimentary Systems, Alice Springs, Australia, 2002, IAHS Publ. no. 276, 113–119, 2002.

Thorp, J. H., Thoms, M. C., and Delong, M. D.: The Riverine Ecosystem Synthesis. Toward Conceptual Cohesiveness in River Science, Elsevier, London, UK, 208 pp., 2008.

Uddin, F. M. J., Asaeda, T., and Rashid, M. H.: Factors affecting the changes of downstream forestation in the South American river channels, Environment and Pollution, 3, 24–40, https://doi.org/10.5539/ep.v3n4p24, 2014a.

Uddin, F. M. J., Asaeda, T., and Rashid, M. H.: Large-Scale Changes of the Forestation in River Channel Below the Dams in Southern African Rivers: Assessment Using the Google Earth Images, Pol. J. Ecol., 62, 607–624, https://doi.org/10.3161/104.062.0407, 2014b.

Van Looy, K., Tormos, T., Ferréol, M., Villeneuve, B., Valette, L., Chandesris, A., Bougon, N., Oraison, F., and Souchon, Y.: Benefits of riparian forest for the aquatic ecosystem assessed at a large geographic scale, Knowl. Managt. Aquatic Ecosyst., 408, 1–16, https://doi.org/10.1051/kmae/2013041, 2013.

Van Rijn, L. C.: Principles of sediment transport in rivers, estuaries and coastal seas, Aqua Publications, Delft, NLD, 1993.

Viera, A. J. and Garrett, J. M.: Understanding interobserver agreement: the Kappa statistic, Fam. Med., 37, 360–363, 2005.

Wainwright, J. and Mulligan, M.: Environmental Modelling: Finding Simplicity in Complexity, John Wiley & Sons, Ltd, London, UK, 430 pp., 2004.

Wootton, J. T.: River Food Web Response to Large-Scale Riparian Zone Manipulations, PLOS ONE, 7, e51839, https://doi.org/10.1371/journal.pone.0051839, 2012.

Wu, R. and Mao, C.: The assessment of river ecology and habitat using a two-dimensional hydrodynamic and habitat model, J. Mar. Sci. Technol., 15, 322–330, 2007.

Yasi, M., Hamzepouri, R., and Yasi, A. R.: Uncertainties in Evaluation of the Sediment Transport Rates in Typical Coarse-Bed Rivers in Iran, Journal of Water Sciences Research, 5, 1–12, 2013.

6

Why is the Arkavathy River drying? A multiple-hypothesis approach in a data-scarce region

V. Srinivasan[1], **S. Thompson**[2], **K. Madhyastha**[1], **G. Penny**[2], **K. Jeremiah**[1], **and S. Lele**[1]

[1]Ashoka Trust for Research in Ecology and the Environment, Royal Enclave Sriramapura, Jakkur Post, Bangalore, Karnataka, India

[2]Department of Civil and Environmental Engineering, University of California, Berkeley, Berkeley, California, USA

Correspondence to: V. Srinivasan (veena.srinivasan@atree.org)

Abstract. Water planning decisions are only as good as our ability to explain historical trends and make reasonable predictions of future water availability. But predicting water availability can be a challenge in rapidly growing regions, where human modifications of land and waterscapes are changing the hydrologic system. Yet, many regions of the world lack the long-term hydrologic monitoring records needed to understand past changes and predict future trends.

We investigated this "predictions under change" problem in the data-scarce Thippagondanahalli (TG Halli) catchment of the Arkavathy sub-basin in southern India. Inflows into TG Halli reservoir have declined sharply since the 1970s. The causes of the drying are poorly understood, resulting in misdirected or counter-productive management responses.

Five plausible hypotheses that could explain the decline were tested using data from field surveys and secondary sources: (1) changes in rainfall amount, seasonality and intensity; (2) increases in temperature; (3) groundwater extraction; (4) expansion of eucalyptus plantations; and (5) fragmentation of the river channel. Our results suggest that groundwater pumping, expansion of eucalyptus plantations and, to a lesser extent, channel fragmentation are much more likely to have caused the decline in surface flows in the TG Halli catchment than changing climate.

The multiple-hypothesis approach presents a systematic way to quantify the relative contributions of proximate anthropogenic and climate drivers to hydrological change. The approach not only makes a meaningful contribution to the policy debate but also helps prioritize and design future research. The approach is a first step to conducting use-inspired socio-hydrologic research in a watershed.

1 Introduction

Freshwater has been identified as one of the gravest challenges of the twenty-first century (Wagener et al., 2010; Vörösmarty et al., 2010; Srinivasan et al., 2012a). Human demands for water have increased while annual freshwater available globally has remained more or less constant through history. To make sound policy choices, water managers need to know how water availability is changing. They must reconcile the ability to meet the needs of their populations and economies with the potential impacts on the well-being of downstream users, ecosystems and/or future generations. But predicting water availability is particularly challenging in rapidly growing regions, which are undergoing population growth, agricultural intensification and industrialization. Human modifications of land and waterscapes are changing the dynamics of the water cycle at unprecedented rates. Many of these regions also lack the long-term hydrologic monitoring records needed to make such analyses possible. As a result, water managers lack the scientific basis to articulate trade-offs. This often leads to policies that address only part of the problem at best or, at worst, have negative or paradoxical outcomes (Sivapalan et al., 2014).

1.1 Challenges in rapidly growing, data-scarce regions

Making hydrological predictions is a non-trivial problem in any context, but it is confounded by three issues encountered in rapidly changing, data-scarce regions: (i) non-stationarity arising from anthropogenic drivers, (ii) the sparse availability of historical data, and (iii) lack of original, place-based scientific research leading to oversimplified assumptions. The prediction challenges arise both from the nature of the system (point i) and researcher constraints (points ii and iii), but the net result is that water managers are forced to rely on conceptual models that poorly represent the underlying system.

Multiple drivers of change: traditional water resources management is based on the assumption of stationarity – the idea that natural systems fluctuate within an unchanging envelope of variability (Milly et al., 2008). However, the impact of humans on the water cycle, either directly through modification of landscapes and waterscapes, or indirectly via climate change, has been identified as a defining challenge for hydrology (Thompson et al., 2013). The potential impacts of climate change on the hydrologic cycle have received enormous attention from researchers and decision makers in recent years (Stocker and Raible, 2005; Huntington, 2006). The role of other direct human interventions like groundwater extraction, small dams and urbanization is also coming under increased scrutiny (Arrigoni et al., 2010; Wang and Hejazi, 2011; Cai and Zeng, 2013; Zeng and Cai, 2014; Hu et al., 2015).

Data sparseness: the task of determining cause and effect relationships with respect to hydrologic behaviour is complicated in regions with sparse or recent hydrologic records (Maeda and Torres, 2012; Systematics, 2010). Despite long-standing global calls to improve data sharing and transparency (Arzberger et al., 2004; Sivapalan et al., 2003; Bonell et al., 2006; Reichman et al., 2011; Dunne and Leopold, 1978), data scarcity remains an impediment to research in many parts of the world. Data scarcity has motivated concerted research efforts such as the predictions in ungauged basins (PUB) effort of IAHS (Wagener et al., 2004; Sivapalan, 2003; Sivapalan et al., 2003). However, these efforts are generally not suitable for predictions in non-stationary, human-impacted basins (Srinivasan et al., 2012b; Sivapalan et al., 2012). In such cases, lack of data confounds both conceptual understanding and building of quantitative models that explain how the water system works.

Over-simplifying assumptions: investments in the water sector must be made even in the absence of long-term records. In the absence of reliable data, modellers are then often forced to make many simplifying assumptions. The choices seem too often to be dictated by what *can be modelled* rather than *what matters*, leading to so-called "modeller myopia" (Buytaert, 2015). For instance, Gosain et al. (2006) predict water availability in space and time in several Indian river basins under climate change, but they do not incorporate man-made structures like dams or diversions into their basic model or trend analyses. Mujumdar and Ghosh (2008) modelled flows in the Mahanadi of eastern India; their model assumed that recent declines in streamflow reflect a "climate signal", without considering the possible influence of more proximate factors like groundwater pumping. Similarly, numerous water resources modelling projects in India decouple the effects of groundwater depletion from surface water responses, even where groundwater overexploitation is known

to be a problem (Kelkar et al., 2008; Gosain et al., 2011; Garg et al., 2013).

1.2 Use-inspired science in data-scarce regions

The mismatch between the needs of water managers and what off-the-shelf models can generate is not a sufficient reason for inaction or ad hoc decision making in regions with rapidly increasing water demand. There is an urgent need to formulate new approaches to frame and conduct hydrologic investigations in human-dominated, data-scarce situations. The conventional response would be to initiate primary data collection and to build new site-specific models from scratch. However, hydrologic data collection is expensive and takes many years. In contrast, information is often needed quickly and projects are limited by time and resource constraints.

How should hydrologists proceed in these circumstances? First, as Thompson et al. (2013) suggest, hydrologists should adopt a "use-inspired science" approach by pursuing scientific understanding while also addressing policy and management goals. This requires identifying the most pressing societal problems and working backwards from them. Second, as Buytaert et al. (2014) suggest, knowledge may be dispersed amongst multiple parties. While researchers and managers may hold some expert knowledge, citizens who have lived through change in the basin may also have useful insights. Third, it should be possible to use this knowledge to identify working hypotheses (Chamberlin, 1965) that might explain the hydrological phenomenon of interest, and then use the sparse data to accept or reject at least some of them. This approach would then guide the choice of future data collection and sophisticated modelling efforts, targeting the most critical knowledge gaps. We use the above approach to narrow down possible causal mechanisms of hydrologic change in the Arkavathy watershed in southern India. Five possible hypotheses that link anthropogenic and climatic changes to the water scarcity in the watershed are outlined and investigated.

2 The problem: drying of TG Halli reservoir

2.1 Description of study area

The Arkavathy River is located in the state of Karnataka in southern India (Fig. 1). The river's catchment overlaps with the western portion of the rapidly growing metropolis of Bengaluru (Bangalore). The region is seasonally monsoonal, receiving approximately 830 mm of precipitation annually. The main stem of the Arkavathy River has its headwaters in the Nandi Hills north of Bengaluru and is joined by its first major tributary, the Kumudvathy River at Thippagondanahalli (TG Halli) village, where a reservoir was constructed in 1935 to supply water to Bengaluru. This reservoir has a catchment area of approximately 1447 km^2.

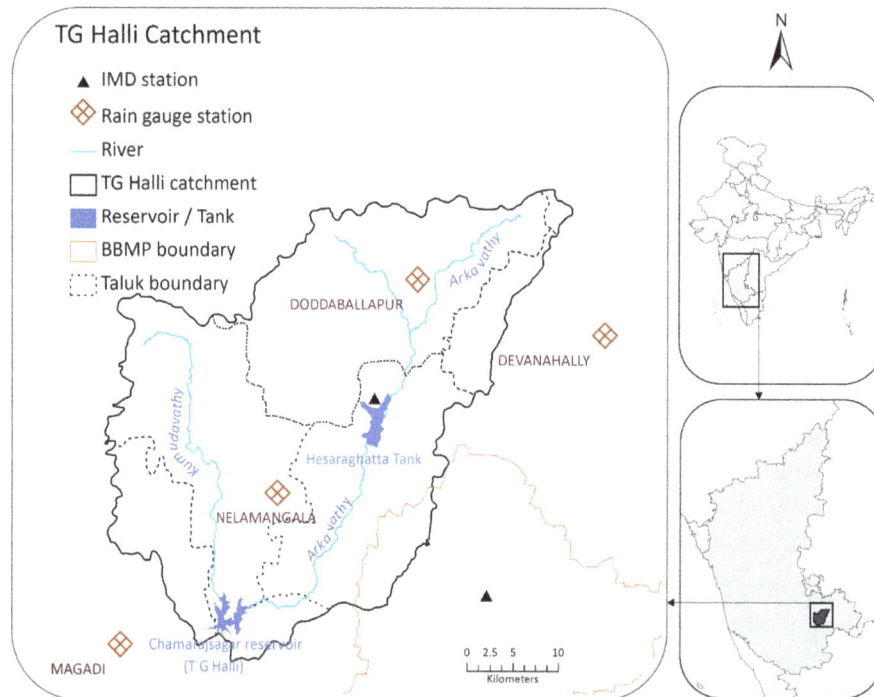

Figure 1. The TG Halli catchment with major features. BBMP is the Greater Bangalore Municipal Corporation Boundary (data source: Survey of India toposheets at 1 : 50 000 scale; ASTER DEM imagery, maps prepared at the ATREE EcoInformatics Lab).

The TG Halli reservoir catchment also contains an older water supply reservoir at Hesaraghatta, as well as an estimated 617 small surface storage structures called "tanks". Tanks are traditional in-stream water harvesting systems that were commonly built in southern India and Sri Lanka over the last 6 centuries to store monsoon runoff for post-monsoon irrigation (Vaidyanathan, 2001; Shah, 2003). The cumulative storage of all these tanks (297 of which are more than 50 ha in size) and Hesaraghatta reservoir is estimated to be 143 million cubic metres, i.e. about 1.5 times the storage capacity of TG Halli reservoir (ISRO and IN-RIMT, 2000).

Most of the TG Halli catchment is underlain by gneissic and granitic aquifers. Highly weathered soils extend to about 20 m below grade level (b.g.l.), and form a shallow aquifer in which seasonal perched water tables can develop. Between about 20 and 60 m b.g.l. lies a fractured rock zone with considerable jointing and cracking, acting as a deeper aquifer. Groundwater yields decline beyond 60 m b.g.l., although fractures continue to be encountered down to 300 m.

2.2   The problem

From 1937 up to the 1980s, the TG Halli reservoir was a major source of water for Bengaluru. However, inflow to the reservoir has steadily declined since the early 1980s (Fig. 2a), and today it supplies only 0–25 % of its design capacity. Average inflows into the TG Halli reservoir have decreased from 385 millions of litres (ML) per day (140 000 ML year^{-1}) pre-

1975 to about 65 ML day^{-1} post-2000 (24 000 ML year^{-1}), a decline of 320 ML day^{-1}. The cascading irrigation tanks dotting the catchment are also mostly dry, indicating that the loss of surface runoff has occurred throughout the catchment (Lele et al., 2013).

The drying of flows into the TG Halli reservoir and tanks in the catchment has clear implications for the 800 000 people that live in the catchment, both in terms of current water availability and because the declining flows may be an indicator of the overall unsustainability of water use in the basin jeopardizing future populations and economic growth.

2.3   The debate about causes and solutions

Given the urgency of the problem, several uncoordinated and often contradictory actions have been undertaken. One reason for this is that the causes of the inflow reductions to the TG Halli reservoir remain unclear. In order to formulate hypotheses that could be investigated systematically, we consulted a range of sources to understand the positions and perceptions of different groups: one-on-one meetings with government officials, written policy documents and reports, a comprehensive literature review (Lele et al., 2013) and an expert consultation meeting held at ATREE (Bangalore) in November 2012. Additionally, at the launch of the research project in early 2013, a meeting was convened by the chief secretary of the state which included the research team and the heads of all government agencies engaged in water is-

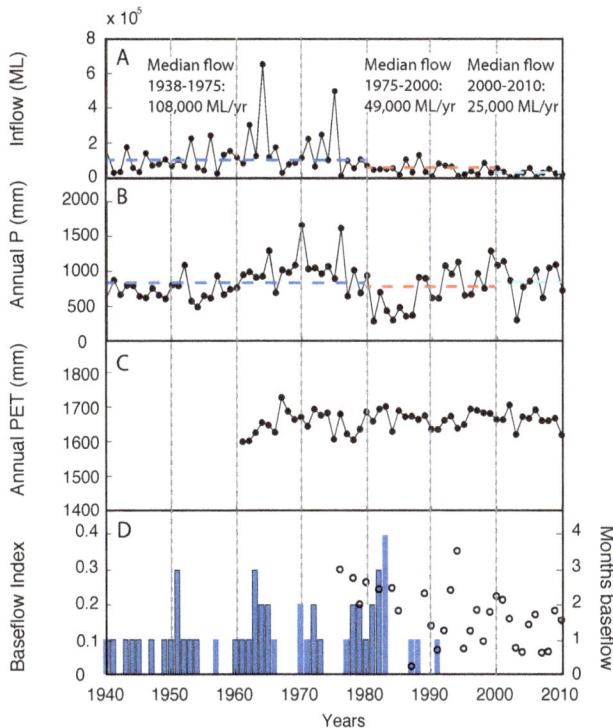

Figure 2. Changes in hydrology and hydrometeorology of the Arkavathy Basin, 1970–2010. (**a**) Annual inflows into the TG Halli reservoir. The 1938–1975, 1975–2000 and 2000–2010 median and mean annual inflows illustrate the decline in inflow that has occurred in recent decades. (**b**) Areally averaged annual rainfall over the seventh taluk (local government areas) comprising the TG Halli catchment. Potential evapotranspiration as estimated from the Hargreaves equation for the TG Halli catchment. (**d**) Two estimates of baseflow contribution to the TG Halli inflows: the number of months/year when 100 % of flow was derived from baseflow (bars) and the baseflow index computed from daily inflow data (dots).

sues. The research team also made several reconnaissance visits, attended over a dozen stakeholder meetings hosted by other groups and held more than 60 "water literacy meetings" in the TG Halli catchment villages in 2014 and 2015, which were collectively attended by over 500 farmers. Finally, the research team conducted over two dozen focus group discussions in 2013 and 2014, targeting specific stakeholder groups.

This initial review identified several policy positions that reflect different perceptions on the drying of the river:

– The Bangalore Water Supply and Sewerage Board (BWSSB), which owns and operates the TG Halli reservoir, commissioned a study (ISRO and IN-RIMT, 2000). This study identified several possible causes for the decline of inflows into the TG Halli – declines in rainfall, groundwater pumping and obstructions in streams. However, the study did not quantify the relative magnitudes of these factors and did not recommend

actions to directly address them. BWSSB has taken no specific actions of its own to address the problem.

– The Cauvery Neeravari Nigam Ltd. (CNNL) was made responsible by the state government for "rejuvenating" the Arkavathy River. CNNL commissioned its own study (CNNL, 2010), which concluded that the primary reason for reduced inflow into TG Halli is obstructions in the channels. Accordingly, the agency response has been to bulldoze the obstructions and desilt the channels.

– Meanwhile, local rural development programmes have focused on constructing check dams to recharge the shallow aquifer and ostensibly restore baseflow in the stream.

– A number of urban based citizen's groups have emerged with the objective of rejuvenating the river or saving Bangalore's water bodies (see http://www.artofliving.org/kumudvathi-river-rejuvenation-project and https://www.facebook.com/arkavathi.rejuvenation). These groups are focused on removal of eucalyptus trees, desilting lake beds and diverting treated wastewater into lakes to recharge the shallow aquifer.

– The state Water Resources Development Organization (WRDO) has argued that climate change, via declining rainfall and rising temperatures, is responsible for the drying of the river. This perception was also held by most farmers we interacted with during the water literacy meetings, many of whom favour inter-basin imports from west-flowing rivers.

2.4 The multiple-hypothesis approach

By examining the different explanations of the causes of streamflow decline and plausible runoff generation mechanisms, we identified and investigated *all plausible hypotheses* that could explain the observed changes in the Arkavathy Basin:

– *Hypothesis 1: changes in rainfall:* changes in rainfall as the primary driver of streamflow could induce changes in surface runoff generation. The climate change literature for this part of Karnataka mentions a possible shift in the monsoon, such that the south-west monsoon June-July-August-September (JJAS) season rainfall would probably decline, and post-monsoon October-November-December (OND) rainfall could increase. A change in the seasonality of precipitation could result in a change in rainfall partitioning to runoff, because a greater fraction is partitioned to evaporation and transpiration. Additionally, if both seasonal and annual rainfall patterns are unchanged, a reduction in the mean storm intensity or depth could result in a failure to trigger infiltration-excess or saturation-excess runoff.

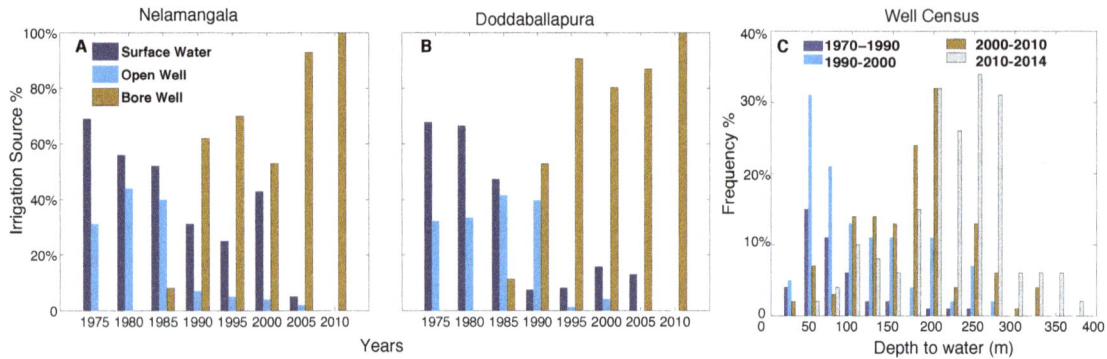

Figure 3. Water use in the TG Halli catchment. **(a, b)** Sources of irrigation water used over time in two taluks (local government areas), indicating the reduction in surface and shallow well use and their replacement by deep borehole over time. Data Source: Department of Economics and Statistics, Karnataka. **(c)** Depths at which water was encountered at time of drilling. Data from well census conducted by ATREE field hydrology team in summer of 2014 covering 482 boreholes in a 26 sq km area in TG Halli catchment.

The previous study commissioned by BWSSB found that rainfall in excess of 20 mm day^{-1} is needed to generate significant inflows into TG Halli reservoir (ISRO and IN-RIMT, 2000).

– *Hypothesis 2: increasing potential evapotranspiration due to climate change:* increases in potential evapotranspiration could result in an increase in actual evapotranspiration, reducing recharge and therefore baseflow. Of the major drivers of potential evapotranspiration (temperature, humidity, solar radiation and wind speed), it is known that temperature has been increasing in southern India over the past century, with most stations reporting temperature increases on the order of 1°C/100 years (Arora et al., 2005; Hingane et al., 1985). On the other hand, solar radiation trends in the region have been negative, in association with the formation of atmospheric brown clouds (Ramanathan et al., 2005), and wind speed trends in India also appear to also be declining (McVicar et al., 2012). Empirical evidence from evaporation gauges throughout India also suggests that pan evaporation has declined over the twentieth century (Chattopadhyay and Hulme, 1997). Nonetheless, temperature-driven increases in evaporative demand could have altered rainfall-runoff partitioning in the catchment and contributed to the reduced streamflow.

– *Hypothesis 3: declining baseflow due to groundwater overexploitation:* previous studies in well-monitored basins have shown that groundwater depletion can reduce baseflow contributions to streamflow, reducing overall flows (Cai and Zeng, 2013; Zeng and Cai, 2014). Our hypothesis is that reduction in groundwater storage induced by pumping lowers the seasonal water table, resulting in the water table intersecting the river channel less frequently and for shorter periods of time, ulti-

mately reducing the baseflow contribution to the Arkavathy. The decline in the Arkavathy River flow has occurred concurrently with an expansion of groundwater extraction in the basin and across Karnataka. Although groundwater monitoring in the region is minimal, irrigation data clearly show a shift from surface to groundwater and open wells to deep boreholes (Fig. 3a, b) (DES, 1970–2012).

– *Hypothesis 4: increasing actual evapotranspiration due to expansion of plantations:* numerous studies indicate that where a catchment area is converted from rain-fed agriculture to deep-rooted perennial vegetation, it can result in decreases in flow (Brown et al., 2005). Eucalyptus cultivation was actively promoted among farmers by the state government under its farm forestry programme in the 1980s (Shiva et al., 1981). Field surveys within the TG Halli catchment indicate a significant increase in eucalyptus plantation area in the past 40 years. Several studies have documented that eucalyptus plantations create unsaturated conditions over a deep root zone and can thus reduce subsurface contributions to streamflow (Calder et al., 1993; Farley et al., 2005).

– *Hypothesis 5: million puddle theory:* the final hypothesis is that the construction of (largely illegal) structures in the channel – along with construction of check dams and unculverted roads – has resulted in the channels in the upper catchment becoming disconnected. In other words, a once-connected, flowing river has been replaced by a "million puddles". A portion of the water in these puddles evaporates or is transpired by riparian vegetation and becomes unavailable.

Table 1. Details of various data sets used.

Parameter	Type of data	Source of data
Precipitation	Daily rainfall from four rain gauges (1934–2010)	Indian Meteorological Department
Temperature	Monthly max., min. and mean temperature from two weather stations (1901–2010)	Indian Meteorological Department
Surface flow	Daily inflows (1975–2010) and Monthly inflows (1937–2010) into TG Halli reservoir	Bangalore Water Supply and Sewerage Board
Area under eucalyptus plantations in 1973–1979	Topographical sheet	Survey of India
Area under eucalyptus in 2001	Land use map	Karnataka State Remote Sensing Applications Centre
Groundwater levels	Well census of 472 wells in two milli-watersheds covering 26 km^2	ATREE hydrology team field survey
Groundwater extraction	Irrigated area (1970–2012)	KA State Annual Season Crop Report
	Irrigated area (1981, 1991, 2001)	Census of India, Village Amenities Dataset
Channel obstructions	Number of check dams and unculverted roads	Primary Survey by Zoomin Tech.
	Number of check dams in two milli-watersheds covering 26 km^2	ATREE hydrology team field survey

3 Methods

3.1 Data sources and quality assurance

To test these hypotheses, we collected available secondary data within and around the Arkavathy Basin. Data were quality-checked and triangulated against other sources and supplemented with field surveys when needed (Table 1).

Monthly inflow data for the TG Halli reservoir were obtained from Bangalore Water Supply and Sewerage Board (BWSSB) for the period 1937 to 2010. Additionally, daily records of inflows from 1970 onwards were obtained from the local BWSSB offices and digitized. The daily and monthly data were cross-validated and any errors were corrected.

3.2 Analysis techniques

The goal of the analysis was two-fold: (i) to determine whether the perceived changes in hydrological drivers have occurred and (ii) whether the magnitude of changes in the drivers could explain the magnitude of the change in flow in the Arkavathy Basin (i.e. consistent with the observed 320 ML day^{-1} reductions in flow).

Hypothesis 1: declining rainfall

Data from four long-term rainfall gauges located in Devanahalli, Doddaballapura, Magadi and Nelamangala towns within the TG Halli catchment were available for analysis (see Fig. 1 for gauge locations). These gauges provide daily rainfall data over 75 years (1934–2010). Although 18 additional rainfall gauges, operated by various government agencies, exist within the catchment, these gauges do not provide continuous data over a sufficient time period to allow trend analysis. As a quality control procedure, we performed double mass plots, compared the total number of rainy days between the gauges, and excluded years where the total number of rainy days represented a low outlier (indicating a likelihood of missing data). Outlier years were determined to be those where the number of recorded rain days was less than $f_{25} - 1.5(f_{75} - f_{25})$, where f_{75} represents the 75th percentile and f_{25} the 25th percentile of the total number of rain days.

Annual rainfall was computed over the water year (June to May). Seasonal rainfall totals were computed in terms of pre-monsoon (January-February-March-April-May: JFMAM), monsoon (June-July-August-September: JJAS) and post-monsoon (October-November-December: OND) rainfall totals. To identify changes in rainfall depths at daily timescales, the number of days per year in which rainfall volumes exceeded 10, 25 and 50 mm were determined for the 1934–2009 period. Trend detection was undertaken for each of the above data sets in two ways. First, we determined whether a trend was present over the full time series. As the data generally did not conform to the assumptions for least-squares regression, we evaluated the trends using a non-parametric Mann–Kendall test. Second, we evaluated whether a change in the mean values of the meteorological parameters had occurred from the pre-1970 and post-1970 period, taking 1970 as a point after which the Arkavathy River flows obviously declined. Where the data were normally distributed we made these comparisons with t tests; otherwise non-parametric Mann–Whitney–Wilcoxon tests were used.

Hypothesis 2: increasing potential evaporation due to climate change

In the absence of detailed meteorological data in the Arkavathy Basin, we estimated changes in the mean daily potential evaporation rate as a function of temperature (PET) using the modified 1985 Hargreaves evapotranspiration equation (Hargreaves and Samani, 1985):

$$PET = 0.0023 \times R_a \times (TC + 17.8) \times TR^{0.5}, \quad (1)$$

where R_a is the extraterrestrial solar radiation (mm day^{-1}), and TC is the average daily temperature (°C) calculated as $(T_{Max} + T_{Min})/2$. TR is the temperature range (TR $= T_{Max} - T_{Min}$, T_{Max} is the maximum daily temperature and T_{Min} is the minimum daily temperature). All results were averaged to the annual timescale. The resulting PET time series was analysed to determine the presence of trends or step changes in the mean PET. Temperature data from 1901 to 2001 were obtained for one station each in Bangalore Urban and Bangalore Rural districts from the Indian Meteorological Department. The data were checked to ensure that there were no missing data and that temperatures were within expected ranges. Extraterrestrial solar radiation was computed based on the weather station latitudes (Maurer, 2014) using the method of Spencer (1971), with an accuracy of 0.01 % (Duffie and Beckman, 2013).

The modified Hargreaves equation relies on the diurnal temperature range to provide a surrogate for solar radiation and is widely used to estimate potential evaporation when only limited ground data (temperature) are available. The resulting PET estimates are typically within 10 % or better of those derived from lysimeter or Penman–Monteith methods, when results are averaged over 5-day or greater time periods (Hargreaves and Allen, 2003). Limitations of the method lie in the fact that the relationship between diurnal temperature range and other drivers of potential evaporation (e.g. net radiation and vapour pressure deficit) may not be stationary over long time periods. In southern India, such non-stationarity

is likely to be associated with so-called "solar dimming" due to increased upper atmospheric pollution (Chattopadhyay and Hulme, 1997). We anticipate that errors due to non-stationarity are likely to lead to an over-estimation of potential evaporation via the Hargreaves equation.

Hypothesis 3: declining baseflow due to groundwater overexploitation

Long-term groundwater level data (> 10 years) existed for only two shallow wells within a 5 km buffer of the TG Halli catchment. These reported stable water levels of 10–30 m b.g.l. However, in the course of extensive field visits, no water was seen in any other open well in the region. We concluded that the two monitoring wells are not representative of surrounding conditions. There are also no deep borehole piezometers with long-term water level data in the catchment area. To infer potential changes in groundwater levels, we conducted a comprehensive census of boreholes in a 26 km^2 area in the TG Halli catchment in the summer of 2014. Data for a total of 472 boreholes were recorded. For each borehole, the owner was interviewed to obtain details of the year of construction, use, status, depths of yielding fractures and year of failure (if applicable). Together, these data provide an understanding of how groundwater levels have changed in the last 4 decades.

We undertook two different analyses to explore whether changes in groundwater were compatible with the observed changes in surface flow. In one analysis we used a baseflow recession technique to benchmark the changes in mobile subsurface water storage that would be needed to account for the decline in annual flows and then estimated how these changes might manifest as a decline in groundwater levels. If this change in storage greatly exceeds observed well declines in the catchment, then the hypothesis that lower groundwater levels have led to streamflow reductions could be rejected. In a second analysis, we performed a baseflow separation on the daily runoff data from 1970 onwards to determine how the trends in total streamflow were reflected by changes in quick-flow and baseflow.

Recession analysis: we follow Brutsaert and Nieber (1977) in positing a nonlinear relationship between storage (S, [ML]) and discharge (Q, [ML day^{-1}]) of the form

$$S = aQ^b. \quad (2)$$

A mass balance during periods of flow recession (i.e. when rainfall P is negligible) would be given by

$$\frac{dS}{dt} = -ET - Q, \quad (3)$$

where recharge to groundwater and inter-basin transfers are assumed negligible and ET represents evapotranspiration. If Eqs. (2) and (3) are coupled and differentiated, then the following expression is obtained relating flow to its rate of change:

$$ab\, Q^{b-1} \frac{dQ}{dt} = -\text{ET} - Q, \tag{4}$$

Under the assumptions that ET is slow in comparison to flow, so that ET $\rightarrow$ 0, Eqs. (3) and (4) simplify to

$$\frac{dQ}{dt} = -\frac{1}{ab} Q^{2-b}. \tag{5}$$

Taking the logarithms of the absolute values of this expression, one obtains

$$\log\left(\frac{dQ}{dt}\right) = \log\left(\frac{1}{ab}\right) + (2-b)\log(Q). \tag{6}$$

That is, a plot of the logarithms of the rate of change of the discharge against the logarithms of the actual discharge at any point in time contains sufficient information (in the form of an intercept and slope) to estimate the parameters of the original storage–discharge expression. To do this, the lower envelope of the expression must be fitted in order to minimize the effects of neglecting evaporation and to focus the analysis on the groundwater response (Brutsaert and Nieber, 1977). No significant changes in the recession behaviour over time were identified from this analysis.

This methodology was applied to the monthly flow data from the Arkavathy at TG Halli, focusing on the seasonal recessions from 1937 to 1970 (i.e. prior to the discernible reductions in river flow). There are two major limitations to using monthly data for this analysis. First, the estimation of the rate of change of the flow is coarse. Second, the contribution of rainfall to runoff events is unlikely to be negligible, even during the seasonal recession. However, because the daily flow data were only available for the post-1970 period, the monthly analysis provides the only opportunity to evaluate the storage–discharge relationship when the river was flowing "normally". As outlined in the results, the calibrated model had an exponent $b = 1.43$, very close to the theoretical value of 1.5, offering some reassurance that the results are reasonable. Using the parameterized storage–discharge equation, we estimated the mobile storage averaged over the catchment area, needed to produce the mean of the peak monthly flows for all years prior to 1970. The resulting storage volume can be normalized by the mobile porosity of the aquifer sediments to generate an estimate of the drop in the surface water table depth required to explain the "missing" flow volume after 1970.

Baseflow trends: using the daily data from 1970 to 2010, we undertook a baseflow separation using a digital filter (Nathan and McMahon, 1990) and computed the annual baseflow and the baseflow index for each water year. Again, we analysed trends in these indices using the methods described previously. Additionally, we analysed baseflow trends in both the monthly data from 1937 to 2008 and in the daily data available for 1970 onward. For the monthly

data, we defined a "baseflow month" as a month in which there was streamflow. This definition implies that 100 % of the flow in these months is from baseflow, a much higher standard than the baseflow index which indicates the proportion of baseflow that occurred through time.

Hypothesis 4: increasing actual evapotranspiration due to expansion of plantations

We calculated the change in eucalyptus plantation area from 1973 to 2001 by comparing the mapped land uses in both years. We used two sources: a land use map provided by Karnataka State Remote Sensing Application Centre (KSRSAC) and Survey of India topographic sheets. The KSRSAC land use map was derived from Indian Remote Sensing (IRS) LISS-3 merged with PAN satellite imagery with an effective 6 m resolution. The map reported the area under eucalyptus plantations in 2001. For other years, no such maps were readily available. So we digitized 1 : 50 000 scale topographic maps prepared by the Survey of India during the 1970s (1973 to 1979), which show eucalyptus plantations on public lands.

We made three assumptions about water use by eucalyptus plantations (which are typically unirrigated). First, the plantations could not themselves have led to groundwater mining (as has been claimed in other parts of Karnataka (Calder et al., 1993)), because shallow groundwater in the region had largely disappeared by time eucalyptus plantations were promoted under the social forestry programme in the early 1980s. Second, we assumed that eucalyptus transpires at a rate of 830 mm yr^{-1} (the annual average rainfall). In effect the trees were perfectly efficient in utilization of rainwater, given that potential evaporation of 1650 mm yr^{-1} greatly exceeds annual rainfall and that many plantations implement practices to limit surface runoff. Third, we assumed the plantations displaced rain-fed coarse cereal crops such as maize or millet, which have a seasonal ET of about 290 mm yr^{-1} for a single crop and 540 mm yr^{-1} for a double crop (Allen et al., 1998).

Hypothesis 5: million puddle theory

Data on the number of channel obstructions in the TG Halli catchment were available in a report commissioned by Cauvery Neeravari Nigam Limited (CNNL) (CNNL, 2010). A total of 344 obstructive structures were recorded including roads, bridges and unculverted roads, of which 277 were small check dams (Table 2). The density of check dams estimated from the report is 0.2 km^{-2} of watershed.

To validate the CNNL data, we conducted a comprehensive survey of all stream obstructions in two milli-watersheds covering a 26 km^2 area within in TG Halli catchment. Over 40 check dams were found in the 26 km^2 area, indicating a check-dam density of 1.35 km^{-2}. Even after discounting 20 % that were leaky or silted, it appears that the CNNL data

Table 2. Number and type of stream encroachments in each section TG Halli catchment source: Zoomin Tech Report to CNNL, 2011.

Type	Hesaraghatta	Kumudvathy	Arkavathy
Check dam	70	65	142
Bridge	4	23	31
Road	0	2	7

are an underestimate of the number of check dams. We therefore assumed the higher density of $1.35\,km^{-2}$ for our analysis.

The volumes of typical obstructions were estimated based on stream profiles made using a dumpy-level instrument on seven check dams. Interpolation of the stream profiles allowed us to estimate the maximum storage volumes as ranging between 100 and $1500\,m^3$ with an average of $325\,m^3$. We multiplied this storage by the basin area and density of obstructing structures to obtain a peak storage volume for the whole basin.

We then plotted the cumulative density function of the daily inflow events into the TG Halli for 15 years from 1976 to 1990 (the period before check dams and unculverted roads were constructed for which we had daily inflow records). We took all flow events less than or equal to the peak storage volume and assumed that the entire flow would be impounded. For events that generated inflows greater than the peak storage, the volume impounded was capped by the peak storage the catchment; anything higher would have overflowed. The volumes impounded were summed to estimate the total loss downstream. This calculation is likely to overestimate the fraction of daily runoff that is impounded behind check dams and unculverted roads, since the structures are unlikely to be empty at the beginning of every rain event.

4 Results

Results are presented separately for each of the hypotheses.

Hypothesis 1: declining rainfall

Annual rainfall trends: Fig. 2b shows the area-averaged monthly and annual rainfall over the basin for the years 1934–2010. With an average of $830\,mm\,yr^{-1}$ and standard deviation of $210\,mm\,yr^{-1}$, the monthly rainfall time series does not show any trend, and no statistically significant trend emerges in the annual rainfall. Similarly, no significant changes are visible in the pre- and post-1970 in mean annual and monthly rainfall totals. The data do exhibit high decadal variability in rainfall, and it is clear that the 1970–1980 period was exceptionally wet. However, there is no evidence that total rainfall volumes have changed in the region.

Seasonal rainfall trends: with the exception of Devanahalli, we did not identify any statistically significant shifts

in the timing of the rainfall over the last 80 years. The observed trend in Devanahalli was for an increase in JJAS rainfall, contrary to the predictions of climate models. Moreover, more rain in JJAS cannot explain the decline in flow production at other times of the year.

Change in rainfall intensity: no statistically significant trends in daily rainfall volumes exceeding threshold values of 10, 25 or 50 mm could be identified at the 95th percentile level at any of the four gauges. Although we cannot exclude the possibility of changes in sub-daily rainfall intensities, analysis of rainfall data in the TG Halli catchment area shows no meaningful historical trends in precipitation volumes, timing or storm characteristics. We find no evidence that rainfall-driven changes could be responsible for the change in flow in the TG Halli catchment.

Hypothesis 2: increasing ET due to increase in temperature

The rise in temperature of about 0.6 to $1\,°C/100$ years was within the range predicted by other studies (Kothawale and Rupa Kumar, 2005; Arora et al., 2005). The estimated PET from the Hargreaves equation averaged to the annual scale is shown in Fig. 2c. As indicated in the figure, there is no statistically significant trend in PET within the basin. We conclude that there is no evidence to support the hypothesis that increasing temperature is increasing potential evaporation and leading to a decline in streamflow.

Hypothesis 3: declining baseflow due to groundwater overexploitation

From the recession analysis, the fitted storage discharge relationship for the pre-1970 period was

$$S = 595Q^{0.57}, \tag{7}$$

where S and Q are given in units of ML per month, consistent with the monthly time step. The slope of the lower envelope was 1.43, very close to the 1.5 slope predicted by the nonlinear Dupuit–Boussinesq theory and found by Brutsaert and Nieber (1977) in their original analysis. We estimated S for the mean of the peak monthly flows from the years prior to 1970 (65 000 ML) using Eq. (7) and normalized this total stored volume by the catchment area. This leads to a prediction that, on average, mobile storage would need to decline by 0.24 m across the catchment to reduce the peak monthly flow rates to zero. We can then use porosity estimates of 20 % for the unconfined sediments and 1 % for the fractured rock to estimate the order of magnitude of the groundwater declines that could effectively remove 0.24 m of mobile water from being in connection with the surface channels. This works out to a decline of approximately 1.25 m in the surficial aquifer, or a decline of approximately 25 m in the fractured rock aquifer (Fig. 3).

Why is the Arkavathy River drying? A multiple-hypothesis approach in a data-scarce region

93

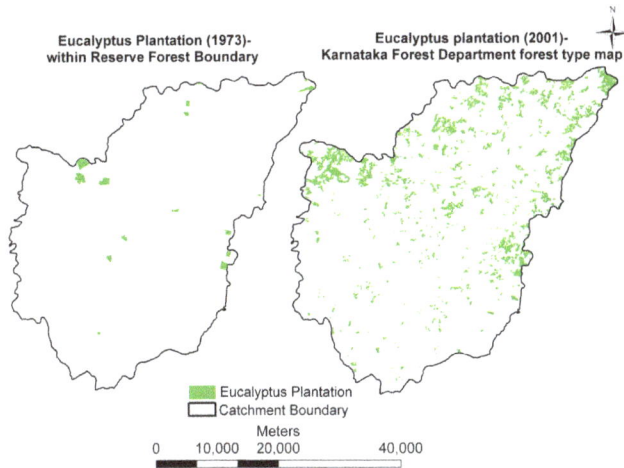

Figure 4. Change in eucalyptus area in Arkavathy Basin between 1973 and 2001.

As can be seen from Fig. 2d, baseflow started declining after the early 1980s, but after 1992 there was not a single month when there was baseflow into the reservoir. The baseflow index, which was computed from daily data and indicates the share of baseflow in the total annual inflow, also declined consistently from 1970 to 2010. The loss of below-ground storage observed in the Arkavathy Basin is of the correct order of magnitude to explain the contemporary absence of surface flow and the hypothesis that loss of groundwater storage in the surface aquifer should be retained for further investigation.

Hypothesis 4: increasing actual evapotranspiration due to expansion of plantations

The area under eucalyptus plantations in 1973, as indicated by Survey of India toposheets, was only $11 \, km^2$, all of it within the boundaries of state reserve forests. By 2001, the area under eucalyptus plantations had increased to $104 \, km^2$ (Fig. 4).

Conversion of $93 \, km^2$ of rain-fed crops to eucalyptus plantations would thus translate into a loss of runoff of $75–135 \, ML \, day^{-1}$ by the year 2001. This figure is significant compared to the observed runoff decline of about $320 \, ML \, day^{-1}$, suggesting that expansion of eucalyptus could be a significant contributor.

Hypothesis 5: million puddle theory

Based on the assumptions about check dam and encroachment water storage, the total loss in runoff at the basin scale attributable to channel encroachment is on the order of $18–54 \, ML \, day^{-1}$. While there are substantial uncertainties associated with this number (for example, other sources of surface water storage such as within-farm impoundments or impoundments in housing plots are ignored leading to an un-

derestimate of the total volume, while the assumption that all storages empty prior to each rainfall event undoubtedly represents an overestimate), the order of magnitudes unequivocally indicate that, while the million puddle theory could have contributed to a fraction of the loss in runoff, it cannot account for the entire loss of inflow into the TG Halli.

5 Discussion

Our analysis indicates that rainfall changes or temperature increases cannot account for any significant fraction of the decline in inflows into TG Halli reservoir. The causes found to be plausible are groundwater extraction, expansion of eucalyptus plantations and to some extent increased obstructions in the stream course. Importantly, many policy approaches currently under consideration do not reflect the major underlying causes of the drying of the Arkavathy River, and in some cases (check dam construction) they are clearly counter-productive. In the future, climate change could play a critical role in exacerbating water stress, but climate stressors will only add to existing local stresses.

Although the hypotheses have been framed as independent, the mechanisms undoubtedly interact with each other, so their inter-relations should be considered in formulating a conceptual model of the catchment and in attributing the effects of each mechanism in terms of the change in river flow. For example, check dams not only impound flow, but also locally elevate recharge. Check dams may thus facilitate high levels of groundwater extraction locally. Spatial heterogeneity in water table levels and eucalyptus root zone access to saturated conditions may vary throughout the catchment, meaning that the assumption that eucalyptus plantations do not contribute to groundwater mining and reduced baseflow should be relaxed in future studies.

The analysis presented here is preliminary. Further work is needed to understand the hydrological processes in the catchment, including the contemporary and historical flow generation pathways and their changes. There are, however, suggestive clues of timing that suggest a potential working hypothesis for the flow generation mechanisms. Expansion of electricity and installation of wells began to increase in the late 1960s – although this period also coincided with a period of relatively high rainfall and streamflow in the 1970s. Flow declines began to emerge in the early 1980s, with baseflow indices and numbers of "baseflow months" plummeting in the early 1990s, approximately at the same time that open wells went dry and deeper boreholes become more prevalent (Fig. 3a, b). During this period, the baseflow index declined, suggesting that less and less of the streamflow entering the TG Halli reservoir was associated with groundwater inputs. These trends are highly reminiscent of those projected by models of the Republican River basin (Zeng and Cai, 2014) as a function of increasing groundwater extraction-reduced baseflow and an increasingly erratic quick-flow response.

Inflows continued to decline after 1992, suggesting that additional mechanisms beyond the decline in baseflow must be considered. Possible additional mechanisms include the conversion of the Arkavathy River into a "losing" river, which provides a source of recharge to the local aquifers, the continued expansion of eucalyptus plantations and increasing implementation of management techniques that prevent surface runoff from leaving farm fields and increasing obstruction of the stream channels. Based on these observations, further research targeting runoff generation mechanisms, establishing the pathways for surface–groundwater connections, evaluating the effect of land use on water balance and estimating groundwater extraction rates has now been initiated in the catchment (see www.atree.org/accuwa).

Finally, from a policy perspective, the fuzzy perception of the causes of streamflow decline and the lack of coordination between agencies have resulted in contradictory policies. The range of policy responses observed reflect both different *stakeholder interests* and *different explanations* of the hydrologic causes of the declining river flow. For instance, even as CNNL is removing encroachments and blockages under the programme to rejuvenate the Arkavathy River, new check dams continue to be authorized under the Mahatma Gandhi Rural Employment Guarantee Scheme (MNREGA). Interestingly, the case study illustrates that the reluctance to acknowledge human feedback is not limited to hydrologists. Even farmers living in the catchment often do not fully acknowledge their role in altering the hydrology. The actors involved have not made any substantive effort to scientifically validate and reconcile these views, resulting in significant wasted investment.

6 Conclusions

The TG Halli catchment case study shows that humans can play a significant role in altering the hydrology of watersheds (Wang and Cai, 2009; Grafton et al., 2013) in a variety of ways. Indeed, the results presented in this paper suggest that proximate drivers like groundwater pumping and land use change, rather than just climate change, are the most likely causes of the drying of the Arkavathy River. The article strengthens the case for use-inspired sociohydrology as a science of water and society that explicitly includes human feedback on hydrologic processes (Sivapalan et al., 2012). In particular, the paper makes three contributions to this nascent field. First, the study highlights the importance of accounting for multiple anthropogenic drivers of change. There has been a tendency within the hydrology community to understate the role of humans in altering hydrology beyond large structures like dams or, more recently, climate change. The dominant conceptualization remains that of the hydrologic system as being separate from society. This case study shows why attention to direct and dispersed human modifications of this system is needed. Second, the study offers guidance on how

human feedback ought to be addressed in a region where data are scarce and unreliable. By adopting a multiple-hypothesis approach, we illustrate how even limited data sources can be marshalled to eliminate some of them and identify critical knowledge gaps. This approach can inform primary data collection efforts and lead to the development of better models of the catchment.

Third, the hypotheses themselves are derived not just from the academic literature but also from perceptions of all stakeholders in the debate. This ensures the legitimacy and usefulness of the research.

Acknowledgements. This research is part of a larger study titled "Adapting to Climate Change in Urbanizing Watersheds (ACCUWa) in India" (www.atree.org/accuwa).

We are grateful to Jayalakshmi from ATREE's EcoInformatic Lab for RS/GIS support and Sowmyashree for help with the 1973 Eucalyptus area calculation. We thank NIT Suratkal student D. N. Shilpa and ATREE Administrative Assistant H. Usha for translation and data entry support on the annual season crop reports. We thank our outreach coordinators K. Janardhan and G. Manjunath for the insights obtained from water literacy meetings. We are grateful to Sekhar Muddu and the other ACCUWa advisory committee members for input on the hydrology study. We are grateful to P. N. Ballukraya for sharing his borehole data and advice on the hydrogeology investigations.

Financial support for most of this research comes from grant no. 107086-001 from the International Development Research Centre (IDRC), Canada. In addition, S. Thompson acknowledges NSF CNIC IIA-1427761 for support of ATREE-UC Berkeley collaborations. G. Penny acknowledges support from the NSF Graduate Research Fellowship Program under grant no. DGE 1106400, the NSF and USAID Research and Innovation Fellowship Program and NSF International Research Experience for Students (OISE-1031194).

Edited by: M. Sivapalan

References

Allen, R. G., Pereira, L. S., Raes, D., and Smith, M.: Crop evapotranspiration – Guidelines for computing crop water requirements, FAO Irrigation and drainage paper 56, FAO, Rome, 300, 6541, 1998.

Arora, M., Goel, N., and Singh, P.: Evaluation of temperature trends over India/Evaluation de tendances de température en Inde, Hydrol. Sci. J., 50, 81–93, 2005.

Arrigoni, A. S., Greenwood, M. C., and Moore, J. N.: Relative impact of anthropogenic modifications versus climate change on the natural flow regimes of rivers in the Northern Rocky Mountains, United States, Water Resour. Res., 46, W12542, doi:10.1029/2010WR009162, 2010.

Arzberger, P., Schroeder, P., Beaulieu, A., Bowker, G., Casey, K., Laaksonen, L., Moorman, D., Uhlir, P., and Wouters, P.: An international framework to promote access to data, Science, American Association for the Advancement of Science, 1777–1778, 2004.

Bonell, M., McDonnell, J., Scatena, F., Seibert, J., Uhlenbrook, S., and van Lanen, H.: HELPing FRIENDs in PUBs: charting a course for synergies within international water research programmes in gauged and ungauged basins, Hydrol. Process., 20, 1867–1874, 2006.

Brown, A. E., Zhang, L., McMahon, T. A., Western, A. W., and Vertessy, R. A.: A review of paired catchment studies for determining changes in water yield resulting from alterations in vegetation, J. Hydrol., 310, 28–61, doi:10.1016/j.jhydrol.2004.12.010, 2005.

Brutsaert, W. and Nieber, J. L.: Regionalized drought flow hydrographs from a mature glaciated plateau, Water Resour. Res., 13, 637–643, 1977.

Buytaert, W.: Interactive comment on "Why is the Arkavathy River drying? A multiple hypothesis approach in a data scarce region", HESS Interactive Comment, 12, 2015.

Buytaert, W., Zulkafli, Z., Grainger, S., Acosta, L., Bastiaensen, J., De Bièvre, B., Bhusal, J., Chanie, T., Clark, J., Dewulf, A., Foggin, M., Hannah, D. M., Hergarten, C., Isaeva, A., Karpouzoglou, T., Pandeya, B., Paudel, D., Sharma, K., Steenhuis, T., Tilahun, S., Zhumanova, M., and Van Hecken, G.: Citizen science in hydrology and water resources: opportunities for knowledge generation, ecosystem service management, and sustainable development, Front. Earth Sci., 2, 26, doi:10.3389/feart.2014.00026, 2014.

Cai, X. and Zeng, R.: Assessing the Stream Flow Effects of Groundwater Pumping and Return Flow from Irrigation, Chap. 40, 416–425, doi:10.1061/9780784412947.040, 2013.

Calder, I. R., Hall, R. L., and Prasanna, K.: Hydrological impact of *Eucalyptus* plantation in India, J. Hydrol., 150, 635–648, doi:10.1016/0022-1694(93)90129-W, 1993.

Chamberlin, T. C.: The method of multiple working hypotheses, Science, 15, 754–759, doi:10.1126/science.148.3671.754, 1965.

Chattopadhyay, N. and Hulme, M.: Evaporation and potential evapotranspiration in India under conditions of recent and future climate change, Agr. Forest Meteorol., 87, 55–73, 1997.

CNNL: Study of Obstructions to flow of natural water in the course of the Arkavathy River and its tributaries from the origin upto the Thippagondanahalli reservoir, Cauvery Neeravari Nigam Limited, Bangalore 560061, 2010.

DES: Annual season crop report for 1970–2012, Government of Karnataka, Bangalore, India, 1970–2012.

Duffie, J. and Beckman, W.: Solar Engineering of Thermal Processes, Hoboken, New Jersey, 3 Edn., 2013.

Dunne, T. and Leopold, L.: Water in environmental planning, W. H. Freeman and Co., San Francisco, CA, 1978.

Farley, K. A., Jobbagy, E. G., and Jackson, R. B.: Effects of afforestation on water yield: a global synthesis with implications for policy, Global Change Biol., 11, 1565–1576, doi:10.1111/j.1365-2486.2005.01011.x, 2005.

Garg, K. K., Wani, S. P., Barron, J., Karlberg, L., and Rockstrom, J.: Up-scaling potential impacts on water flows from agricultural water interventions: opportunities and trade-offs in the Osman Sagar catchment, Musi sub-basin, India, Hydrol. Process., 27, 3905–3921, 2013.

Gosain, A., Rao, S., and Basuray, D.: Climate change impact assessment on hydrology of Indian river basins, Curr. Sci., 90, 346–353, 2006.

Gosain, A., Rao, S., and Arora, A.: Climate change impact assessment of water resources of India, Curr. Sci. (Bangalore), 101, 356–371, 2011.

Grafton, R. Q., Pittock, J., Davis, R., Williams, J., Fu, G., Warburton, M., Udall, B., McKenzie, R., Yu, X., Che, N., Connell, D., Jiang, Q., Kompas, T., Lynch, A., Norris, R., Possingham, H., and Quiggin, J.: Global insights into water resources, climate change and governance, Nat. Clim. Change, 3, 315–321, doi:10.1038/nclimate1746, 2013.

Hargreaves, G. H. and Allen, R. G.: History and evaluation of Hargreaves evapotranspiration equation, J. Irrig. Drain. Eng., 129, 53–63, doi:10.1061/(ASCE)0733-9437(2003)129:1(53), 2003.

Hargreaves, G. H. and Samani, Z. A.: Reference crop evapotranspiration from ambient air temperature, Am. Soc. Agr. Eng., 1, 96–99, 1985.

Hingane, L., Rupa Kumar, K., and Ramana Murty, B. V.: Long-term trends of surface air temperature in india, J. Climatol., 5, 521–528, doi:10.1002/joc.3370050505, 1985.

Hu, Z., Wang, L., Wang, Z., Hong, Y., and Zheng, H.: Quantitative assessment of climate and human impacts on surface water resources in a typical semi-arid watershed in the middle reaches of the Yellow River from 1985 to 2006, Int. J. Climatol., 35, 97–113, 2015.

Huntington, T. G.: Evidence for intensification of the global water cycle: review and synthesis, J. Hydrol., 319, 83–95, 2006.

ISRO and IN-RIMT: Tippagondanahalli Reservoir (TGR): A remote sensing based evaluation, Technical report for Bangalore Metropolitan Region Development Authority, Bengaluru, 2000.

Kelkar, U., Narula, K. K., Sharma, V. P., and Chandna, U.: Vulnerability and adaptation to climate variability and water stress in Uttarakhand State, India, Global Environ. Change, 18, 564–574, 2008.

Kothawale, D. and Rupa Kumar, K.: On the recent changes in surface temperature trends over India, Geophys. Res. Lett., 32, L18714, doi:10.1029/2005GL023528, 2005.

Lele, S., Srinivasan, V., Jamwal, P., Thomas, B., Eswar, M., and Zuhail, T. M.: Water Management in Arkavathy Basin: A Situation Analysis, Environ. Develop., Discussion Paper 1, 2013.

Maeda, E. and Torres, J.: Open Environmental Data in Developing Countries: Who Benefits?, AMBIO, 41, 410–412, doi:10.1007/s13280-012-0283-4, 2012.

Maurer, E.: Extraterrestrial solar radiation calculator, available at: http://www.engr.scu.edu/~emaurer/tools/calc_solar_cgi.pl (last access: 13 April 2015), 2014.

McVicar, T. R., Roderick, M. L., Donohue, Randall J.and Li, L. T., Van Niel, T. G., Thomas, A., Grieser, J., Jhajharia, D., Himri, Y., Mahowald, N. M., Mescherskaya, A. V., Kruger, A. C., Rehman, S., and Dinpashoh, Y.: Global review and synthesis of trends in observed terrestrial near-surface wind speeds: Implications for evaporation, J. Hydrol., 416, 182–205, doi:10.1016/j.jhydrol.2011.10.024, 2012.

Milly, P., Betancourt, J., Falkenmark, M., Hirsch, R., Kundzewicz, Z., Lettenmaier, D., and Stouffer, R.: Stationarity is dead: Whither water management?, Science, 319, 573–574, doi:10.1126/science.1151915, 2008.

Mujumdar, P. and Ghosh, S.: Modeling GCM and scenario uncertainty using a possibilistic approach: Application to the Mahanadi River, India, Water Resour. Res., 44, W06407, doi:10.1029/2007WR006137, 2008.

Nathan, R. and McMahon, T.: Evaluation of Automated Techniques for Baseflow and Recession Analysis, Water Resour. Res., 26, 1465–1473, doi:10.1029/WR026i007p01465, 1990.

Ramanathan, V., Chung, C., Kim, D., Bettge, T., Buja, L., Kiehl, J., Washington, W., Fu, Q., Sikka, D., and Wild, M.: Atmospheric brown clouds: Impacts on South Asian climate and hydrological cycle, P. Natl. Acad. Sci. USA, 102, 5326–5333, 2005.

Reichman, O., Jones, M., and Schildhauer, M.: Challenges and opportunities of open data in ecology, Science (Washington), 331, 703–705, 2011.

Shah, E.: Social designs: Tank irrigation technology and agrarian transformation in Karnataka, South India, Vol. 4, Orient Longman Hyderabad, 2003.

Shiva, V., Sharatchandra, H., and Bandyopadhyay, J.: Social, economic, and ecological impact of social forestry in Kolar, Indian Institute of Management, 1981.

Sivapalan, M.: Prediction in ungauged basins: a grand challenge for theoretical hydrology, Hydrol. Process., 17, 3163–3170, doi:10.1002/hyp.5155, 2003.

Sivapalan, M., Takeuchi, K., Franks, S., Gupta, V., Karambiri, H., Lakshmi, V., Liang, X., McDonnell, J., Mendiondo, E., O'Connell, P., Oki, T., Pomeroy, J., Schertzer, D., Uhlenbrook, S., and Zehe, E.: IAHS Decade on Predictions in Ungauged Basins (PUB), 2003–2012: Shaping an exciting future for the hydrological sciences, Hydrol. Sci. J., 48, 857–880, 2003.

Sivapalan, M., Savenije, H. H., and Blöschl, G.: Socio-hydrology: A new science of people and water, Hydrol. Process., 26, 1270–1276, doi:10.1002/hyp.8426, 2012.

Sivapalan, M., Konar, M., Srinivasan, V., Chhatre, A., Wutich, A., Scott, C., Wescoat, J., and Rodríguez-Iturbe, I.: Socio-hydrology: Use-inspired water sustainability science for the Anthropocene, Earth's Future, 2, 225–230, doi:10.1002/2013EF000164, 2014.

Spencer, J.: Solar position and radiation tables for Brisbane (latitude 27. 5/sup 0/S), Tech. Rep., Commonwealth Scientific and Industrial Research Organization, Melbourne (Australia), 1971.

Srinivasan, V., Lambin, E., Gorelick, S., Thompson, B., and Rozelle, S.: The nature and causes of the global water crisis: Syndromes from a meta-analysis of coupled human-water studies, Water Resour. Res., 48, 10, doi:10.1029/2011WR011087, 2012a.

Srinivasan, V., Seto, K. C., Emerson, R., and Gorelick, S. M.: The impact of urbanization on water vulnerability: A coupled human–environment system approach for Chennai, India, Global Environ. Change, doi:10.1016/j.gloenvcha.2012.10.002, 2012b.

Stocker, T. F. and Raible, C. C.: Climate change: water cycle shifts gear, Nature, 434, 830–833, doi:10.1038/434830a, 2005.

Systematics, C.: Data & Capacity Needs for Transportation NAMAs Report 1: Data Availability, 2010.

Thompson, S. E., Sivapalan, M., Harman, C. J., Srinivasan, V., Hipsey, M. R., Reed, P., Montanari, A., and Blöschl, G.: Developing predictive insight into changing water systems: use-inspired hydrologic science for the Anthropocene, Hydrol. Earth Syst. Sci., 17, 5013–5039, doi:10.5194/hess-17-5013-2013, 2013.

Vaidyanathan, A.: Tanks of South India, Centre for Science and Environment New Delhi, India, 2001.

Vörösmarty, C. J., McIntyre, P., Gessner, M. O., Dudgeon, D., Prusevich, A., Green, P., Glidden, S., Bunn, S. E., Sullivan, C. A., Liermann, C. R., Reidy, C., and Davies, P. M.: Global threats to human water security and river biodiversity, Nature, 467, 555–561, 2010.

Wagener, T., Sivapalan, M., McDonnell, J., Hooper, R., Lakshmi, V., Liang, X., and Kumar, P.: Predictions in ungauged basins as a catalyst for multidisciplinary hydrology, Eos, Trans. Am. Geophys. Union, 85, 451–457, 2004.

Wagener, T., Sivapalan, M., Troch, P. A., McGlynn, B. L., Harman, C. J., Gupta, H. V., Kumar, P., Rao, P. S. C., Basu, N. B., and Wilson, J. S.: The future of hydrology: An evolving science for a changing world, Water Resour. Res., 46, doi:10.1029/2009WR008906, 2010.

Wang, D. and Cai, X.: Detecting human interferences to low flows through base flow recession analysis, Water Resour. Res., 45, W07426, doi:10.1029/2009WR007819, 2009.

Wang, D. and Hejazi, M.: Quantifying the relative contribution of the climate and direct human impacts on mean annual streamflow in the contiguous United States, Water Resour. Res., 47, W00J12, doi:10.1029/2010WR010283, 2011.

Zeng, R. and Cai, X.: Analyzing streamflow changes: irrigation-enhanced interaction between aquifer and streamflow in the Republican River basin, Hydrol. Earth Syst. Sci., 18, 493–502, doi:10.5194/hess-18-493-2014, 2014.

7

On the appropriate definition of soil profile configuration and initial conditions for land surface–hydrology models in cold regions

Gonzalo Sapriza-Azuri[1], Pablo Gamazo[1], Saman Razavi[2,3,4], and Howard S. Wheater[2,3,4]

[1]Departamento del Agua, Centro Universitario Regional Litoral Norte, Universidad de la República, Salto, Uruguay
[2]Global Institute for Water Security, University of Saskatchewan, Saskatoon, SK, Canada
[3]School of Environment and Sustainability, University of Saskatchewan, Saskatoon, SK, Canada
[4]Department of Civil and Geological Engineering, University of Saskatchewan, Saskatoon, SK, Canada

Correspondence: Gonzalo Sapriza-Azuri (gsapriza@gmail.com)

Abstract. Arctic and subarctic regions are amongst the most susceptible regions on Earth to global warming and climate change. Understanding and predicting the impact of climate change in these regions require a proper process representation of the interactions between climate, carbon cycle, and hydrology in Earth system models. This study focuses on land surface models (LSMs) that represent the lower boundary condition of general circulation models (GCMs) and regional climate models (RCMs), which simulate climate change evolution at the global and regional scales, respectively. LSMs typically utilize a standard soil configuration with a depth of no more than 4 m, whereas for cold, permafrost regions, field experiments show that attention to deep soil profiles is needed to understand and close the water and energy balances, which are tightly coupled through the phase change. To address this gap, we design and run a series of model experiments with a one-dimensional LSM, called CLASS (Canadian Land Surface Scheme), as embedded in the MESH (Modélisation Environnementale Communautaire – Surface and Hydrology) modelling system, to (1) characterize the effect of soil profile depth under different climate conditions and in the presence of parameter uncertainty; (2) assess the effect of including or excluding the geothermal flux in the LSM at the bottom of the soil column; and (3) develop a methodology for temperature profile initialization in permafrost regions, where the system has an extended memory, by the use of paleo-records and bootstrapping. Our study area is in Norman Wells, Northwest Territories of Canada, where measurements of soil temperature profiles and historical re-

constructed climate data are available. Our results demonstrate a dominant role for parameter uncertainty, that is often neglected in LSMs. Considering such high sensitivity to parameter values and dependency on the climate condition, we show that a minimum depth of 20 m is essential to adequately represent the temperature dynamics. We further show that our proposed initialization procedure is effective and robust to uncertainty in paleo-climate reconstructions and that more than 300 years of reconstructed climate time series are needed for proper model initialization.

1 Introduction

Arctic and subarctic regions are amongst the most susceptible on Earth to climate change (IPCC, 2013; Hinzman et al., 2005). For example, shrub expansion into the tundra regions (Sturm et al., 2001), permafrost thaw (Connon et al., 2014; Rowland et al., 2010), and glacier retreat (Marshall, 2014) are some of the current manifestations of climate change. All these changes are triggered by the interaction of climate, the carbon cycle, and hydrology in response to global warming (Schuur et al., 2015). These effects are expected to be exacerbated due to global warming trends in the coming years (IPCC, 2013; Slater and Lawrence, 2013; Lawrence and Slater, 2005). Therefore, being able to evaluate and assess the impact of climate change in cold regions is a primary concern for the scientific community, stakeholders, and

First Nations communities in northern regions. The significance of this problem in Canada has led to the creation of the Changing Cold Regions Network (DeBeer et al., 2015; http://www.ccrnetwork.ca, last access: 5 September 2017), which aims to provide improved science and modelling to address these concerns.

Earth system models are essential tools for evaluating the impacts of climate change. At global and regional scales, general circulation models (GCMs) and regional climate models (RCMs) are used to simulate climate change evolution. Land surface models (LSMs) are used with GCMs and RCMs (coupled or offline) to represent the hydrological processes associated with the lower boundary condition of the atmosphere. These models typically represent the coupled energy and water balance in the soil, based on numerical solution of the Richards equation and using a relatively coarse vertical discretization.

In general, a standard soil configuration with a depth of no more than 4 m is used in all LSMs that are commonly implemented in GCMs and RCMs (see for example the comparison made by Slater and Lawrence, 2013, for the soil configuration depth (SCD) in LSMs implemented in some GCMs). The typical boundary conditions to solve the energy and water balance in the soil column are (1) the exchanges with atmosphere at the top, (2) no lateral exchange of water or energy with the surrounding grids (only vertical fluxes), and (3) no heat flux at the bottom of the soil.

For moderate climate conditions and at the spatial scales on which these models are commonly applied, the above depth and boundary conditions are commonly deemed to be sufficient to capture the intra-annual variability in the energy and water balance. However, for cold regions, where the energy balance is closely related to the water balance through the phase change (Woo, 2012), deeper soil configurations and more representative boundary conditions are needed. A deeper soil profile in a model can result in a more accurate process representation as it allows the heat signal to propagate to deeper soil layers and hence avoids erroneous near-surface states and fluxes, such as overheating or over-freezing during summer and winter, respectively (e.g, Lawrence et al., 2008; Stevens et al., 2007). An alternative to modelling a deeper soil profile is the incorporation of a rigorous lower boundary condition that adaptively changes with time and includes a geothermal heat flux (Hayashi et al., 2007). Developing and incorporating a dynamic lower boundary condition is, however, impractical in most cases due to lack of adequate data; in addition, the geothermal heat flux is usually ignored in LSMs, as its effects on temperature dynamics within the upper 20–30 m of soil are considered negligible on century timescales (Nicolsky et al., 2007).

The aforementioned challenges and shortcomings have been recognized by the climate, permafrost, and hydrology community. For climate models, Slater and Lawrence (2013), Alexeev et al. (2007), Nicolsky et al. (2007), and Stevens et al. (2007) have disputed the validity of GCM future projec-

tions due to the shallow soil profile depth in LSMs for the reasons stated above. There have been studies of how the spatial distribution of permafrost is improved by including deeper soil configurations in an LSM. For example, Paquin and Sushama (2015) considered a 65 m deep soil configuration for the arctic region with a spin-up period of 200 years through recycling the 1970–1999 period in the Canadian RCM, which uses Canadian Land Surface Scheme (CLASS) (Verseghy, 1991) as the LSM, and showed an improved spatial distribution of permafrost. Zhang et al. (2003, 2006, 2008) used a thermal soil model that includes soil water balance and showed the importance of considering deep soil configurations. In the context of LSMs, Troy et al. (2012) simulated river basins in northern Eurasia using a 50 m soil configuration with a spin-up of 500 years by recycling the 1901–2001 period 5 times. Decharme et al. (2013), who applied the ISBA model to the whole of France, concluded that an 18 m depth was needed to properly simulate the energy and water balance.

In addition, deeper soil/rock configurations possess extended system memories, and, as such, particular care should be taken to properly define the initial conditions for the subsurface system. The presence of significant non-stationarity in climate and hydrology further complicates the process of model initialization, as it leads to significant changes to the statistical properties and envelope of variability of forcings (Razavi et al., 2015). Due to such non-stationarity, it may be inadvisable to initialize a model by recycling the (typically short) historical records (i.e., repeating the simulation over the same period multiple times and using the final model state of one run as the initial state of the next run), as implemented in Troy et al. (2012) or Paquin and Sushama (2015); such practice, in particular, may result in serious misrepresentation of soil processes, because the significant warming trend in the historical records of cold regions leads to unrealistically warmer soil states after each cycle. Together, these reasons highlight the pressing need for multi-century-long hydroclimatic records to include past non-stationarity that may affect the present state and flux variables. Proxy records such as tree rings can provide a vehicle to reconstruct long hydroclimatic time series, typically at annual to multi-year timescales (Razavi et al., 2016).

The sensitivity of LSMs to initial conditions and the initialization methods has been the focus of several studies (e.g., Yang et al., 1995; Rodell et al., 2005; Shrestha and Houser, 2010). However, most of these works have focused on relatively shallow soil profiles located in areas other than cold regions. An exception is the work of Ednie et al. (2008) that illustrated the need for a suitable model initialization procedure to properly simulate soil thermal profiles in permafrost regions and applied a simplified thermal model of soil by using reconstructed past climate variables.

Despite significant advances, as briefly outlined above, the appropriate soil configuration depth in land surface modelling of cold regions remains an open question. This question is further complicated by the fact that parameter uncertainty is typically ignored in LSMs, and parameter values are usually collected from look-up tables based on land cover and soil maps (Mendoza et al., 2015). Related to this, there have been some previous efforts for "sensitivity analysis" of model outputs to parameters (Razavi and Gupta, 2015) but these have been mainly limited to comparisons of different cover types (e.g., Paquin and Sushama, 2015; Yang et al., 1995) with some few exceptions (e.g., Bastidas et al., 2006).

In this paper, we focus on the three interrelated aspects of LSMs, namely soil depth, parameter uncertainty, and initializations, together to address the above question. Unlike the previous studies that focus on each aspect in isolation, this study looks at their joint and individual effects. We set up a series of systematic modelling experiments with the following three objectives to (1) identify the appropriate SCD for a given LSM and location in the presence of uncertainty in model parameter values and climate conditions, (2) assess the significance of including or excluding geothermal flux as the lower boundary condition in an LSM, and (3) develop an initialization procedure for LSMs in cold regions based on paleo-reconstructions of climate variables and statistical bootstrapping.

2 Methods

To advance our understanding and modelling capability of soil moisture and energy dynamics in permafrost regions, we developed two series of numerical experiments for a study area located in the Northwest Territories, Canada, where observations of soil temperature at several depths and historical reconstructed climate data are available.

2.1 Study area and data

The experimental test case is located at Norman Wells, in the Mackenzie Valley, Northwest Territories, Canada (Fig. 1). Based on the Permafrost Map of Canada (Geological Survey of Canada, 2000), the area is located in a zone of extensive discontinuous permafrost. The land cover is characterized by moss lichen groundcover, ericaceous shrubs, and black spruce and tamarack trees (Smith et al., 2004). The subsurface is formed by ice-rich silt clays. The climate of the region is subarctic, according to the Köppen climate classification (Peel et al., 2007), with an average annual mean daily temperature of $-5\,°C$ and average annual precipitation of $295\,mm\,yr^{-1}$.

This area is selected due to the availability of both soil temperature at several depths down to $20\,m$ (Smith et al., 2004) and dendroclimatic reconstructions of summer air temperature (Szeicz and MacDonald, 1995). These data will be used to test the proposed methodology to define the SCD and the initialization approach.

2.1.1 Soil temperature profiles

Annual soil temperature profiles are available based on the maximum and minimum daily average of soil temperature at several borehole locations in the Mackenzie Valley, administrated by the Geological Survey of Canada (Smith et al., 2004). Figure 2 shows the temperature profiles for the borehole 84-1-T5 selected for our analysis. The soil temperatures were measured at the following depths (in metres) for the period 1985–2001: {1.0, 2.0, 3.0, 4.0, 6.0, 8.0, 10.0, 12.0, 15.0, 18.0, 19.6}. The active layer thickness, defined as the soil depth that encapsulates the seasonal freeze-and-thaw cycle (Woo, 2012), was also reported and varied from $1.5\,m$ at the beginning of the period of record (1985) up to $3.0\,m$ to the end of the period (2000), showing an increasing trend in the active layer thickness over time.

2.1.2 Reconstructed summer air temperature

Szeicz and MacDonald (1995) generated proxy climate records of average summer (June–July) air temperature based on tree rings for the period 1638–1988 in northwestern Canada near to Norman Wells (Fig. 3). These proxy data have been previously used by other authors (Ednie et al., 2008; Esper et al., 2002). For example, Ednie et al. (2008) showed that the linear trend of proxy summer air temperature can be used as an approximation of the linear trend of the mean annual air temperature for the region. Following this approach, we generate a stochastic climate time series (Sect. 2.5.1) that follows the historical reconstructions of mean annual air temperature based on the proxy data of Szeicz and MacDonald (1995).

2.2 Design of experiments

The methodology and experiments were designed to be carried out in two stages. In the first stage, we focus on the characterization of the adequate soil profile depth for land surface–hydrologic modelling in the permafrost regions, in relation to climate condition and model parameterization. For this purpose, we run a 1-D model under a variety of soil profile, parameter, climate configurations, and lower boundary conditions. This stage is referred to as "Experiment 1" in this paper.

In the second stage, "Experiment 2", we propose a method to handle the presence of non-stationarity in climate and hydrology, in order to include effects of past non-stationarity on the present state and flux variables. This method utilizes paleo-climate reconstructions to generate long, synthetic time series of climate variables for model initialization.

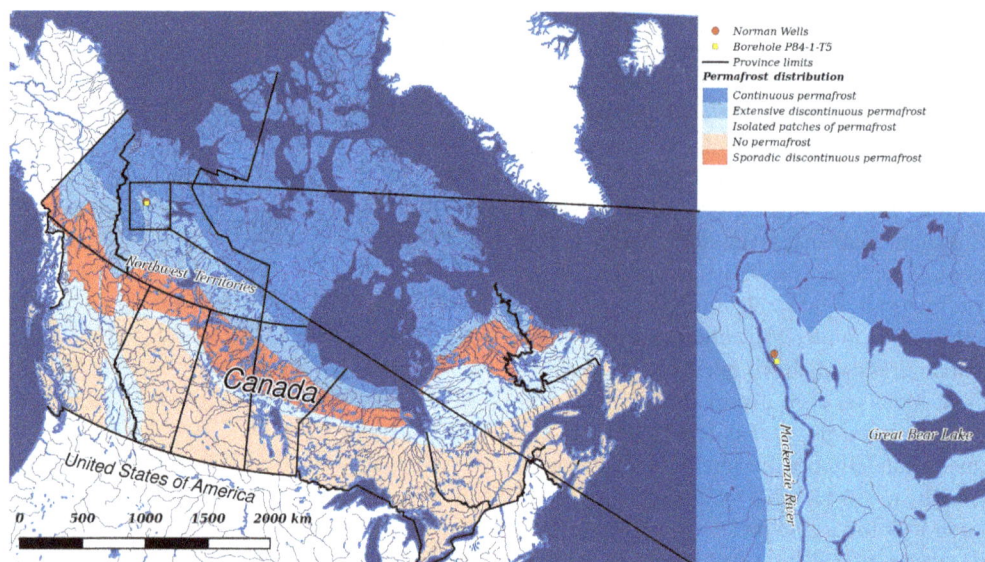

Figure 1. Permafrost Map of Canada and location of the area of study. Temperature soil profiles are available at the borehole P84-1-T5 (yellow dot).

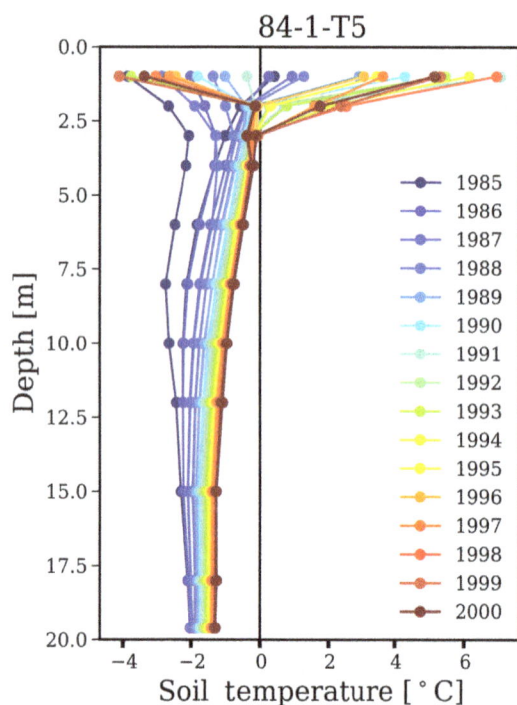

Figure 2. Permafrost Annual maximum and minimum soil temperature profiles for the borehole 84-1-T5 located in Normal Wells. Each colour represent an individual year (1985–2000).

2.3 The 1-D model

The core of the experiments is a 1-D model implemented in MESH, Environment and Climate Change Canada's community model (Pietronero et al., 2007). This integrates the

CLASS LSM (Verseghy et al., 1993; Verseghy, 1991), which solves coupled energy and water balance equations for vegetation, snow and soil and their exchange of heat and moisture with the atmosphere, and WATROF (Soulis et al., 2000) or PDMROF (Mekonnen et al., 2014) to solve the horizontal flow processes for basin-scale integration. MESH discretizes the spatial domain based on regular grid cells and each individual cell is then subdivided in grouped response units (GRUs) based on land cover and/or soil types. MESH has been commonly used to simulate land surface–hydrology processes in many cold regions (e.g., Yassin et al., 2017; Haghnegahdar et al., 2017). The 1-D CLASS model is implemented here at one grid cell, and a unique GRU was used. The upper boundary condition of the model is formed by atmospheric forcings. At the lower boundary condition, in terms of heat, we include two cases: no heat flux and geothermal flux (only in Experiment 1) and, in terms of mass, we assume the water flux that reaches the bottom of the soil profile drains to generate base flow. The climate forcings needed are temperature, precipitation, shortwave radiation, longwave radiation, specific humidity, wind velocity, and atmospheric pressure.

2.4 Experiment 1

A schematic representation of the modelling experiment is illustrated in Fig. 4. Several 1-D model set-ups were implemented by a combination of (1) various SCDs, (2) several climate conditions selected to spin up the model, (3) different values for the parameters that control hydrological processes (water and energy balance), and (4) the inclusion or exclusion of the geothermal flux as the lower boundary condition.

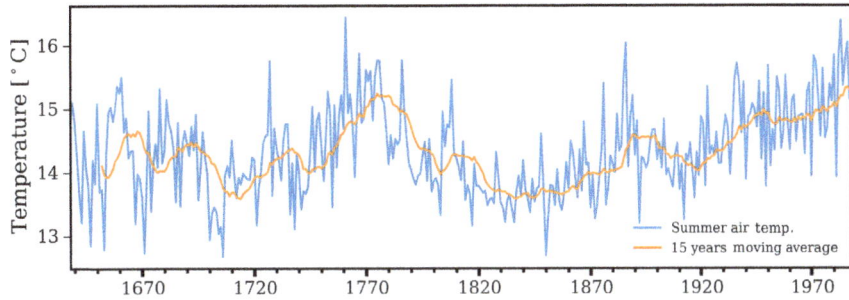

Figure 3. Reconstructed summer (June–July) air temperature based on tree rings for the period 1638–1988 along with its 15-year moving average.

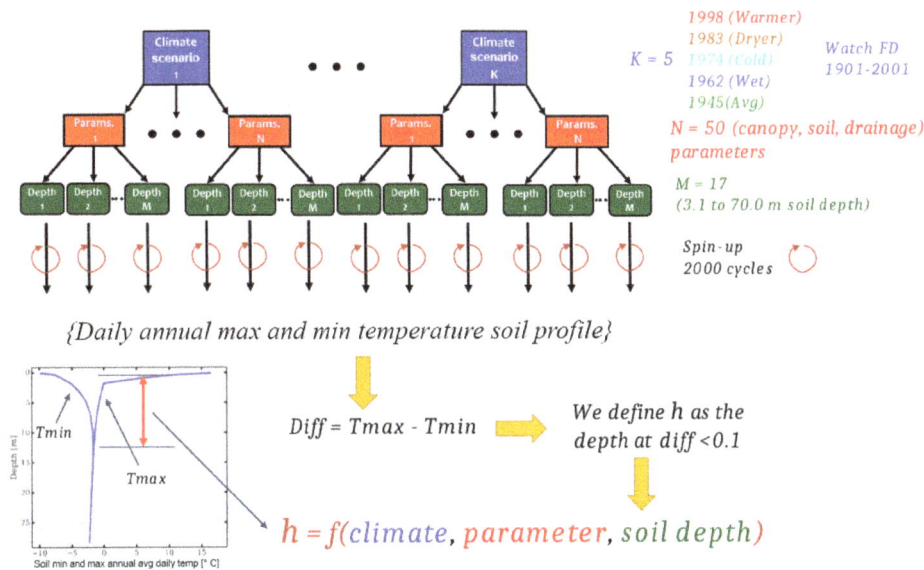

Figure 4. Schematic representation of the model experiment for Experiment 1. The model set-ups are defined as combinations of 5 different climate conditions, 50 randomly selected sets of parameter values within their uncertainty ranges, and 17 different soil configurations. Each model is then run in a spin-up mode for 2000 cycles. The last year of spin-up is taken to compute the daily annual maximum and minimum soil temperature profiles and their difference is computed. The depth at which this difference becomes less than 0.1 is referred to as the "non-oscillation depth" or h_T-non-oscillation.

2.4.1 Variable soil depth configuration

For this experiment, a series of 1-D models with incremental numbers of soil layers (corresponding to different total soil depths) are defined. The soil configurations of the 1-D models are illustrated in Fig. 5, as well as the range from the standard CLASS configuration of three layers with a 4.1 m depth to 20 layers corresponding to a depth of 71.59 m. The thickness of each layer is increased exponentially for deeper soil layers. A total of 17 different soil configurations are tested.

2.4.2 Climate conditions

To account for the effect of climate conditions, years 1998 (warm), 1983 (dry), 1974 (cold), 1962 (wet), and 1945 (average) (Table 1) are used with every model configuration. Each model was run five times (for the 5 years) over 2000-year-

long sequences, each of which comprised 2000 back-to-back repetitions of 1 of the above years. These five climate conditions are defined based on temperature and precipitation obtained from the WATCH FD (WCH-FD) gridded data base of climate forcing (Weedon et al., 2011) for the period 1901–2001 at the location of our study area. We do not use the historical sequence of years 1901–2001 to avoid overheating effects that could be introduced due to the warming trend of the last century.

2.4.3 Parameter uncertainty

Three groups of parameters representing canopy, soil, and drainage processes are perturbed within their ranges of uncertainty to analyze their influence on SCD. Table 2 describes all the parameters considered along with their lower and upper bounds of variation. Monte Carlo sampling with a uni-

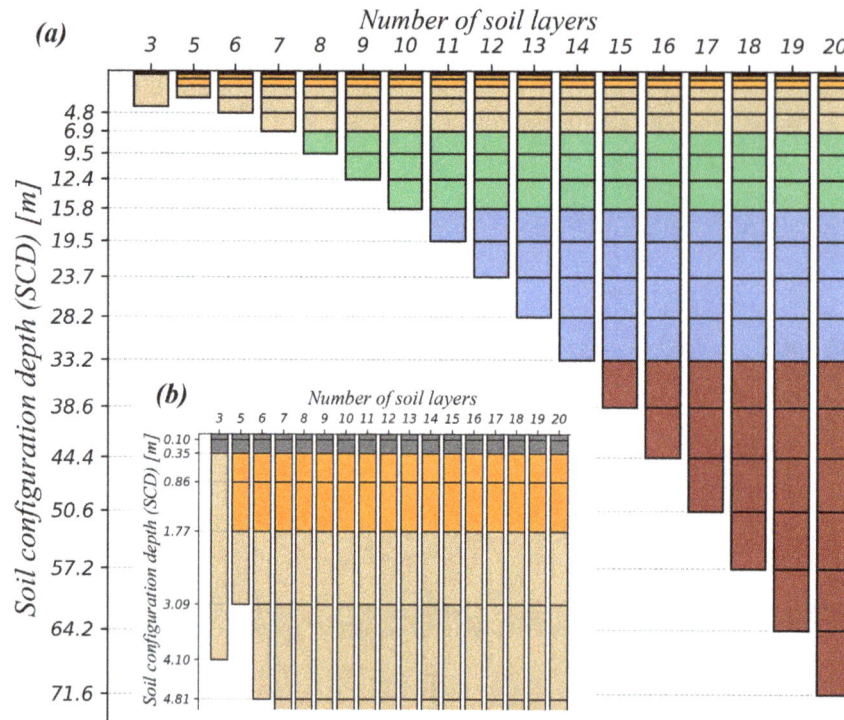

Figure 5. The variable soil configuration profiles defined for the 1-D model: number of soil layers, depth of each layer, and total depth. Each colour represents a group of layers that are assigned the same parameter values. Panel (a) shows all the configurations and panel (b) shows a zoom-in window of the previous panel for the first few layers. The first soil configuration (three layers) represents the standard CLASS soil model configuration.

Table 1. Climate conditions of the 5 representative years used in this study.

Year	Precipitation ($mm\,yr^{-1}$)	Temperature (°C)	Climate condition
1945	396	−6.5	Average
1962	667	−5.6	Wet
1974	534	−8.3	Cold
1983	252	−7.1	Dry
1998	363	−3.6	Warm

form distribution is applied to generate a collection of 50 samples for each parameter. The range of the canopy parameter values used represents different vegetation covers that are present in the area based on the look-up table from the CLASS user manual (Versegey, 2009). To set a consistent parametrization scheme for the soil texture across the models with different numbers of layers, we grouped layers and assigned the same values to the parameters of the layers in each group. These groups are represented with different colours in Fig. 5.

2.4.4 Lower boundary conditions: the geothermal flux

To assess the effect of the lower boundary condition on the energy balance and soil temperature profile, an analysis was made to compare two scenarios: (1) no heat flow at the bottom of the lowest soil layer and (2) a constant geothermal flow (called ggeo flux in CLASS). The comparative analysis was carried out for the average climatic condition (year 1945). All the 17 different soil configurations and 50 sets of parameter values were tested, resulting in a total of 850 model configurations to be run for scenario 2 above. For this scenario, the geothermal heat flow was set to be $0.083\,W\,m^{-2}$, based on measurements made in a borehole in Norman Wells (Garland and Lennox, 1962).

2.4.5 Non-oscillation depth

In Experiment 1, we ran a total of {(17 SCD) × (5 climates) × (50 parameters)+850 (with geothermal flux)} = 5100 model combinations. In each of these model set-ups, a 2000-year model run was performed. All the models were set with the same initial conditions and constant temperature and liquid/ice saturation soil profiles. The soil thermal profile was defined at −3.0 °C and all the soil water was defined as ice

Table 2. List, description, and ranges of model parameters perturbed in this study. The values of soil texture parameters SAND, CLAY, and ORG (denoted by *) sampled such that they sum to 100 %.

Id	Name	Units	Lower bound	Upper bound	Description
1	LAMX	(–)	2.0	4.0	Annual maximum leaf-area index
2	LAMN	(–)	2.0	4.0	Annual minimum leaf-area index
3	ALVC	(–)	0.03	0.06	Average visible albedo of the vegetation when fully leafed
4	ALIC	(–)	0.2	0.34	Average near-infrared albedo of the vegetation when fully leafed
5	ROOT	(m)	0.2	1.55	Root depth
6	SDEP	(m)	2.0	Maximum depth	Permeable depth
7	GRKF	(–)	0.001	1.0	Fraction of the saturated surface soil conductivity moving in the horizonal direction
8	KSAT	$(m\,s^{-1})$	0.0001	5.5	Saturated surface soil hydraulic conductivity
9	SAND*	(%)	0.0	100	% sand texture
10	CLAY*	(%)	0.0	100	% clay texture
11	ORG*	(%)	0.0	100	% material organic texture
12	ZSNL	(m)	0.05	0.5	Minimum depth to consider 100 % cover of snow on the ground surface
13	ZPLS	(m)	0.05	0.5	Maximum depth of liquid water allowed to be stored on the ground surface for snow-covered areas
14	ZPLG	(m)	0.05	0.5	Maximum depth of liquid water allowed to be stored on the ground surface for snow-free areas

content. We assume that after the spin-up a quasi-equilibrium between the climate conditions and the ground thermal state was reached. The last cycle, a complete 1-year simulation, was used to compute the annual soil temperature profiles based on the maximum (maxTsp) and minimum (minTsp) daily average of soil temperature (Fig. 4). Next, we computed the difference between maxTsp and minTsp and defined a depth (h) at which this difference was less than 0.1 °C. We named this depth h as the "non-oscillation depth" of annual soil temperature. Therefore, h, which is a function of climate condition, parameter values, and simulated soil depth, represents the depth at which the soil thermal response remains invariant over seasons. In other words, the non-oscillation depth indicates the depth at which the SCD no longer has a significant effect on the energy balance computed by the model.

2.5 Experiment 2

To be able to simulate the hydrology using LSMs in cold regions in the last century (period of record) and in the future, it is necessary to correctly set the initial conditions of the models. When the SCD of the model is considered to be shallow (no more than 4 m), the initialization can be easily carried out with a relatively short spin-up period (Yang et al., 1995). However, with deeper SCDs, the memory of the system is longer, and it remembers the past climate regimes and trends. Therefore, it is necessary to run the model over an extended period of time to diminish the effect of uncertainty in initial conditions on model predictions. This is a major challenge, however, as the typical length of periods of records (say ~ 100 years) is not sufficient.

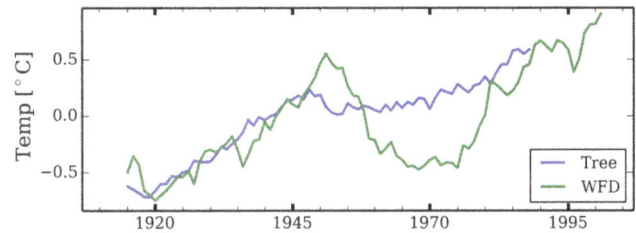

Figure 6. Trend comparison of the annual average air temperature data (15-year moving average) based on WCH-FD and tree-ring-based reconstructions.

2.5.1 Methodology of reconstruction

To overcome the above challenge, we stochastically generated past climate variables, back to year 1678 based on proxy data of reconstructed summer air temperature described in Sect. 2.1.2. To this end, we applied a block-bootstrapping technique (Razavi et al., 2015; Politis and Romano 1994).

The stochastic time series of climate variables were generated as follows:

1. First, we assumed that the reconstructed summer air temperature by Szeicz and MacDonald (1995) can be used as proxy data to derive the past trends in air temperature. The historical temperature trend back to 1678 (T_{Htrend}) was estimated by first computing the moving average with a window of 15 years and then subtracting the moving average from the annual time series. Figure 6 compares both temperature trends (15-year moving average) obtained from WCH-FD data and tree rings for the same period, showing a reasonable agreement, with a Pearson correlation coefficient of 0.66. The exist-

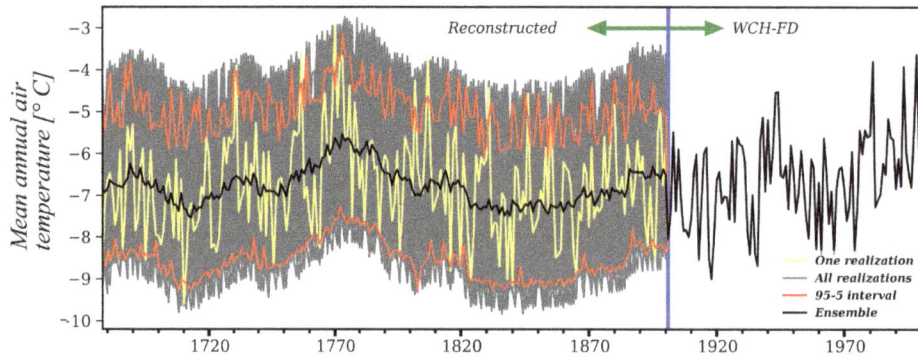

Figure 7. Combined air temperature time series generated using the block-bootstrapping technique and WCH-FD. The time series is divided in two periods. From 1678–1900 the temperature and the other six climate variables were generated using the block-bootstrapping technique with a block of 5 years assembled on tree-ring-based reconstructions. The 100 realizations (grey lines), the 5–95 % confidence interval (red lines), and the average of the ensemble (black line) are shown. In the second period (1901–2000) the climate variables are used directly from the WCH-FD database.

ing discrepancy may be in part due to a lack of consideration of longer-term variability (longer than annual) in the reconstruction of the time series, an issue explained in Razavi et al. (2016).

2. Then, we decomposed the WCH-FD temperature time series (6-hourly time resolution) for the period 1901–2001 into its trend (based on the 15-year moving average) and its seasonality component (T_{seas}).

3. Next, we applied the block-bootstrapping technique with a block size of 5 years to T_{seas}. We sampled 45 blocks of 5 years so as to generate a time series long enough to cover the 1678–1901 period.

4. To finish the reconstruction of the 6-hourly temperature data, we added T_{seas} to the T_{Htrend} from step (1).

5. The other six climate variables needed by MESH to run were precipitation, shortwave and longwave radiation, specific humidity, wind, and atmospheric pressure. They were generated by applying the block-bootstrapping technique with the same time indexes of the temperature blocks (step 3). In this way, we maintained the interdependence between all the climate variables.

6. Finally, we generated 100 realizations of the climate variables for the period 1678–1901. The complete climate time series of 1678–2000 was finally obtained by combining the generated ones and the WCH-FD data for 1901–2000. Figure 7 shows the mean annual temperature of these 6-hourly time series generated with the methodology presented.

2.5.2 Evaluation procedure

We used the 100 realizations of the climate variables of Sect. 2.5 to run the models with the 50 parameter sets and 17

SCDs used before. For the initial conditions, we used the stabilized model outputs obtained from the 2000 cycles for year 1945 (average with respect to temperature and precipitation). Finally, the simulated soil temperature profiles obtained were compared with the observed data (see Sect. 2.1.1) by computing the root mean squared error (RMSE). To evaluate the model performance in reproducing the observations (1985–2000), individual maximum and minimum soil temperature profiles of simulated and observed data were used to compute RMSE for each individual year. Then all the values of RMSE obtained, one for each year, were averaged to obtain the overall RMSE of the corresponding simulation.

3 Results

3.1 Soil configuration depth

Using the experiments proposed in Experiment 1, we explored the combined and individual effects of climate, parameters, and SCD on the non-oscillation depth of the soil temperature profile. Figures 8, 9, and 10 summarize these analyses as 2-D histograms: SCD, h_T-non-oscillation (Fig. 8); years, h_T-non-oscillation (Fig. 9); and parameter sample group, h_T-non-oscillation (Fig. 10). Notably, Fig. 8 shows that for SCDs less than 15 m, there is a high probability that the h_T-non-oscillation condition is never reached, regardless of the parameter values and the climate conditions (year). For SCDs greater than 20 m, the h_T-non-oscillation condition is always reached, with this condition occurring at a higher frequency at a depth between 13 and 16 m.

The variability observed in h_T-non-oscillation depth for each SCD is, in general, mainly explained by the variation in parameter values rather than the year selected (i.e., climate condition) for spinning up the model (Figs. 9 and 10).

From the previous results, it seems clear that we need at least an SCD of greater than 20 m to adequately repre-

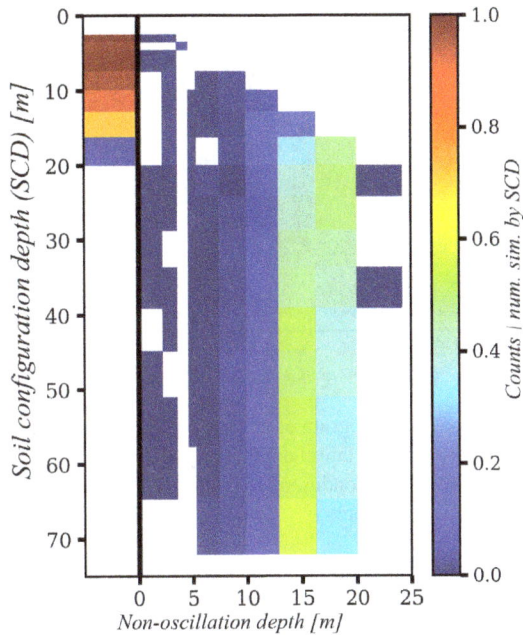

Figure 8. The 2-D histogram of SCD and h_T-non-oscillation depth. Counts are normalized by the number of simulations per SCD. The thick black line separates the frequency of reaching and not reaching the h_T-non-oscillation conditions; bins to the left of this line are for simulations that never reached the h_T-non-oscillation condition.

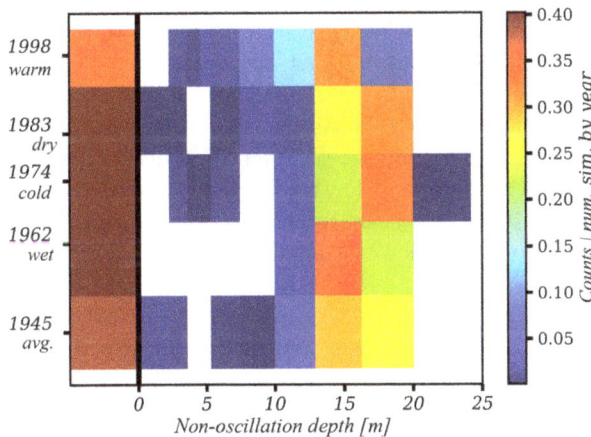

Figure 9. The 2-D histogram of climate condition (years) and h_T-non-oscillation depth. Counts are normalized by the number of simulations per year. The thick black line separates the frequency of reaching and not reaching the h_T-non-oscillation conditions; bins to the left of this line are for simulations that never reached the h_T-non-oscillation condition.

sent the temperature dynamics of permafrost. This conclusion is supported by the fact that the soil temperature at which h_T-non-oscillation condition is reached remains invariant throughout the annual cycle. The distribution of this "non-oscillating temperature" is shown using 2-D histograms

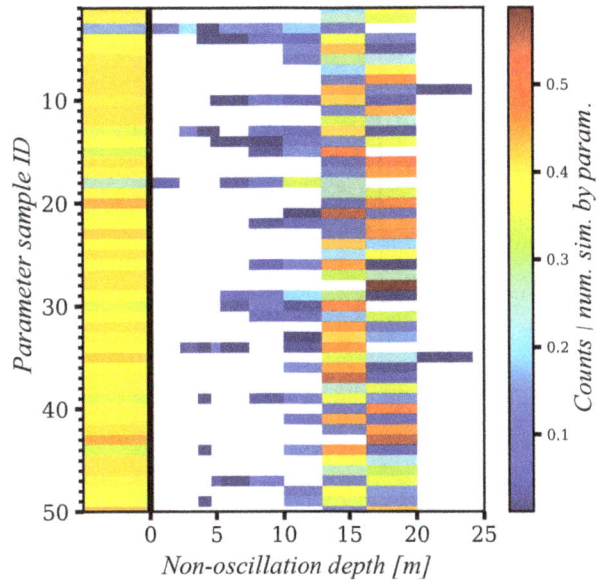

Figure 10. The 2-D histogram of parameter and h_T-non-oscillation depth. Counts are normalized by the number of simulations by parameter sample. The thick black line separates the frequency of reaching and not reaching the h_T-non-oscillation conditions; bins to the left of this line are for simulations that never reached the h_T-non-oscillation condition.

in Figs. 11 and 12 with respect to the SCD and the climate conditions (years), respectively.

Figure 11 shows that for shallow SCDs, from 3.1 to 16 m, there is a tendency to obtain a warmer soil temperature such that the permafrost is thawed. In the SCDs with the depth of 16 m and deeper, there is much more variability in the soil temperature (between −6 and 0 °C), but with a high probability that the soil temperature at the h_T-non-oscillation condition is between −3 and −2.5 °C. In Fig. 12 the effect of the climate condition can be appreciated. The main behavioural difference is for the warmest year (1998) when, as expected, the warmest soil temperatures at the h_T-non-oscillation condition occur. As for the other climate conditions, the behaviours are quite similar and in general have a range of variation between −7 and 0.5 °C. As before (Fig. 11), the probability distribution for each climate condition is quite symmetrical with a peak value around −2.5 °C. A slightly cooler soil temperature is obtained for the coldest year (1974).

3.2 Lower boundary conditions

Figure 13 shows the 2-D histograms (SCD, h_T-non-oscillation) for simulations where the geothermal flux is not included (Fig. 13a) and with the geothermal flux (Fig. 13b) in the lower boundary condition. On both experiments the same number of models are run ({(17 SCD) × (1 climate year) × (50 parameters)} = 850). The visual comparison indicates that the histogram differences are negligible in most cases. Some marginal differences suggest, as expected, that the

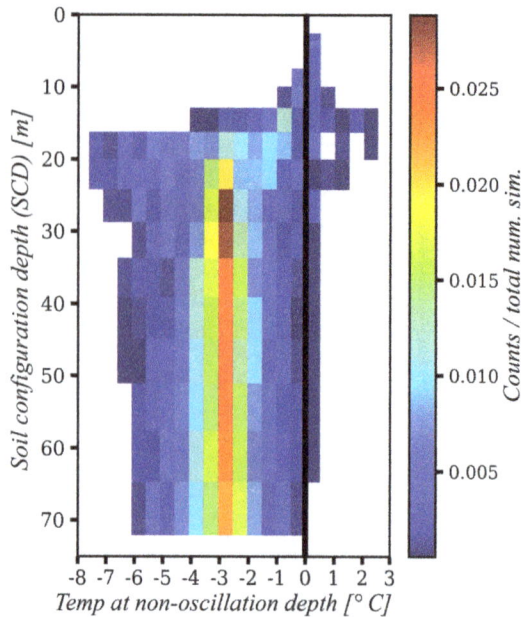

Figure 11. The 2-D histogram of SCD and temperature at h_T-non-oscillation depth. Only SCDs that have reached the h_T-non-oscillation condition are included. The black line represents the 0 °C temperature.

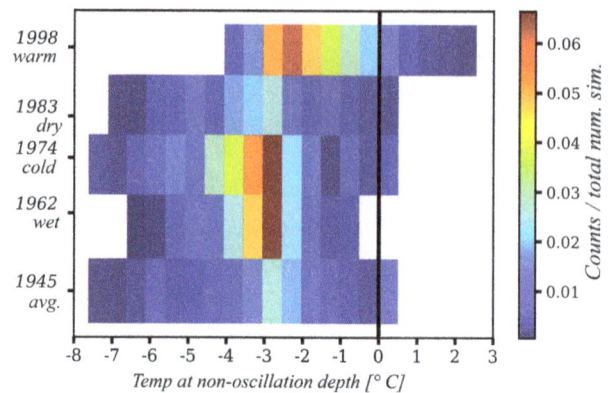

Figure 12. The 2-D histogram of climate condition (years) and temperature at h_T-non-oscillation depth. Only SCDs that have reached the h_T-non-oscillation condition are included. The black line represents the 0 °C temperature.

models with a constant geothermal flux result in slightly warmer soil profiles and slightly deeper non-oscillation depth compared with no-heat-flow counterparts. These differences are small, and the results confirm that more than 20 m of soil depth are needed to adequately represent the temperature dynamics. To further compare the two scenarios, Fig. 14 shows the cumulative distribution function (CDF) of the differences in soil temperature at the non-oscillation depth of the two simulation scenarios (with and without geothermal flux at the bottom). As shown, the temperature difference of the two scenarios is small in most simulations and is within ±0.15 °C in approximately 60 % of simulations.

3.3 Initialization by paleo-reconstructions

The previous sections have shown evidence that, regardless of the climate conditions, parameter uncertainty, and lower boundary conditions, we need to have an SCD that is deeper than 20 m. However, such depths make the model initialization problem challenging. Here, we show the results of our 1-D models with different SCDs and parameter values when driven by a set of 100 tree-ring, bootstrap-based reconstructed climate forcing realizations for the period 1678–2001.

Figure 15a shows a general overview of the model's ability to reproduce the observed soil thermal behaviour between years 1985 and 2000, by plotting the 2-D histogram of SCD and RMSE. The colours for a specific SCD represent the frequency distribution of RMSE values; the variability in this

distribution includes the effects of different parameter values and climate forcing realizations. The RMSE was calculated as described in Sect. 2.5.2. In general, for the shallower SCDs (say less than 15 m), the RMSE tends to be larger with a higher variability (1.5 to 9.0 °C). The frequency distributions for deeper SCDs become, however, quite similar regardless of the depth, with an RMSE range between 1 and 5 °C with a high density around 1.5 to 3.0 °C.

In the histogram of Fig. 15a, we included all the simulations, even if in a simulation from Experiment 1 the non-oscillation conditions had not been reached. Figure 15b, however, presents only the simulations that have reached the non-oscillation condition. As can be seen, the histograms of Fig. 15a and b become the same for SCDs deeper than 16 m similar. This is explained by the fact that almost all the simulations with SCDs that are sufficiently deep (> 16.0 m) reach the h_T-non-oscillation condition.

Figure 16 shows a series of histograms of RMSE values generated by the different reconstructed climate series, each of which is for a different set of parameter values. This figure is designed to assess the relative effects of the variation in the different reconstructed climate time series (a manifestation of data uncertainty) and variation in model parameters (a manifestation of parameter uncertainty) on the variability of RMSE. Here, we only take into account the simulations that have reached the h_T-non-oscillation condition. As can be seen, the range of variation in RMSE for each set of parameter values is quite narrow compared to the union of all the ranges across the different sets of parameter values. Therefore, two points can be made here: (1) the variability observed in RMSE can be attributed mainly to the parameter variations, indicating the significant role of parameter uncertainty; and (2) the effect of stochasticity in the reconstructed time series for the period preceding the period of record is minimal on the model performance in the evaluation period.

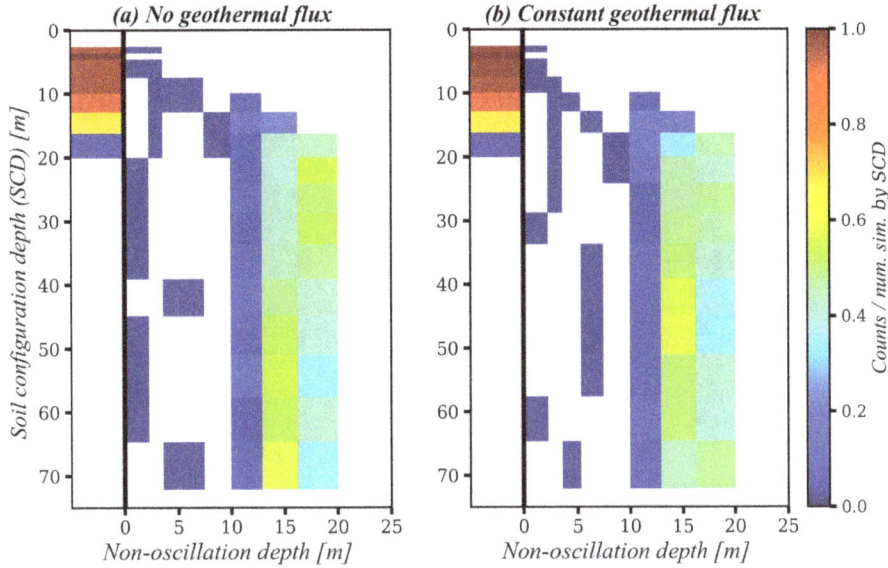

Figure 13. The 2-D histogram of SCD and h_T-non-oscillation depth. Counts are normalized by the number of simulation by SCD. The black line represents the limit at which the conditions reach or do not the h_T-non-oscillation conditions. Bins to the left represent SCDs that never reach the h_T-non-oscillation condition. **(a)** no geothermal flux; **(b)** constant geothermal flux as lower boundary condition at the bottom of the soil layers.

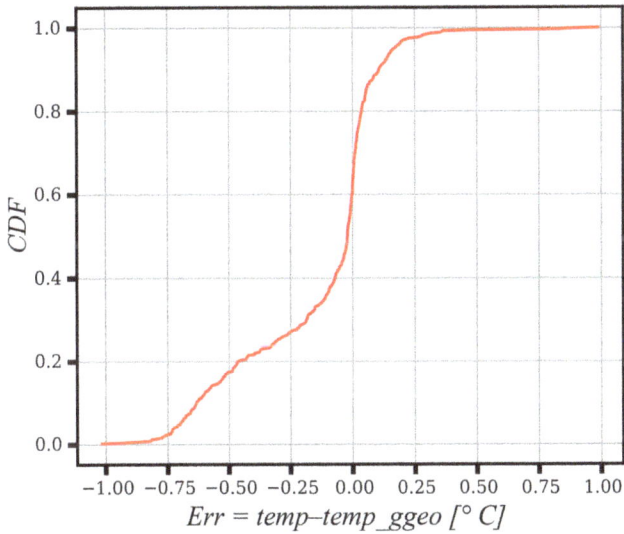

Figure 14. Cumulative distribution function (CDF) of the soil temperature difference at the h_T-non-oscillation depth between simulations with and without geothermal flux.

4   Discussion and conclusions

This study concludes that for permafrost regions, deeper soil configurations in LSMs are needed than commonly adopted to be able to correctly simulate the coupled energy and water balance in the subsurface. This conclusion can be extended to all Earth system models that incorporate an LSM with permafrost representation. While this conclusion has also been pointed out by other authors, this work investigated the in-

dividual and joint effects of parameter uncertainty, total soil depth, lower boundary conditions (geothermal flux), and climate conditions. Further, this work addresses the uncertainty in the reconstructions of past climate for model initialization and also the question of how the initialization should be carried out.

Our analysis shows that the minimum total soil depth should be around 20 m. This value is the reliable depth considering uncertainty and variability in model parameters and climate conditions used to initialize the model and whether or not the geothermal flux is included as lower boundary condition. The metric defined to assess this depth was based on a depth at which the annual maximum and minimum of daily soil temperature are equal, referred to as h_T-non-oscillation condition in this paper. This depth represents a thermally stable condition and ensures that the lower boundary condition is deep enough to accommodate a no-heat-flux or constant-heat-flux boundary condition at the bottom of the soil configuration.

The variability observed in the value of h_T-non-oscillation across the many simulations we conducted was mainly explained by parameter perturbations rather than climate conditions. This assessment was the case for the both sets of analyses in Experiment 1 and 2. This emphasizes the importance of recognizing and addressing parameter uncertainty and raises serious issues with the common practice in using LSMs with GCMs, where model capabilities are constrained by using hard coded parameters determined based on look-up tables (Mendoza et al., 2015).

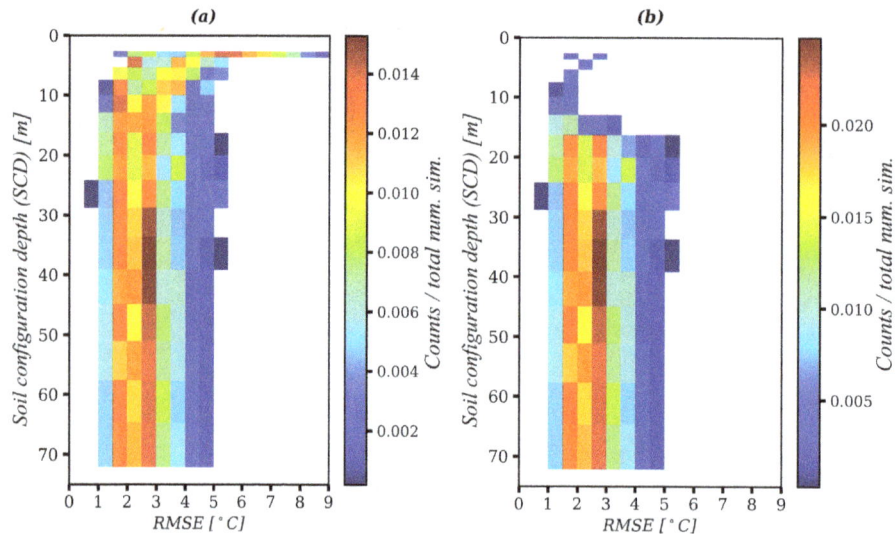

Figure 15. The 2-D histograms of SCD and RMSE for the period 1985–2000 when initialized by the bootstrap-based paleo-reconstructions. In plot **(a)** all the simulations are included, but in plot **(b)** only the simulations that have reach the h_T-non-oscillation condition are included.

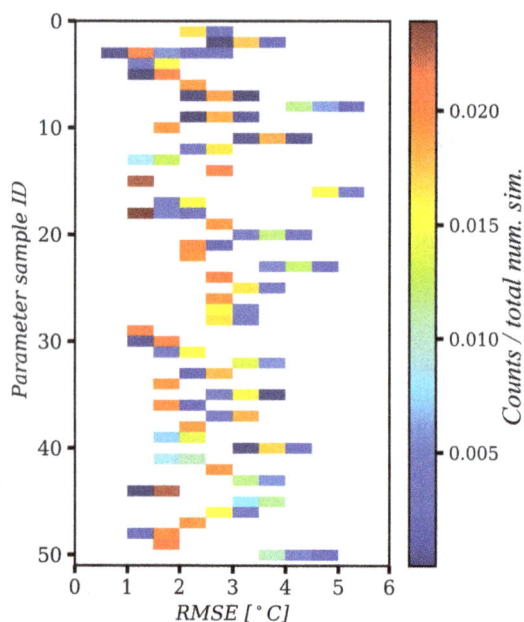

Figure 16. The 2-D histogram of parameter sample and RMSE. Only the simulations that reached the h_T-non-oscillation condition are included.

We argued that model spin-ups that are based on recycling the 20th century data should be avoided, as simulations on back-to-back repetitions of any sequence of years with a warming trend will result in an unrealistically warm soil temperature profile. Instead, we recommend a two-stage procedure to set the initial conditions of the model: in stage 1 (as conducted in Experiment 1), we spin up the model on an "average" year and then, in stage 2, we further run the model on a multi-century-long bootstrap-based paleo-reconstruction to

the beginning of the period of record. The first phase has a stabilizing effect and assures that coherent state variables and fluxes are set before subsequent initialization of the model. This is an important step, as the majority of the LSMs have a large number of state variables to initialize (e.g., CLASS has 17). For the first step, we recommend selecting an average year in terms of air temperature and precipitation, and we recycle that year in simulation until the soil temperature profile is stabilized. Then, in the second step, we recommend using multi-century-long time series of climate variables generated based on the procedure proposed in this study. The proposed procedure reconstructs the time series of temperature using proxy records of summer temperatures derived from tree rings and generates the concurrent time series of other climate variables such as precipitation by applying block bootstrapping on historical records. We were able to reproduce quite well the past trends of summer temperature and we included the effect of uncertainty in the climate time series by generating 100 realizations. An important remark here is that the effect of short-timescale (e.g., annual) fluctuations in the reconstructed time series used for initialization was minimal, while low frequency trends were important. The length of reconstructions required for proper initialization is longer for deeper SCDs.

Finally, we envision our future work being directed to generalize the results obtained here by extending the analyses to other locations where observations of soil profile temperature and past climate are available. Furthermore, implementing a variable SCD in regional and global models may be investigated, as also proposed by Brunke et al. (2016) for the Community Land Model version 4.5. However, the computational burden is a bottleneck for large-scale simulations. To address the computational issues, surrogate modelling strategies that develop cheaper-to-run statistical or mechanistic surrogates of the original models may be explored (Razavi et al., 2012), and also an endeavour may be made by the cryosphere community to generate a unified gridded data set for the last millennium or so (Jungclaus et al., 2017; Landrum et al., 2013; Schmidt et al., 2011) that approximates soil temperature profiles with adequate soil depth, considering the effect of parameter uncertainty via generating ensembles of approximations.

Competing interests. The authors declare that they have no conflict of interest.

Special issue statement. This article is part of the special issue "Understanding and predicting Earth system and hydrological change in cold regions". It is not associated with a conference.

Acknowledgements. This research was undertaken as part of the Changing Cold Region Network, funded by Canada's Natural Science and Engineering Research Council and by the Canada Excellence Research Chair in Water Security at the University of Saskatchewan.

Edited by: Sean Carey

References

Alexeev, V. A., Nicolsky, D. J., Romanovsky, V. E., and Lawrence, D. M.: An evaluation of deep soil configurations in the CLM3 for improved representation of permafrost, Geophys. Res. Lett., 34, L09502, https://doi.org/10.1029/2007GL029536, 2007.

Bastidas, L. A., Hogue, T. S., Sorooshian, S., Gupta, H. V., and Shuttleworth, W. J.: Parameter sensitivity analysis for different complexity land surface models using multicriteria methods, J. Geophys. Res., 111, D20101, https://doi.org/10.1029/2005JD006377, 2006.

Brunke, M., Broxton, P., Pelletier, J., Gochis, D., Hazenberg, P., Lawrence, D., Leung, L., Niu, G., Troch, P., and Zeng, X.: Implementing and Evaluating Variable Soil Thickness in the Community Land Model, Version 4.5 (CLM4.5). J. Climate, 29, 3441–3461, https://doi.org/10.1175/JCLI-D-15-0307.1, 2016.

Connon, R. F., Quinton, W. L., Craig, J. R., and Hayashi, M.: Changing hydrologic connectivity due to permafrost thaw in the lower Liard River valley, NWT, Canada, Hydrol. Process., 28, 4163–4178, https://doi.org/10.1002/hyp.10206, 2014.

DeBeer, C. M., Wheater, H. S., Quinton, W., Carey, S. K., Stewart, R., MacKay, M., and Marsh, P.: The changing cold regions network: Observation, diagnosis, and prediction of environmental change in the Saskatchewan and Mackenzie River basins, Canada, Sci. China-Earth Sci., 58, 46–60, 2015.

Decharme, B., Martin, E., and Faroux, S.: Reconciling soil thermal and hydrological lower boundary conditions in land surface models, J. Geophys. Res.-Atmos., 118, 7819–7834, https://doi.org/10.1002/jgrd.50631, 2013.

Ednie, M., Wright, J. F., and Duchesne, C.: Establishing initial conditions for transient ground thermal modelling in the Mackenzie Valley: A paleo-climatic reconstruction approach, in: Proceedings, 9th International Conference on Permafrost, Vol. 29, 403–408, 2008.

Esper, J., Cook, E. R., and Schweingruber, F. H.: Low-frequency signals in long tree-ring chronologies for reconstructing past temperature variability, Science, 295, 2250–2253, https://doi.org/10.1126/science.1066208, 2002.

Garland, G. D. and Lennox, D. H.: Heat Flow in Western Canada, Geophys. J. Roy. Astr. S., 6, 245–262, 1962.

Geological Survey of Canada: Canadian Permafrost. Government of Canada, Natural Resources Canada, Earth Sciences Sector; Canada Centre for Mapping and Earth Observation, available at: http://geogratis.gc.ca/api/en/nrcan-rncan/ess-sst/092b663d-198b-5c8d-9665-fa3f5970a14f.html (last access: 5 September 2017), 2000.

Haghnegahdar A., Razavi S., Yassin F., and Wheater H.: Multi-criteria sensitivity analysis as a diagnostic tool for understanding model behavior and characterizing model uncertainty, Hydrol. Process., 31, 4462–4476, https://doi.org/10.1002/hyp.11358, 2017.

Hayashi, M., Goeller, N. T., Quinton, W. L., and Wright, N.: A simple heat-conduction method for simulating the frost-table depth in hydrological models, Hydrol. Process., 21, 2610–2622, 2007.

Hinzman, L. D., Bettez, N. D., Bolton, W. R., Chapin, F. S., Dyurgerov, M. B., Fastie, C. L., Griffith, B., Hollister, R. D., Hope, A., Huntington, H. P., Jensen, A. M., Jia, G. J., Jorgenson, T., Kane, D. L., Klein, D. R., Kofinas, G., Lynch, A. H., Lloyd, A. H., McGuire, A. D., Nelson, F. E., Oechel, W. C., Osterkamp, T. E., Racine, C. H., Romanovsky, V. E., Stone, R. S., Stow, D. A., Sturm, M., Tweedie, C. E., Vourlitis, G. L., Walker, M. D., Walker, D. A., Webber, P. J., Welker, J. M., Winker, K. S., and Yoshikawa, K.: Evidence and Implications of Recent Climate Change in Northern Alaska and Other Arctic Regions, Climatic Change, 72, 251–298, https://doi.org/10.1007/s10584-005-5352-2, 2005.

IPCC: Climate Change 2013: The Physical Science Basis, Contribution of Working Group I to the Fifth Assessment Report of the Intergovernmental Panel on Climate Change, edited by: Stocker, T. F., Qin, D., Plattner, G.-K., Tignor, M. M. B., Allen, S. K., Boschung, J., Nauels, A., Xia, Y., Bex, V., and Midgley, P. M., Cambridge Univ. Press, 1535 pp., 2013.

Jungclaus, J. H., Bard, E., Baroni, M., Braconnot, P., Cao, J., Chini, L. P., Egorova, T., Evans, M., González-Rouco, J. F., Goosse, H., Hurtt, G. C., Joos, F., Kaplan, J. O., Khodri, M., Klein Goldewijk, K., Krivova, N., LeGrande, A. N., Lorenz, S. J., Luterbacher, J., Man, W., Maycock, A. C., Meinshausen, M., Moberg, A., Muscheler, R., Nehrbass-Ahles, C., Otto-Bliesner, B. I., Phipps, S. J., Pongratz, J., Rozanov, E., Schmidt, G. A., Schmidt, H., Schmutz, W., Schurer, A., Shapiro, A. I., Sigl, M., Smerdon, J. E., Solanki, S. K., Timmreck, C., Toohey, M., Usoskin, I. G., Wagner, S., Wu, C.-J., Yeo, K. L., Zanchettin, D., Zhang, Q., and Zorita, E.: The PMIP4 contribution to CMIP6 – Part 3: The last millennium, scientific objective, and experimental design for the PMIP4 past1000 simulations, Geosci. Model Dev., 10, 4005–4033, https://doi.org/10.5194/gmd-10-4005-2017, 2017.

Landrum, L., Otto-Bliesner, B., Wahl, E., Conley, A., Lawrence, P., Rosenbloom, N., and Teng, H.: Last Millennium Climate and Its Variability in CCSM4, J. Climate, 26, 1085–1111, https://doi.org/10.1175/JCLI-D-11-00326.1, 2013.

Lawrence, D. M. and Slater, A. G.: A projection of severe near-surface permafrost degradation during the 21st century, Geophys. Res. Lett., 32, L24401, https://doi.org/10.1029/2005GL025080, 2005.

Lawrence, D. M., Slater, A. G., Romanovsky, V. E., and Nicolsky, D. J.: Sensitivity of a model projection of near-surface permafrost degradation to soil column depth and representation of soil organic matter, J. Geophys. Res., 113, F02011, https://doi.org/10.1029/2007JF000883, 2008.

Marshall, S.: Glacier retreat crosses a line, Science, 345, p. 872, https://doi.org/10.1126/science.1258584, 2014.

Mekonnen, M. A., Wheater, H. S., Ireson, A. M., Spence, C., Davison, B., and Pietroniro, A.: Towards an Improved Land Surface Scheme for Prairie Landscapes, J. Hydrol., 511, 105–116, https://doi.org/10.1016/j.jhydrol.2014.01.020, 2014.

Mendoza, P. A., Clark, M. P., Barlage, M., Rajagopalan, B., Samaniego, L., Abramowitz, G., and Gupta, H.: Are we unnecessarily constraining the agility of complex process-based models?, Water Resour. Res., 51, 716–728, https://doi.org/10.1002/2014WR015820, 2015.

Nicolsky, D. J., Romanovsky, V. E., Alexeev, V. A., and Lawrence, D. M.: Improved modeling of permafrost dynamics in a GCM land-surface scheme, Geophys. Res. Lett., 34, L08501, https://doi.org/10.1029/2007GL029525, 2007.

Paquin, J. P. and Sushama, L.: On the Arctic near-surface permafrost and climate sensitivities to soil and snow model formulations in climate models, Clim. Dynam., 44, 203–228, https://doi.org/10.1007/s00382-014-2185-6, 2015.

Peel, M. C., Finlayson, B. L., and McMahon, T. A.: Updated world map of the Köppen-Geiger climate classification, Hydrol. Earth Syst. Sci., 11, 1633–1644, https://doi.org/10.5194/hess-11-1633-2007, 2007.

Pietroniro, A., Fortin, V., Kouwen, N., Neal, C., Turcotte, R., Davison, B., Verseghy, D., Soulis, E. D., Caldwell, R., Evora, N., and Pellerin, P.: Development of the MESH modelling system for hydrological ensemble forecasting of the Laurentian Great Lakes at the regional scale, Hydrol. Earth Syst. Sci., 11, 1279–1294, https://doi.org/10.5194/hess-11-1279-2007, 2007.

Politis, D. N. and Romano, J. P.: The stationary bootstrap, J. Am. Stat. Assoc., 89, 1303–1313, 1994.

Razavi, S. and Gupta, H. V.: What do we mean by sensitivity analysis? The need for comprehensive characterization of "global" sensitivity in Earth and Environmental systems models, Water Resour. Res., 51, 3070–3092, https://doi.org/10.1002/2014WR016527, 2015.

Razavi, S., Tolson, B. A., and Burn, D. H.: Review of surrogate modelling in water resources, Water Resour. Res., 48, W07401, https://doi.org/10.1029/2011WR011527, 2012.

Razavi, S., Elshorbagy, A., Wheater, H., and Sauchyn, D.: Toward understanding nonstationarity in climate and hydrology through tree ring proxy records, Water Resour. Res., 51, 1813–1830, https://doi.org/10.1002/2014WR015696, 2015.

Razavi, S., Elshorbagy, A., Wheater, H., and Sauchyn, D.: Time scale effect and uncertainty in reconstruction of Paleo-hydrology, Hydrol. Process., 30, 1985–1999. https://doi.org/10.1002/hyp.10754, 2016.

Rodell, M., Houser, P. R., Berg, A. A., and Famiglietti, J. S.: Evaluation of 10 Methods for Initializing a Land Surface Model, J. Hydrometeorol., 6, 146–155, https://doi.org/10.1175/JHM414.1, 2005.

Rowland, J. C., Jones, C. E., Altmann, G., Bryan, R., Crosby, B. T., Geernaert, G. L., Hinzman, L. D., Kane, D. L., Lawrence, D. M., Mancino, A., Marsh, P., McNamara, J. P., Romanovsky, V. E., Toniolo, H., Travis, J., Trochim, E., and Wilson, C. J. : Arctic landscapes in transition: Responses to thawing permafrost, Eos, 91, 229–230, 2010.

Schmidt, G. A., Jungclaus, J. H., Ammann, C. M., Bard, E., Braconnot, P., Crowley, T. J., Delaygue, G., Joos, F., Krivova, N. A., Muscheler, R., Otto-Bliesner, B. L., Pongratz, J., Shindell, D. T., Solanki, S. K., Steinhilber, F., and Vieira, L. E. A.: Climate forcing reconstructions for use in PMIP simulations of the last millennium (v1.0), Geosci. Model Dev., 4, 33–45, https://doi.org/10.5194/gmd-4-33-2011, 2011.

Schuur, E. A. G., McGuire, A. D., Schädel, C., Grosse, G., Harden, J. W., Hayes, D. J., Hugelius, G., Koven, C. D., Kuhry, P., Lawrence, D. M., Natali, S. M., Olefeldt, D., Romanovsky, V. E., Schaefer, K., Turetsky, M. R., Treat, C. C., and Vonk, J. E.: Climate change and the permafrost carbon feedback, Nature, 250, 171–178, 2015.

Shrestha, R. and Houser, P.: A heterogeneous land surface model initialization study, J. Geophys. Res., 115, D19111, https://doi.org/10.1029/2009JD013252, 2010.

Slater, A. G. and Lawrence, D. M.: Diagnosing present and future permafrost from climate models, J. Climate, 26, 5608–5623, 2013.

Smith, S. L., Burgess, M. M., Riseborough, D., Coultish, T., and Chartrand, J.: Digital summary database of permafrost and thermal conditions - Norman Wells Pipeline study sites; Geological Survey of Canada, Open File 4635, 104 pp., 2004.

Soulis, E. D., Snelgrove, K. R., Kouwen, N., Seglenieks, F., and Verseghy, D. L. Towards closing the vertical water balance in Canadian atmospheric models: Coupling of the land surface scheme class with the distributed hydrological model watflood, Atmos. Ocean, 38, 251–269, https://doi.org/10.1080/07055900.2000.9649648, 2000.

Stevens, M. B., Smerdon, J. E., González-Rouco, J. F., Stieglitz, M., and Beltrami, H.: Effects of bottom boundary placement on subsurface heat storage: Implications for climate model simulations, Geophys. Res. Lett., 34, L02702, https://doi.org/10.1029/2006GL028546, 2007.

Sturm, M., Racine, C., and Tape, K.: Climate change: increasing shrub abundance in the Arctic, Nature, 411, 546–547, 2001.

Szeicz, J. M. and MacDonald, G. M.: Dendroclimatic reconstruction of Summer Temperatures in northwestern Canada Since AD 1638 Based on Age-Dependent Modeling, Quaternary Res., 44, 257–266, 1995.

Troy, T. J., Sheffield, J., and Wood, E. F.: The role of winter precipitation and temperature on northern Eurasian streamflow trends, J. Geophys. Res., 117, D05131, https://doi.org/10.1029/2011JD016208, 2012.

Verseghy, D.: CLASS – The Canadian Land Surface Scheme (Version 3.4), Technical Documentation (Version 1.1), Climate Research Division, Science and Technology Branch, Environment Canada, 180 pp., 2009.

Verseghy, D. L.: CLASS – A Canadian land surface scheme for GCMs, I. Soil model, Int. J. Climatol., 11, 111–133, 1991.

Verseghy, D. L., McFarlane, N. A., and Lazare, M.: Class – A Canadian land surface scheme for GCMS, II. Vegetation model and coupled runs, Int. J. Climatol., 13, 347–370, https://doi.org/10.1002/joc.3370130402, 1993.

Weedon, G., Gomes, S., Viterbo, P., Shuttleworth, W., Blyth, E., Österle, H., Adam, J., Bellouin, N., Boucher, O., and Best, M.: Creation of the WATCH Forcing Data and Its Use to Assess Global and Regional Reference Crop Evaporation over Land during the Twentieth Century, J. Hydrometeorol., 12, 823–848, https://doi.org/10.1175/2011JHM1369.1, 2011.

Woo, M.: Permafrost Hydrology, Springer, New York, 572 pp., 2012.

Yang, Z.-L., Dickinson, R. E., Henderson-Sellers, A., and Pitman, A. J.: Preliminary study of spin-up processes in land surface models with the first stage data of Project for Intercomparison of Land Surface Parameterization Schemes Phase 1(a), J. Geophys. Res., 100, 16553–16578, https://doi.org/10.1029/95JD01076, 1995.

Yassin, F., Razavi, S., Wheater, H., Sapriza-Azuri, G., Davison, B., and Pietroniro, A.: Enhanced identification of a hydrologic model using streamflow and satellite water storage data: A multicriteria sensitivity analysis and optimization approach, Hydrol. Process., 31, 3320–3333, https://doi.org/10.1002/hyp.11267, 2017.

Zhang, Y., Chen, W., and Cihlar, J.: A process-based model for quantifying the impact of climate change on permafrost thermal regimes, J. Geophys. Res., 108, 4695, https://doi.org/10.1029/2002JD003354, 2003.

Zhang, Y., Chen, W., and Riseborough, D. W.: Temporal and spatial changes of permafrost in Canada since the end of the Little Ice Age, J. Geophys. Res., 111, D22103, https://doi.org/10.1029/2006JD007284, 2006.

Zhang, Y., Chen, W., and Riseborough, D. W.: Disequilibrium response of permafrost thaw to climate warming in Canada over 1850–2100, Geophys. Res. Lett., 35, L02502, https://doi.org/10.1029/2007GL032117, 2008.

8

Ecohydrological optimality in the Northeast China Transect

Zhentao Cong[1,2]**, Qinshu Li**[1,2]**, Kangle Mo**[1,2]**, Lexin Zhang**[1,2]**, and Hong Shen**[1,2]

[1]Department of Hydraulic Engineering, Tsinghua University, Beijing, 100084, China

[2]State Key Laboratory of Hydroscience and Engineering, Beijing, 100084, China

Correspondence to: Zhentao Cong (congzht@tsinghua.edu.cn)

Abstract. The Northeast China Transect (NECT) is one of the International Geosphere-Biosphere Program (IGBP) terrestrial transects, where there is a significant precipitation gradient from east to west, as well as a vegetation transition of forest–grassland–desert. It is remarkable to understand vegetation distribution and dynamics under climate change in this transect. We take canopy cover (M), derived from Normalized Difference Vegetation Index (NDVI), as an index to describe the properties of vegetation distribution and dynamics in the NECT. In Eagleson's ecohydrological optimality theory, the optimal canopy cover (M^*) is determined by the trade-off between water supply depending on water balance and water demand depending on canopy transpiration. We apply Eagleson's ecohydrological optimality method in the NECT based on data from 2000 to 2013 to get M^*, which is compared with M from NDVI to further discuss the sensitivity of M^* to vegetation properties and climate factors. The result indicates that the average M^* fits the actual M well (for forest, $M^* = 0.822$ while $M = 0.826$; for grassland, $M^* = 0.353$ while $M = 0.352$; the correlation coefficient between M and M^* is 0.81). Results of water balance also match the field-measured data in the references. The sensitivity analyses show that M^* decreases with the increase of leaf area index (LAI), stem fraction and temperature, while it increases with the increase of leaf angle and precipitation amount. Eagleson's ecohydrological optimality method offers a quantitative way to understand the impacts of climate change on canopy cover and provides guidelines for ecorestoration projects.

1 Introduction

Transect studies play an important role in understanding the role of the terrestrial biosphere in global change (Koch et al., 1995a). The Global Change and Terrestrial Ecosystems (GCTE) project of International Geosphere-Biosphere Program (IGBP) has chosen 15 transects along with environmental or land-use gradients, aiming at understanding how these factors influence terrestrial ecosystem and the interaction between biosphere and atmosphere (Koch et al., 1995b; Canadell et al., 2002; Austin and Sala, 2002). The Northeast China Transect (NECT) was identified as one of the IGBP transects in 1993, with precipitation/moisture as the main driving climate factor (Ni and Zhang, 2000; Zhang and Zhou, 2011). Along with the moisture gradient, vegetation types vary gradually from forest in the east, to the cropland in the middle, and grassland and bare soil in the west.

Vegetation plays an important role in terrestrial ecosystems. It strongly influences the exchange of energy, substances and moisture between land and atmosphere through photosynthesis, respiration and transpiration (Graetz, 1991; Mcpherson, 2007). At the same time, the vegetation growth condition is largely affected by climate factors, such as precipitation, air temperature and greenhouse gases (Füssler and Gassmann, 2000; Lotsch et al., 2003; Liu et al., 2006).

The most common indices to describe vegetation performance include Normalized Difference Vegetation Index (NDVI) and vegetation canopy cover. NDVI is a linear combination of remotely sensed near-infrared reflectance and red reflectance. It is an index reflecting the greenness of vegetation canopy and photosynthetic activity (Dorman et al., 2013; Fontana et al., 2008; Hmimina et al., 2013). Vegetation canopy cover is defined as the fraction of total ground surface covered by vegetation. Semi-empirical relationships between NDVI and canopy cover were used to derive the possible arithmetic expression of canopy cover (Baret et al., 1995; Carlson and Ripley, 1997; Gutman and Ignatov, 1998; Jiang et al., 2006). With the rising attention of the climate change issue, studies on the relationship between canopy cover and climate factors have been conducted in different regions of the world (Zhou et al., 2001; Schultz and Halpert, 1993; Piao et al., 2011; Park and Sohn, 2010; Li et al., 2002; Wang et al., 2003). Nie et al. (2012) used correlation analysis to check the relationship between NDVI and climate factors in the NECT, with regression equations given for different timescales. NDVI driven by climate changes varied differently between vegetation types and seasons (Piao et al., 2006). Duan et al. (2011) illustrated that precipitation was the most important factor in affecting the temporal NDVI patterns over semi-arid and arid regions of China. Peng et al. (2012) found that more than 70 % of the temporal variations in NDVI were contributed by precipitation during the growing season in typical and desert steppes in northeast China. Mao et al. (2012), however, discovered that the correlation between NDVI and temperature was higher than with precipitation over most parts of northeast China for all vegetation covers; NDVI presented a downward trend with increased temperature and remarkably decreased precipitation. Further, Yuan et al. (2015) suggested diverse responses of grasslands to precipitation intensities.

Although the statistical models have been established to describe the response of vegetation to climate factors, they cannot express the underlying mechanism of the response quantitatively. Vegetation models were developed to detect how vegetation reacts to climate change based on the biophysical and physiological processes, including plant life cycle, carbon and nitrogen cycles, but many data and parameters were required (Myoung et al., 2011). It is a big challenge to build a simplified model that can describe the mechanism of vegetation response to climate change with relatively few parameters. Fortunately, Eagleson (2002) presented a theory and method: ecohydrological optimality (Eagleson, 1978a–g, 1982; Eagleson and Tellers, 1982). In Eagleson's ecohydrological optimality theory, vegetation characteristics, such as leaf angle, leaf area index (LAI) and canopy cover, are determined by the light, energy, water and soil conditions in long-term average state. Different from the models above, Eagleson's ecohydrological optimality theory can not only explore the mechanism of canopy cover distribution, mainly from the water balance perspective but is also easy to conduct. The optimality theory provides a new way to explore the quantitative relationship between vegetation and climate factors. Despite the fact that Eagleson's work is regarded as the basis for ecohydrology and is of great importance (Hatton et al., 1997; Kerkhoff et al., 2004), limited studies have been conducted using the theory in practice (Shao et al., 2011; Mo et al., 2015), which is partly due to the limitation of long-term temporal scale, partly due to the difficulty to measure vegetation characteristics.

In Eagleson's ecohydrological optimality theory, the optimal canopy cover (M^*) is determined by the trade-off between water supply depending on water balance and water demand depending on canopy transpiration. The NDVI data offer us a method to estimate actual canopy cover. If we can verify Eagleson's ecohydrological optimality theory by comparing the optimal canopy cover and remote sensing canopy cover, we can discuss the impacts of climate factors and vegetation properties on vegetation cover. From this framework, we can certainly provide some insights in terms of the understanding of climate change impacts on canopy cover dynamics and therefore can provide useful guidelines for ecorestoration projects, especially for the selection of vegetation species and plant density. Mo et al. (2015) applied this method in Horqin Sand, China, just for one kind of vegetation. In this study, we apply Eagleson's ecohydrological optimality method in the NECT based on data from 2000 to 2013 to get M^* and then compare with M determined by NDVI to discuss the sensitivity of M^* to vegetation properties and climate factors.

2   Study area and data

2.1   Study area

The NECT is one of the midlatitude IGBP terrestrial transects. It ranges from 42 to 46° N and from 106 to 134° E. The major change gradient is precipitation, which decreases gradually from the eastern mountainous region to the middle farmland and western steppes (Fig. 1). In the east, the annual precipitation is over 600 mm yr^{-1}; meanwhile, in the west, the annual precipitation is under 200 mm yr^{-1}. The land cover types show a significant zonal distribution from east to west: temperate evergreen coniferous–deciduous broadleaf mixed forests, deciduous broadleaf forests and woodlands in the east, shrublands and crop in the middle, grassland and bare soil in the west (Fig. 1). In this study, we just focus on the growing season from May to September. The input data and parameters include remote sensing data, meteorological data, vegetation data and soil data. The remote sensing data are the vegetation cover and LAI. The main meteorological data are length of growing season, potential evaporation, air temperature and storm duration. The main vegetation data are surface retention depth, leaf angle and stem height. The main soil data are soil porosity and hydraulic conductivities. The input values are listed in Table 1.

2.2   Remote sensing data

Monthly NDVI (MOD13A3), yearly land cover types (MCD12Q1) and 8-day LAI (MCD15A2) datasets derived from the Moderate-resolution Imaging Spectroradiometer (MODIS) aboard the Aqua and Terra satellites are applied. These satellite data are available on the NASA website (http://reverb.echo.nasa.gov/).

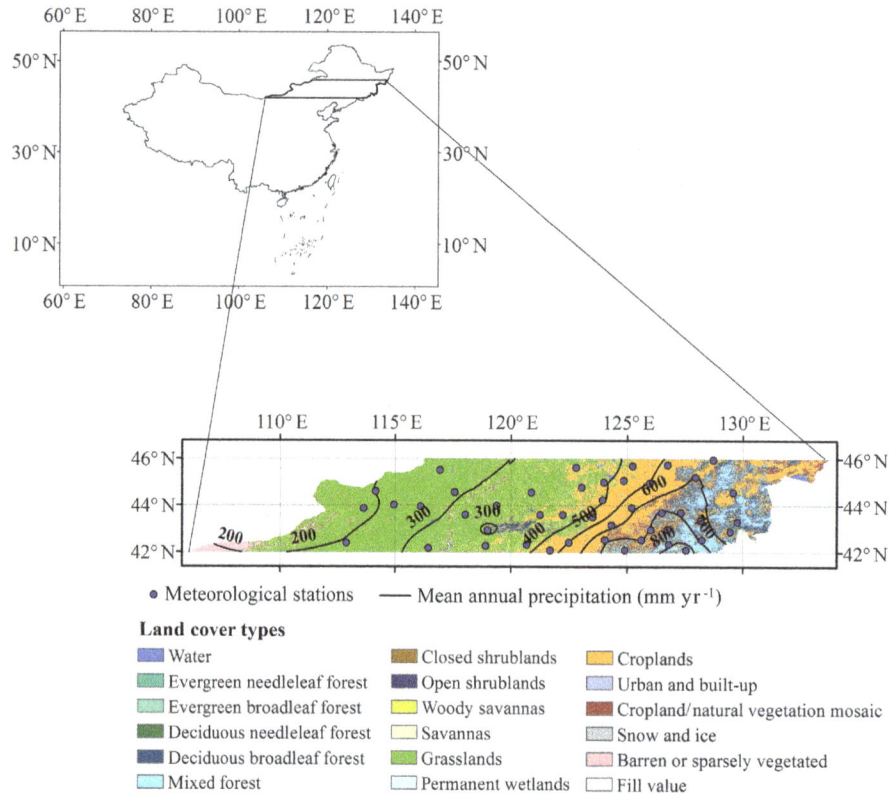

Figure 1. The geographic location, land cover, spatial distribution of precipitation and meteorological station locations of the NECT.

The spatial resolution of NDVI, land cover types and LAI dataset are 1 km, 500 m and 1 km, respectively. Considering the wide longitudinal and latitudinal extends of the NECT, these remote-sensed data are resampled to be $10\,\mathrm{km} \times 10\,\mathrm{km}$. The MODIS Reprojection Tool (MRT) is applied to define coordinate systems for the images. MRT is also used to generate NDVI and LAI data for growing season of each year.

Canopy cover is defined as the fraction of total ground surface covered by vegetation (Eagleson, 2002). Usually a linear transformation of remote-sensed NDVI is used to calculate actual canopy cover (M) (Gutman and Ignatov, 1998; Jiang et al., 2006):

$$M = \frac{\mathrm{NDVI} - \mathrm{NDVI_{min}}}{\mathrm{NDVI_{max}} - \mathrm{NDVI_{min}}}, \tag{1}$$

in which $\mathrm{NDVI_{min}}$ is the NDVI of barren soil and $\mathrm{NDVI_{max}}$ is the NDVI of forests. Since the land cover of a fixed grid may be changed in different years, it was hard to define the real barren soil or the forest areas. We considered the area sensed as barren soil for every year as the barren soil area, and the $\mathrm{NDVI_{min}}$ is the spatial average of barren area NDVI. Similarly, the area sensed as forests every year is considered as the forests area, and the spatial average of forests NDVI is $\mathrm{NDVI_{max}}$. In this study, $\mathrm{NDVI_{min}}$ and $\mathrm{NDVI_{max}}$ are 0.05 and 0.63, respectively, which means the canopy cover can be regarded as 1 if the NDVI is above 0.63 and as 0 if the NDVI is below 0.05.

2.3 Meteorological data

The meteorological data used in this study from 2000 to 2013 are provided by China Meteorological Data Sharing Service System (http://cdc.cma.gov.cn). The spatial distribution of the 45 meteorological stations is shown in Fig. 1. Atmospheric pressure (P_a), wind speed (W_{nd}), average air temperature (T_a), net radiation (R_n) (estimated by sunshine hours, S_h, and air temperature; Allen, 1998), relative humidity (RH), minimum air temperature (T_n) and maximum air temperature (T_m) are required to calculate the potential evapotranspiration by Penman–Monteith equation (Ni and Zhang, 2000; Eagleson, 2002). A kriging interpolation method is applied to generate the spatial distribution of the meteorological factors and potential evapotranspiration. The spatial resolution is 10 km in order to be consistent with that of remote sensing data.

3 Methodology

Eagleson proposed three hypotheses in his ecohydrological optimality theory. He considered that climate and vegetation can influence and adapt to each other on different timescales.

Table 1. The terminology, interpretation, units and values of inputs.

	Terminology	Interpretation and units	Value
Remote sensing data	f_c	average vegetation cover of growing season	0.00–1.00
	M_d	average vegetation cover of non-growing season	0.00–1.00
	l_t	leaf area index in growing season (dimensionless)	0.00–4.70
	l_{td}	leaf area index in dormant season (dimensionless)	0.00–1.70
Meteorological data	m_t	length of the growing season (days)	153
	m_d	length of the non-growing season (days)	212
	E_{pst}	free water surface potential evaporation during growing season (mm day^{-1})	3.6–4.4
	E_{psd}	free water surface potential evaporation during dormant season (mm day^{-1})	0.7–1.0
	P_τ	precipitation in growing season (mm)	149.7–624.3
	P_d	precipitation in dormant season (mm)	26.2–226.3
	t_0	average temperature in growing season (°)	16.12–21.24
	m_{tb}	mean time between storms (days)	4.65–6.35
	m_{tr}	mean storm duration (days)	0.37–0.64
	γ_0	surface psychrometric constant (Pa K^{-1})	0.06
Vegetation data	m	exponent relating shear stress on foliage to horizontal wind velocity (dimensionless)	0.5
	n	number of sides of each foliage element producing surface resistance to wind (dimensionless)	2
	η_0	stomatal leaf area/illuminated leaf area (dimensionless)	2.50
	h_0	surface retention depth (mm)	1.00
	β	cosine of leaf angle (dimensionless)	0.45
Soil data	h_s	stem height (i.e., height of crown base above substrate) (m)	
	h	height of tree from ground surface to top of crown (m)	
	m	soil pore size distribution index (dimensionless)	0.50
	n_e	effective soil porosity (dimensionless)	0.45
	d	diffusivity index of soil (dimensionless)	4.30
	ψ	saturated matrix potential of soil (mm)	900.0
	k	effective saturated hydraulic conductivity of soil (mm day^{-1})	29.4
	s_0	space–time average soil moisture concentration in the root zone (dimensionless)	0.30–0.62

First, when climate and soil change in a short time period, the canopy cover will adjust its value to maximize the soil moisture. Second, as the timescales get longer, the species whose potential transpiration efficiency make the soil moisture highest will be selected through natural selection. Third, the soil properties will be altered to ensure the species get their maximum canopy cover (Eagleson, 2002; Hatton et al., 1997). These hypotheses mentioned two important canopy state variables, i.e., the canopy cover (M) and canopy conductance (k_v). Canopy conductance is defined as the ratio of potential rates of transpiration E_v and soil surface evaporation E_{ps} (Eagleson, 1978d, 2002):

$$k_v = \frac{E_v}{E_{ps}}. \tag{2}$$

The potential evapotranspiration E_{ps} is calculated by Penman equation:

$$\lambda E_{ps} = \frac{\Delta R_n + \rho c_p [e_s(T) - e]/r_a}{\Delta + \gamma_0}. \tag{3}$$

The canopy transpiration rate E_v is calculated by Penman–Monteith equation:

$$\lambda E_v = \frac{\Delta R_n + \rho c_p [e_s(T) - e]/r_a}{\Delta + \gamma_0 (1 + r_c/r_a)}, \tag{4}$$

where E_{ps} is the potential rate of evaporation from a wet simple surface (mm day^{-1}); E_v is the rate of canopy transpiration (mm day^{-1}); λ is the latent heat of vaporization of water $= 2500$ J g^{-1} (at 0 °C); Δ is the slope of the saturation vapor pressure vs. temperature curve (Pa K^{-1}); R_n is the net solar radiation, J/(mm^2 day^{-1}); ρ is the fluid mass density (g mm^{-3}); c_p is the specific heat of air at constant pressure, J/(g K^{-1}); $e_s(T)$ is the saturation vapor pressure at the temperature of the evaporation site (Pa); e is the partial pressure of water vapor (Pa); r_a is the lumped atmospheric resistance over the 2 m above the canopy top (day mm^{-1}); r_c is the lumped resistance to flow through the canopy which does not vary with water supply (day mm^{-1}); and γ_0 is the surface psychrometric constant (Pa K^{-1}).

When the stomas fully open, the canopy transpiration rate E_v will reach its maximum value – potential canopy transpiration E_{pv} – thus making k_v its maximum value as well, which is called the potential canopy conductance k_v^*:

$$k_v^* = \frac{E_{pv}}{E_{ps}} = \frac{1 + \Delta/\gamma_0}{1 + \Delta/\gamma_0 + (1-M)\left(\frac{r_c}{r_a}\right)_{M\to 0} + M\left(\frac{r_c}{r_a}\right)_{M=1}}, \quad (5)$$

where $(r_c/r_a)_{M\to 0}$ is the resistance ratio for open canopies, which is related to the exponent relating shear stress on foliage to horizontal wind velocity and horizontal leaf area index. $(r_c/r_a)_{M=1}$ is the resistance ratio for closed ($M=1$) canopies whose main influence factor is the ratio of stem height h_s and tree height h. According to Eq. (5) and explanations above, the resistance ratio can be fixed once the vegetation specie is given. The potential canopy conductance k_v^* is inversely proportionate to the canopy cover M. The $k_v^* - M$ curve is called the water demand curve.

The relationship between k_v^* and M can also be described by the water balance equation. In the growing season, the average inflows and outflows of the soil column can be described as

$$P_\tau - m_v E[E_r] - \Delta S = m_v E[R_{sj}] + E[E_{T\tau}] + m_\tau v - m_\tau w, \quad (6)$$

where P_τ is the growing season precipitation (mm); m_v is the number of independent storm times (dimensionless); E_r is the storm surface retention depth (mm); R_{sj} is the storm rainfall excess (mm); ΔS is the average carryover (from dormant season to growing season) soil moisture storage (mm); m_τ is the growing season length (days); v is the percolation to water table (mm day^{-1}); w is the capillary rise from the water table (mm day^{-1}); and $E[\ ]$ indicates the expected value of $[\ ]$.

Some assumptions are made to describe each item (Eagleson, 1978a–g). Thus, the water balance of the growing season can be expressed as

$$Mk_v^* = \frac{V_e}{m_{tb} E_{ps}}, \quad (7)$$

where V_e is the volume of soil moisture (per unit of surface area) available for exchange with atmosphere during average inter-storm period (mm), and m_{tb} is mean time between storms (h); more details can be found in the Appendix.

Equation (7) describes water supply in naturally selected canopy moisture state, while Eq. (5) describes water demand of fixed vegetation species. By drawing these two lines in a figure (Fig. 2), we can notice that the water demand grows with the increase of M, but as water is limited, water supply decreases under M enhancement. The intersection point of these two lines is the theoretical optimal canopy cover and potential canopy conductance in the vegetation state space. This method is applied in each grid (10 km × 10 km) of the NECT area.

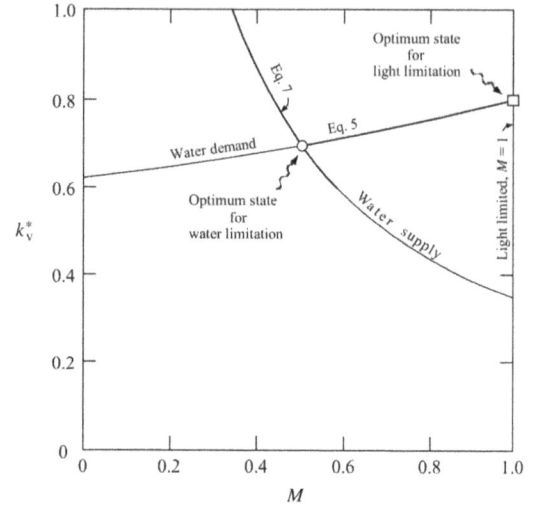

Figure 2. Optimum canopy state (from Eagleson, 2002).

4 Results and discussion

4.1 Canopy cover of the NECT

The observed canopy cover M shows a significant gradient ranging from 1 in the eastern forests to 0 in the western desert (Fig. 3a).

The dark blue area is mainly the forests of the Changbai Mountains, where the average canopy cover reaches up to 0.83. The light blue area is the Songnen Plain. The Songnen Plain is one of the most famous commodity grain bases, rich in corn, sorghum, soybean, wheat and paddies (Zhou and Wang, 2003), with the average canopy cover of 0.55. Farther westward, there is Horqin Sand, in which most of the vegetation is grass. Then, there is a narrow northeast–southwest-oriented band with a relatively higher value (blue color) at around 120° E, which is mainly caused by the elevation. The band is the location of Greater Khingan Range. The east slope of the Greater Khingan Range is very steep; thus, the maritime monsoon can bring a lot of rainfall, causing the existence of the forest ecosystem. However, most of the vegetation on the west slope is grass, mainly because of the gentle gradient and dry climate (Guo and Zhang, 2013). The grassland is the Inner Mongolia steppe.

The ecohydrological optimality theory is applied in this study to simulate theoretic optimal canopy cover (M^*) of the NECT. As shown in Fig. 3b, the modeled canopy cover has the same trend with the actual M but transitions more smoothly, which is mainly caused by the interpolation of meteorological data. The blank grids in the simulation result are due to the missing data of LAI. The result indicates that the average M^* fits the average M well (for forest, $M^* = 0.822$ while $M = 0.826$; for grassland, $M^* = 0.353$ while $M = 0.352$; the correlation coefficient between M and M^* is 0.81). The corresponding areas are highlighted in the

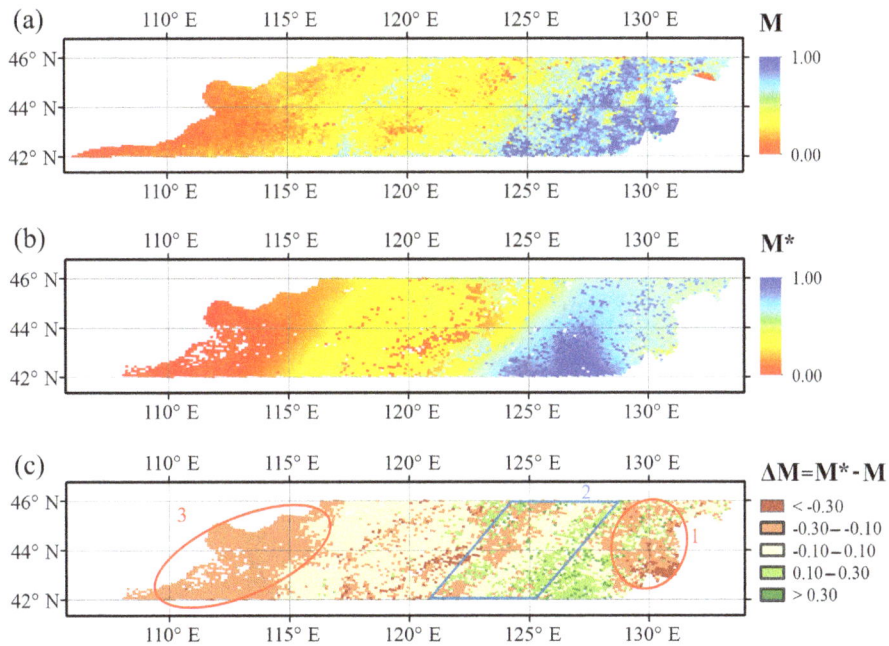

Figure 3. Spatial distribution of mean canopy cover from MODIS NDVI (**a**), optimal canopy cover (**b**) and their difference (**c**).

figure of spatial distribution of ΔM, defined as M^* minus M (Fig. 3c). The spatial average ΔM is only -0.050 for the whole NECT area; meanwhile, there are 45.7 % pixels of the NECT area where the ΔM value is between -0.1 and 0.1. There are three regions where the differences between M and M^* are relatively large. Region 1 is the Yanbian Korean Autonomous Prefecture. The simulation result is relatively small mainly because of the forest protection project. The Natural Forest Protection Project (NFPP) has been conducted in northeastern China since 1998, aiming at protecting the natural forest resources (Wei et al., 2014). Yanbian forest acreage has increased by $800\,km^2$ during the first stage of NFPP. The dark red area is the city of Hunchun. Hunchun is a representative nature reserve, and the forest acreage has increased by 9009 ha from 1999 to 2012 (Li, 2014). Region 2 is the southern Xilin Gol grassland. In the past decades, Xilin Gol grassland has been extremely dry and has been suffering from severe degradation (Tong et al., 2004). The Beijing–Tianjin Sand Source Control Project is undertaken to improve the canopy cover of degraded grassland. Over 66 000 water source projects and 47 000 water-saving irrigation projects increased the water supply of this area, thus contributing to the increase of vegetation activities (Yu et al., 2010). The irrigation part is not considered in Eagleson's water balance system, which leads to the deviation of the modeled results. In the crop region (the blue frame in Fig. 3c), some M^* are higher than M, while some are lower. This is because of the close relationship between canopy cover and crop growth stage. The growth process of various crops is different, and the timing of plantation and harvesting is mainly affected by human intervention rather than natural processes

(Liu et al., 2013; Kim and Wang, 2005). Meanwhile, the water supply for the crop is not only from the natural hydrological cycle but also from agricultural irrigation, which is not considered in the theory.

The correlation coefficient R between M and M^* is high, which indicates the ecohydrological optimality theory is applied well in the NECT during the long-term period. Previous studies suggest there are lagged relationships between NDVI and climate factors, and the time lags are different in different regions or different biomes (Braswell et al., 1997; Piao et al., 2003; Li et al., 2011; Hu et al., 2011; Bao et al., 2015). Figure 4 shows that, in the grassland area, the variation amplitude of M is smaller than M^*, and the delay usually happens within a year, while in the forest area, there is a trend delay across the years. For example, M^* increases from 2007 to 2009, but the increasing trend of M does not appear until 2009 to 2010. This can be explained by the vegetation adaptation strategy to climate changes. For example, the canopy cover might not increase immediately with the increasing precipitation but might increase in the next year. Eagleson's theory describes how vegetation adapts to climate change in a relatively long term (Eagleson, 2002). Once climate changes, it takes years for vegetation to reach its optimum canopy cover.

4.2 Water balance components

As the NECT spans a wide range from west to east, and the vegetation and climate vary significantly, the NECT is divided into three parts according to land cover types: forest, cropland and grassland (Ni and Zhang, 2000). The propor-

Table 2. Water balance components of different land cover types.

Result		Grassland		Cropland		Forests	
	M	0.352		0.548		0.826	
	M^*	0.353		0.557		0.822	
Water balance		mm	$/P$	mm	$/P$	mm	$/P$
component	Precipitation	253	100.00 %	414	100.00 %	478	100.00 %
	Interception	29	11.61 %	39	9.29 %	68	14.24 %
	Runoff	1	0.19 %	119	28.77 %	119	24.92 %
	ΔS	91	36.29 %	14	3.34 %	−74	−15.47 %
	Evaporation	131	51.90 %	243	58.60 %	365	76.31 %

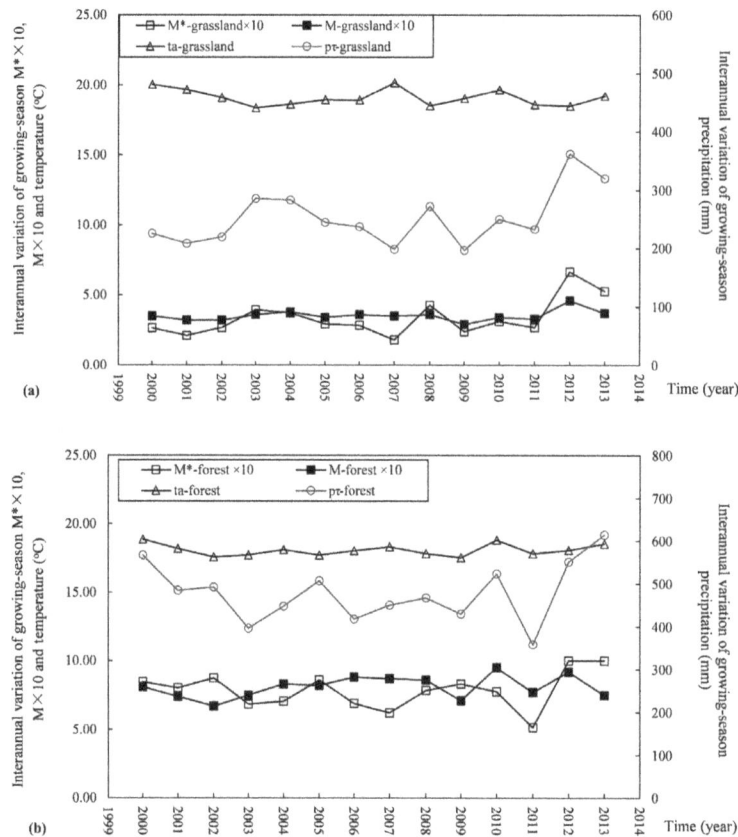

Figure 4. Variation of M^*, M, precipitation (P_τ) and air temperature (t_a) during 2000–2013: **(a)** grassland; **(b)** forest.

tions of the water balance components for the annual average growing season are calculated for each part based on Eq. (6), as shown in Table 2. According to the research conducted before, in the grassland area, the interception was 20.86 and 7.88 % for shrub and grass, respectively (Peng et al., 2014), and the runoff of Xilin Gol grassland occupied around 0.046–1.8 % (Wang et al., 2008a; Miao et al., 2008). In the forest area, the dominant tree species are *Pinus koraiensis* (Pk), *Quercus mongolica* (Qm), *Populus davidiana* (Pd) and *B. platyphylla* (Bp) (Chen, 2001; Zhang and Zhou, 2009). The interception consists of 19.61 % for Pk and 14.97 % for Bp in the Greater Khingan Range and 10.20 %

for Pk in the Changbai Mountains (Cai et al., 2008; Wang et al., 2006). The runoff coefficients of the Suifen River and Second Songhua River, both of which are located in the forest area, are around 20–30 % (Huang, 1999; Song, 2010). The simulated interception and runoff for both grassland and forest areas are consistent with previous studies, which demonstrates the reasonability of this theory. The negative value of ΔS in the forest area means a recharge of soil moisture. As the air temperature in the non-growing season is low, most of the precipitation is snow rather than rain, so the water is frozen in the soil and melts in the next spring (Fan et al., 2006; Yang et al., 2006). Therefore, most of the water is

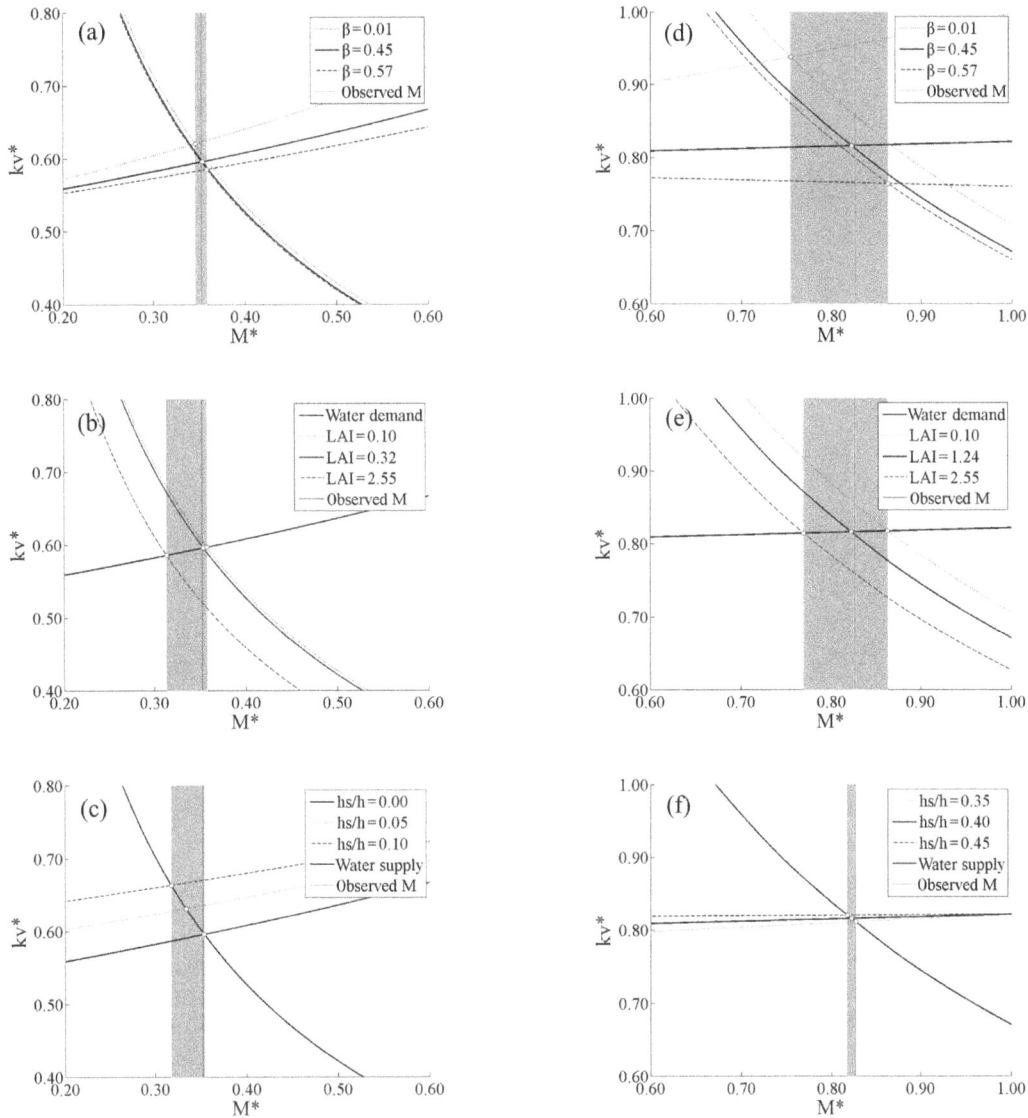

Figure 5. M^* changes with β, LAI and h_{s}/h: **(a)–(c)** grassland; **(d)–(f)** forest. The shaded areas indicate the range of M^* with the change of climate factors.

stored in the dormant season for the vegetation growth in the next growing season.

The rationality of the calculated proportions of water balance components for each part demonstrates the applicability of the optimality theory. By adapting this method, it is much easier to figure out the allocation of precipitation if the vegetation and soil conditions are known.

4.3 Sensitivity of M^* to vegetation properties

LAI, β and h_{s}/h control the physical and biological processes of plant canopies, such as interception and evaporation (Chen and Black, 1992; Asner, 1998; Huete et al., 2002). β and LAI are the dominant parameters for the interception calculation, thus leading to the variance of water supply for

vegetation growth. β and h_{s}/h affect the plant evaporation by affecting the resistance ratio, which influences the water demand curve (Eagleson, 2002). The thresholds of the three parameters are from the experiments conducted before (Du, 2004; Wang et al., 2008b; Rauner, 1976; Eagleson, 2002). Figure 5 shows the different reactions of optimal canopy cover to vegetation species change between grassland and forest areas. M^* increases with the increase of leaf angle and decrease of stem fraction and LAI. Mo et al. (2015) studied the relationship between vegetation properties and optimal canopy cover in Horqin Sand, China, and reached the same conclusion. In the grassland area, the water demand curve is more sensitive to the variation of h_{s}/h compared to β. M^* decreases by 0.037 as h_{s}/h increases by 0.10. The water supply curve changes a lot with the change of LAI

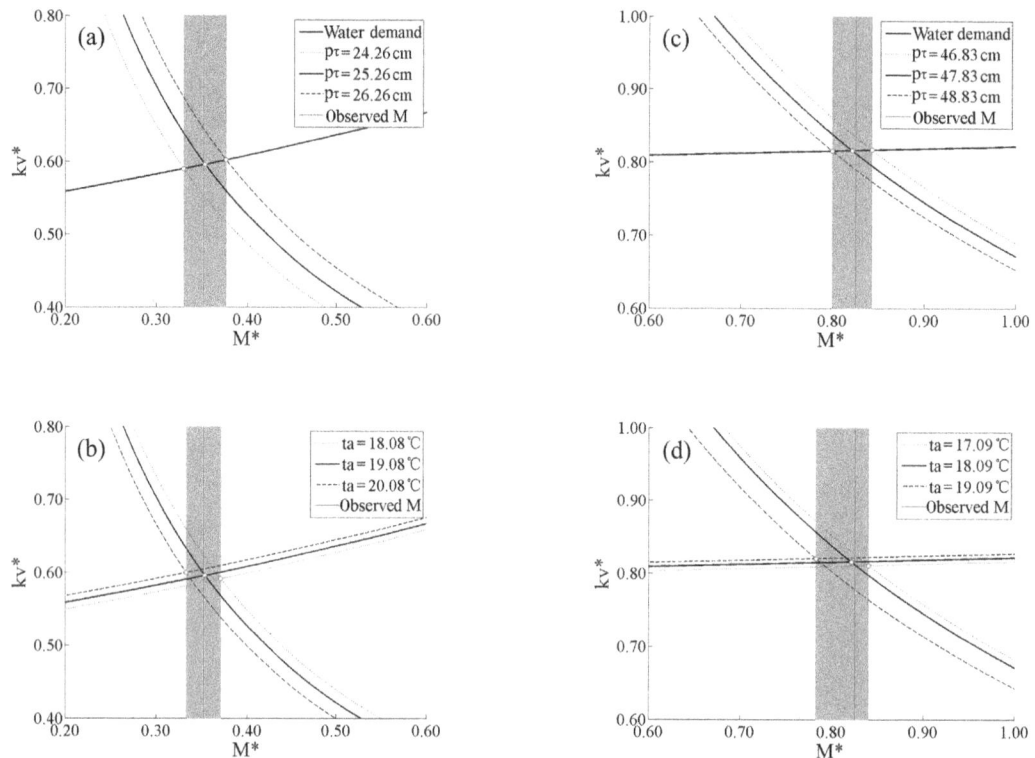

Figure 6. M^* changes with precipitation (P_τ) and air temperature (t_a): **(a, b)** grassland; **(c, d)** forest. The shaded areas indicate the range of M^* with the change of vegetation properties.

but slightly with β or h_s/h. In the forest area, M^* is less sensitive to h_s/h than β or LAI. Because the average stem fraction of trees (0.4–0.5) is usually larger than grasslands or shrub (0.0–0.1), the water demand curve is much gentler (Eagleson, 2002) and changes little with h_s/h. The forest interception takes up 14.24 % of precipitation during the growing season, which is larger than that of grassland. M^* increases by 0.108 and 0.094 with a decrease of 0.56 in β and 2.45 in LAI, respectively.

The sensitivity of M^* to vegetation properties can be used to offer advice about species choice and plant density to ecorestoration projects. If the purpose is to increase canopy cover, different strategies should be conducted in different areas. For the grassland area, shrubby or herbaceous plants with low h_s/h value are more welcome. Nevertheless, in the forest area, as h_s/h does not affect canopy cover that much, more considerations should be taken when choosing the species with relatively lower β and LAI values. However, vegetation with a larger canopy cover always requires more water to maintain functions (Woodward and Mckee, 1991; Zhang and Zhou, 2011). If the plant species are determined, the optimum canopy cover can be calculated and the upper limit for plant density can be estimated.

4.4 Sensitivity of M^* to climate factors

Studies of the relationship between climate factors and vegetation growth conditions reveal that precipitation and temperature are the two dominant factors that affect M^* (Ichii et al., 2002; Liu and Li, 2015). Under this framework, the variation of precipitation (P_τ) affects the availability of water, thus changing water supply curve; air temperature (t_a) affects not only water supply but also water demand, through changing resistance ratio and evaporation (Fig. 6). In the grassland area, M^* exhibits a positive relationship with precipitation but a negative relationship with air temperature (Fig. 6a and b), which is consistent with studies conducted before (He et al., 2015; Peng et al., 2012). This can be explained by the limited water supply in arid and semi-arid regions, and the increase of air temperature enhances transpiration and evaporation intensity (Duan et al., 2011; Mao et al., 2012). The variation of grassland M^* during 2000–2013 (Fig. 4a) also shows a similar trend of precipitation, while the trend of air temperature is different from that of precipitation in most years. In the forest area, M^* increases with the increase of precipitation and decrease of air temperature, but the variance of grassland M^* is less than that of forest with the same range of air temperature, which indicates the forest plants are more sensitive to air temperature than grassland. However, the result is different from previous studies. Most correlation anal-

Table 3. The variance of inputs and their corresponding M^*.

Inputs	Variation range	Grassland M^*	Forest M^*
β	0.01–0.57	0.346–0.358	0.755–0.863
LAI	0.10–2.55	0.313–0.357	0.770–0.864
h_s/h	0.00–0.10 (grassland); 0.35–0.45 (forest)	0.317–0.354	0.817–0.827
P_t	24.26–26.26 (grassland); 46.83–48.83 (forest)	0.330–0.377	0.800–0.844
t_a	18.08–20.08 (grassland); 17.09–19.09 (forest)	0.333–0.371	0.783–0.841

yses of NDVI with air temperature and precipitation show that, in the forest area, NDVI increases with the increase of air temperature and decrease of precipitation, because air temperature is the dominant factor in humid areas, and the light-use efficiency increases under elevated air temperatures (Peng et al., 2012; Wang et al., 2014; Liu et al., 2011). The difference may be caused by the deficient hypotheses of the theory. Under this framework, the surface runoff is assumed to be Hortonian, but in most humid areas, the runoff is saturation excess. The improper hypothetical runoff mechanism leads to the deviation of runoff in water-sufficient areas, thus causing the deviation of the water supply curve.

5 Conclusion

In this study, remote-sensed NDVI is used to generate the actual canopy cover of the NECT, while the ecohydrological optimality method has been applied to calculate the optimal canopy cover. The proportions of water balance components have been explored, as well as the influence of vegetation properties and climate factors on optimal canopy cover. The main conclusions are summarized as follows:

1. The observed canopy cover M shows a significant decreasing gradient from eastern forests to the west. The modeled canopy cover M^* has the same trend with M but transitions more smoothly, which is mainly caused by the interpolation of meteorological data. The relatively lower M^* in Yanbian Korean Autonomous Prefecture and Xilin Gol grassland is mainly because of human activity. The correlation coefficient R between M and M^* is 0.81, which indicates the ecohydrological optimality theory is applied well in the NECT during the long-term period. There is a 2-year lag between M^* and M during 2002–2012, due to the long-term adaptation strategy of vegetation to climate change.

2. The proportions of the water balance components are calculated for three parts: forest, cropland and grassland. The simulated results are within the observed range, which demonstrates the reasonability of this theory. By adapting this method, it is much easier to figure out the allocation of precipitation with fixed vegetation and soil conditions.

3. M^* has a positive relationship with β and a negative relationship with h_s/h and LAI. Grassland plants are more sensitive to h_s/h and LAI compared to β, while forest plants are more sensitive to β and LAI than h_s/h. The sensitivity of M^* to vegetation properties can be used to offer advice about species choice and plant density for ecorestoration projects.

4. Precipitation and temperature are the two dominant climate factors that affect M^*. M^* increases with the increase of precipitation and decrease of air temperature. Eagleson's ecohydrological optimality theory offers an opportunity to explore the quantitative relationship between vegetation and climate factors from the mechanism, but the runoff mechanism description in the wet region still needs improvement.

Appendix A: Algorithm of optimal canopy cover

Eagleson made several assumptions for each item of Eq. (4). The Poisson precipitation model was used to simulate the precipitation process by random storm depth and duration (Eagleson, 1978b). The probability density functions of storm depth and storm duration are incomplete gamma and exponential distributions, respectively. The growing season precipitation can be expressed as

$$P_\tau = m_v m_h, \tag{A1}$$

where m_h is the mean storm depth (mm).

Surface retention $m_v E_r$ is the water held on the surface during the rainstorm of duration. The total surface retention is proportioned by bare soil and vegetation canopy:

$$E[E_r] = (1 - M)E[E_{rs}] + ME[E_{rv}], \tag{A2}$$

respectively. $E[E_r]$ can be further expressed as

$$E[E_r] = (1 + M\eta_0\beta L_t)h_0 = \overline{h_0}, \tag{A3}$$

where η_0 is the ratio of stomatal leaf area to illuminated leaf area (dimensionless); β is the cosine of leaf angle (dimensionless); L_t is the foliage area index (dimensionless);

and h_0 is the surface retention depth (mm). The interception depth retained on the horizontal projection of leaves is assumed to be 1.00 cm (Eagleson, 1978d).

Average carryover soil moisture storage ΔS (mm) is determined by the soil profile and seasonality (Eagleson, 2002):

$$\Delta S = -\left[P_d - (1-M)E_{psd}m_d - Y_d\right], \tag{A4}$$

where P_d (mm), E_{psd} (mm day^{-1}), m_d (days) and Y_d (days) are the precipitation, evaporation, mean length and runoff in the non-growing season, respectively.

Assume that there is no surface inflow from outside of the region, and the surface runoff is Hortonian (Eagleson, 1978e). When the storm intensity m_i and storm duration m_{tr} are independent random variables, the storm surface runoff $m_v R_{sj}$ is

$$E\left(R_{sj}\right) = m_h e^{-G-2\sigma^{3/2}}, \tag{A5}$$

where

$$G \equiv \omega K(1)\left(\frac{1+s_0^c}{2}\right) \tag{A6}$$

$$\sigma \equiv \left[\frac{5n_e\lambda_0^2 K(1)\psi(1)(1-s_0)^2\phi_i(d,s_0)}{6\pi\delta m\kappa_0^2}\right], \tag{A7}$$

where s_0 is the space–time average soil moisture in the root zone (dimensionless); $\omega = 1/m_i$; $K(1)$ is the effective saturated hydraulic conductivity of soil (cm day^{-1}); n_e is effective soil porosity (dimensionless); λ_0 is scale parameter of probability density function of storm depth (cm^{-1}); $\psi(1)$ is the saturated matrix potential of soil (cm); ϕ_i is the sorption diffusivity (dimensionless); $\delta = 1/m_{tr}$ (day^{-1}); m is the soil pore size distribution index (dimensionless); and κ_0 is the shape parameter or distribution index of storm depth (dimensionless).

Evapotranspiration consists of bare soil evaporation and vegetal transpiration:

$$E[E_{T\tau}] = m_v m_{tb} E_{ps\tau}\left[(1-M)\beta_s + Mk_v^*\beta_v\right], \tag{A8}$$

where m_{tb} is the mean time between storms (days); $E_{ps\tau}$ is the potential free water surface potential evaporation during the growing season (mm day^{-1}); and β_s and β_v are the bare soil evaporation efficiency and canopy transpiration efficiency, respectively (Eagleson, 1978d).

The percolation rate is mainly affected by s_0 (Eagleson, 1978f):

$$v(s_0) = K(1)s_0^c. \tag{A9}$$

The capillary rise is considered to be 0 due to the deep water table in the NECT.

Using Eqs. (A1)–(A9), Eq. (4) gives the water balance of growing season as

$$1 - e^{-G-2\sigma^{3/2}} - \frac{\overline{h_0}}{m_h} + \frac{\Delta S}{m_v m_h}$$
$$= \frac{m_{tb}E_{ps}}{m_h}\left[(1-M)\beta_s + Mk_v^*\beta_v\right] + \frac{m_\tau K(1)}{P_\tau}s_0^c. \tag{A10}$$

β_v is equal to 1.0 when the water condition reaches optimal state. When the bare soil evaporation is ignored, Eq. (A10) can be simplified into

$$Mk_v^* = \frac{V_e}{m_{tb}E_{ps}}, \tag{A11}$$

where

$$V_e = m_h - m_h e^{-G-2\sigma^{3/2}} - \overline{h_0} + \frac{\Delta S}{m_v} - \frac{m_\tau K(1)}{P_\tau}s_0^c. \tag{A12}$$

Competing interests. The authors declare that they have no conflict of interest.

Acknowledgements. This work was supported by National Natural Science Foundation (grant nos. 51479088, 41630856).

Edited by: L. Wang

References

Allen, R. G.: Crop evapotranspiration: guidelines for computing crop water requirements, xxvi, Food and Agriculture Organization of the United Nations, Rome, 300 pp., 1998.

Asner, G. P.: Biophysical and Biochemical Sources of Variability in Canopy Reflectance, Remote Sens. Environ., 64, 234–253, 1998.

Austin, A. T. and Sala, O. E.: Carbon and nitrogen dynamics across a natural precipitation gradient in Patagonia, Argentina, J. Veg. Sci., 13, 351–360, 2002.

Bao, G., Bao, Y., Sanjjava, A., Qin, Z., Zhou, Y., and Xu, G.: NDVI-indicated long-term vegetation dynamics in Mongolia and their response to climate change at biome scale, Int. J. Climatol., 35, 4293–4306, 2015.

Baret, F., Clevers, J. G. P. W., and Steven, M. D.: The robustness of canopy gap fraction estimates from red and near-infrared reflectances: A comparison of approaches, Remote Sens. Environ., 54, 141–151, 1995.

Braswell, B. H., Schimel, D. S., Linder, E., and Moore, B.: The Response of Global Terrestrial Ecosystems to Interannual Temperature Variability, Science, 278, 870–873, 1997.

Cai, T., Sheng, H., and Cui, X.: Rainfall redistribution of a virgin Pinus koraiensis forest and secondary Betula platyphylla forest in Northeast China, Front. Forest. China, 3, 189–193, doi:10.1007/s11461-008-0022-y, 2008.

Canadell, J. G., Steffen, W. L., and White, P. S.: IGBP/GCTE terrestrial transects: dynamics of terrestrial ecosystems under environmental change– introduction, J. Veg. Sci., 13, 298–300, 2002.

Carlson, T. N. and Ripley, D. A.: On the relation between NDVI, fractional vegetation cover, and leaf area index?, Remote Sens. Environ., 62, 241–252, 1997.

Chen, J. M. and Black, T. A.: Defining leaf area index for non-flat leaves, Plant Cell Environ., 15, 421–429, 1992.

Chen, X.: Change of tree diversity on Northeast China Transect (NECT), Biodivers. Conserv., 10, 1087–1096, 2001.

Dorman, M., Svoray, T., and Perevolotsky, A.: Homogenization in forest performance across an environmental gradient – The interplay between rainfall and topographic aspect, Forest Ecol. Manage., 310, 256–266, 2013.

Du, J.: Approach to the Phenotypic Plasticity of Leymus chinensis Population Responding to Grazing[D]. Northeast Normal University, 2004. (in Chinese)

Duan, H., Yan, C., Tsunekawa, A., Song, X., Li, S., and Xie, J.: Assessing vegetation dynamics in the Three-North Shelter Forest region of China using AVHRR NDVI data (SCI), Environ. Earth Sci., 64, 1011–1020, 2011.

Eagleson, P. S.: Climate, soil, and vegetation: 1. Introduction to water balance dynamics, Water Resour. Res., 14, 705–712, 1978a.

Eagleson, P. S.: Climate, soil, and vegetation: 2. The distribution of annual precipitation derived from observed storm sequences, Water Resour. Res., 14, 713–721, 1978b.

Eagleson, P. S.: Climate, soil, and vegetation: 3. A simplified model of soil moisture movement in the liquid phase, Water Resour. Res., 14, 722–730, 1978c.

Eagleson, P. S.: Climate, soil, and vegetation: 4. The expected value of annual evapotranspiration, Water Resour. Res., 14, 731–739, 1978d.

Eagleson, P. S.: Climate, soil, and vegetation: 5. A derived distribution of storm surface runoff, Water Resour. Res., 14, 741–748, 1978e.

Eagleson, P. S.: Climate, soil, and vegetation: 6. Dynamics of the annual water balance, Water Resour. Res., 14, 749–764, 1978f.

Eagleson, P. S.: Climate, soil, and vegetation: 7. A derived distribution of annual water yield, Water Resour. Res., 14, 765–776, 1978g.

Eagleson, P. S.: Ecological optimality in water-limited natural soil–vegetation systems: 1. Theory and hypothesis, Water Resour. Res., 18, 325–340, 1982.

Eagleson, P. S.: Ecohydrology: Darwinian expression of vegetation form and function, Cambridge University Press, Cambridge, 2002.

Eagleson, P. S. and Tellers, T. E.: Ecological optimality in water-limited natural soil-vegetation systems: 2. Tests and applications, Water Resour. Res., 18, 341–354, 1982.

Fan, S., Gao, Y., Cheng, Y., and Bai, Q.: Research of plot testing for effects of the woods and grass vegetation on runoff, J. Shandong Agr. Univers., 37, 43–47, 2006.

Fontana, F., Rixen, C., Jonas, T., Aberegg, G. and Wunderle, S.: Alpine Grassland Phenology as Seen in AVHRR, VEGETATION, and MODIS NDVI Time Series – a Comparison with In Situ Measurements, Sensors, 8, 2833–2853, 2008.

Füssler, J. S. and Gassmann, F.: On the role of dynamic atmosphere–vegetation interactions under increasing radiative forcing, Global Ecol. Biogeogr., 9, 337–349, 2000.

Graetz, R. D.: The nature and significance of the feedback of changes in terrestrial vegetation on global atmospheric and climatic change, Climatic Change, 18, 147–173, 1991.

Guo, X. Y. and Zhang, H. Y.: The Vegetation Dynamic Research Under of Eco-geographical Region Framework on Greater Khingan Mountains, Scient. Geogr. Sin., 33, 181–188, 2013.

Gutman, G. and Ignatov, A.: The derivation of the green vegetation fraction from NOAA/AVHRR data for use in numerical weather prediction models, Int. J. Remote Sens., 19, 1533–1543, 1998.

Hatton, T. J., Salvucci, G. D., and Wu, H. I.: Eagleson's optimality theory of an ecohydrological equilibrium: quo vadis?, Funct. Ecol., 11, 665–674, 1997.

He, B., Chen, A., Wang, H., and Wang, Q.: Dynamic Response of Satellite-Derived Vegetation Growth to Climate Change in the Three North Shelter Forest Region in China, Remote Sensing, 7, 9998–10016, 2015.

Hmimina, G., Dufrêne, E., Pontailler, J. Y., Delpierre, N., Aubinet, M., Caquet, B., De Grandcourt, A., Burban, B., Flechard, C., and Granier, A.: Evaluation of the potential of MODIS satellite data to predict vegetation phenology in different biomes: An investigation using ground-based NDVI measurements, Remote Sens. Environ., 132, 145–158, 2013.

Hu, M., Mao, F., Sun, H., and Hou, Y.: Study of normalized difference vegetation index variation and its correlation with climate factors in the three-river-source region, Int. J. Appl. Earth Obs. Geoinf., 13, 24–33, 2011.

Huang, G.: Impact of climate change on water resources of international rivers in northeastern China, Acta Geogr. Sin., 66, 152–156, doi:10.11821/xb1999S1020, 1999.

Huete, A., Didan, K., Miura, T., Rodriguez, E. P., Gao, X., and Ferreira, L. G.: Overview of the radiometric and biophysical performance of the MODIS vegetation indices, Remote Sens. Environ., 83, 195–213, 2002.

Ichii, K.: Global correlation analysis for NDVI and climatic variables and NDVI trends: 1982–1990, Int. J. Remote Sens., 23, 3873–3878, 2002.

Ichii, K., Kawabata, A., and Yamaguchi, Y.: Global correlation analysis for NDVI and climatic variables and NDVI trends: 1982–1990, Int. J. Remote Sens., 23, 3873–3878, 2002.

Jiang, Z., Huete, A. R., Chen, J., Chen, Y., Li, J., Yan, G., and Zhang, X.: Analysis of NDVI and scaled difference vegetation index retrievals of vegetation fraction, Remote Sens. Environ., 101, 366–378, 2006.

Kerkhoff, A. J., Martens, S. N., and Milne, B. T.: An ecological evaluation of Eagleson's optimality hypotheses, Funct. Ecol., 18, 404–413, 2004.

Kim, Y. and Wang, G.: Modeling seasonal vegetation variation and its validation against Moderate Resolution Imaging Spectroradiometer (MODIS) observations over North America, J. Geophys. Res., 110, D04106, doi:10.1029/2004JD005436, 2005.

Koch, G. W., Scholes, R. J., Steffen, W. L., Vitousek, P. M. and Walker, B. H.: The IGBP terrestrial transects: science plan. International Geosphere–Biosphere Report 36, Programme, Stockholm, https://getsemaberri.files.wordpress.com/2017/04/20172718.pdf (last access: May 2017), 1995a.

Koch, G., Vitousek, P., Steffen, W., and Walker, B.: Terrestrial transects for global change research, Vegetatio, 121, 53–65, 1995b.

Li, B., Tao, S., and Dawson, R. W.: Relations between AVHRR NDVI and ecoclimatic parameters in China, Int. J. Remote Sens., 23, 989–999, 2002.

Li, S., Zhao, Z., Wang, Y., and Wang, Y.: Identifying spatial patterns of synchronization between NDVI and climatic determinants using joint recurrence plots, Environ. Earth Sci., 64, 851–859, 2011.

Li, X.: Study on demonstration and construction of NFPP effect evaluation system in Yanbian Forest Region, PhD thesis, Northeast Forestry University, Harbin, China, 2014.

Liu, C., Shang, J., Vachon, P., and Mcnairn, H.: Multiyear Crop Monitoring Using Polarimetric RADARSAT-2 Data, IEEE T. Geosci. Remote, 51, 2227–2240, 2013.

Liu, W., Cai, T., Ju, C., Fu, G., Yao, Y., and Cui, X.: Assessing vegetation dynamics and their relationships with climatic variability in Heilongjiang Province, northeast China, Environ. Earth Sci., 64, 2013–2024, 2011.

Liu, Y. and Li, S.: Spatial and Temporal Patterns of Global NDVI Trends: Correlations with Climate and Human Factors, Remote Sensing, 7, 13233–13250, 2015.

Liu, Z., Notaro, M., Kutzbach, J., and Liu, N.: Assessing Global Vegetation–Climate Feedbacks from Observations, J. Climate, 19, 787–814, doi:10.1175/JCLI3658.1, 2006.

Lotsch, A., Friedl, M., Anderson, B., and Tucker, C.: Coupled vegetation–precipitation variability observed from satellite and climate records, Geophys. Res. Lett., 30, 107–218, 2003.

Mao, D., Wang, Z., Luo, L., and Ren, C.: Integrating AVHRR and MODIS data to monitor NDVI changes and their relationships with climatic parameters in Northeast China, Int. J. Appl. Earth Obs. Geoinf., 18, 528–536, 2012.

Mcpherson, R. A.: A review of vegetation–atmosphere interactions and their influences on mesoscale phenomena, Prog. Phys. Geogr., 31, 261–285, 2007.

Miao, B. L., Liang, C. Z., Wang, W., Wang, L. X., and Yun, W. L.: Effects of vegetation on degradation surface runoff of typical steppe, Res. Soil Water Conserv., 22, 10–14, 2008.

Mo, K., Cong, Z., and Lei, H.: Optimal vegetation cover in the Horqin Sands, China, Ecohydrology, 9, 700–711, doi:10.1002/eco.1668, 2015.

Myoung, B., Choi, Y. S., and Park, S. K.: A review on vegetation models and applicability to climate simulations at regional scale, Asia-Pac. J. Atmos. Sci., 47, 463–475, 2011.

Ni, J. and Zhang, X. S.: Climate variability, ecological gradient and the Northeast China Transect (NECT), J. Arid Environ., 46, 313–325, 2000.

Nie, Q., Xu, J., Ji, M., Cao, L., Yang, Y., and Hong, Y.: The vegetation coverage dynamic coupling with climatic factors in Northeast China Transect, Environ. Manage., 50, 405–417, doi:10.1007/s00267-012-9885-7, 2012.

Park, H. S. and Sohn, B. J.: Recent trends in changes of vegetation over East Asia coupled with temperature and rainfall variations, J. Geophys. Res.-Atmos., 115, 1307–1314, 2010.

Peng, H., Li, X., and Tong, S.: Effects of shrub (Caragana microphalla Lam.) encroachment on water redistribution and utilization in the typical steppe of Inner Mongolia, Acta Ecol. Sin., 34, 2256–2265, 2014.

Peng, J., Dong, W., Yuan, W., and Zhang, Y.: Responses of Grassland and Forest to Temperature and Precipitation Changes in Northeast China, Adv. Atmos. Sci., 29, 1063–1077, 2012.

Piao, S., Fang, J., Zhou, L., Guo, Q., Henderson, M., Ji, W., Li, Y., and Tao, S.: Interannual variations of monthly and seasonal normalized difference vegetation index (NDVI) in China from 1982 to 1999, J. Geophys. Res.-Atmos., 108, 4401, doi:10.1029/2002JD002848, 2003.

Piao, S., Mohammat, A., Fang, J., Cai, Q., and Feng, J.: NDVI-based increase in growth of temperate grasslands and its responses to climate changes in China, Global Environ. Change, 16, 340–348, 2006.

Piao, S., Wang, X., Ciais, P., Zhu, B., Wang, T., and Liu, J.: Changes in satellite-derived vegetation growth trend in temperate and boreal Eurasia from 1982 to 2006, Global Change Biol., 17, 3228–3239, 2011.

Rauner, J. L.: Vegetation and the Atmosphere, in: vol. 2, Case Studies, edited by: Monteith, J. L., Academic Press, New York, 241–264, 1976.

Schultz, P. A. and Halpert, M. S.: Global correlation of temperature, NDVI and precipitation, Adv. Space Res., 13, 277–280, 1993.

Shao, W., Yang, D., Sun, F., and Wang, J.: Analyzing the Regional Soil–Vegetation–Atmosphere Interaction Using Both the Eagleson and Budyko's Water Balance Models, Proced. Environ. Sci., 10, 1908–1913, 2011.

Song, X.: Precipitation, streamflow and sediment transport changes and its response to human activities in Songhua basin, Graduate University of Chinese Academy of Sciences, Beijing, China, 2010.

Tong, C., Wu, J., Yong, S., Yang, J., and Yong, W.: A landscapescale assessment of steppe degradation in the Xilin River basin, Inner Mongolia, China, J. Arid Environ., 59, 133–149, 2004.

Wang, A., Pei, T., Jin, C., and Guan, D.: Estimation of rainfall interception by broad-leaved Korean pine forest in Changbai Mountains, Chinese J. Appl. Ecol., 17, 1403–1407, 2006.

Wang, J., Rich, P. M., and Price, K. P.: Temporal responses of NDVI to precipitation and temperature in the central Great Plains, USA, Int. J. Remote Sens., 24, 2345–2364, 2003.

Wang, Q., Zhang, B., Zhang, Z., Zhang, X., and Dai, S.: The Three-North Shelterbelt Program and Dynamic Changes in Vegetation Cover, J. Resour. Ecol., 5, 53–59, 2014.

Wang, Y. L., Yun, W. L., Miao, B. L., Liang, C. Z., and Wang, W.: The Pattern and Dynamics of Surface Runoff in the Typical Steppe of Inner Mongolia, Res. Soil Water Conserv., 15, 114–115, 2008a.

Wang, Y. F., Mo, X. G., Hao, Y. B., Guo, R. P., and Huang, X. Z.: Simulating seasonal and interannual variations of ecosystem evapotranspiration and its components in Inner Mongolia steppe with VIP model, J. Plant Ecol., 32, 1052–1060, 2008b.

Wei, Y., Yu, D., Lewis, B. J., Zhou, L., Zhou, W., Fang, X., Zhao, W., Wu, S., and Dai, L.: Forest Carbon Storage and Tree Carbon Pool Dynamics under Natural Forest Protection Program in Northeastern China, Chinese Geogr. Sci., 24, 397–405, 2014.

Woodward, F. I. and Mckee, I. F.: Vegetation and climate, Environ. Int., 17, 535–546, 1991.

Yang, H., Pei, T., Guan, D., Jin, C., and Wang, A.: Soil moisture dynamics under broad-leaved Korean pine forest in Changbai Mountains, Chinese J. Appl. Ecol., 17, 587–591, 2006.

Yu, X. X., Gu, J. J., and Yue, Y. J.: Benefit evaluation on forestry ecological projects, Science Press, Beijing, China, 2010.

Yuan, X., Li, L., Chen, X., and Shi, H.: Effects of Precipitation Intensity and Temperature on NDVI-Based Grass Change over Northern China during the Period from 1982 to 2011, Remote Sensing, 7, 10164–10183, 2015.

Zhang, Y. and Zhou, G.: Exploring the effects of water on vegetation change and net primary productivity along the IGBP Northeast China Transect, Environ. Earth Sci., 62, 1481–1490, 2011.

Zhou, G. S. and Wang, Y. H.: Global Ecology, China Meteorological Press, Beijing, China, 2003.

Zhou, L., Tucker, C., Kaufmann, R., Slayback, D., Shabanov, N., and Myneni, R.: Variations in northern vegetation activity inferred from satellite data of vegetation index during 1981 to 1999, J. Geophys. Res.-Atmos., 106, 20069–20083, 2001.

How does initial soil moisture influence the hydrological response? A case study from southern France

Magdalena Uber[1,2,a], **Jean-Pierre Vandervaere**[1], **Isabella Zin**[1], **Isabelle Braud**[3], **Maik Heistermann**[2], **Cédric Legoût**[1], **Gilles Molinié**[1], and **Guillaume Nord**[1]

[1]Univ. Grenoble Alpes, CNRS, IRD, Grenoble-INP, IGE Grenoble, 38000, France

[2]Institute of Earth and Environmental Science, University of Potsdam, Potsdam, 14476, Germany

[3]Irstea, UR RiverLy, Lyon-Villeurbanne Centre, Villeurbanne, 69625, France

[a]now at: Univ. Grenoble Alpes, CNRS, IRD, Grenoble-INP, IGE Grenoble, 38000, France

Correspondence: Magdalena Uber (magdalena.uber@univ-grenoble-alpes.fr)

Abstract. The Cévennes–Vivarais region in southern France is prone to heavy rainfall that can lead to flash floods which are one of the most hazardous natural risks in Europe. The results of numerous studies show that besides rainfall and physical catchment characteristics the catchment's initial soil moisture also impacts the hydrological response to rain events. The aim of this paper is to analyze the relationship between catchment mean initial soil moisture $\widetilde{\theta}_{ini}$ and the hydrological response that is quantified using the event-based runoff coefficient ϕ_{ev} in the two nested catchments of the Gazel ($3.4\,km^2$) and the Claduègne ($43\,km^2$). Thus, the objectives are twofold: (1) obtaining meaningful estimates of soil moisture at catchment scale from a dense network of in situ measurements and (2) using this estimate of $\widetilde{\theta}_{ini}$ to analyze its relation with ϕ_{ev} calculated for many runoff events. A sampling setup including 45 permanently installed frequency domain reflectancy probes that continuously measure soil moisture at three depths is applied. Additionally, on-alert surface measurements at ≈ 10 locations in each one of 11 plots are conducted. Thus, catchment mean soil moisture can be confidently assessed with a standard error of the mean of $\leq 1.7\,vol\,\%$ over a wide range of soil moisture conditions.

The ϕ_{ev} is calculated from high-resolution discharge and precipitation data for several rain events with a cumulative precipitation P_{cum} ranging from less than $5\,mm$ to more than $80\,mm$. Because of the high uncertainty of ϕ_{ev} associated with the hydrograph separation method, ϕ_{ev} is calculated with several methods, including graphical methods, digital filters and a tracer-based method. The results indicate that the

hydrological response depends on $\widetilde{\theta}_{ini}$: during dry conditions ϕ_{ev} is consistently below 0.1, even for events with high and intense precipitation. Above a threshold of $\widetilde{\theta}_{ini} = 34\,vol\,\%$ ϕ_{ev} can reach values up to 0.99 but there is a high scatter. Some variability can be explained with a weak correlation of ϕ_{ev} with P_{cum} and rain intensity, but a considerable part of the variability remains unexplained.

It is concluded that threshold-based methods can be helpful to prevent overestimation of the hydrological response during dry catchment conditions. The impact of soil moisture on the hydrological response during wet catchment conditions, however, is still insufficiently understood and cannot be generalized based on the present results.

1 Introduction

The Cévennes–Vivarais region in southern France is prone to intense rainfall that can lead to the occurrence of flash floods in catchments of various scales ranging from small headwater catchments to ones of several thousand kilometers squared (Boudevillain et al., 2011; Braud et al., 2014). Flash floods are one of the most destructive natural hazards in Europe, both in terms of number of fatalities and economic damage (Gaume et al., 2009). Striking examples are the October 2015 flash flood of the Brague river that hit the French Riviera and the 2002 flash flood of the Gard river with 23 deaths and an estimated direct tangible damage of EUR 1.2 billion (Huet et al., 2003).

Despite the recognition of their high damage potential, the hydrological processes leading to the generation of flash floods are still insufficiently understood at a scale that is important for prediction and management (Gaume et al., 2009; Braud et al., 2014). One of the main problems that hinders flash flood prediction is the ignorance of the water retention capacity of the soil (Creutin and Borga, 2003). Other issues concern the lack of high-resolution data measured during flash flood events as well as the variety of catchment characteristics that influence their occurrence. The high degree of nonlinearity in the hydrological response of catchments hinders the predictability of flash floods (Braud et al., 2014). This has motivated the installation of several measurement networks in first-order catchments, especially in the USA and Australia (Slaughter et al., 2001; Renard et al., 2008; Moran et al., 2008; Baffaut et al., 2013) and – at the mesoscale and in a Mediterranean context – the FloodScale project in the Cévennes–Vivarais region (Braud et al., 2014; Nord et al., 2017).

Flash floods are usually associated with intense rainfall of > 100 mm in a few hours or long-lasting rainfall (≈ 24 h) with moderate intensities (Braud et al., 2014) often generated by mesoscale convective systems and/or orographic precipitation (Marchi et al., 2010; Molinié et al., 2012; Panziera et al., 2015). However, the hydrological response to rain events varies greatly between catchments and between events. It can be quantified using the event-based runoff coefficient ϕ_{ev}, i.e., the ratio of event runoff volume to total event rainfall volume. The major drawback of this quantity is the lack of standard procedures for obtaining event runoff volumes and for defining the beginning and end of an event, which impedes comparisons between studies (Blume et al., 2007). Yet, event-based runoff coefficients of flash-flood events have been found to differ substantially, spanning nearly the full range of values from zero to one, with a high positive skewness in their frequency distribution (Merz et al., 2006; Blume et al., 2007; Norbiato et al., 2008; Merz and Blöschl, 2009; Marchi et al., 2010). They were shown to differ considerably between regions (Marchi et al., 2010), seasons (Li et al., 2012) and flood types (Merz et al., 2006); to increase with mean annual precipitation; and event rainfall depth (Merz et al., 2006; Norbiato et al., 2009) and to depend on rain intensity, soil types and antecedent soil moisture conditions (Wood et al., 1990; Crow et al., 2005; Marchi et al., 2010; Hrachowitz et al., 2011; Penna et al., 2011; Li et al., 2012; Huza et al., 2014). Furthermore, a multitude of catchment characteristics also determine runoff generation and concentration, namely topography, geology, hydraulic routing and geomorphological controls (Braud et al., 2014).

Soil moisture is known to govern overland flow generation (Zehe and Sivapalan, 2009). As it controls threshold behavior, it implies qualitative changes of hydrological processes and the hydrologic system's response to rain events (Zehe and Sivapalan, 2009; Hardie et al., 2011). Initial soil moisture θ_{ini}, i.e., the soil water content at the onset of a rain event, is

a crucial factor that influences the water storage capacity of the catchment as well as soil hydraulic properties and thus the hydrological response to rainfall events. It controls the soil moisture deficit and, consequently, in the interplay with rainfall forcing, it determines whether soil saturation and saturation excess overland flow (Dunne and Black, 1970) occur during a rain event or not. Moreover, soil moisture controls the unsaturated hydraulic conductivity and thus the occurrence of infiltration excess overland flow (Horton, 1933). Given the high spatial and temporal variability of soil moisture and the incoherence of scale of the measurements with the catchment size, it remains challenging to obtain meaningful estimates of θ_{ini} at the catchment scale (Brocca et al., 2009a; Vereecken et al., 2014; Korres et al., 2015). There are multiple controls on soil moisture such as soil texture, topography and vegetation as well as small-scale random variability (Jawson and Niemann, 2007; Garcia-Estringana et al., 2013; McMillan and Srinivasan, 2015). This problem is addressed by Vachaud et al. (1985) by introducing the concept of temporal stability, based on the finding that deviations of point measurements from the catchment mean can be persistent in time. Thus, optimum sampling locations can be identified and the number of samples required can be reduced (e.g., Brocca et al., 2009a; Huza et al., 2014).

Several studies consider the impact of initial soil moisture on the hydrological response of catchments on heavy rain events.

Seasonality in the occurrence of flash floods (Gaume et al., 2009) and discharge magnitudes (Borga et al., 2007; Li et al., 2012) have been attributed to initial soil moisture conditions. Numerous modeling studies have shown the high sensitivity of the modeled runoff response to θ_{ini} and the importance of estimates of θ_{ini} at the catchment scale as initial conditions in event-based models (e.g., Castillo et al., 2003; Huang et al., 2007; Le Lay and Saulnier, 2007; Berthet et al., 2009; Brocca et al., 2009b; Tramblay et al., 2010, 2012; Li et al., 2012; Massari et al., 2014a, b, 2015; Grillakis et al., 2016). The dependence of catchment responses to initial soil moisture is also observed by Marchi et al. (2010) in a dataset comprising data for 25 flood events in 60 basins across Europe and on this study's site by Huza et al. (2014). This relationship is characterized by high nonlinearity and threshold effects (Zehe et al., 2005; Huza et al., 2014). There is no consent on the importance of initial soil moisture during extreme events. Wood et al. (1990) conclude that catchment characteristics are important only for flood events with a low return period (up to ca. 10 years), whereas rainfall characteristics dominate those with a higher return period. On the other hand this finding is rejected in analyses of historic flash floods (Gaume et al., 2004; Borga et al., 2007; Le Lay and Saulnier, 2007) or flash flood databases (Marchi et al., 2010) whose authors conclude that soil moisture plays an important role in the hydrological response, also under extreme conditions.

The aim of this study is to assess how soil moisture controls the hydrological response in a flash-flood-prone area in southern France. The study is conducted in the two nested catchments of the Claduègne (43 km^2) and Gazel (3.4 km^2), Ardèche, France. Thanks to an exceptionally good database, it is possible to obtain reliable estimates of the two catchments' initial soil moisture states for several rain events and to quantify the hydrological response with the event-based runoff coefficient. To this end, the spatiotemporal heterogeneity is assessed to obtain reliable estimates for mean initial soil moisture at the catchment scale as well as its uncertainty. Other studies results suggest a dependence of ϕ_{ev} on initial soil moisture (e.g., Merz et al., 2006; Blume et al., 2007; Merz and Blöschl, 2009; Norbiato et al., 2009). However, most of these studies use indirect information such as remote sensing data, antecedent precipitation indices, initial baseflow, continuous soil moisture accounting models or the ratio of actual evaporation to precipitation. These approaches offer many advantages, such as the global availability of remote sensing data and the easier acquisition of these data (e.g., Brocca et al., 2009c). Numerous studies found good agreement of soil moisture data obtained from in situ measurements and remote sensing (e.g., Brocca et al., 2009c, 2013; Huza et al., 2014). Nonetheless, case studies are important to confirm the results obtained with indirect data as well as the results from modeling exercises and to thoroughly understand the hydrologic functioning of local sites. At this study's site the impact of θ_{ini} on ϕ_{ev} was already considered by Huza et al. (2014). However, these authors used soil moisture data obtained from ASCAT satellite data which is fitted to in situ measurements of topsoil moisture that were conducted on grasslands only. They considered five rain events only, so this relation could not be quantified unambiguously. Thus, this study's novelty is to analyze the relation between ϕ_{ev} and θ_{ini} when both are obtained from a comprehensive, high-resolution data set allowing the assessment of the uncertainty of the two variables. Relying solely on in situ data, we aim to (i) obtain a meaningful estimate of catchment-scale soil moisture and its uncertainty and (ii) answer the research question how does soil moisture at the event onset affect the hydrological response?

2 Methods

2.1 Study site

For this study two nested subcatchments of the Ardèche river in the Cévennes–Vivarais region of southern France are considered: the catchments of the intermittent Gazel stream and the perennial Claduègne river, with areas of 3.4 and 43 km^2 respectively (Fig. 1).

Both catchments can be clearly divided into two distinct geologies: the northern part is constituted by the Coiron basaltic plateau that is bounded by a steep cliff of basaltic columns in the south, whereas the southern part of both catchments is a landscape of piedmont hills underlain by sed-

imentary limestone lithology (Nord et al., 2017). The basaltic plateau covers 51 % of the Claduègne catchment, whereas its fraction of the Gazel catchment is only 23 %. Thus, the northern part is dominated by silty and stony soils on pebble deposit of basaltic component, while the soils in the southern part are predominantly rendzinas or other clay–stony soils, cultivated soils of loam and clay-loam and in the south of the Claduègne catchment lithosols and regosols (Nord et al., 2017). The terrain is hilly, and has a mean slope of about 20 %. The area is characterized by extensive agriculture and natural vegetation. Hence, the main land use and land cover types are grasslands, pastures, vineyards, forests and Mediterranean open woodlands. The vineyards are predominantly found on the finer textured soils in the southern part of the Claduègne catchment while the other land use types are found throughout the catchments. The average annual precipitation at Le Pradel at the outlet of the Gazel catchment is 1030 mm (Huza et al., 2014; original data: daily rain gauge data for 1958–2000 from Météo-France). For further details see Nord et al. (2017).

2.2 Data availability

As part of the HyMex (Hydrological Cycle in the Mediterranean Experiment, Ducrocq et al., 2014) and FloodScale (Braud et al., 2014) projects and the Cévennes–Vivarais Mediterranean Hydrometeorological Observatory (OHM-CV; Boudevillain et al., 2011), the area is exceptionally well monitored; thus, high-resolution spatiotemporal data on rainfall, discharge and soil moisture is available. The data used for this study were published in Nord et al. (2017) and the link to download the data can be found at the publishers website: https://www.earth-syst-sci-data.net/9/221/2017/essd-9-221-2017-assets.html (last access: 19 November 2018).

2.2.1 Soil moisture θ

Two different sets of soil moisture data are available: continuous measurements and on-alert measurements. Soil moisture is continuously measured with 45 fixed soil moisture probes at nine plots (two vineyards, one fallow, six grasslands) within the Claduègne catchment since June 2013. Six of the plots are located in the piedmont hills and three on the basaltic plateau (Fig. 1). Concerning topography, most of the sensors are located on hillslopes which is the dominant topographic zone according to Savenije (2010) in the catchment. Only two plots are located in the riparian area and are potentially connected to the stream during rain events. At each plot, five frequency domain reflectometry (FDR) probes (Decagon 10HS soil moisture sensors) are installed at different depths: 10 cm ($n = 2$), 20–25 cm ($n = 2$) and in the subsoil (33–50 cm, $n = 1$). The temporal resolution is 15–20 min (Nord et al., 2017). The sensors in the vineyards were installed between two vine plants in a row, which is a compro-

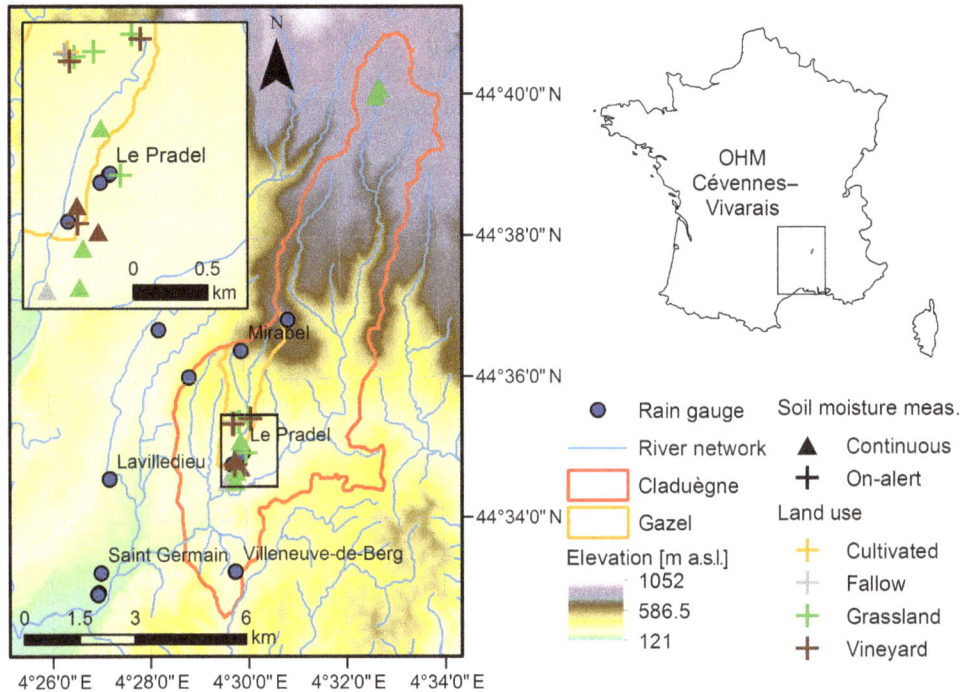

Figure 1. Location of the study site and measurement network. At every on-alert site measurements were taken at about 10 randomly chosen locations. At every continuous measurement sites two sensors were installed at 10 cm depth, two sensors at 20–25 cm depth and one sensor at a depth of 33–50 cm.

mise between feasibility and representativeness of soil moisture in the vineyards, which is heterogeneous due to transpiration. The accuracy and the range of the probe as provided by the manufacturer are ± 3 vol % and 0–57 vol %. The data are available from June 2013–November 2014 in the dataset (OHMCV, 2013) presented in Nord et al. (2017).

In addition, following forecasts of heavy rain events, on-alert measurements of soil moisture in 0–5 cm depth were conducted at 11 plots within the Gazel catchment with a hand-held FDR soil moisture sensor (Delta-T SM200). The accuracy and the range of the probe are ± 3 vol % and 0–50 vol %. The plots comprised four vineyards, five grasslands, one fallow and one cultivated field. All of these sites are located on hillslopes. The sampling sites were selected for reasons of accessibility, congruence with other measurements conducted during the FloodScale project and representativeness of the catchments' landscapes. All on-alert measurements were conducted in about 1 h at ≈ 10 randomly chosen measurement points within each plot. The distance between the measurement points was at least 1 m to ensure spatial independence (Huza et al., 2014). In the vineyards the measurements were conducted in between the rows of vine plants because this is where surface runoff started (visual inspection). On-alert measurements were taken before and after 11 heavy rain events during the special observation periods of the HyMex Project in autumn (September–December)

of the period 2012–2015. The dataset is found in the Supplement of this article (Supplement S1).

2.2.2 Precipitation P

Rainfall data were obtained from the HPiconet rain gauge network at a resolution of 1 min. The network consists of 19 tipping bucket rainfall gauges with a sampling surface of $1000\,\mathrm{cm}^2$ and a resolution of 0.2 mm, out of which 12 are located in the Claduègne catchment or its close vicinity (OHMCV, 2010; Nord et al., 2017, Fig. 1).

2.2.3 Discharge Q

Water level is continuously measured at the outlets of the two catchments with water level gauges at 2 min resolution (Gazel) and 10 min resolution (Claduègne) respectively (OHMCV, 2011; Nord et al., 2017). The water level is converted to discharge with a stage–discharge relationship established using the BaRatin framework (Le Coz et al., 2014) that also gives the uncertainty of the rating curve that is quantified as the 90 % confidence interval of discharge. The rating curve is based on numerous discharge measurements performed in 2012–2014 (Nord et al., 2017).

2.2.4 Additional data

Spatial data used for this study include a digital elevation model with a resolution of 5 m (Nord, 2015) and the Ardèche soil database by the French National Institute for Agricultural Research, Bureau de Recherches Géologiques et Minières and the French Department of Agriculture (Braud 2015; Nord et al., 2017). Furthermore, a 0.5 m resolution land use map of the Claduègne catchment based on QuickBird satellite images is available (Andrieu, 2015). Data on soil properties such as porosity, texture and saturated hydraulic conductivity were obtained during a measurement campaign in 2012 by Braud and Vandervaere (2015). Electrical conductivity (EC) of streamflow is continuously measured at the outlets of both catchments (OHMCV, 2011), and measurements of EC of overland flow from two runoff and erosion plots in a vineyard in the south of the Gazel catchment are available (OHMCV, 2009; Cea et al., 2015).

2.3 Precipitation data processing

The catchment mean hyetographs for both catchments are calculated from the HPiconet rain gauge data with the method of Thiessen polygons. Rain events are separated by using a threshold of 12 h without precipitation being recorded at any rain gauge. The onset of an event was defined as the first time rain occurred after a dry period of at least 12 h, the end as the last time with rain being recorded by at least one rain gauge before the next dry period. The threshold of 12 h provides a good compromise between having a high number of events and excluding two separate events that are not independent from each other. Averaged catchment rainfall is then summed over the whole period of the rain event to calculate cumulative event precipitation P_{cum}. Furthermore, mean rain intensity I_μ over the whole event as well as maximum rain intensity I_{max} at 2, 10, 20, 30 and 60 min are calculated using the averaged catchment rainfall.

2.4 Soil moisture analysis

From both data sets (continuous and on-alert measurements) plot and catchment mean values are calculated for the initial and final state of each rain event. From the continuous data, mean values are calculated for all three depths and the profile mean value is calculated.

$\theta(x_{i,j}, t_{ev})$ refers to a spatially and temporally discrete on-alert soil moisture measurement, with the subscript i denoting the index of the n_i (usually 10) measurements within the plot, j denoting the plot and ev the event and the state (initial or final).

Mean soil moisture was calculated at the plot scale ($\overline{\theta}_j(t_{ev})$), for each land use class ($\overline{\overline{\theta}}_{lu}(t_{ev})$) and at the catchment scale ($\overline{\overline{\theta}}_{ev}$). See Table 1 for the formulas. Plot means and catchment means obtained from the continuously measured data are computed for all three layers l. Here, the plot mean is obtained by averaging not only the probes installed at the same depth and the same location, but also all measurements in a dry period of 2 h before the onset or after the end of the rain event in order to diminish noise. The catchment

mean averaged over the three layers $\widetilde{\theta}_{ev}$, i.e., considering the topmost 60 cm, is calculated from the continuously measured data (Eq. 4, Table 1). Finally, for all events the soil moisture storage change ΔS (mm event^{-1}) in the upper 60 cm is computed from the difference between initial and final soil moisture (Eq. 5, Table 1). It is assessed whether significant differences between the four land use classes exist by performing a visual inspection of box plots or histograms and Student t tests. Moreover, standard deviations σ as measures of spatial variability are calculated at the plot scale (σ_j^{inner}, Eq. 6, Table 1) and at the catchment scale ($\sigma_{cat.}^{inter}$, Eq. 7, Table 1). Furthermore, σ is calculated between plots of the same land use ($\sigma_{lu}^{inter}(t_{ev})$, Eq. 8, Table 1) and between land use classes ($\sigma^{betw}(t_{ev})$, Eq. 9, Table 1). As an estimate of the uncertainty of the calculated plot and catchment mean values, the standard error of the plot mean SEM$_j^{inner}$ and the one of the catchment mean SEM$_{cat.}^{inter}$ are calculated. It should be noted that SEM$_{cat.}^{inter}$ is calculated from the on-alert measurements in the topsoil as well as from the continuous measurements over the soil profile, SEM$_j^{inner}$ only from the on-alert measurements. The SEM is used as a measure of the confidence that the sample mean corresponds to the universal mean; it increases with the standard deviation and decreases with the number of sampling points.

Moreover, it is assessed whether temporal stability, i.e., consistency of soil moisture patterns at the catchment scale at different times of measuring (Vachaud et al., 1985), as reported by Huza et al. (2014) for six grassland plots in the Gazel catchment, is also found in the present on-alert data set: the relative spatial difference $\delta_{j,ev}$ of each plot corresponds to the relative difference between the plot mean and the catchment mean (Eq. 1); its temporal mean $\overline{\delta}_j$ is calculated with Eq. (2):

$$\delta_{j,ev} = \frac{\overline{\theta}_j(t_{ev}) - \overline{\overline{\theta}}_{ev}}{\overline{\overline{\theta}}_{ev}}, \tag{1}$$

$$\overline{\delta}_j = \frac{1}{n_{ev}} \sum_{ev=1}^{n_{ev}} \delta_{j,ev}. \tag{2}$$

The plot with the smallest $\delta_{j,ev}$ is the one that agrees best with the catchment mean at a given time of measurement. The temporal variability of the spatial difference σ_{δ_j} serves as an auxiliary variable to assess whether this behavior is stable in time:

$$\sigma_{\delta_j} = \sqrt{\frac{1}{n_{ev}-1} \sum_{ev=1}^{n_{ev}} \left(\delta_{j,ev} - \overline{\delta}_j\right)^2}. \tag{3}$$

2.5 Hydrological response

2.5.1 Event-based runoff coefficients

In order to quantify the hydrological response of the catchment to different rainfall events, the dimensionless event-

How does initial soil moisture influence the hydrological response? A case study from southern France

131

Table 1. Calculated measures of averaged soil moisture and its variability at different scales.

Name	Formula	Eq. no.	Purpose and abbreviations
Plot mean soil moisture	$\bar{\theta}_j(t_{ev}) = \frac{1}{n_i} \sum_{i=1}^{n_i} \theta(x_{i,j}, t_{ev})$	(1)	Best estimate at plot scale; n_i: number of measurements in plot j.
Land use mean soil moisture	$\bar{\bar{\theta}}_{lu}(t_{ev}) = \frac{1}{n_{j_{lu}}} \sum_{j_{lu}=1}^{n_{j_{lu}}} \bar{\theta}_j(t_{ev})$	(2)	Best estimate for land use classes; $n_{j_{lu}}$: number of measurements in plots of land use lu; g: grassland, v: vineyard, c: cultivated field, f: fallow.
Catchment mean soil moisture	$\bar{\bar{\theta}}_{ev} = \frac{1}{n_p} \sum_{lu}^{n_{c_{lu}}} \bar{\bar{\theta}}_{lu} \cdot n_{j_{lu}}$	(3)	Best estimate at catchment scale; n_p: number of plots ($n_p = 11$); $n_{c_{lu}}$: number of land use classes ($n_{c_{lu}} = 4$).
Profile mean soil moisture	$\tilde{\theta}_{ev} = \dfrac{\sum_{l=1}^{n_l} \bar{\bar{\theta}}_{ev,l} \cdot m_l}{\sum_{l=1}^{n_l} m_l}$	(4)	Best estimate at catchment scale, integrated over the soil profile; n_l: number of layers ($n_l = 3$); m_l: thickness of layer l^{a}.
Soil moisture storage change	$\Delta S = \sum_{l=1}^{n_l} \frac{1}{100} \left(\bar{\bar{\theta}}_{fin,l} - \bar{\bar{\theta}}_{ini,l} \right) \cdot m_l$	(5)	Soil water retention during events; $\bar{\bar{\theta}}_{fin,l}$; $\bar{\bar{\theta}}_{ini,l}$: final and initial soil moisture in layer l.
Inner plot SD	$\sigma_j^{inner}(t_{ev}) = \sqrt{\frac{1}{n_i-1} \sum_{i=1}^{n_i} \left(\theta(x_{i,j}, t_{ev}) - \bar{\theta}_j(t_{ev}) \right)^2}$	(6)	Estimate of spatial variability at the plot scale.
Interplot SD (catchment)	$\sigma_{cat.}^{inter}(t_{ev}) = \sqrt{\frac{1}{n_p-1} \sum_{j=1}^{n_p} \left(\bar{\theta}_j(t_{ev}) - \bar{\bar{\theta}}(t_{ev}) \right)^2}$	(7)	Estimate of spatial variability at the catchment scale.
Interplot SD (g/v)[b]	$\sigma_{lu}^{inter}(t_{ev}) = \sqrt{\frac{1}{n_{j_{lu}}-1} \sum_{j_{lu}=1}^{n_{j_{lu}}} \left(\bar{\theta}_j(t_{ev}) - \bar{\bar{\theta}}_{lu}(t_{ev}) \right)^2}$	(8)	Estimate of interplot variability in the grasslands and vineyards.
Between land use SD	$\sigma^{betw}(t_{ev}) = \sqrt{\frac{1}{n_{c_{lu}}-1} \sum_{lu}^{n_{c_{lu}}} \left(\bar{\bar{\theta}}_{lu}(t_{ev}) - \bar{\bar{\theta}}_{ev} \right)^2}$	(9)	Estimate variability between the grasslands and vineyards.

[a] Thicknesses of layers 1–3 are assumed to be 175, 150 and 275 mm. [b] Only calculated for grasslands and vineyards, because the number of plots is 1 for land use classes fallow and cultivated field.

based runoff coefficient ϕ_{ev} is calculated for all events:

$$\phi_{ev} = \frac{Q_{ev,cum}}{P_{cum}}, \tag{4}$$

To obtain cumulative event discharge $Q_{ev,cum}$, the time series of stream discharge Q_{tot} has to be separated into baseflow Q_b and event flow Q_{ev}. Q_{ev} is defined here to be the fast-responding part of discharge that occurs during or directly after the rain event. It usually encompasses surface runoff or overland flow and fast subsurface flow. Q_b on the other hand is the slow responding part of discharge that lasts long after the rain event and feeds the stream between rain events. To obtain $Q_{ev,cum}$, Q_{ev} is summed up over the whole period of the event. The onset of a discharge event is defined as the first increase of discharge in response to a rain event. Defin-

ing the end of the eventdischarge is more complicated and depends on the hydrograph separation method (Blume et al., 2007). Usually, the end of the event flow is defined as the moment when Q_b equals Q_{tot}, but for some events the onset of a second event impedes this procedure which causes errors. Taking into consideration that there is no standard method for hydrograph separation and that results obtained with different methods can differ substantially (Blume et al., 2007), seven different hydrograph separation techniques are applied and compared (Sect. 2.5.2). The uncertainty of Q_{tot} associated with the stage–discharge relation can be important, especially for high-flow conditions. This was taken into account by calculating ϕ_{ev} with the upper and the lower limit of the 90 % confidence interval of discharge obtained with the BaRatin framework.

2.5.2 Hydrograph separation

Straight line method

The straight line (SL) method is a simple, graphical method where baseflow during the event is interpolated by connecting the point in the event hydrograph at which discharge first increases as a response to the rain event with the first point on the falling limb of the hydrograph with the same discharge value. As this condition is often never met, the end of the event flow is often determined by the onset of the next event or discharge below a threshold.

Constant-k method

The constant-k (CK) method proposed by Blume et al. (2007) is based on the assumption that baseflow recession behaves similarly to the outflow of a linear storage. Thus, baseflow at time step t can be described as exponential recession with the recession parameter k and initial flow Q_0:

$$Q_b(t) = Q_0 e^{-kt}. \tag{5}$$

The value of k is calculated at each time step by differentiating Eq. (5) and division by $Q_b(t)$:

$$k = -\frac{dQ}{dt} \frac{1}{Q_b(t)}. \tag{6}$$

Event flow is assumed to terminate at time step t_e, which is defined as the end of the event runoff, once k becomes approximately constant. Baseflow is assumed to be equal to the discharge before the onset of event flow up to t_e when it equals Q_{tot}.

Electrical conductivity method

Hydrograph separation is also conducted based on EC, which serves as a natural tracer (Miller et al., 2014; Pellerin et al., 2008). The method relies on the assumption that streamflow Q_{tot} with electrical conductivity EC_{tot} is a mixture of subsurface flow Q_{sb} and surface flow Q_s, which have significantly different EC signals EC_{sb} and EC_s (Nakamura, 1971):

$$Q_{tot} = Q_{sb} + Q_s, \tag{7}$$

$$Q_{tot} \cdot EC_{tot} = Q_{sb} \cdot EC_{sb} + Q_s \cdot EC_s. \tag{8}$$

Thus, with given values for EC_{sb} (interpolated EC between values before the onset and after the end of the event discharge) and EC_s (measured in overland flow collected at the outlet of four erosion plots representative of the signature of the rainfall flowing at the surface of soils developed on sedimentary rocks), a time series of Q_{sb} can be calculated. As no EC_s values were available for overland flow occurring on soils developed on basalts, covering half of the Claduègne catchment, the method could only be applied to the Gazel catchment, where the proportion of basaltic geology to total catchment surface is much smaller.

It should be noted that this method considers only the surficial part of event flow and is not able to separate the fast responding subsurficial flow occurring in the unsaturated zone. Thus, event flow is likely to be underestimated, especially in conditions under which the latter plays an important role.

Recursive digital filter

The RDF method proposed by Lyne and Hollick (1979) uses a low-pass filter to separate high-frequency event flow signals from low frequency baseflow signals:

$$f(t) = af(t-1) + \frac{1+a}{2}(Q_{tot}(t) - Q_{tot}(t-1)), \tag{9}$$

where $f(t)$ is filtered event flow at time t, a is a filter parameter that is usually in the range of $0.00 < a < 0.95$ (Nathan and McMahon, 1990) and $Q_{tot}(t)$ is original streamflow at time t. The data are passed through the filter several times, forwards and backwards. Recommendations for the number of passes vary depending on the temporal resolution of the discharge series (Ladson et al., 2013). The method is implemented in the R function BaseflowSeparation of the package EcoHydRology (Archibald, 2014).

Hysep filters (HS1–HS3)

Three further filtering approaches are implemented in the Unites Stated Geological Survey's (USGS's) HySep program (Sloto and Crouse, 1996). It is based on finding minima in the discharge time series. The minima are determined either within fixed (HS1) or sliding (HS2) intervals or with a local minima algorithm (HS3). The interval width is adjusted according to Gonzales et al. (2009). It is applied using the R code of the USGS (2015).

3 Results

3.1 Spatial and temporal variability of soil moisture

The variability of soil moisture at the plot scale, determined from the on-alert measurements in the topsoil, is very high: the median range between the highest and the lowest measurement in one plot is 7.8 vol %, but maximum values can get up to > 30 vol %. The mean of the inner-plot standard deviation $\sigma_j^{inner}(t_{ev})$ is 2.7 vol %. Values range from 1 vol % to 8 vol % with no significant difference between the land use classes. There is no significant correlation between plot means and standard deviations (Fig. 2a). The inner-plot standard deviation in the deeper layers, determined with the continuously measuring probes, cannot be confidently assessed because of the low number of probes installed in each plot at the same depth. However, the difference of two sensors installed at the same depth indicate that the variability is in the order of the one derived from the on-alert measurements (Table 2).

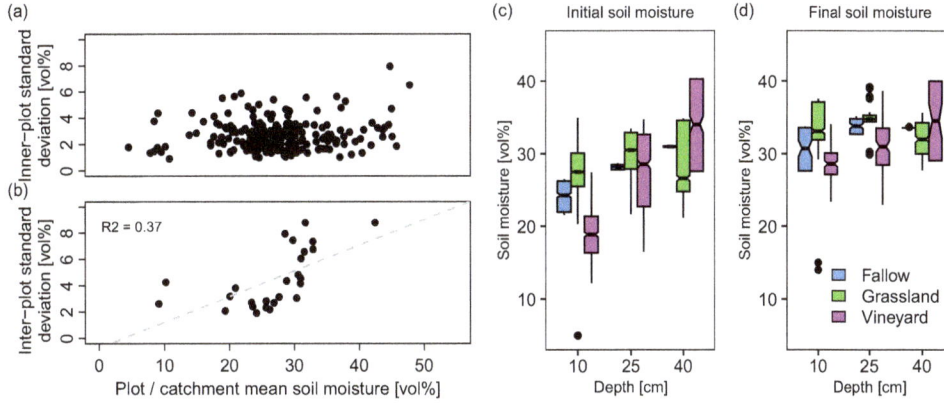

Figure 2. (a) Relationship between plot mean soil moisture $\bar{\theta}_j(t_{ev})$ and inner-plot standard deviation $\sigma_j^{inner}(t_{ev})$, as calculated with Eqs. (1) and (6) in Table 1. Panel (b) shows the same at the catchment scale for catchment mean soil moisture $\bar{\bar{\theta}}_{ev}$ and interplot standard deviation $\sigma_{cat.}^{inter}(t_{ev})$ as calculated with Eqs. (3) and (7) in Table 1. The right panels show initial (c) and final (d) soil moisture profile in plots of different land use during event 27 (6–9 September 2013).

Table 2. Spatial variability of soil moisture at the plot scale (mean of all events calculated for all plots: mean $\sigma_j^{inner}(t_{ev})$; for the grassland plots: mean $\sigma_{j\in g}^{inner}(t_{ev})$; and the vineyard plots: mean $\sigma_{j\in v}^{inner}(t_{ev})$ calculated with Eq. 6 in Table 1) and at the catchment scale (mean interplot variability of the grassland plots: mean $\sigma_g^{inter}(t_{ev})$, and the vineyard plots: mean $\sigma_v^{inter}(t_{ev})$, calculated with Eq. 8 in Table 1 as well as between land use class variability mean $\sigma^{betw.}(t_{ev})$, calculated with Eq. 7 in Table 1) determined at different depth with the two measuring schemes.

	Initial states				Final states			
	0–5 cm	10 cm	25 cm	40 cm	0–5 cm	10 cm	25 cm	40 cm
Mean $\sigma_j^{inner}(t_{ev})$	2.62	2.77	1.85	NA	2.91	2.71	1.88	NA
Mean $\sigma_{j\in g}^{inner}(t_{ev})$	2.77	2.18	1.77	NA	2.80	2.18	1.78	NA
Mean $\sigma_{j\in v}^{inner}(t_{ev})$	2.49	3.15	2.24	NA	2.84	3.15	2.24	NA
Mean $\sigma_g^{inter}(t_{ev})$	3.48	4.49	2.07	4.40	3.70	4.26	2.04	4.30
Mean $\sigma_v^{inter}(t_{ev})$	2.63	2.26	5.36	6.81	2.11	2.39	5.31	6.65
Mean $\sigma^{betw}(t_{ev})$	2.12	2.20	1.14	1.71	3.78	1.98	1.10	1.55

The mean $SEM_j^{inner}(t_{ev})$ is 0.8 vol % with only 3 out of 228 data sets exceeding 2.0 vol %. Thus, the confidence that the population plot mean lies within the sample mean $\bar{\theta}_j(t_{ev})$ ± 2.0 vol % is very high. This accuracy was achieved with about 10 measurements per plot. The variability at catchment scale is also high (Table 2), the catchment mean can be confidently assessed nonetheless. The mean of the standard error of the Claduègne catchment $SEM_{cat.}^{inter}$ is 1.5 vol %, the maximum is 1.7 vol %. The mean $SEM_{cat.}^{inter}$ of the Gazel catchment is 1.3 vol %, the maximum is 1.7 vol %.

Both the on-alert and the continuous measurements were analyzed for differences between the land use classes. In the present study site the plot means of grasslands, vineyards and a fallow are not significantly different from each other (Student t test, significance level $\alpha = 0.05$). Only in the culti-

vated field mean soil moisture is significantly and systematically different from the one in the grasslands. A comparison of the variability between the four land use classes expressed as $\sigma^{betw}(t_{ev})$ to the one within land use classes $\sigma_{lu}^{inter}(t_{ev})$ or within plots $\sigma_j^{inner}(t_{ev})$ also reveals that it is smaller than both other standard deviations (Table 2). The initial and final soil moisture profile of the first major event in 2013 (event 27) shows, nonetheless, that there are differences in the profile shape and in the wetting behavior between grasslands and vineyards (Fig. 2c and d). While the grasslands have a nearly homogenous profile before the rain event, the vineyards have a much more pronounced vertical soil moisture profile with higher values in the deeper layers. In response to the rain event, the profile of the grasslands shifts towards higher soil moisture, with similar differences in each depth. In the vineyards, mainly the moisture in the topmost layer increases, whereas soil moisture in the subsoil hardly changes.

Figure 3 shows the relative spatial difference $\delta_{j,\text{ev}}$ of all plots for all on-alert measurements conducted from 2012 to 2015. It can be seen that temporal stability is found to some degree. The mean values of some plots are (nearly) consistently below the catchment mean (v4, v3, g5, f1, c1), others above (g4, g3, g1). This is also the case if deviations were related to the land use mean instead of the catchment mean (see for example the noticeable difference between grasslands g2 and g5 on the one hand and g1, g3, g4 on the other hand). However, there are also plots with above-average soil moisture for a certain period of time and below-average soil moisture during other periods, indicated by a change in signs of $\delta_{j,\text{ev}}$ between events (v1, v2, g2). The plots with the lowest mean spatial difference $\bar{\delta}_j$ are v1, v2 and c1 (3 %, 4 % and 8 % respectively). The one with the lowest temporal variability of the relative spatial difference ($\sigma_{\delta_j} = 5$ %) is the cultivated field c1.

Figure 4 shows the evolution of soil moisture in 10, 25 and 40 cm depth in autumn 2013. Due to several large rain events in July and August 2013 (not shown here), soil moisture at the beginning of the season is already relatively high. In the topsoil, however, soil moisture is much lower at the beginning of the season (Fig. 5). After the first major rain events, it remains constantly above 30 vol % at 10 cm depth and above 36 vol % in the deeper layers, with maximum values of around 42 vol % reached after major rain events. This value is not exceeded, even after rain events that occur during wet initial conditions.

Temporal variability of soil moisture varies considerably between wet and dry conditions. Soil moisture in all continuously sampled depths increases rapidly as a response to rain events (Fig. 4). Differences between initial and final state in the topsoil can be even larger (Fig. 5). The rapid response is evident from the small lag between the peak of rainfall and the peak of soil moisture (usually less than 2 h for all soil layers).

3.2 Event-based runoff coefficients

The event-based runoff coefficients ϕ_{ev} calculated for the Gazel catchment with seven hydrograph separation methods for 54 events range from 0 to 0.99, with large differences between the methods and a high positive skewness (Fig. 6a). In the Claduègne catchment, ϕ_{ev} was only calculated with the recursive digital filter method RDF and the HySep filter methods, but values still range from 0 to 0.97. The electric conductivity and constant-k methods result in the lowest values for ϕ_{ev} while the three HySep filters yield considerably larger values than all other methods. Apart from the HySep filter methods, the other four methods correlate well with each other (Fig. 6b). The HySep filters correlate very well with each other ($R^2 \geq 0.96$ for all three pairs; not shown here), but to a lesser degree with the other methods (Fig. 6b).

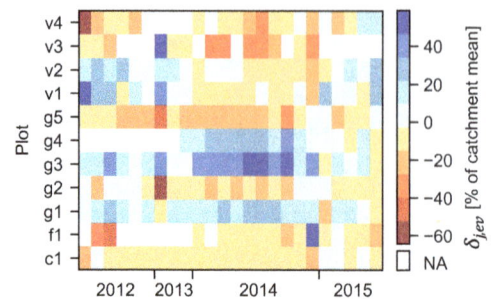

Figure 3. Temporal stability (Vachaud et al., 1985): the relative difference between the plot mean and the catchment mean $\delta_{j,\text{ev}}$ for each on-alert measurement and each plot where on-alert measurements were conducted (four vineyards v1–v4, five grasslands g1–g5, one fallow f1 and one cultivated field c1). Note that the time axis represents a sequence of events, no equidistant time line. Blue squares show plots with plot means that exceed the catchment mean, red squares are those with plot means below the catchment mean, and white squares indicate plots that were not sampled during the respective measurement.

For the following sections, values of ϕ_{ev} determined with the RDF method are used for reasons explained in the discussion (Sect. 4.4).

The uncertainty of ϕ_{ev} associated with the stage–discharge relation is shown in Fig. 7 as black vertical error bars. In both catchments this uncertainty is very small for events with low ϕ_{ev} while it can get up to 0.28 (difference between ϕ_{ev} calculated with the 5 % and the 95 % confidence interval of discharge) for event 40, which is the one with the highest ϕ_{ev} and highest discharge in both catchments (Tables 3 and 4). The uncertainty due to different ϕ_{ev} obtained by different hydrograph separation methods is visualized as gray vertical error bars in Fig. 7. It can be very high for any event regardless of the ϕ_{ev} and is often due to the discordance of the HySep methods with the other methods. The mean standard deviation of ϕ_{ev} calculated with different hydrograph separation methods is 0.03, when the HySep methods are excluded it decreases to 0.02. However, these measures are biased by the important positive skewness of the distribution of ϕ_{ev}.

Factors that are suggested to influence ϕ_{ev} include rainfall depth and rain intensity. Figure 7 shows the correlation of the meteorological forcing quantified as cumulative catchment rainfall depth P_{cum}, mean rain intensity I_μ and maximum 20 min rain intensity $I_{\text{max},20}$ with ϕ_{ev}. In the present data set there is a weak correlation between ϕ_{ev} and the meteorologic variables P_{cum}, I_μ and $I_{\text{max},20}$ (Fig. 7). The correlations of ϕ_{ev} with maximum rain intensity calculated at 2, 10, 30 and 60 min time steps were worse that the one at 20 min. None of these variables can, therefore, explain more than 30 % of the variability of ϕ_{ev}. Figure 7 shows that events with similar rainfall characteristics (events 30 and 40, similar intensity) can have very different ϕ_{ev}. Additionally, similar ϕ_{ev} are obtained for events with very different rainfall characteristics

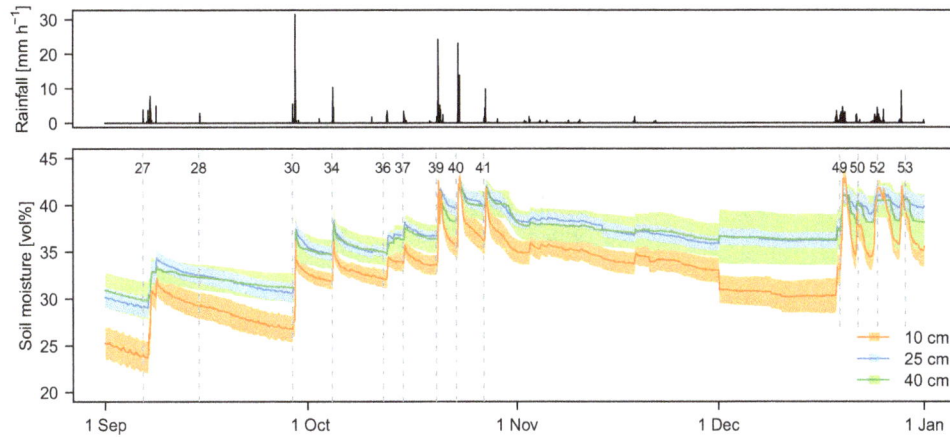

Figure 4. Rainfall and soil moisture in autumn 2013 measured at 10, 25 and 40 cm depth. The line represents the catchment mean calculated with Eq. (3) in Table 1 in the respective depth and the shaded area the mean $\pm$SEM$_{\text{cat.}}^{\text{inter}}$. The dashed lines represent the onsets of several rain events and the labels refer to the event numbers as in Tables 3 and 4. Note that on 27 November 2013 one of the probes stopped working, which is the reason for the step in the data and the higher SEM$_{\text{cat.}}^{\text{inter}}$.

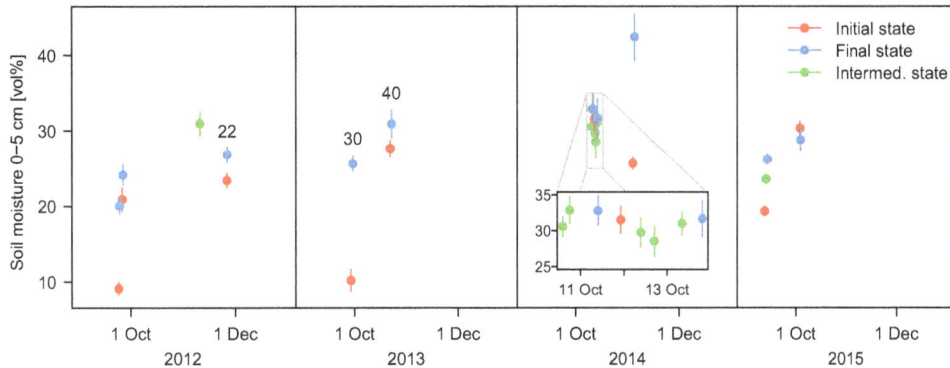

Figure 5. Soil moisture in the topsoil (0–5 cm) measured on alert basis before and after major rain events in autumn seasons in 2012–2015. The points show catchment mean values, the lines the range of the mean $\pm$SEM$_{\text{cat.}}^{\text{inter}}$. The numbers above selected events give the event number as in Tables 3 and 4 and as referred to in the text.

(events 22 and 39). These striking differences in catchment behavior can partly be attributed to differences in initial soil moisture as shown in the following section.

3.3 Soil moisture's impact on runoff generation

The hydrological responses concerning the temporal dynamics of soil moisture and discharge in reaction to rain events of the two catchments vary greatly. In Tables 3 and 4 the characteristics of all rain events in autumn 2013 that generate event flow at the river gauges of the Gazel and/or Claduègne are given. The hyetographs, hydrographs and time series of catchment mean soil moisture of four of these events with very different behavior are exemplarily shown in Fig. 8: event 27 and event 30 occur at the beginning of the season when soil moisture is still relatively low. Rainfall leads to a considerable increase in soil moisture in all three layers and to a storage change ΔS in the topmost 50 cm of the soil

profile that constitutes a notable share of cumulative precipitation. For event 30 on-alert surface soil moisture measurements show a sharp increase from 10.2 vol % before the event to 25.7 vol % afterwards (Fig. 5). The runoff coefficients ϕ_{ev} of both events are very low. The within-event temporal dynamic of rainfall, soil moisture and runoff during event 27 is also noteworthy: the discharge peak does not follow the rainfall peak, which is closely followed by the steepest increase in soil moisture, but the second rainfall pulse that occurs when soil moisture is considerably higher than at the beginning of the event. As a response to this much smaller rainfall impulse, soil moisture rises only slightly. This behavior is also observed during event 40, when the first rainfall impulse leads to a sharp increase in soil moisture and only a small discharge peak, while the second rainfall pulse generates a substantial discharge peak and only a slight increase in soil moisture. Event 40 and event 53 both occur during wet initial soil moisture conditions, but event 53 has a much smaller

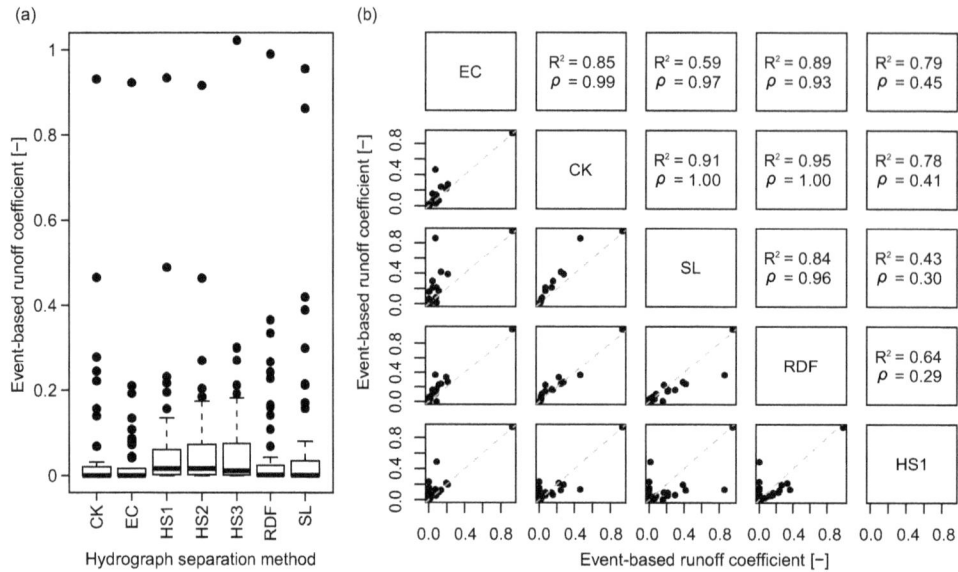

Figure 6. (a) Differences of event-based runoff coefficients ϕ_{ev} calculated for 54 rain events in the Gazel catchment in autumn 2012 and autumn 2013 derived with different methods for hydrograph separation. **(b)** Correlation between different methods. The upper panel gives the coefficient of determination (R^2) and Spearman's rank correlation coefficient ρ, the dashed line in the lower panel is the line of identity. The methods used for hydrograph separation are described in Sect. 2.5.2: constant-k (CK), electric conductivity (EC), Hysep filter with fixed or sliding interval (HS1–HS2), Hysep filter with local minima algorithm (HS3), recursive digital filter (RDF) and straight line (SL).

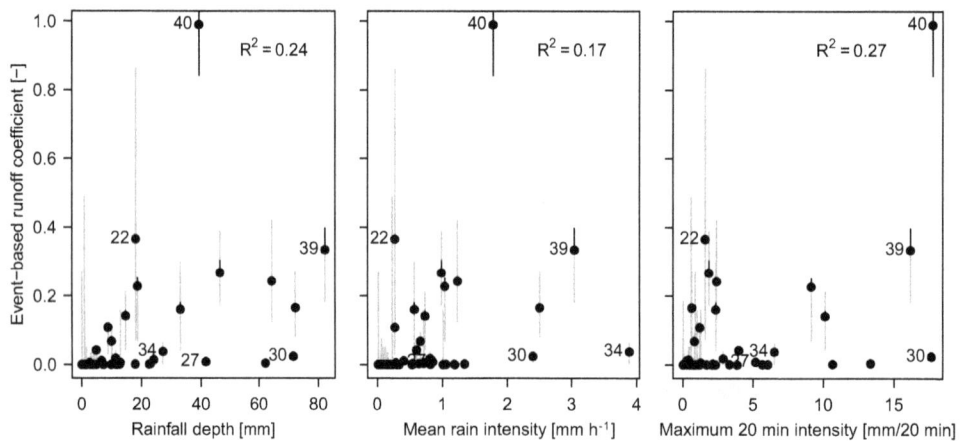

Figure 7. Correlation of three variables describing meteorological forcing with event-based runoff coefficients ϕ_{ev} of the Gazel catchment calculated with the recursive digital filter method. The lines represent the uncertainty associated with the hydrograph separation method (gray vertical lines: range of ϕ_{ev} calculated with the seven different methods) and the stage–discharge relation (black vertical lines: range between ϕ_{ev} calculated with the 5 % and the 95 % confidence interval of discharge obtained with the BaRatin framework). The point labels give the numbers of selected events as in Tables 3 and 4 and described in the text.

ϕ_{ev} than event 40. During these two events an inversion of the vertical soil moisture profile, i.e., temporally higher soil moisture at the topsoil than in deeper layers, can also be observed at approximately the time of peak discharge. This inversion is an indicator of Hortonian overland flow. Overland flow was indeed observed during event 40 in vast areas in the north of the Claduègne catchment (Supplement S2).

The large range of ϕ_{ev} of events with high $\widetilde{\theta}_{ini}$ can also be seen in Fig. 9. While the three events with low $\widetilde{\theta}_{ini}$ consis-

tently have very low ϕ_{ev}, above a threshold of approximately 34 vol %, ϕ_{ev} can have a value anywhere between zero and one. An examination of Fig. 9 shows that both high $\widetilde{\theta}_{ini}$ and high P_{cum} are necessary but not sufficient criteria for high ϕ_{ev} and that the relation between $\widetilde{\theta}_{ini}$ and ϕ_{ev} is characterized by strong nonlinearity and threshold effects. This is observed in both catchments and the threshold value is very similar for the Gazel and the Claduègne catchment. Further analysis of the relation between $\widetilde{\theta}_{ini}$ and ϕ_{ev} for events with high P_{cum}

Table 3. Rainfall, soil moisture and discharge characteristics of selected rain events in autumn 2013 in the Gazel catchment: beginning of the rain event, cumulative precipitation (P_{cum}), maximum 20 min rain intensity ($I_{max,20}$), mean intensity (I_μ), initial soil moisture ($\widetilde{\theta}_{ini}$), final soil moisture ($\widetilde{\theta}_{fin}$), standard error of the catchment mean ($SEM_{cat.}^{inter}$) during initial and final stage, soil storage change at depth 0–50 cm (ΔS), peak discharge (Q_p), cumulative total discharge ($Q_{tot,cum}$), cumulative event discharge ($Q_{ev,cum}$) and event-based runoff coefficient calculated with the recursive digital filter method (ϕ_{ev}).

	Rainfall				Soil moisture					Discharge		
Ev. no.	Beg. rain DD-MM hh:mm	P_{cum} mm	$I_{max,20}$ mm h^{-1}	I_μ mm h^{-1}	$\widetilde{\theta}_{ini}$ vol %	$SEM_{cat.}^{inter}$ vol %	$\widetilde{\theta}_{fin}$ vol %	$SEM_{cat.}^{inter}$ vol %	ΔS mm	Q_p L s^{-1}	$Q_{ev,cum}$ mm	ϕ_{ev} –
27	06-09 16:06	41.59	33.30	0.85	27.99	1.73	32.69	1.07	28.17	25	0.34	0.01
28	15-09 00:23	4.82	24.00	0.60	31.40	1.31	31.34	1.38	0.00	14	0.20	0.04
30	28-09 17:09	71.73	59.70	2.39	29.80	1.57	35.20	1.45	32.43	324	1.73	0.02
34	04-10 15:39	27.20	25.50	3.89	34.33	1.29	36.43	1.25	12.62	58	1.03	0.04
36	12-10 05:12	12.86	4.50	0.71	34.55	1.15	35.91	1.02	8.12	18	0.07	0.01
37	15-10 03:46	11.29	9.00	0.81	35.76	1.07	36.54	1.07	4.72	23	0.21	0.02
39	20-10 02:41	82.20	53.40	3.04	35.65	1.10	38.12	1.41	14.83	8660	27.49	0.33
40	23-10 01:01	39.31	60.00	1.79	37.34	1.44	38.40	1.54	6.33	30096	38.91	0.99
41	27-10 03:37	14.64	37.20	0.73	37.17	1.46	38.09	1.52	5.54	361	2.07	0.14
49	18-12 06:52	33.09	8.70	0.57	34.26	1.20	37.90	1.32	21.82	402	5.30	0.16
50	21-12 03:11	8.73	4.20	0.27	38.39	1.23	37.78	1.12	0.00	175	0.95	0.11
52	24-12 01:15	46.33	7.50	0.99	37.09	1.17	39.83	1.08	16.46	1047	12.38	0.27
53	28-12 04:51	18.70	32.70	1.04	37.52	1.40	39.25	1.04	10.41	654	4.25	0.23

Table 4. Rainfall, soil moisture and discharge characteristics of selected rain events in autumn 2013 in the Claduègne catchment; abbreviations as in Table 3.

	Rainfall				Soil moisture					Discharge		
Ev. no.	Beg. rain DD-MM hh:mm	P_{cum} mm	$I_{max,20}$ mm h^{-1}	I_μ mm h^{-1}	$\widetilde{\theta}_{ini}$ vol%	$SEM_{cat.}^{inter}$ vol%	$\widetilde{\theta}_{fin}$ vol%	$SEM_{cat.}^{inter}$ vol%	ΔS mm	Q_p m^3 s^{-1}	$Q_{ev,cum}$ mm	ϕ_{ev} –
27	06-09 15:41	43.26	33.3	0.88	27.00	1.38	32.62	1.16	29.05	0.19	0.38	0.01
28	15-09 00:23	3.61	26.7	0.45	31.09	1.32	31.00	1.32	0.00	NA	NA	NA
30	28-09 17:04	77.71	63	2.59	29.09	1.49	34.49	1.24	29.93	16.39	3.16	0.04
34	04-10 15:33	28.09	27	4.01	33.48	1.53	35.46	1.44	11.39	0.59	0.66	0.02
36	12-10 05:12	12.81	4.5	0.71	33.58	1.53	35.20	1.46	8.42	0.13	0.10	0.01
37	15-10 03:46	11.51	9	0.82	34.98	1.55	35.67	1.52	4.34	0.16	0.19	0.02
39	20-10 02:41	83.67	53.4	3.10	34.83	1.63	37.54	1.51	15.33	54.64	20.82	0.25
40	23-10 01:01	51.01	60	2.32	36.92	1.69	38.75	1.60	11.37	60.76	36.37	0.93
41	27-10 03:37	22.06	52.5	1.10	37.62	1.67	38.47	1.64	4.40	12.08	5.44	0.37
49	18-12 06:52	67.91	8.7	1.17	34.40	1.61	38.22	1.44	21.59	NA	NA	NA
50	21-12 03:11	10.98	4.2	0.34	38.70	1.61	38.40	1.63	0.00	NA	NA	NA
52	24-12 01:15	51.24	7.5	1.09	37.70	1.68	40.29	1.58	14.30	NA	NA	NA
53	28-12 04:51	19.65	32.7	1.09	38.09	1.66	39.68	1.60	9.56	NA	NA	NA

and high $\widetilde{\theta}_{ini}$ is limited because of the low number of events fulfilling these criteria and the uncertainty of both $\widetilde{\theta}_{ini}$ and ϕ_{ev}.

Consideration of single events shows that $\widetilde{\theta}_{ini}$ can partly explain the high scatter in Fig. 7. The contrary behavior of events 30 and 40 can be explained by different initial soil moisture conditions. It can also be hypothesized that the high ϕ_{ev} of event 22 despite low P_{cum}, I_μ and $I_{max,20}$ is due to high initial soil moisture. For this event, that started on 26 Novem-

ber, 2012, only on-alert soil moisture is available and initial topsoil moisture is relatively high at 23.5 vol %. The event occurred late in the season (26 November 2012, Fig. 5) 2 weeks after a heavy rain event with $P_{cum} = 72.5$ mm. However, event 39, which has a similar ϕ_{ev} higher P_{cum}, I_μ and $I_{max,20}$, also occurred during high initial moisture conditions, which indicates that the relation between ϕ_{ev}, $\widetilde{\theta}_{ini}$ and rainfall characteristics cannot easily be generalized.

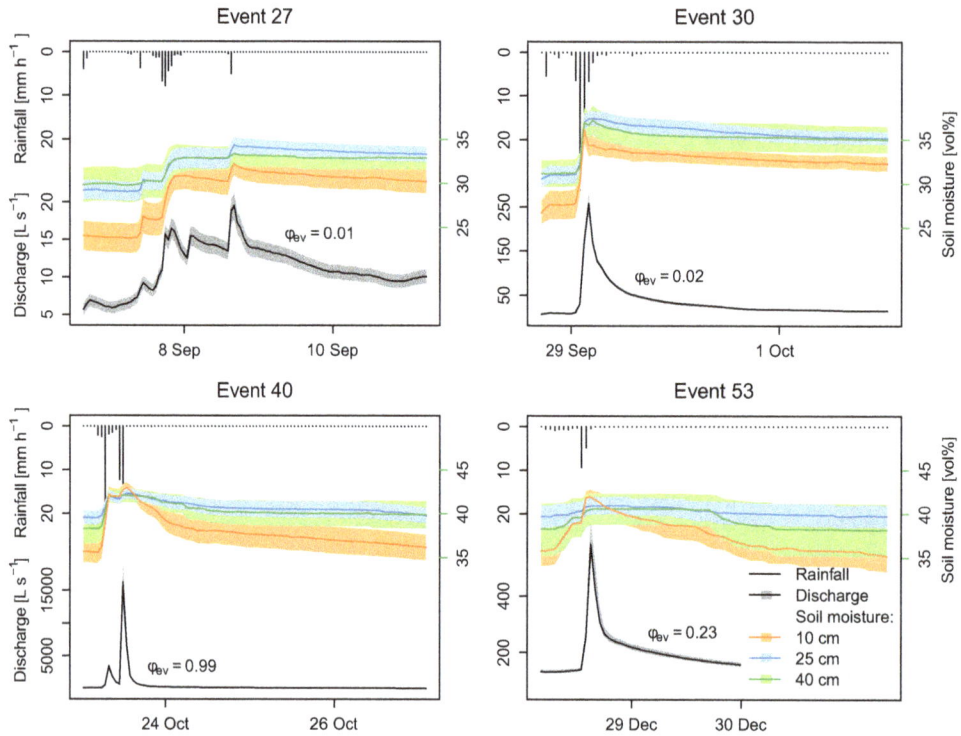

Figure 8. Hyetographs, hydrographs of total discharge and evolution of soil moisture in the Gazel catchment during four different events in 2013. The event-based runoff coefficient ϕ_{ev} is also given for all events. The representation of soil moisture gives the mean $\pm \mathrm{SEM}_{cat.}^{inter}$ in the respective depth and that of discharge the best estimate $\pm$ the uncertainty of the stage–discharge relation.

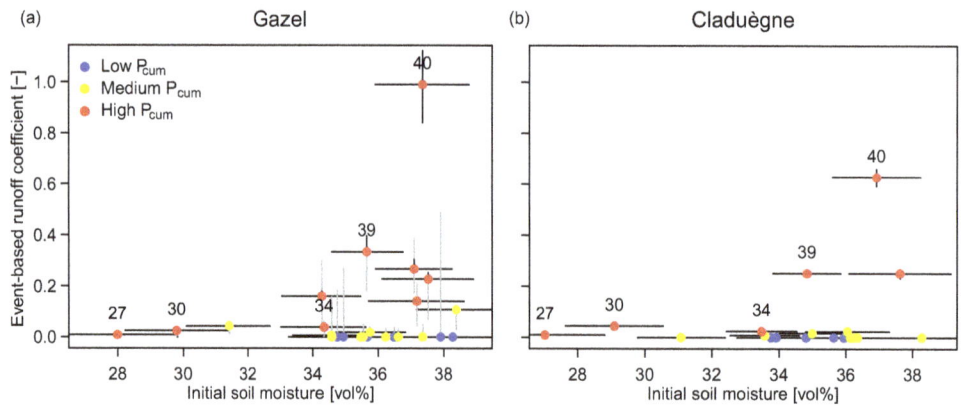

Figure 9. Relationship between initial soil moisture $\widetilde{\theta}_{ini}$ and event-based runoff coefficients ϕ_{ev} in the Gazel (**a**) and Claduègne (**b**) catchments. On the x axis the point represents the profile mean initial soil moisture and the horizontal line the range of the mean $\pm \mathrm{SEM}^{inter}$. On the y axis the point represents ϕ_{ev} calculated with the recursive digital filter method and the line the uncertainty as in Fig. 7. The color of the points indicates whether cumulative precipitation is low ($P_{cum} < 1.5\,\mathrm{mm}$), medium ($1.5\,\mathrm{mm} < P_{cum} < 13\,\mathrm{mm}$) or high ($P_{cum} > 13\,\mathrm{mm}$).

4 Discussion

4.1 Soil moisture estimation at the catchment scale

The sampling design applied in this project proved to be efficient to assess spatial variability of soil moisture across scales, as well as to document temporal dynamics. The on-alert measurements of soil moisture allow a good estimate of the plot mean, with a low mean SEM_j^{inner} of 0.8 vol % as well as an accurate estimate of the inner-plot variability, quantified as $\sigma_j^{inner}(t_{ev})$ during wet and dry conditions. On the other hand, the continuous soil moisture measurements cover a larger extent of the two studied catchments and different depths in the soil profile. Due to the higher variability at the catchment scale, the mean $\mathrm{SEM}_{cat.}^{inter}$ is somewhat higher (1.5 vol %). The values obtained for SEM_j^{inner} and $\mathrm{SEM}_{cat.}^{inter}$

show that at an accepted uncertainty of the mean of ± 2 vol %, the number of 10 measurements per plot is sufficient. This is consistent with the results of Zucco et al. (2014), who found a maximum number of 11 or 20 required samples at the plot scale and catchment scale respectively, and those of Molina et al. (2014), who concluded that plot mean soil moisture in a Mediterranean mountain area was well represented with nine probes. The review by Vereecken et al. (2014) shows that there is a wide range of estimates for these numbers and that they are site specific. The continuous measurements reveal the temporal evolution of soil moisture over the season and within events. The only drawback is the lack of continuous soil moisture estimates in the topsoil. The sampling at the plot scale and in nested catchments is considered to be a good approach to assess heterogeneity across scales and to cope with the change of scale problem (Braud et al., 2014).

In this study the interplot variability within one land use class $\sigma_{lu}^{inter}(t_{ev})$ usually exceeds the inner-plot variability $\sigma_{j}^{inner}(t_{ev})$. This is not consistent with findings of Huza et al. (2014), which may be due to different sampling strategies: whereas Huza et al. (2014) conducted measurements along 50 m transects, for this study random locations were sampled within one field in an area of $\approx$ 20 by 20 m. At scales larger than 10 m, they found a spatial structure revealed by a higher semi-variance at distances of more than 10 m in at least one of their transects.

The results of this study indicate differences between grasslands and vineyards in the vertical soil profile and in the response of the profile to rain events. These differences are most likely due to differences in soil texture, as vineyards are usually found on soils with higher clay content than the ones of the other land use types. However, there are no significant and systematic differences between the plot means of different land use classes. Thus, land use cannot be used as additional information to improve spatially distributed soil moisture estimation in the study site.

The cultivated field c1 shows a remarkable temporal stability of the difference of this plot's mean soil moisture and the catchment mean $\delta_{j,ev}$. This suggests that if the catchment mean has to be approximated by measurements in just one field, this one is the best choice (Vachaud et al., 1985; Vanderlinden et al., 2012). Other fields show, however, that $\delta_{j,ev}$ is not consistent in time. The observation that several sites change the sign of $\delta_{j,ev}$ between measurements was also made at the plot scale on a grassland, a field cultivated with wheat and an olive grove by Vachaud et al. (1985) and at the catchment scale on grasslands by Huza et al. (2014). Here, notably the vineyards v1 and v2 are considerably wetter than the catchment mean throughout the autumn seasons of 2012 and 2013, dryer in 2014 and wetter again in 2015. Possible reasons include changes in cultivation. In particular, tillage practices play an important role in the vineyards (not shown here). Therefore, conclusions based on this finding should be considered carefully. Moreover, the choice of the plot which best represents the catchment mean should include the temporal variability of $\delta_{j,ev}$ and should not be solely based on the minimal mean difference $\bar{\delta}_j$, which is in this case that of v1 and v2.

4.2 Quantification of the hydrological response

The available precipitation and discharge data at a high spatiotemporal resolution is a major asset that is necessary to understand the hydrological processes at small scales and during short time spans that lead to flash flood generation (Nord et al., 2017). It allows the event-based runoff coefficient ϕ_{ev} to be calculated and its uncertainty to be estimated. The main sources of its uncertainty are that of the stage–discharge relation, which is especially important for events with high discharge and which was assessed with the BaRatin framework; the uncertainty associated with the choice of the method used for hydrograph separation; and the uncertainty of the catchment mean precipitation. The latter source of uncertainty is not considered in this study. It stems from the rainfall measurements with tipping buckets and the interpolation between the rain gauges. Tipping buckets are known to underestimate precipitation at high intensities (Marsalek, 1981; Molini et al., 2005); thus, including radar data could improve the estimation of catchment mean rainfall even in relatively well-gauged catchments such as those of the Gazel and Claduègne (e.g., Creutin and Borga, 2003; Delrieu et al., 2014; Abon et al., 2015).

In this study, the uncertainty associated with the hydrograph separation method exceeds that of the stage–discharge relation. The high range and positive skewness of event-based runoff coefficients is consistent with the results of other studies (Merz et al., 2006; Blume et al., 2007; Merz and Blöschl, 2009; Norbiato et al., 2009; Marchi et al., 2010). The dependence of ϕ_{ev} on rain characteristics suggested by other authors (Merz et al., 2006; Norbiato et al., 2009) was not entirely confirmed in this study, as none of the rain characteristics examined here (P_{cum}, I_{μ}, $I_{max,20}$) could explain more than 30 % of the variability in ϕ_{ev}.

Each of the hydrograph separation methods used here has advantages and disadvantages. The method based on electric conductivity has a physically based foundation as it distinguishes components with different EC and represents subsurface flow dynamics. This method could not be applied to both catchments because values for surface flow EC_s were only available from Le Pradel in the south of the Gazel catchment, and it is assumed that EC_s on the basaltic plateau differs considerably as this geology accounts for a large part of the Claduègne catchment. Furthermore, it is not possible to conduct a three-component hydrograph decomposition with the available data, so unlike with the other methods, the fast reacting subsurface flow is considered to be baseflow. Thus, event discharge is underestimated.

Unlike the other methods, the CK method offers a physical explanation for the end of the event flow. The method builds on the assumption of baseflow behaving like the slow responding outflow of a linear reservoir. For the discharge data of the Gazel, this method could not always be applied because of the low discharge that results in "steps" in the data and high noise so the threshold for defining that k is constant as proposed by Blume et al. (2007) was never reached. An adjusted threshold yielded reasonable results for some but not all events.

The straight line method was rejected because it does not consider baseflow dynamics and the end of the event flow has to be determined arbitrarily. The filter methods have the advantage of being easy to apply to all data sets without further data treatment or demand of additional data, but these methods are very sensitive to parameters such as the interval width (HySep filters) or the number of passes (RDF). The HySep filters were discarded because of the disagreement with the other methods. Thus, the RDF method was used for all further analyses because it correlates well with the EC and CK methods and can easily be applied to all events and both catchments. The number of passes had to be calibrated, as suggested by Ladson et al. (2013), in such a manner that ϕ_{ev} is below 1 for all cases and that it is slightly higher than the value obtained with the EC method in order to compensate for the underestimation of event discharge. Nonetheless, underestimation of event discharge is still a source of uncertainty.

4.3   The impact of initial soil moisture on the hydrological response

The relation between ϕ_{ev} and $\widetilde{\theta}_{ini}$ is not as clear as one might have expected from other studies' results. Moreover, both variables are still subject to large uncertainties. Catchment mean initial soil moisture $\widetilde{\theta}_{ini}$ below a threshold of 34 vol % inhibits high ϕ_{ev}. However, only three of the events considered here occur during such dry conditions, so further measurements would be useful to corroborate this finding. Threshold effects in the relation of ϕ_{ev} and $\widetilde{\theta}_{ini}$ are also observed by other authors (e.g., McMillan et al., 2014; Hrachowitz et al., 2011). In the Mediterranean context, the thresholds obtained by Huza et al. (2014) in the Gazel catchment and Braud et al. (2014) in the Valescure catchment (22 vol % and 25 vol % respectively) are lower than the one obtained here. The threshold at 45 vol % observed by Penna et al. (2011) in a 1.9 km^2 headwater catchment in the Italian Dolomites, on the other hand, is higher than the one obtained here. McMillan et al. (2014) show that thresholds in different subcatchments of a 50 km^2 catchment in New Zealand are highly variable: they range between 27 vol % and 58 vol % and are more or less pronounced in different subcatchments. These differences might be due to different soil and land use features, climate, and sampling designs. The values for $\widetilde{\theta}_{ini}$ that Huza et al. (2014) used are obtained from satellite data, while this study uses in situ data from several land use classes. Moreover, a profile mean is considered here, while Huza et al. (2014) used only values of topsoil moisture.

Furthermore, different methods were applied for hydrograph separation. Huza et al. (2014) used a method similar to the HySep 3 filter, which yielded different results than the other methods applied for this study.

The high range of ϕ_{ev} obtained at high $\widetilde{\theta}_{ini}$ also agrees with findings of Huza et al. (2014). It indicates that the hydrological response is influenced by other factors as well. The parameters describing the impact of meteorological forcing (P_{cum}, I_μ and $I_{max,20}$) neither explain that variability. When only events with high cumulative precipitation are considered, the range is still very high. Results obtained in virtual experiments (Merz and Plate, 1997; Bronstert and Bárdossy, 1999; Zehe and Blöschl, 2004) showed that spatial patterns of soil moisture and threshold effects strongly impact the runoff response. The latter authors show that, especially during initial moisture conditions close to the threshold, the runoff response depends strongly on the resampling of spatially distributed soil moisture. Therefore, actual, small-scale soil moisture patterns that control connectivity of pathways but are not reflected in the catchment mean value are a possible explanation for the very diverse runoff behavior. On the other hand, Morbidelli et al. (2016) conclude that spatial heterogeneity of $\widetilde{\theta}_{ini}$ does not affect the runoff response for events that are associated with floods, so more research is needed on this topic. Additionally, subsurface flow along preferential flow paths can contribute to high ϕ_{ev} and Hortonian overland flow is not directly related to $\widetilde{\theta}_{ini}$ but produces a substantial proportion of event flow.

The results of this study partly confirm the suggestions of other authors (i.e., Brocca et al., 2009b; Javelle et al., 2010; Grillakis et al., 2016) to consider estimates of initial soil moisture in flash flood warning based on the dependence of ϕ_{ev} on $\widetilde{\theta}_{ini}$. This offers high potential for predictions in poorly gauged or ungauged basins given the global availability of remote sensing soil moisture data (Crow et al., 2005; Beck et al., 2009; Brocca et al., 2009c; Massari et al., 2014a, b). Threshold-based warning systems are advocated, by Norbiato et al. (2008) for example. Including a threshold value for initial soil moisture could prevent false positive flash flood warnings in cases in which high precipitation is expected under dry initial catchment conditions, while above-threshold soil moisture in combination with high precipitation increases the likelihood of high runoff coefficients. This threshold seems not to be scale dependent. However, the threshold values differ between catchments and depend to a high degree on the methodology to determine it, as indicated by the different values in this study and that by Huza et al. (2014). Furthermore, there are high data requirements to determine such thresholds and it is not known whether they can be transferred from one catchment to another, so it is not applicable for operational flash-flood warning. Moreover, the high scatter of ϕ_{ev} under high initial soil moisture conditions suggests that the relation between ϕ_{ev} and $\widetilde{\theta}_{ini}$ is very complex and depends on other factors and processes that are still insufficiently understood. Thus, the impact of soil moisture on the hydrological response during wet catchment conditions cannot be generalized based on the results obtained here. Further research and instrumentation could include the installation of piezometers in the catchment to understand subsurficial flow in the catchment, using tracers other than EC to differentiate subsurficial stormflow as a third flow component during hydrograph separation as well as the application of multivariate regression analysis methods that systematically examine different controls on ϕ_{ev}, such as meteorological forcing as well as $\widetilde{\theta}_{ini}$ and their interactions.

5 Conclusions

This study aimed at assessing the influence of initial soil moisture on the hydrological response in a flash-flood prone area in southern France. To this end, two issues were addressed and exemplarily examined in the nested Gazel ($3.4\,\mathrm{km}^2$) and Claduègne ($43\,\mathrm{km}^2$) catchments: (1) obtaining a meaningful estimate of soil moisture at the catchment scale and (2) analyzing the relation between initial soil moisture $\widetilde{\theta}_{ini}$ and the hydrological response quantified as the event-based runoff coefficient ϕ_{ev}.

The main findings of this study related to the first objective are as follows:

1. Spatial variability of soil moisture at the plot scale and at the catchment scale is very high. There are differences between land use classes in the vertical soil moisture profile and in wetting behavior, but no significant and systematic differences in catchment mean soil moisture values between land use classes exist. Between land use standard deviation σ_{lu}^{betw} exceeds neither σ^{inter} nor σ^{inner}.

2. There is one plot, c1, with remarkable temporal stability of the spatial difference between plot mean and catchment mean. Thus, this field should be opted for, if the catchment mean has to be assessed from measurements in just one plot. However, not one of the other plots shows this temporal stability.

The sampling design applied for this study allowed a detailed characterization of soil moisture heterogeneity across scales as well as the assessment of temporal dynamics. The catchment mean soil moisture was derived with a mean standard error of the catchment mean of $1.3\,\mathrm{vol}\,\%$ or $1.5\,\mathrm{vol}\,\%$ for the Gazel and Claduègne catchments respectively.

Main findings concerning the impact of initial soil moisture on the hydrological response quantified with the event-based runoff coefficient ϕ_{ev} are as follows:

1. The ϕ_{ev} obtained with different hydrograph separation methods can differ considerably, but results obtained with EC, CK, SL and RDF methods correlate well. The RDF method was preferred for this study because it is easy to apply and because of the good correlation with the more physically based methods EC and CK, which could not be applied to all events and both catchments.

2. There is a weak correlation between ϕ_{ev} and cumulative event precipitation P_{cum}, mean rain intensity I_μ and maximum 20 min rain intensity $I_{max,20}$ ($R^2 = 0.24$, 0.17 or 0.27 respectively).

3. The hydrological response depends on initial soil moisture $\widetilde{\theta}_{ini}$: below a threshold of 34 vol %, ϕ_{ev} remains very low, even during high precipitation events. However, there is a large scatter in ϕ_{ev} above that threshold, indicating that other factors and processes also have an important impact on ϕ_{ev}. The threshold is identical for both catchments, which indicates that at this study's site it might be scale invariant.

4. Analysis of the seasonal and within-event evolution of soil moisture and discharge shows that discharge peaks of two considered events did not follow the peaks in rainfall, but a second, smaller rain impulse, while the rainfall peaks lead to a considerable refilling of soil water storage.

These results indicate that $\widetilde{\theta}_{ini}$ does impact the hydrological response. For single events ϕ_{ev} can be attributed to $\widetilde{\theta}_{ini}$, P_{cum} or $I_{max,20}$. However, these results cannot be generalized and no systematic and unequivocal relationship between $\widetilde{\theta}_{ini}$ and ϕ_{ev} was found. Even though the present data set is exceptionally detailed, there still is substantial uncertainty in the values for $\widetilde{\theta}_{ini}$, P_{cum} and cumulative event flow $Q_{ev,cum}$.

The results of this study support suggestions by other authors to include estimates of initial soil moisture in flash flood warning, based on the dependence of ϕ_{ev} on $\widetilde{\theta}_{ini}$, by including a threshold that could prevent false positive flash flood warnings under dry initial conditions. Further research could focus on the role of subsurface flow and on elaborating multivariate regression analysis methods.

Author contributions. JPV and IZ conceptualized the study and MU carried out data analysis and its interpretation. JPV, IB, CL, GM and GN were principal investigators in the FloodScale project responsible for instrumentation and analyzed and provided the data used for this study. JPV, IZ and IB contributed to soil moisture analyses, MH, CL and GN to hydrograph separation, and GM to precipitation data processing. MU prepared the paper with contributions from all coauthors.

Competing interests. The authors declare that they have no conflict of interest.

Acknowledgements. This work was funded by the French National Research Agency (ANR) via the FloodScale project under contract no. ANR 2011 BS56 027, which contributes to the HyMeX program. OHM-CV is supported by the Institut National des Sciences de l'Univers (INSU/CNRS), the French Ministry for Education and Research, the Environment Research Cluster of the Rhône-Alpes Region, the Observatoire des Sciences de l'Univers de Grenoble (OSUG) and the SOERE Réseau des Bassins Versants (Alliance Allenvi) and belongs to the OZCAR Reasearch Infrastructure. Magdalena Uber received financial support from German Academic Exchange Service (DAAD). The authors want to thank Brice Boudevillain, Sahar Hachani, Simon Gérard and Cindy Nicoud for their help during field work and Luca Brocca, Hongkai Gao and two anonymous referees for their comments on earlier versions of this paper that were very valuable to improve this paper.

Edited by: Roberto Greco

References

Abon, C., Kneis, D., Crisologo, I., Bronstert, A., David, C. P. C., and Heistermann, M.: Evaluating the potential of radar-based rainfall estimates for streamflow and flood simulations in the Philippines, Geomat. Nat. Haz. Risk, 7, 1–16, https://doi.org/10.1080/19475705.2015.1058862, 2015.

Andrieu, J.: Landcover map Claduègne catchment, ESPRI/IPSL, https://doi.org/10.14768/mistrals-hymex.1381, 2015.

Archibald, J.: BaseflowSeparation {EcoHydRology}, available at: https://cran.r-project.org/web/packages/EcoHydRology/ (last access: 19 November 2018), 2014.

Baffaut, C., Ghidey, F., Sudduth, K. A., Lerch, R. N., and Sadler, E. J.: Long-term suspended sediment transport in the Goodwater Creek Experimental Watershed and Salt River Basin, Missouri, USA, Water Resour. Res., 49, 7827–7830, https://doi.org/10.1002/wrcr.20511, 2013.

Beck, H. E., de Jeu, R. A. M., Schellekens, J., van Dijk, A. I. J. M., and Bruijnzeel, L. A.: Improving Curve Number Based Storm Runoff Estimates Using Soil Moisture Proxies, IEEE J. Sel. Top. Appl., 2, 250–259, https://doi.org/10.1109/JSTARS.2009.2031227, 2009.

Berthet, L., Andréassian, V., Perrin, C., and Javelle, P.: How crucial is it to account for the antecedent moisture conditions in flood forecasting? Comparison of event-based and continuous approaches on 178 catchments, Hydrol. Earth Syst. Sci., 13, 819–831, https://doi.org/10.5194/hess-13-819-2009, 2009.

Blume, T., Zehe, E., and Bronstert, A.: Rainfall–runoff response, event-based runoff coefficients and hydrograph separation, Hydrol. Sci. J., 52, 843–862, https://doi.org/10.1623/hysj.52.5.843, 2007.

Borga, M., Boscolo, P., Zanon, F., and Sangati, M.: Hydrometeorological Analysis of the 29 August 2003 Flash Flood in the Eastern Italian Alps, J. Hydrometeorol., 8, 1049–1067, https://doi.org/10.1175/JHM593.1, 2007.

Boudevillain, B., Delrieu, G., Galabertier, B., Bonnifait, L., Bouilloud, L., Kirstetter, P. E., and Mosini, M. L.: The Cévennes-Vivarais Mediterranean Hydrometeorological Observatory database, Water Resour. Res., 47, 1–6, https://doi.org/10.1029/2010WR010353, 2011.

Braud, I.: Soil properties Auzon catchment, https://doi.org/10.6096/MISTRALS-HyMeX.1385, 2015.

Braud, I. and Vandervaere, J. P.: Analysis of infiltration tests and performed in the Claduègne catchment in May–June 2012, Report to the FloodScale project, 66 pp., https://doi.org/10.6096/MISTRALS-HyMeX.1321, 2015.

Braud, I., Ayral, P.-A., Bouvier, C., Branger, F., Delrieu, G., Le Coz, J., Nord, G., Vandervaere, J.-P., Anquetin, S., Adamovic, M., Andrieu, J., Batiot, C., Boudevillain, B., Brunet, P., Carreau, J., Confoland, A., Didon-Lescot, J.-F., Domergue, J.-M., Douvinet, J., Dramais, G., Freydier, R., Gérard, S., Huza, J., Leblois, E., Le Bourgeois, O., Le Boursicaud, R., Marchand, P., Martin, P., Nottale, L., Patris, N., Renard, B., Seidel, J.-L., Taupin, J.-D., Vannier, O., Vincendon, B., and Wijbrans, A.: Multi-scale hydrometeorological observation and modelling for flash flood understanding, Hydrol. Earth Syst. Sci., 18, 3733–3761, https://doi.org/10.5194/hess-18-3733-2014, 2014.

Brocca, L., Melone, F., Moramarco, T., and Morbidelli, R.: Antecedent wetness conditions based on ERS scatterometer data, J. Hydrol., 364, 73–87, https://doi.org/10.1016/j.jhydrol.2008.10.007, 2009a.

Brocca, L., Melone, F., Moramarco, T., and Singh, V. P.: Assimilation of Observed Soil Moisture Data in Storm Rainfall-Runoff Modeling, J. Hydrol. Eng., 14, 153–165, https://doi.org/10.1061/(ASCE)1084-0699(2009)14:2(153), 2009b.

Brocca, L., Melone, F., Moramarco, T., and Morbidelli, R.: Soil moisture temporal stability over experimental areas in Central Italy, Geoderma, 148, 364–374, https://doi.org/10.1016/j.geoderma.2008.11.004, 2009c.

Brocca, L., Melone, F., Moramarco, T., Penna, D., Morga, M., Matgen, A., Gumuzzio, A., Martinez-Fernandez, J., and Wagner, W.: Detecting threshold hydrological response through satellite soil moisture data, Bodenkultur, 64, 7–12, 2013.

Bronstert, A. and Bárdossy, A.: The role of spatial variability of soil moisture for modelling surface runoff generation at the small catchment scale, Hydrol. Earth Syst. Sci., 3, 505–516, https://doi.org/10.5194/hess-3-505-1999, 1999.

Castillo, V. M., Gómez-Plaza, A., and Martínez-Mena, M.: The role of antecedent soil water content in the runoff response of semi-arid catchments: A simulation approach, J. Hydrol., 284, 114–130, https://doi.org/10.1016/S0022-1694(03)00264-6, 2003.

Cea, L., Legout, C., Grangeon, T., and Nord, G.: Impact of model simplifications on soil erosion predictions: application of the GLUE methodology to a distributed event-based model at the hillslope scale, Hydrol. Process., 30, 1096–1113, https://doi.org/10.1002/hyp.10697, 2015.

Creutin, J. D. and Borga, M.: Radar hydrology modifies the monitoring of flash-flood hazard, Hydrol. Process., 17, 1453–1456, https://doi.org/10.1002/hyp.5122, 2003.

Crow, W. T., Bindlish, R., and Jackson, T. J.: The added value of spaceborne passive microwave soil moisture retrievals for forecasting rainfall-runoff partitioning, Geophys. Res. Lett., 32, 1–5, https://doi.org/10.1029/2005GL023543, 2005.

Delrieu, G., Wijbrans, A., Boudevillain, B., Faure, D., Bonnifait, L., and Kirstetter, P. E.: Geostatistical radar-raingauge merging: A novel method for the quantification of rain estimation accuracy, Adv. Water Resour., 71, 110–124, https://doi.org/10.1016/j.advwatres.2014.06.005, 2014.

Ducrocq, V., Braud, I., Davolio, S., Ferretti, R., Flamant, C., Jansa, A., Kalthoff, N., Richard, E., Taupier-Letage, I., Ayral, P. A., Belamari, S., Berne, A., Borga, M., Boudevillain, B., Bock, O., Boichard, J. L., Bouin, M. N., Bousquet, O., Bouvier, C., Chiggiato, J., Cimini, D., Corsmeier, U., Coppola, L., Cocquerez, P., Defer, E., Delanoe, J., Di Girolamo, P., Doerenbecher, A., Drobinski, P., Dufournet, Y., Fourrié, N., Gourley, J. J., Labatut, L., Lambert, D., Le Coz, J., Marzano, F. S., Molinié, G., Montani, A., Nord, G., Nuret, M., Ramage, K., Rison, W., Roussot, O., Said, F., Schwarzenboeck, A., Testor, P., Van Baelen, J., Vincendon, B., Aran, M., and Tamayo, J.: HyMeX-SOP1: The field campaign dedicated to heavy precipitation and flash flooding in the northwestern mediterranean, B. Am. Meteorol. Soc., 95, 1083–1100, https://doi.org/10.1175/BAMS-D-12-00244.1, 2014.

Dunne, T. and Black, R. D.: Partial Area Contributions to Storm Runoff in a Small New England, Water Resour. Res., 6, 1296–1311, 1970.

Garcia-Estringana, P., Latron, J., Llorens, P., and Gallart, F.: Spatial and temporal dynamics of soil moisture in a Mediterranean mountain area (Vallcebre, NE Spain), Ecohydrology, 6, 741–753, https://doi.org/10.1002/eco.1295, 2013.

Gaume, E., Livet, M., Desbordes, M., and Villeneuve, J. P.: Hydrological analysis of the river Aude, France, flash flood on 12 and 13 November 1999, J. Hydrol., 286, 135–154, https://doi.org/10.1016/j.jhydrol.2003.09.015, 2004.

Gaume, E., Bain, V., Bernardara, P., Newinger, O., Barbuc, M., Bateman, A., Blaškovičová, L., Blöschl, G., Borga, M., Dumitrescu, A., Daliakopoulos, I., Garcia, J., Irimescu, A., Kohnova, S., Koutroulis, A., Marchi, L., Matreata, S., Medina, V., Preciso, E., Sempere-Torres, D., Stancalie, G., Szolgay, J., Tsanis, I., Velasco, D., and Viglione, A.: A compilation of data on European flash floods, J. Hydrol., 367, 70–78, https://doi.org/10.1016/j.jhydrol.2008.12.028, 2009.

Gonzales, A. L., Nonner, J., Heijkers, J., and Uhlenbrook, S.: Comparison of different base flow separation methods in a lowland catchment, Hydrol. Earth Syst. Sci., 13, 2055–2068, https://doi.org/10.5194/hess-13-2055-2009, 2009.

Grillakis, M. G., Koutroulis, A. G., Komma, J., Tsanis, I. K., Wagner, W., and Blöschl, G.: Initial soil moisture effects on flash flood generation – A comparison between basins of contrasting hydro-climatic conditions, J. Hydrol., 541, 206–217, https://doi.org/10.1016/j.jhydrol.2016.03.007, 2016.

Hardie, M. A., Cotching, W. E., Doyle, R. B., Holz, G., Lisson, S., and Mattern, K.: Effect of antecedent soil moisture on preferential flow in a texture-contrast soil, J. Hydrol., 398, 191–201, https://doi.org/10.1016/j.jhydrol.2010.12.008, 2011.

Horton, R. E.: The rôle of infiltration in the hydrologic cycle, Eos, Trans. Am. Geophys. Union, 14, 446–460, 1933.

Hrachowitz, M., Bohte, R., Mul, M. L., Bogaard, T. A., Savenije, H. H. G., and Uhlenbrook, S.: On the value of combined event runoff and tracer analysis to improve understanding of catchment functioning in a data-scarce semi-arid area, Hydrol. Earth Syst. Sci., 15, 2007–2024, https://doi.org/10.5194/hess-15-2007-2011, 2011.

Huang, M., Gallichand, J., Dong, C., Wang, Z., and Shao, M.: Use of soil moisture data and curve number method for estimating runoff in the Loess Plateau of China, Hydrol. Process., 21, 1471–1481, https://doi.org/10.1002/hyp.6312, 2007.

Huet, P., Martin, X., Prime, J.-L., Foin, P., Laurain, C., and Cannard, P.: Retour D'Expérience des Crues de Septembre 2002 dans les Departements du Gard de l'Hérault, du Vaucluse, des Bouches-du-Rhône, de l'Ardèche et de la Drôme. Rapport Consolidé après Phase Contradictoire, Paris, available at: https://www.ladocumentationfrancaise.fr/var/storage/rapports-publics/034000547.pdf (last access: 19 November 2018), 2003.

Huza, J., Teuling, A. J., Braud, I., Grazioli, J., Melsen, L. A., Nord, G., Raupach, T. H., and Uijlenhoet, R.: Precipitation, soil moisture and runoff variability in a small river catchment (Ardèche, France) during HyMeX Special Observation Period 1, J. Hydrol., 516, 330–342, https://doi.org/10.1016/j.jhydrol.2014.01.041, 2014.

Javelle, P., Fouchier, C., Arnaud, P., and Lavabre, J.: Flash flood warning at ungauged locations using radar rainfall and antecedent soil moisture estimations, J. Hydrol., 394, 267–274, https://doi.org/10.1016/j.jhydrol.2010.03.032, 2010.

Jawson, S. D. and Niemann, J. D.: Spatial patterns from EOF analysis of soil moisture at a large scale and their dependence on soil, land-use, and topographic properties, Adv. Water Resour., 30, 366–381, https://doi.org/10.1016/j.advwatres.2006.05.006, 2007.

Korres, W., Reichenau, T. G., Fiener, P., Koyama, C. N., Bogena, H. R., Cornelissen, T., Baatz, R., Herbst, M., Diekkrüger, B., Vereecken, H., and Schneider, K.: Spatiotemporal soil moisture patterns – A meta-analysis using plot to catchment scale data, J. Hydrol., 520, 326–341, https://doi.org/10.1016/j.jhydrol.2014.11.042, 2015.

Ladson, A., Bronw, R., Neal, B., and Nathan, R.: A standard approach to baseflow separation using the Lyne and Hollick filter, Aust. J. Water Resour., 17, 25–34, https://doi.org/10.7158/W12-028.2013.17.1, 2013.

Le Coz, J., Renard, B., Bonnifait, L., Branger, F., and Le Boursicaud, R.: Combining hydraulic knowledge and uncertain gaugings in the estimation of hydrometric rating curves: A Bayesian approach, J. Hydrol., 509, 573–587, https://doi.org/10.1016/j.jhydrol.2013.11.016, 2014.

Le Lay, M. and Saulnier, G. M.: Exploring the signature of climate and landscape spatial variabilities in flash flood events: Case of the 8–9 September 2002 Cévennes-Vivarais catastrophic event, Geophys. Res. Lett., 34, 1–5, https://doi.org/10.1029/2007GL029746, 2007.

Li, H., Sivapalan, M., and Tian, F.: Comparative diagnostic analysis of runoff generation processes in Oklahoma DMIP2 basins: The Blue River and the Illinois River, J. Hydrol., 418–419, 90–109, https://doi.org/10.1016/j.jhydrol.2010.08.005, 2012.

Lyne, V. D. and Hollick, M.: Stochastic time-variable rainfall-runoff modeling, Institute of Engineers Australia National Conference Publications, Perth, 10, 89–92, 10–12 September 1979.

Marchi, L., Borga, M., Preciso, E., and Gaume, E.: Characterisation of selected extreme flash floods in Europe and implications for flood risk management, J. Hydrol., 394, 118–133, https://doi.org/10.1016/j.jhydrol.2010.07.017, 2010.

Marsalek, J.: Calibration of the tipping-bucket raingage, J. Hydrol., 53, 343–354, https://doi.org/10.1016/0022-1694(81)90010-X, 1981.

Massari, C., Brocca, L., Moramarco, T., Tramblay, Y., and Didon Lescot, J.-F.: Potential of soil moisture observations in flood modelling: Estimating initial conditions and correcting rainfall, Adv. Water Resour., 74, 44–53, https://doi.org/10.1016/j.advwatres.2014.08.004, 2014a.

Massari, C., Brocca, L., Barbetta, S., Papathanasiou, C., Mimikou, M., and Moramarco, T.: Using globally available soil moisture indicators for flood modelling in Mediterranean catchments, Hydrol. Earth Syst. Sci., 18, 839–853, https://doi.org/10.5194/hess-18-839-2014, 2014b.

Massari, C., Brocca, L., Ciabatta, L., Moramarco, T., Gabellani, S., Albergel, C., De Rosnay, P., Puca, S., and Wagner, W.: The Use of H-SAF Soil Moisture Products for Operational Hydrology: Flood Modelling over Italy, Hydrology, 2, 2–22, https://doi.org/10.3390/hydrology2010002, 2015.

McMillan, H. K., Gueguen, M., Grimon, E., Woods, R., Clark, M., and Rupp, D. E.: Spatial variability of hydrological processes and model structure diagnostics in a $50\,km^2$ catchment, Hydrol. Process., 28, 4896–4913, https://doi.org/10.1002/hyp.9988, 2014.

McMillan, H. K. and Srinivasan, M. S.: Characteristics and controls of variability in soil moisture and groundwater in a headwater catchment, Hydrol. Earth Syst. Sci., 19, 1767–1786, https://doi.org/10.5194/hess-19-1767-2015, 2015.

Merz, B. and Plate, E. J.: An analysis of the effects of spatial variability of soil and soil moisture on runoff, Water Resour. Res., 33, 2909–2922, https://doi.org/10.1029/97WR02204, 1997.

Merz, R. and Blöschl, G.: A regional analysis of event runoff coefficients with respect to climate and catchment characteristics in Austria, Water Resour. Res., 45, 1–19, https://doi.org/10.1029/2008WR007163, 2009.

Merz, R., Blöschl, G., and Parajka, J.: Spatio-temporal variability of event runoff coefficients, J. Hydrol., 331, 591–604, https://doi.org/10.1016/j.jhydrol.2006.06.008, 2006.

Miller, M. P., Susong, D. D., Shope, Ch. L., Heilweil, V. M., and Stolp, B. J.: Continuous estimation of baseflow in snowmelt-dominated streams and rivers in the Upper Colorado River Basin: A chemical hydrograph separation approach, Water Resour. Res., 50, 6986–6999, https://doi.org/10.1002/2013WR014939, 2014.

Molina, A. J., Latron, J., Rubio, C. M., Gallart, F., and Llorens, P.: Spatio-temporal variability of soil water content on the local scale in a Mediterranean mountain area (Vallcebre, North Eastern Spain). How different spatio-temporal scales reflect mean soil water content, J. Hydrol., 516, 182–192, https://doi.org/10.1016/j.jhydrol.2014.01.040, 2014.

Molini, A., Lanza, L. G., and La Barbera, P.: The impact of tipping-bucket raingauge measurement errors on design rainfall for urban-scale applications, Hydrol. Process., 19, 1073–1088, https://doi.org/10.1002/hyp.5646, 2005.

Molinié, G., Ceresetti, D., Anquetin, S., Creutin, J. D., and Boudevillain, B.: Rainfall regime of a mountainous mediterranean region: Statistical analysis at short time steps, J. Appl. Meteorol. Clim., 51, 429–448, https://doi.org/10.1175/2011JAMC2691.1, 2012.

Moran, M. S., Holifield Collins, C. D., Goodrich, D. C., Qi, J., Shannon, D. T., and Olsson, A.: Long-term remote sensing database, Walnut Gulch Experimental Watershed, Arizona, United States, Water Resour. Res., 44, 1–8, https://doi.org/10.1029/2006WR005777, 2008.

Morbidelli, R., Saltalippi, C., Flammini, A., Corradini, C., Brocca, L., and Govindaraju, R. S.: An investigation of the effects of spatial heterogeneity of initial soil moisture content on surface runoff simulation at a small watershed scale, J. Hydrol., 539, 589–598, https://doi.org/10.1016/j.jhydrol.2016.05.067, 2016.

Nakamura, R.: Runoff analysis by electrical conductance of water, J. Hydrol., 14, 197–212, https://doi.org/10.1016/0022-1694(71)90035-7, 1971.

Nathan, R. J. and McMahon, T.: Evaluation of Automated Techniques for Base Flow and Recession Analyses, Water Resour. Res., 26, 1465–1473, https://doi.org/10.1029/WR026i007p01465, 1990.

Norbiato, D., Borga, M., Degli Esposti, S., Gaume, E., and Anquetin, S.: Flash flood warning based on rainfall thresholds and soil moisture conditions: An assessment for gauged and ungauged basins, J. Hydrol., 362, 274–290, https://doi.org/10.1016/j.jhydrol.2008.08.023, 2008.

Norbiato, D., Borga, M., Merz, R., Blöschl, G., and Carton, A.: Controls on event runoff coefficients in the eastern Italian Alps, J. Hydrol., 375, 312–325, https://doi.org/10.1016/j.jhydrol.2009.06.044, 2009.

Nord, G.: Digital Terrain Model (DTM) of the Auzon catchment region, https://doi.org/10.6096/MISTRALS-HyMeX.1389, 2015.

Nord, G., Boudevillain, B., Berne, A., Branger, F., Braud, I., Dramais, G., Gérard, S., Le Coz, J., Legoût, C., Molinié, G., Van Baelen, J., Vandervaere, J.-P., Andrieu, J., Aubert, C., Calianno, M., Delrieu, G., Grazioli, J., Hachani, S., Horner, I., Huza, J., Le Boursicaud, R., Raupach, T. H., Teuling, A. J., Uber, M., Vincendon, B., and Wijbrans, A.: A high space–time resolution dataset linking meteorological forcing and hydro-sedimentary response in a mesoscale Mediterranean catchment (Auzon) of the Ardèche region, France, Earth Syst. Sci. Data, 9, 221–249, https://doi.org/10.5194/essd-9-221-2017, 2017.

OHMCV: Runoff and erosion plots, Pradel, CNRS – OSUG – OREME, available at: http://mistrals.sedoo.fr/?editDatsId=1347 (last access: 19 November 2018), doi10.17178/OHMCV.ERO.PRA.10-13.1, 2009.

OHMCV: Hpiconet rain gauge network, CNRS – OSUG – OREME, available at: http://mistrals.sedoo.fr/?editDatsId=656 (last access: 19 November 2018), https://doi.org/10.17178/OHMCV.RTS.AUZ.10-14.1, 2010.

OHMCV: Gazel and Claduègne hydro-sedimentary stations, CNRS – OSUG – OREME, available at: http://mistrals.sedoo.fr/?editDatsId=993 (last access: 19 November 2018), https://doi.org/10.17178/OHMCV.HSS.CLA.11-14.1, 2011.

OHMCV: Soil moisture sensor network, Gazel and Claduègne catchments, CNRS – OSUG – OREME, available at: http://mistrals.sedoo.fr/?editDatsId=1350 (last access: 19 November 2018), https://doi.org/10.17178/OHMCV.SMO.CLA.13-14.1, 2013.

Panziera, L., James, C. N., and Germann, U.: Mesoscale organization and structure of orographic precipitation producing flash floods in the Lago Maggiore region, Q. J. Roy. Meteor. Soc., 141, 224–248, https://doi.org/10.1002/qj.2351, 2015.

Pellerin, B. A., Wollheim, W. M., Feng, X., and Vörösmarty, C. J.: The application of electrical conductivity as a tracer for hydrograph separation in urban catchments, Hydrol. Process., 22, 1810–1818, https://doi.org/10.1002/hyp.6786, 2008.

Penna, D., Tromp-van Meerveld, H. J., Gobbi, A., Borga, M., and Dalla Fontana, G.: The influence of soil moisture on threshold runoff generation processes in an alpine headwater catchment, Hydrol. Earth Syst. Sci., 15, 689–702, https://doi.org/10.5194/hess-15-689-2011, 2011.

Renard, K. G., Nichols, M. H., Woolhiser, D. A., and Osborn, H. B.: A brief background on the U.S. Department of Agriculture Agricultural Research Service Walnut Gulch Experimental Watershed, Water Resour. Res., 44, 1–11, https://doi.org/10.1029/2006WR005691, 2008.

Savenije, H. H. G.: HESS Opinions "Topography driven conceptual modelling (FLEX-Topo)", Hydrol. Earth Syst. Sci., 14, 2681–2692, https://doi.org/10.5194/hess-14-2681-2010, 2010.

Slaughter, C. W., Marks, D., Flerchinger, G. N., Van Vactor, S. S., and Burgess, M.: Thirty-five years of research data collection at the Reynolds Creek Experimental Watershed, Idaho, United States, Water Resour. Res., 37, 2819–2823, https://doi.org/10.1029/2001WR000413, 2001.

Sloto, R. and Crouse, M. Y.: Hysep: a computer program for streamflow hydrograph separation and analysis, U.S. Geological Survey Water-Resources Investigations Report, 1996–4040, 46 pp., available at: https://pubs.er.usgs.gov/publication/wri964040 (last access: 19 November 2018), 1996.

Tramblay, Y., Bouvier, C., Martin, C., Didon-Lescot, J. F., Todorovik, D., and Domergue, J. M.: Assessment of initial soil moisture conditions for event-based rainfall-runoff modelling, J. Hydrol., 387, 176–187, https://doi.org/10.1016/j.jhydrol.2010.04.006, 2010.

Tramblay, Y., Bouaicha, R., Brocca, L., Dorigo, W., Bouvier, C., Camici, S., and Servat, E.: Estimation of antecedent wetness conditions for flood modelling in northern Morocco, Hydrol. Earth Syst. Sci., 16, 4375–4386, https://doi.org/10.5194/hess-16-4375-2012, 2012.

U.S. Geological Survey (USGS): hysep.R, GitHub Repository, available at: https://github.com/USGS-R/DVstats/blob/master/R/hysep.R (last access: 19 November 2018), 2015.

Vachaud, G., Passerat De Silans, A., Balabanis, P., and Vauclin, M.: Temporal Stability of Spatially Measured Soil Water Probability Density Function, Soil Sci. Soc. Am. J., 49, 822–828, https://doi.org/10.2136/sssaj1985.03615995004900040006x, 1985.

Vanderlinden, K., Vereecken, H., Hardelauf, H., Herbst, M., Martínez, G., Cosh, M. H., and Pachepsky, Y. A.: Temporal Stability of Soil Water Contents: A Review of Data and Analyses, Vadose Zone J., 11, 19, https://doi.org/10.2136/vzj2011.0178, 2012.

Vereecken, H., Huisman, J. A., Pachepsky, Y., Montzka, C., van der Kruk, J., Bogena, H., Weihermüller, L., Herbst, M., Martinez, G., and Vanderborght, J.: On the spatio-temporal dynamics of soil moisture at the field scale, J. Hydrol., 516, 76–96, https://doi.org/10.1016/j.jhydrol.2013.11.061, 2014.

Wood, E. F., Sivapalan, M., and Beven, K.: Similarity and Scale in Catchment Storm Response, Rev. Geophys., 28, 1–18, 1990.

Zehe, E. and Blöschl, G.: Predictability of hydrologic response at the plot and catchment scales: Role of initial conditions, Water Resour. Res., 40, 1–21, https://doi.org/10.1029/2003WR002869, 2004.

Zehe, E. and Sivapalan, M.: Threshold behaviour in hydrological systems as (human) geo-ecosystems: manifestations, controls, implications, Hydrol. Earth Syst. Sci., 13, 1273–1297, https://doi.org/10.5194/hess-13-1273-2009, 2009.

Zehe, E., Becker, R., Bárdossy, A., and Plate, E.: Uncertainty of simulated catchment runoff response in the presence of threshold processes: Role of initial soil moisture and precipitation, J. Hydrol., 315, 183–202, https://doi.org/10.1016/j.jhydrol.2005.03.038, 2005.

Zucco, G., Brocca, L., Moramarco, T., and Morbidelli, R.: Influence of land use on soil moisture spatial–temporal variability and monitoring, J. Hydrol., 516, 193–199, https://doi.org/10.1016/j.jhydrol.2014.01.043, 2014.

10

How to predict hydrological effects of local land use change: How the vegetation parameterisation for short rotation coppices influences model results

F. Richter[1], **C. Döring**[1], **M. Jansen**[1], **O. Panferov**[2,3], **U. Spank**[4], and **C. Bernhofer**[4]

[1]Department of Soil Science of Temperate Ecosystems, Georg-August-Universität Göttingen, Büsgenweg 2, 37077 Göttingen, Germany

[2]Department of Bioclimatology, Georg-August-Universität Göttingen, Büsgenweg 2, 37077 Göttingen, Germany

[3]Institute of Climatology and Climate Protection, University of Applied Sciences, Bingen am Rhein, Berlinstr. 109, 55411 Bingen am Rhein, Germany

[4]Institute of Hydrology and Meteorology, Technische Universität Dresden, Pienner Str. 23, 01737 Tharandt, Germany

Correspondence to: F. Richter (falk.richter@forst.uni-goettingen.de)

Abstract. Among the different bioenergy sources, short rotation coppices (SRC) with poplar and willow trees are one of the promising options in Europe. SRC provide not only woody biomass but also additional ecosystem services. However, a known shortcoming is the potentially lower groundwater recharge caused by the potentially higher evapotranspiration demand compared to annual crops. The complex feedbacks between vegetation cover and water cycle can be only correctly assessed by application of well-parameterised and calibrated numerical models. In the present study, the hydrological model system WaSim (Wasserhaushalts-Simulations-Model) is implemented for assessment of the water balance. The focus is the analysis of simulation uncertainties caused by the use of guidelines or transferred parameter sets from scientific literature compared to "actual" parameterisations derived from local measurements of leaf area index (LAI), stomatal resistance (Rsc) and date of leaf unfolding (LU). The analysis showed that uncertainties in parameterisation of vegetation lead to implausible model results. LAI, Rsc and LU are the most sensitive plant physiological parameters concerning the effects of enhanced SRC cultivation on water budget or groundwater recharge. Particularly sensitive is the beginning of the growing season, i.e. LU. When this estimation is wrong, the accuracy of LAI and Rsc description plays a minor role. Our analyses illustrate that the use of locally measured vegetation parameters, like maximal LAI, and meteorological variables, like air temperature, to estimate LU give better results than literature data or data from remote network stations. However, the direct implementation of locally measured data is not always advisable or possible. Regarding Rsc, the adjustment of local measurements gives the best model evaluation. For local and accurate studies, measurements of model sensitive parameters like LAI, Rsc and LU are valuable information. The derivation of these model parameters based on local measurements shows the best model fit. Additionally, the adjusted seasonal course of LAI and Rsc is less sensitive to different estimates for LU. Different parameterisations, as they are all eligible either from local measurements or scientific literature, can result in modelled ground water recharge to be present or completely absent in certain years under poplar SRC.

1 Introduction

In the context of climate change, mitigation and reduction of greenhouse gas (GHG) emissions, bioenergy is one of the possible alternatives for fossil fuels. Among the different bioenergy sources, short rotation coppices (SRC) with mainly poplar and willow trees are promising options in Europe (Djomo et al., 2011). SRC provide woody biomass as well as additional ecosystem services. Seepage water quality is enhanced due to lower fertiliser requirements and higher nutrient use efficiency (Aronsson et al., 2000; Schmidt-Walter and Lamersdorf, 2012). Compared to conventional annual crops, SRC sequester more carbon and emit less N_2O (Don et al., 2012), which is one of the most important GHGs. As structural landscape elements in rural areas SRC might also contribute positively to biodiversity (Baum et al., 2012).

However, SRC are not without disadvantages. The most quantitatively assessable disadvantage is the potentially lower groundwater recharge (GWR) being caused by higher evapotranspiration of poplar and willow plantations in comparison to annual crops (Lasch et al., 2010; Schmidt-Walter and Lamersdorf, 2012). An assessment using hydrological models can help to minimise negative and maximise positive ecological effects caused by land use change from arable land to SRC at regional and local scales, e.g. to regional climate and/or to adjacent ecosystems. To adequately quantify the effects of this land use change, the hydrological models must correctly reproduce the hydrological feedback effects of vegetation and land use management. Beside the choice of proper modelling approach, a careful parameterisation of land use and vegetation is needed to obtain reliable simulation results. The aim of the present study is to assess the effect of parameter uncertainties of the land use type poplar SRC on modelling results.

The planting of SRC causes the occurrence of new factors and complex factor interactions influencing site water fluxes. One factor is the perennial vegetation cover with higher leaf area index (LAI; $m^2 \, m^{-2}$) combined with a longer growing season compared to annual crops (Petzold et al., 2010).

LAI directly affects canopy interception due to an almost linear correlation between LAI and canopy storage (Rutter et al., 1971). Outside the growing season LAI, more precisely plant area index (PAI), additionally provides canopy storage by woody biomass – i.e. stems and branches. Furthermore, the transpiration is positively correlated with LAI; i.e. higher LAI causes higher transpiration rates. However, LAI is negatively correlated with soil evaporation as higher LAI results in more shadowing, less solar radiation input and consequently less evaporation below the vegetation cover. Other important parameters controlling the water balance are the stomatal resistance (Rsc), rooting depth and root distribution. The structural and biophysical parameters do not remain constant during the year, but have a seasonal or even annual variability. The largest variations occur during the growing period. Thus, the beginning and length of the growing season should be known for an adequate description of the seasonal dynamic of vegetation parameters.

The smaller the scale of interest, the more time-resolved parameterisation of land use and vegetation is necessary to capture the spatial and temporal variability of effects. There are two possibilities to obtain the required information: the first one is to measure the parameters like LAI or Rsc directly at the investigated site, which is labour consuming and time-consuming. A second possibility is to use parameters from

scientific literature. This information is quite rare for SRC, because this land use scarcely came into focus of investigation as a part of the renewable energy discussion during the last years (Surendran Nair et al., 2012). Typically only one value, rather than the annual course, is given in literature, e.g. the maximum value for LAI, or the minimum value for Rsc, which are related to the maximum transpiration. In such cases the annual course for these parameters has to be estimated for hydrological modelling. Hence, the question on transferability of published results obtained in a certain area and in a specific year to other regions and years has to be solved in each study separately. However, it is well-known that neither literature values nor direct measurements provide the true values of model parameters, but more or less representative approximations, because of spatial and temporal heterogeneity of vegetation stands including SRC.

The overarching aim of our research is the evaluation of land use change effects. However, this study does not focus on land use change effects in any way but rather on the evaluation of a suitable tool. In this study we used the results of our own measurements of LAI, Rsc and the estimation of leaf unfolding date (LU), determining the beginning of growing season, for a poplar SRC to parameterise the hydrological model system WaSim (Wasserhaushalts-Simulations-Model) (Schulla and Jasper, 2013). The aim of the present study is to assess the effect of parameterisation uncertainties of the land use type poplar SRC on modelling results. The hypothesis that the parameterisation of LAI, Rsc and LU based on measurements shows a better model fit than the use of values reported in literature in combination with approximation about the annual course should be proofed. Our specific objectives are (1) to quantify the WaSim response (sensitivity) to variations of following parameters: LAI, Rsc and LU, caused by different measurement methods and modelling approaches; (2) to estimate the most sensitive parameter; and (3) to evaluate quantitatively whether it is advisable to implement locally point-measured values of sensitive parameters directly. We used GWR and plant available water as indicators.

2 Material and methods

2.1 Study site

The study site Reiffenhausen (51.67° N, 10.65° E; 325 m a.s.l.) is located south of Göttingen, central Germany. According to the meteorological data provided by the German Weather Service (Deutscher Wetterdienst, DWD), for the station Göttingen (DWD station-ID: 01691), nearest to the study site, the climate is characterised by an average temperature of 9.1 °C (±0.7 °C) and a mean annual precipitation sum of 635 mm (±122 mm) for the period 1971–2010. The site Reiffenhausen was established as part of the interdisciplinary investigations of SRC by the joint integrated project BEST ("Bioenergie-Regionen stärken"

– Boosting Bioenergy Regions), which ran from 2010 to 2014 and was funded by the German Ministry of Education and Research (BMBF). The aim of BEST was to develop regionally appropriate concepts and innovative solutions for the production of biomass, with focus on SRC, and to evaluate ecological and economic impacts.

The soil in Reiffenhausen is characterised by a sedimentary deposits of Middle and Upper Buntsandstein, like sandstone, siltstone and clay stone. The main soil types present at the field level are Stagnic Cambisol and Haplic Stagnosol with a soil quality (Ackerzahl) of approximately 45 points. The maximum available points are 100 for very good agriculture fields (Blume et al., 2010). The soil texture is dominated by loamy sand or silty clay.

The plantation Reiffenhausen was established at a former arable field in March 2011 with the poplar clone *Populus nigra x Populus maximowiczii*, hereafter Max1. The poplar SRC were planted with 0.2 m long cuttings on 0.4 ha in a double row system with alternating inter-row distances of 0.75 and 1.50 m, and a spacing of 1.0 m within the rows, yielding an overall planting density of 8800 cuttings per hectare. A detailed site description, including soil chemistry and biomass information is given by Hartmann et al. (2014). In 2011 the weather conditions at Reiffenhausen were unfavourably dry for the initial growth after planting. During the first months in 2011, from February to May, the precipitation sum was unusually low: only 42 % of the long-term (1971–2010) precipitation (78 of 188 mm for the long-term mean value of the same period of the year). This led to dry soil conditions, especially in the upper soil, and resulting in a survival rate of only 63 % for the mono-stem cycle of the poplar SRC. In 2013, when the most measurements and investigations took place the poplar SRC reached a height of 5–6 m, indicating that the unfavourable initial conditions were somewhat improved. All parameters influencing the hydrology in the model are of the order of a fully developed SRC, although it is in mono-stem cycle. The development of the rooting system was eventually enforced by the dry conditions during the initial phase (Broeckx et al., 2013). Rooting depth was more than 1 m in the second year after planting, exploiting the main part of soil layer above the bedrock (Kalberlah, 2013). The LAI reached 5 in 2012 and 6–7 in 2013, which is typical for a fully grown poplar SRC (Schmidt-Walter et al., 2014).

Observations for LU are also used from the site Großfahner, which is also part of the BEST-research project. Großfahner is located near Erfurt, Thuringia, Germany, and was also established in 2011 with the poplar clone Max1 (Hartmann et al., 2014).

2.2 Measurements
2.2.1 Meteorological and local soil measurements
In Reiffenhausen, the micrometeorological measurements were carried out in the centre of the SRC stand; the instruments were installed above the vegetation on a 10 m mast. The air temperature and humidity were measured using Hmp45C (Campbell Sci.; Loughborough, UK); wind speed and wind direction (wind sensor compact and wind direction sensor compact, both ThiesClima; Göttingen, Germany), atmospheric pressure (pressure sensor, Theodor Friedrichs& Co.; Schenefeld, Germany) and solar radiation (CMP3, Kipp&Zonen, Delft, the Netherlands) were measured continuously with 1 Hz frequency and averaged over 15 min. 5 min precipitation sums were obtained using an ombrometer (precipitation transmitter, ThiesClima; Göttingen, Germany). The values were averaged and stored by a CR1000 data logger (Campbell Sci.; Loughborough, UK). In addition to the meteorological measurements in the centre of the poplar SRC, a reference station was similarly equipped and installed approximately 500 m to the north from the stand in the open place (short-grass meadow) to measure the climate variables unaffected by the poplar SRC.

The soil moisture was measured continuously every 15 min using tensiometers and soil water content probes at 20, 60 and 120 cm depth, by six (tensiometers) and three (probes) sensors at every depth. Tensiometers were constructed in the Department of Soil Science for the study using the PCFA6D pressure sensor (Honeywell; Morristown, NJ, USA) and a P-80 ceramic (CeramTec AG; Marktredwitz, Germany). Volumetric soil water content and soil temperature were measured using SM-300 probes (Delta-T Devices Ltd; Cambridge, UK). Additionally, descriptions of soil horizons and soil texture were assessed using a soil pit near the SRC.

In addition to the data of Reiffenhausen, meteorological and phenological data are used from the Tharandter Wald (Tharandt Forest) region located in the federal state Saxony (Germany), 15 km south-west of city of Dresden. As climate characteristics of this region are comparable with Reiffenhausen, a proper set of comparison data are provided. Detailed information about measurement programs of Tharandter Wald can be found, i.a. in Bernhofer (2002) and Spank et al. (2013). In the frame of this study, phenological observation data from the International Phenological Garden (IPG) Tharandt–Hartha and meteorological measurement data (air temperature, air humidity and precipitation) from climate stations Grillenburg and Wildacker have special importance. Grillenburg and Wildacker are the nearest meteorological long-term measurements sites from IPG and are situated about 3 km away. Both stations provide meteorological and climatological information since 1958. Grillenburg represents a standard meteorological station fulfilling all guidelines and standards of the World Meteorological Organisation (WMO) for large-scale representativeness of climatological observations. However, measurements on this site sometimes do not describe local climatic characteristic of the region, particularly related to the daily minimum and maximum of air temperature. In contrary, the climate station Wildacker, not fulfilling WMO standards of fetch and horizon heightening, better represents the local climatic situation.

2.2.2 Leaf area index

For the present study we use the definition of leaf area index by Watson (1947) cited in Breda (2003) as the total one-sided area of leaf tissue per unit ground surface area with the dimension of $m^2 \, m^{-2}$ (or dimensionless). There are numerous ground-based as well as remote sensing-based techniques to estimate LAI. An extensive overview of ground-based methods is given by Breda (2003). Direct methods, such as allometric, litter collection and harvesting, are based on statistically significant sampling of phytoelements and phytoelement dimensions. Among these method, only the harvesting can provide the information on the seasonal dynamic of LAI for the whole season or year. The obvious disadvantages of harvesting as a destructive method, however, are that it is very time-consuming and labour consuming, the canopy is irreversibly damaged and further statistically representative LAI measurements for seasonal dynamics are affected.

Indirect ground-based methods are non-destructive and based on the inversion of the Beer–Lambert law, i.e. on measurements of the extinction of short-wave solar radiation by the canopy. The extinction is related to the vegetation structure parameters including LAI (Eq. 1).

$$\text{LAI} = \frac{\ln(I/I_0)}{k} \quad \text{with} \quad k = \frac{G(\Theta, \alpha)}{\cos\Theta} \quad (1)$$

where LAI is the leaf area index for the vegetation layer, I_0 is the radiation intensity incident to the vegetation layer, I radiation intensity at the lower bound of vegetation layer and k is the extinction coefficient (Breda, 2003). The function G is the projection of unit foliage area on the plane normal to the direction Θ, Θ zenith angle and α the leaf angle distribution. It should be also noted that indirect methods do not estimate LAI rather PAI as the light attenuation is caused not only by leaves but also by branches and tree stems as well. To derive LAI, either the share of woody material is subtracted from PAI, or it is assumed that the attenuation is caused for the most part by leaves (especially for dense canopies). The underlying assumptions, e.g. on stand homogeneity and small black opaque phytoelements that have to be considered to ensure the applicability of indirect methods, as well as advantages and disadvantages of various methods, are presented, e.g. in the LAI-2000 Manual (LI-COR INC, 1992), in Breda (2003) and (Jonckheere et al., 2004).

For the present study the data of one direct and two indirect methods for the estimation of LAI of the poplar SRC in Reiffenhausen were used. For the indirect method we used two different types of instruments. First, two LI-191 SA Line Quantum Sensors (LI-COR Inc., USA) were used to measure incident ($I_{0,\text{PAR}}$) and within-stand photosynthetic active radiation (I_{PAR}) to calculate the LAI using Eq. (1). The $k = 0.5$ for mixed broadleaved species was accepted in our study (Breda, 2003). Second, two plant canopy analysers LAI-2000 (LI-COR Inc., USA) were implemented in two-sensor mode (LI-COR INC, 1992) to obtain LAI and k. Measurements were performed weekly whenever possible from May to November 2013 under homogenous illumination, i.e. at days with overcast conditions or during morning or evening hours. Sensor pairs were cross-validated at the beginning of each measurement day.

In the homogeneous poplar SRC 10 evenly distributed plots were selected. To account for the double row planting of the SRC, 3 m × 3 m square grids with 1 m distance between grid points were marked at every plot so that 16 grid points per plot were obtained for measurements. At each grid point two measurements were performed with instruments oriented along and perpendicularly to SRC rows. Thus, 32 measurements were performed at each of 10 plots during every measurement day. The LAI-2000 was used in two-instrument mode with 25 % view restriction caps to eliminate the influence of operator. The measurements with line quantum sensors LI-191SA were also carried out in two-instrument mode; the measurement design was identical to LAI-2000.

To obtain the reference values for leaf area, the direct destructive sampling, i.e. harvesting, was carried out. All phytoelements within the square column of 1 m^2 surface area were collected and measured with leaf area meter (LI-3100; LI-COR Inc., USA). The sampling was carried out on 26 August 2013 at three plots within the investigated stand.

2.2.3 Stomatal resistance

The dominant factor controlling both the water loss from plant leaves and the uptake of CO_2 for photosynthesis is the resistance of stomata, regulated by the plant in response to environmental conditions. Stomatal resistance, or the reciprocal stomatal conductance, is an important parameter in hydrological modelling, as it controls the transpiration rate for different vegetation types. The version of WaSim applied in the present study uses the Penman–Monteith approach for calculating evapotranspiration and requires a parameter of minimal surface resistance for a state when plants are fully supplied with water (Schulla, 1997; Schulla and Jasper, 2013). The real transpiration modelled is further influenced by meteorological boundary conditions and the available soil water.

For the Rsc measurements in poplar SRC, we used the SC-1 leaf porometer (Decagon Devices Inc.; Pullman, WA, USA). The measurements took place in Reiffenhausen in 2013 and were carried out weekly or fortnightly from May to September only under favourable weather conditions promising minimal resistances: preferable sunny, but at least without rain and with dry leaves. The same 10 plots in the poplar SRC as for the LAI measurements were used, where three sun leaves were marked to be measured at different times. All 10 locations were measured during 1 h to minimise the effects of changing weather conditions. Measurements were started in the morning and when leaves are dry and continued till afternoon, or as long as weather conditions were appropriate.

2.2.4 Phenology – start of growing season

The phenological phases of plants, e.g. leaf unfolding, leaf colouring and the falling of leaves, are controlled by environmental conditions and internal genetic characteristics of plants. Thus, the site and species-specific phenological state is a result of complex interference between length of day, meteorological drivers (mainly temperature and radiation), soil properties, plant provenance, age and height (Menzel, 2000).

Within WaSim a modified approach for estimating LU according to Cannell and Smith (1983) is implemented and used here. A detailed description of this model as presented in Eqs. (2–5), as well as parameterisation examples (Table 1) is given by Menzel (1997).

The model has four parameters: T_0, T_1, a and b which are the threshold temperatures for chilling units and for forcing units and two tree-specific regression parameters, respectively. The starting day for leaves unfolding is calculated according to Eqs. (2), (3), (4) and (5).

$$T_S = \sum_{i=0}^{n} \left\{ \begin{array}{ll} T(t_1 + i\Delta t) - T_1 & T(t_1 + i\Delta t) \geq T_1 \\ 0 & T(t_1 + i\Delta t) < T_1 \end{array} \right. \tag{2}$$

where T_S is the temperature sum, T the daily mean temperature for a day $t_1 + i\Delta t$, t_1 set as 1 February in present study and time step, Δt, as 1 day. The daily mean temperature is calculated according to Eq. (3).

$$T = \frac{T_{\min} + T_{\max}}{2}, \tag{3}$$

where $T_{\min}$ is the daily minimum temperature and $T_{\max}$ the daily maximum temperature. The LU occurs when T_S reaches the critical value $T_{S,\text{crit}}$ (Eqs. 4 and 5).

$$T_{S,\text{crit}} = a + b \ln(CD_n) \tag{4}$$

with

$$CD_n = \sum_{i=0}^{n} \left\{ \begin{array}{ll} 1 & T(t_0 + i\Delta t) \leq T_0 \\ 0 & T(t_0 + i\Delta t) > T_0 \end{array} \right. \tag{5}$$

CD is the number of chilling days, i.e. when $T < T_0$, between days t_0 and t_1. The date t_0 was set as 1 November in present study. Values for T_0, T_1, a and b for *Populus tremula* (IPG235) are given by Menzel (1997; Table 1)

Table 1. Parameters of the modified approach for estimating leaf unfolding (LU) according to Cannell and Smith (1983), which is used in WaSim. Estimated day of years (DOY) of LU for Reiffenhausen (2012 and 2013) are used to calibrate the Max1 parameters T_0, T_1, a and b, i.e. threshold temperature for chilling units and forcing units and regression parameters of two poplars, i.e. Max1 and IPG235 (*Populus tremula*), respectively. Additionally, the parameter set for IPG235 according to (Menzel, 1997, Appendix A7) is shown.

Parameter	Max1 (present study)	IPG235 (Menzel, 1997)
T_0 [°C]	10	8
T_1 [°C]	2	5
a	2200	1693.4161
b	−403	−301.9361

Using these numbers as initial values we fitted the parameters T_0, T_1, a and b to observed LU for the poplar clone Max1 in Reiffenhausen for the years 2012 and 2013 using a least squares method. Finally we evaluated the obtained model parameters against the independent observations in Reiffenhausen (for 2014), and observation in Großfahner for the years 2012 and 2013 (Lorenz and Müller, 2013) and 2014 (K. Lorenz, personal communication, 2014). The observed LU in Reiffenhausen and Großfahner is comparable to the recommendations according to Volkert and Schnelle (1966). Because the estimation of LU, as used in this study is based on meteorological measures, the parameterisation should hold true for the same poplar clone in the same age, if other environmental factors are of minor importance. The results will confirm this.

For a long-term comparison, the data from the international phenological observation networks (IPG) are used (Chmielewski et al., 2013), namely LU of *Populus tremula* (IPG235) at the IPG station Tharandt–Hartha. For poplar clone Max1 we could not find parameter sets in the literature; therefore, for comparison we used the IPG235 parameters of Menzel (1997) for *Populus tremula*, which has been more extensively investigated. The IPG data are used for long-term comparison, because there were no long-term investigations of LU available on the research plots of the BEST project or nearby. IPG235 is the acronym of the parameterisation for *Populus tremula* used by (Menzel, 1997). We decided to retain this acronym to make it comparable to published results, and also because it is an acronym used in the data provided by the phenological garden network. The phenological phase of leaf unfolding is defined as the stage UL, according to the IPG web page (International Phenological Gardens of Europe, 2014) and is obtained by daily observations of plant's development state. The IPG station Tharandt–Hartha is located at the eastern border of the Tharandter Wald. It is the nearest IPG station to the Reiffenhausen site, and comparable in climate and altitude.

2.3 Modelling approach

For simulation, the deterministic spatially distributed hydrological catchment model system WaSim (version 9.05.04) was used. Complete and comprehensive descriptions of this model and its internal structure can be found inter alia in (Schulla, 1997; Schulla and Jasper, 2013). The set up of physically based parameters, such as LAI and Rsc as well as phenological state (date of leaf unfolding – LU here), are predicated on direct measurements and observations. Thus, physical nexus between model image and reality is reproduced as best as possible. The SRC described with these measurements and observations represents a poplar SRC in the third growing season of its mono-stem cycle, which can be seen as a hydrological fully developed canopy, concerning LAI, Rsc and root development. The simulated local soil water contents were compared and evaluated to the measurements.

Different model simulations are done to show the suitability of the direct use of specific plant physiological measurements, as well as the effects of an approximated parameter description in the model, i.e. the annual course and the quantity of LAI, Rsc and phenology.

All these model approaches were done on a plot model domain, which are 3 × 3 raster cells based on a digital elevation model with a spatial resolution of 12.5 m (LGLN – Landesbetrieb Landesvermessung und Geobasisinformation, 2013), provided by the project partner NW-FVA[1]. All topographic information needed by the model is derived by the model itself. The research site, providing the measured soil water contents for model calibration is located in the centre of the domain. A retention curve required in hydrological modelling for the description of soil physical properties was taken from Van Genuchten (1980). The Van Genuchten retention parameters from Blume et al. (2010) were accepted based on a characterisation of soil texture and soil horizons in Reiffenhausen.

The meteorological forcing data were taken from our own measurements for the period 2011–2013, whereas the first 2 years were used for the model spin-up. Analyses and the comparison to measured local soil water contents were done just for the year 2013. To show the effects of different parameterisations under various climate conditions, the simulations were performed for the period from 1969 to 2013 using the forcing meteorological data from the DWD station Göttingen. The period was chosen as the longest period without missing values. The parameterisation of land use is kept constant for the whole period. A WaSim control file including all information about parameterisation and model set up is provided as supplementary material.

2.4 Data analysis

All measured and applied meteorological, soil physical and eco-physiological parameters have been checked for plausibility and measurement errors.

The data have been numerically analysed and graphically presented with the free software package GNU Octave, version 3.6.2 (Octave community, 2012). Parts of the statistical analysis were performed using the hydroGOF package (Mauricio Zambrano-Bigiarini, 2014) within the R software environment (R-Studio under Windows, version 0.98.501) for statistical computing and graphics (R Development Core Team, 2011).

The evaluation of model performance was done according to the objective criteria of Moriasi et al. (2007). Important quality criterions of simulation runs are the Nash–Sutcliffe model efficiency criterion (NSC), the percent bias (PBIAS), and the ratio of the root mean square error to the standard deviation of measured data (RSR).

3 Results

3.1 Measurements

3.1.1 Leaf area index

Figure 1 shows the annual course of LAI as derived from two different indirect optical methods and one direct destructive method. LU started shortly before the first measurement on 1 May 2013. Until 1 August there was almost a linear increase of LAI up to 7.3 and 5.5 for the LI191SA and the LAI2000 measurements, respectively. After that LAI started to decrease, with a more rapid decline toward the end of August 2013. Leaf fall was almost finished at 25 October.

Differences between devices, i.e. LI191SA vs. LAI2000, are large. The LAI values obtained with the LI191SA are systematically higher ($\approx 2\,m^2\,m^{-2}$). The values obtained by direct destructive sampling at 26 August are rather on the level of the LAI2000 estimates.

3.1.2 Stomatal resistance

Figure 2 shows the stomatal resistance (Rsc) as measured on well-illuminated leaves in 2013. The values ranged from 100 to $300\,s\,m^{-1}$ until August 2013. On 18 June Rsc was higher with larger standard deviations as the previous measurement on 14 June. Soil water supply was sufficient on both days. The 2 days significantly differ in temperature, although June 14 was relatively colder with a daily maximum temperature of approximately 17 °C, while 18 June was quite hot, reaching a maximum temperature of 33 °C. This shows the effect of local environmental conditions to measurements, possibly influencing derived model parameters. From August on, both mean Rsc and standard deviation steadily increased. This period was characterised by decreasing soil water availability

---

[1] Nordwestdeutsche Forstliche Versuchsanstalt (NW-FVA), North-west German Forest Research Station

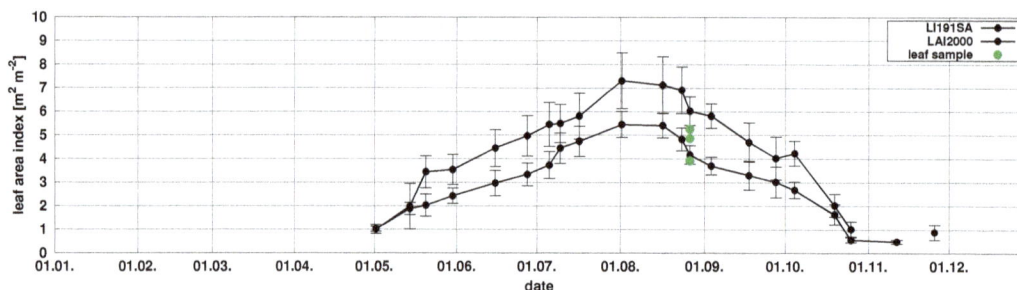

Figure 1. Means and standard deviations of leaf area index of the poplar SRC Reiffenhausen in 2013. Measurements of two optical devices: LI191SA calculated with constant extinction coefficient $k = 0.5$ and LAI2000 are shown. LAI values obtained by destructive harvesting at 26 August on three plots are shown as green dots.

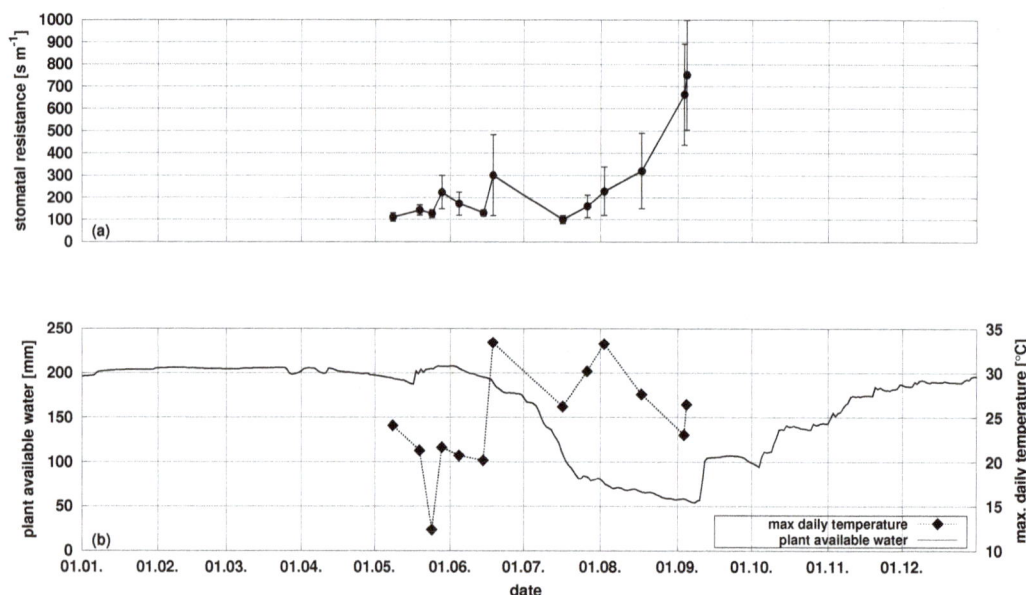

Figure 2. **(a)** Means and standard deviations of stomatal resistance of sun leaves derived from 10 to 11 repetitions every day at 10 measurement plots and **(b)** the plant available water, calculated from soil water content measurements until 1 m soil depth and daily maximum temperature at the poplar SRC Reiffenhausen in 2013. High temperatures affecting stomatal resistance (18 June); starting from August drought stress occurred, increasing the stomatal resistances.

leading to severe drought stress conditions. Due to higher Rsc the trees counteracted the drought stress to avoid water loss and xylem damage, e.g. embolism of xylem vessels. The increase of standard deviation is an expression of stand heterogeneity, single trees still have access to water, and others may already be limited or stressed. In September 2013 we stopped measurements because leaves were visibly affected by the drought stress event. The correlation of plant available soil water and plant regulation via stomata seems to be consistent, increasing confidence in the distinct measurements. The minimum observed stomatal resistance is $80 \, \mathrm{s \, m^{-1}}$.

3.1.3 Phenology – start of growing season

We used in situ phenological observations of the 2 years 2012 and 2013 in Reiffenhausen to calibrate the modified approach

for estimating LU according to Cannell and Smith (1983), which is used in WaSim. In 2012 LU for Max1 in Reiffenhausen started at day of year (DOY) 88, i.e. 28 March 2012 (Table 2). In 2013 LU was delayed by approximately 4 weeks due to low temperatures in spring and started at DOY 115 (25 April 2013). Our calibration resulted in values of 10 and $2 \, ^\circ \mathrm{C}$ for T_0 and T_1. The regression parameters a and b are 2200 and -403, respectively (Table 1). Estimates of LU using these values for T_0, T_1, a and b show deviations from the observed dates of $+3$ and -3 days for Reiffenhausen in 2012 and 2013 (Table 2). Then the phenological model results with the obtained parameter set and local temperatures were compared to phenological observations in Reiffenhausen in 2014 and in Großfahner in 2012 and 2013. Observed LU in Großfahner were almost equal to that in Reiffenhausen, also showing the delay of about 4 weeks in

Table 2. Observed and estimated day of years (DOY) of leaf unfolding (LU) for Reiffenhausen and Großfahner. The Max 1 parameters are calibrated at the observations in Reiffenhausen using local temperatures (2012 and 2013) and evaluated with Reiffenhausen (2014) and Großfahner (2012–2014). Additionally the DOY of LU is compared to estimates using the IPG235 parameter set as well as the temperatures of the nearest DWD climate station (Göttingen for Reiffenhausen, distance approx. 17 km; Dachwig for Großfahner, distance approx. 3.5 km) are presented.

	Observed	Max 1	Max 1	IPG235	IPG235
Temperature data		local	nearest DWD	local	nearest DWD
Reiffenhausen DOY 2012	88	91	89	119	121
Reiffenhausen DOY 2013	115	112	112	125	126
Reiffenhausen DOY 2014	83	82	85	111	113
Großfahner DOY 2012	88	89	88	111	112
Großfahner DOY 2013	114	113	112	121	121
Großfahner DOY 2014	89	–	86	–	103

2013 compared to 2012. The phenological model using the Max 1 parameters results in differences of −1 day for Reiffenhausen in 2014 and of +1 and −1 days for Großfahner in 2012 and 2013 compared to observations. Table 2 also shows the application of the IPG235 parameter set provided by Menzel (1997) for *Populus tremula*. Parameters of a and b for IPG235 are both smaller in magnitude and threshold temperatures for chilling and forcing units, T_0 and T_1 show smaller differences. Due to this widespread between T_0 and T_1, the Max 1 model is able to describe extreme values and therefore a higher variability of LU, which was observed in 2012 and 2013. The model estimations of LU with Max 1 and with the IPG235 parameters differ considerably. The IPG235 set produces systematically later dates. Differences in observations are +31 (2012), +10 (2013) and +28 (2014) days for Reiffenhausen and +23 (2012) and +7 (2103) days for Großfahner using the local temperatures (Table 2, column: local).

To assess the effects of non-local micrometeorological data sources, the model was driven by temperature measurements from the nearest DWD stations, namely, Göttingen for Reiffenhausen and Dachwig for Großfahner. As expected, the use of DWD data instead of the local measurements produces mostly larger estimation errors for both the Max 1 and the IPG235 parameter set (Table 2, column: nearest DWD).

We used the varying parameter sets, i.e. our Max 1 model and the IPG235 parameter set (Table 1) to analyse the effect for hydrological modelling for the year 2013, where soil hydrological measurements are available to evaluate the hydrological model results.

To analyse the species dependence of LU estimations, the model with Max 1 and IPG235 parameters was also driven by temperature measurements during 2012–2014 at the phenological station Tharandt. Figure 3a–d illustrate that the parameter sets better fit with the observations at species for which they were calibrated: Max 1 parameters to Reiffenhausen and the IPG235 parameters to Tharandt observations,

which were part of its calibration data set. The differences between estimated and observed DOY of LU are smaller when local temperature measurements are used (Fig. 3a vs. b and Fig. 3c vs. d).

Figure 3e–f show the long-term courses of estimated DOY of LU for Reiffenhausen and Tharandt using the temperatures of the nearest DWD stations (Göttingen and Wildacker, respectively). The model with IPG235 is systematically later and shows less variability than with Max 1 parameters. For Reiffenhausen no long-term phenological observations are available. However, the average DOY of LU in Reiffenhausen is DOY 97 ± 9 days using Max 1 and DOY 124 ± 5 days for IPG235. The long-term phenological observations in Tharandt fit well with the IPG235 estimates, but showing less variability than observed. The average DOY of LU in Tharandt as observed is DOY 123 ± 10 days, estimated using Max 1 DOY 101 ± 9 days and with IPG235 DOY 124 ± 7 days. In general, the estimations fit best to observations when the corresponding parameters are used, i.e. Max 1 for Reiffenhausen and IPG235 for Tharandt. But variability is underestimated by IPG235 compared to observations.

3.2 Hydrological model simulations

Several model simulations were performed with different parameterisations of LAI, Rsc and LU. Table 3 summarises the eight performed model simulations and introduces their abbreviations. The detailed descriptions of model simulations are given in the text.

First the measured values of LAI, Rsc and LU are implemented for hydrological modelling (LAI2000 Rsc80 and LI191SA Rsc80). Starting from here we changed the parameter sets: (i) to improve the model fit, (ii) to adjust the suitability of applied parameterisations and (iii) to show the effects of different parameterisations on hydrological model results.

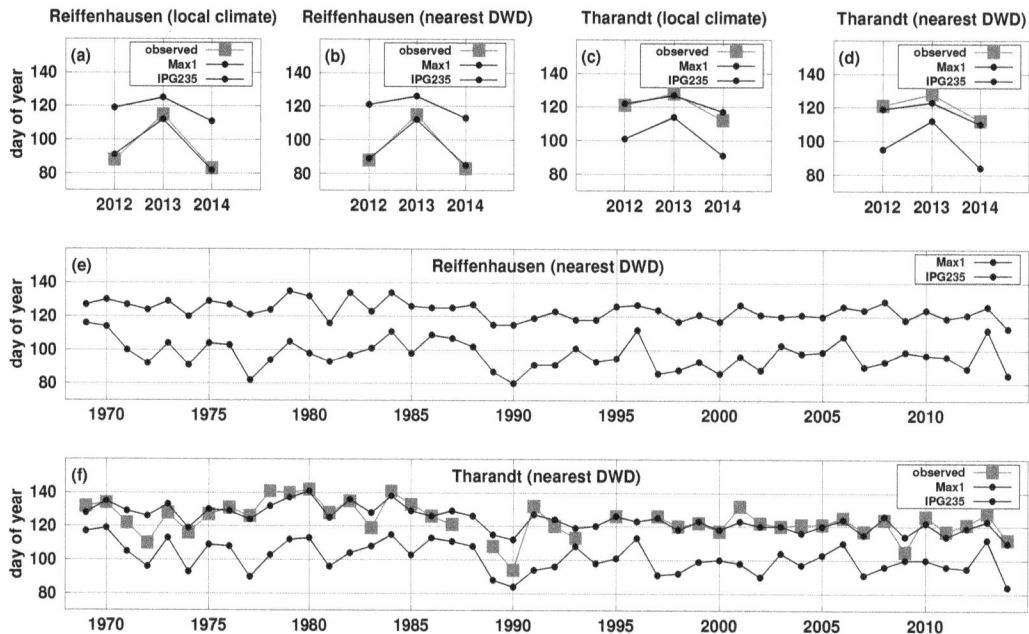

Figure 3. Estimated day of year (DOY) of leaf unfolding (LU) using the Max1 and IPG235 parameters for the site Reiffenhausen with local temperature measurements (**a**) and temperatures from the nearest DWD station Göttingen (**b**) for the years 2012 to 2014; same for Tharandt, local temperatures (**c**) and nearest climate station Wildacker (**d**); with observations. The lower subplots show long-term estimates for DOY of LU using the DWD temperatures for Reiffenhausen (**e**) and for Tharandt (**f**), where also long-term observations are available.

Table 3. Description of the eight performed model simulations. All model parameters are constant, except leaf area index (LAI), stomatal resistance (Rsc) and the date of leaf unfolding (LU) for the two parameter sets Max1 and IPG235.

Version	LAI	Rsc	LU
LAI2000 Rsc80	LAI-2000 measurements	minimum $80 \, \text{s} \, \text{m}^{-1}$ (LAI > 1)	defined by measured LAI
LAI2000 Rsc40	LAI-2000 measurements	minimum $40 \, \text{s} \, \text{m}^{-1}$ (LAI > 1)	defined by measured LAI
LI191SA Rsc80	LI-191 SA measurements	minimum $80 \, \text{s} \, \text{m}^{-1}$ (LAI > 1)	defined by measured LAI
LI191SA Rsc40	LI-191 SA measurements	minimum $40 \, \text{s} \, \text{m}^{-1}$ (LAI > 1)	defined by measured LAI
LAIstep Rsc40 Max1	step function (6 in growing season; else 1)	minimum $40 \, \text{s} \, \text{m}^{-1}$ (LAI > 1)	Max1 model
LAIstep Rsc40 IPG235	step function (6 in growing season; else 1)	minimum $40 \, \text{s} \, \text{m}^{-1}$ (LAI > 1)	IPG235 model
LAIadjusted Rsc40adjusted Max1	course calibrated to improve model fit (max. = 6)	minimum $40 \, \text{s} \, \text{m}^{-1}$ (LAI > 1)	Max1 model
LAIadjusted Rsc40adjusted IPG235	course calibrated to improve model fit (max. = 6)	minimum $40 \, \text{s} \, \text{m}^{-1}$ (LAI > 1)	IPG235 model

3.2.1 Simulation using observed parameters and adaptation of stomatal resistance

First we used the measured annual courses of LAI for hydrological modelling. Rsc is set to the measured minimum of $80 \, \text{s} \, \text{m}^{-1}$ when LAI is larger than 1. LU is not calculated for measured LAI from air temperature using the approach of Cannell and Smith (1983), because this information is already imprinted in LAI measurements and therefore fixed for the year 2013.

Figure 4 shows the applied model parameterisations for LAI and Rsc, as well as the plant available water (PAW), calculated for 1 m soil depth from measured and modelled soil water contents.

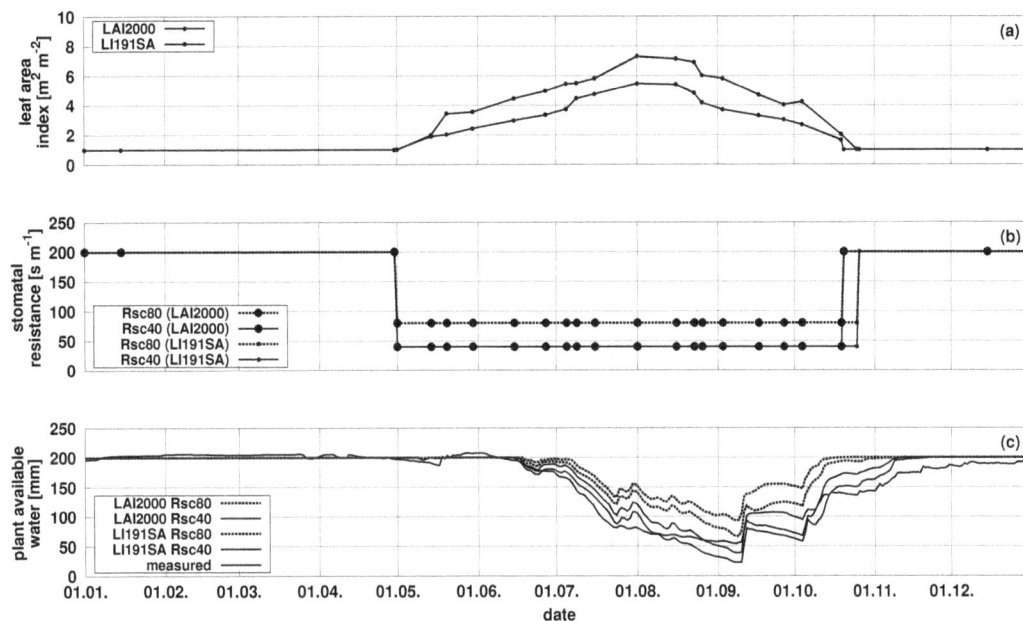

Figure 4. Leaf area index and stomatal resistance as parameterised from measurements. Leaf area index is used as measured, LAI2000 (red) and LI191SA (blue). Stomatal resistance is set to the measured minimum, i.e. $80 \, \text{s m}^{-1}$ (dashed line) and to $40 \, \text{s m}^{-1}$ (solid line) as leaf area index is larger than 1. Length of growing season is determined by the leaf area observations. Simulation results using the four combinations of leaf area index and stomatal resistance are shown as plant available soil water, calculated until 1 m soil depth and compared to values based on soil water content measurements.

For all simulations using the measured value for Rsc (i.e. $80 \, \text{s m}^{-1}$), the modelled soil water contents were higher than measured values, resulting in larger PAW values than observed. This is also reflected by the Nash–Sutcliffe criterion (NSC) calculated from PAW (Table 4). The annual course of PAW is captured quite well by the model, but the drying up in summer is not sufficient, neither for LI191SA, nor for LAI2000 measurements. The NSC is better for the experiments with LAI measured by LI191SA (0.69) than with LAI2000 (0.44) because of the higher LAI values (Table 4).

As the maximum LAI measurements using LAI2000 showed better agreement with direct destructive measurements, we halved the value of Rsc from 80 to $40 \, \text{s m}^{-1}$ to reach the low PAW values observed. This decision is justified by two reasons: first, it can be assumed that the measured Rsc is always higher than the minimum value needed for parameterisation, because the conditions by measuring Rsc are not satisfying the requirements for the parameter to be used in the model – i.e. optimal conditions for transpiration and no water stress. Another way to get lower PAW values would be to increase the LAI, as can be seen by comparing the results for LI191SA and LAI2000. However, in our experiment LAI has to be increased to unrealistically high values to minimise the differences to observed PAW. Additionally LAI is also affecting other processes in hydrological models, like interception evaporation and soil evaporation. Together the decrease of Rsc is a consistent way to minimise the deviations to observations and to improve the model fit.

The reduction of Rsc, from 80 to $40 \, \text{s m}^{-1}$, improved the NSC from 0.69 and 0.44 to 0.89 and 0.87 for LI191SA and LAI2000, respectively (Table 4).

3.2.2 Approximation and adaptation of annual course of leaf area and stomatal resistance

In many cases when hydrological models should be applied for analyses involving vegetation, there are no locally measured data on LAI and/or Rsc. Often only the literature data for the maximum and minimum values of LAI and Rsc are available. Then the annual course for these parameters has to be derived or approximated for modelling. The simplest approximation is a stepwise function, where the increase from minimum to maximum or decrease from maximum to minimum occurs within one time step. We applied this form to the LAI and Rsc as shown in Fig. 5. Here the maximum of LAI is set to $6 \, \text{m}^2 \, \text{m}^{-2}$, which is the observed maximum plus standard deviation of LAI2000 measurements. The minimum of Rsc is set to $40 \, \text{s m}^{-1}$. For this kind of approximation the start and length of the growing season become important, because the maximum transpiration rate occurs immediately after LU. In Fig. 5 we compare two different parameterisations for dynamical estimating LU, i.e. the Max1 and the IPG235 parameter set (Table 1).

The NSCs for both simulations are 0.89 (Table 4), which is slightly better than for applying the direct LAI measurements. However, PBIAS values are negative for the step-

Table 4. Statistical parameters for model evaluation in terms of the accuracy of simulated data compared to measured values. Nash–Sutcliffe efficiency criterion (NSC), percent bias (PBIAS), and ratio of the root mean square error to the standard deviation of measured data (RSR) are calculated from plant available soil water till 1 m soil depth as derived from model simulations and soil water content measurements for the period from April to December 2013, to cover the period of most variability.

	NSC	RSR	PBIAS [%]
Recommended as satisfactory by (Moriasi et al., 2007)	> 0.5	≤ 0.7	±25
LAI2000 Rsc80	0.44	0.75	19.2
LAI2000 Rsc40	0.87	0.37	5.5
LI191SA Rsc80	0.69	0.56	13.9
LI191SA Rsc40	0.89	0.33	0.0
LAIstep Rsc40 Max1	0.89	0.33	−2.0
LAIstep Rsc40 IPG235	0.89	0.33	−1.7
LAIadjusted Rsc40adjusted Max1	0.90	0.31	1.3
LAIadjusted Rsc40adjusted IPG235	0.90	0.31	1.6

function simulations, where they are positive for the simulations using LAI measurements. Negative PBIAS values indicate a stronger drying signal.

Differences in PAW are only visible in the period when parameterisation is different (Fig. 5). The year 2013 shows no drought event in spring, where effects would be more obvious. There are small differences in May, where the step-function simulation using the Max1 parameters are closer to observations.

However, the abrupt increase to the maximum transpiration rate immediately after LU is rather unrealistic as already shown by the LAI measurements. Unfolding of leaves in nature can happen very quickly, as everybody can observe when spring comes late in the year followed by favourable growth conditions. When spring starts early the full leaf development can take much longer. To account for this and to further improve our model fit, we changed the annual development of LAI and Rsc by using these parameters for manual model calibration, guided by the course of LAI measurements mainly (Fig. 6). Major changes are higher LAI and lower Rsc values at the date of leaf unfolding, i.e. $2\,m^2\,m^{-2}$ and $150\,s\,m^{-1}$, respectively. LU is estimated with the dynamic approach like in the step-function simulations. This resulted in modelled higher transpiration rates in spring. Nevertheless, the annual course of LAI and Rsc is described more detailed and more similar to the observed LAI dynamics – the LAI increase and decline is smoother but also starts a bit earlier in the year and last a bit longer in autumn. Due to that smoother increase in spring the sensitivity to deviations in estimating LU is reduced.

Due to these changes the NSC increased to 0.90 for both Max1 and IPG235 parameter sets. This is the best fit obtained in manual calibration procedure (Table 4). PBIAS values are positive for the adjusted models, which is a slightly too small drying signal. However, the magnitudes of PBIAS and RSR values are smaller than for the step-function simulations, indicating better agreement with observations and lower root mean square errors or residual variations (Moriasi et al., 2007).

3.2.3 Long-term simulations

In all simulations shown for the year 2013, the effects on PAW caused by changes in estimated LU are quiet low due to the high soil water contents in spring 2013. Therefore, we applied all simulations for the years 1969–2013, which was the longest meteorological period without missing data. A focus is set to the year 2012, which was characterised by an early drought event in May. Because of missing data there is no complete set of soil water content information available for this year and there is no information about LAI and Rsc for 2012 that can be used to parameterise the hydrological model. So no evaluation of model fit is possible for 2012 or the other years of the period 1969–2013, like it is done for 2013.

To illustrate the effects for the different courses of LAI and Rsc development as well as the estimation of LU, Fig. 7 shows the precipitation, the plant available water and the GWR for the step function and for the adjusted simulations combined with the estimates of LU, i.e. the Max1 and IPG235 parameter set, respectively. Results in Fig. 7 show the last 2 years from the long-term simulations 1969–2013, mean values for evapotranspiration (ETR) and GWR for the whole period 1969–2013 are presented in Table 5. In 2012 and 2013 as well as for both estimations of LU, the adjusted simulation shows the highest GWR and the step-function simulations result in the lowest GWR. The reason is the change in transpiration in spring, as described due to the different parameterisations of the step function and adjusted course for LAI and Rsc. However, the largest effects on GWR are caused by the different estimation of LU (Fig. 7). In the step-function experiments, GWR is zero in 2012 with both the Max1 and the IPG235 parameterisations of LU. Plant available water is reduced more strongly for the Max1

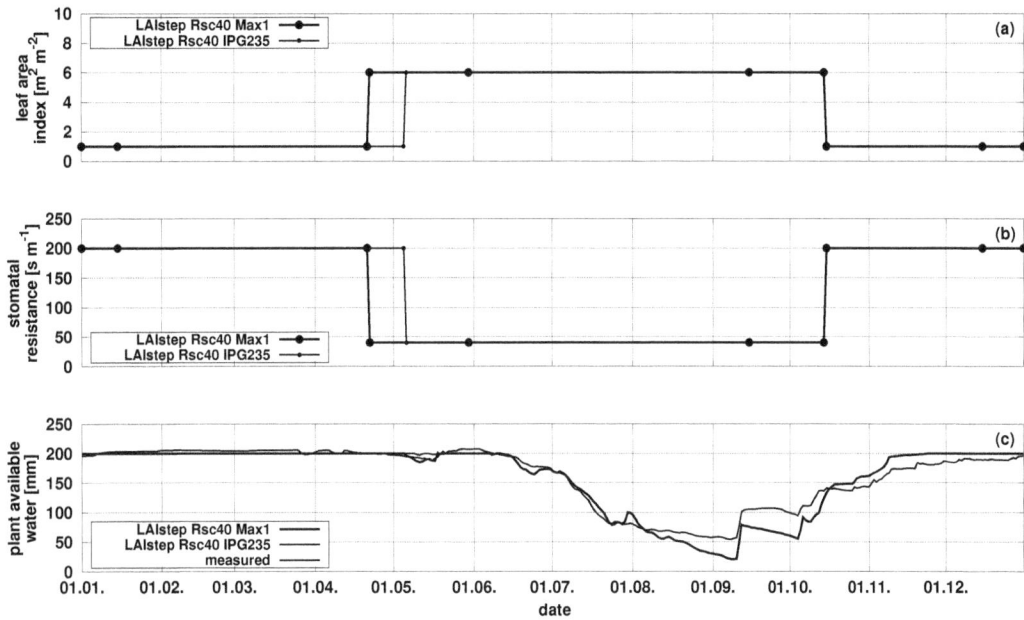

Figure 5. Leaf area index and stomatal resistance parameterised as step function, using maximum and minimum values in the growing season 2013, respectively. LAI is set to 6, stomatal resistance is set to $40\,\mathrm{s\,m^{-1}}$ when LAI is larger than 1. Leaf unfolding (LU) is determined by the dynamic phenology approach implemented in WaSim, using the Max1 parameterisation and IPG235. Simulations results using the two combinations of LAI and stomatal resistance are shown as plant available soil water, calculated till 1 m soil depth and compared to values based on soil water content measurements.

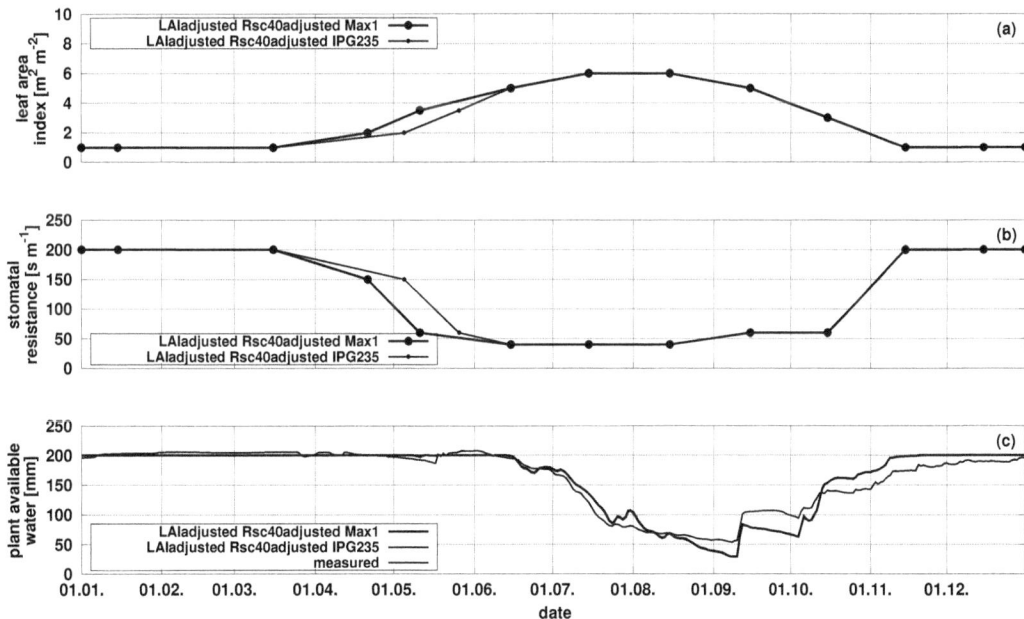

Figure 6. Leaf area index and stomatal resistance as parameterised from adjusted values for 2013. Maximum of leaf area index is set to 6, minimum of stomatal resistance is set to $40\,\mathrm{s\,m^{-1}}$. Leaf unfolding (LU) is determined by the dynamic phenology approach implemented in WaSim, using the Max1 and IPG235parameterisation. The annual course of leaf area index and stomatal resistance is orientated on measurements for leaf area and used as calibration parameter for stomatal resistance. Simulations results using the two combinations of leaf area index and stomatal resistance are shown as plant available soil water, calculated till 1 m soil depth and compared to values based on soil water content measurements.

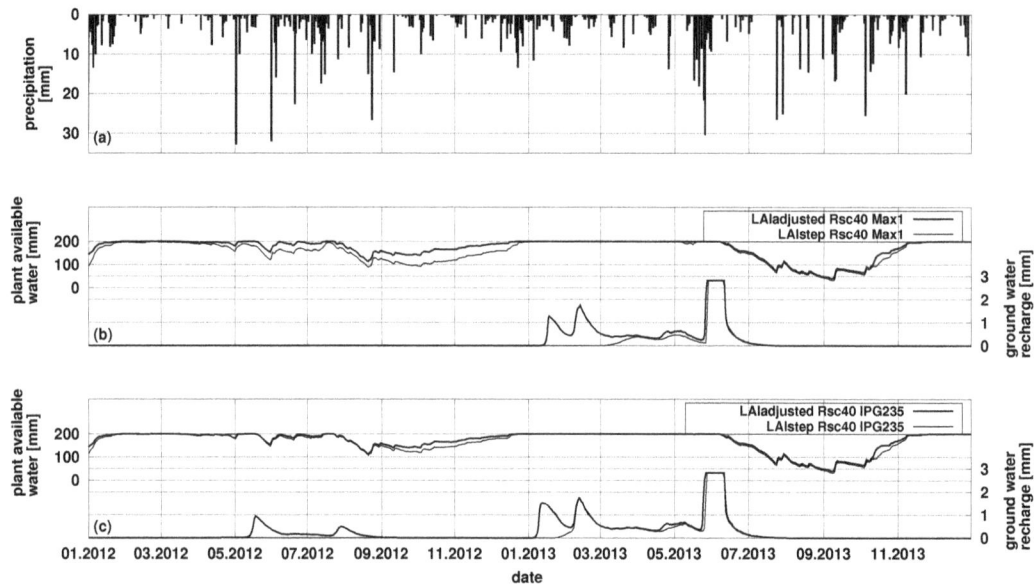

Figure 7. Measured daily precipitation (**a**). Plant available water and ground water recharge as simulated for the step function and adjusted course for leaf area index and stomatal resistance, using the leaf unfolding (LU) parameters Max1 (**b**) and IPG235 (**c**) for simulating LU. The parameterisation for poplar is equal for 2012 and 2013, i.e. the same vegetation hydrological modelled driven by different weather conditions, i.e. a drier year 2012 with an earlier dry period in May.

parameterisation, due to the early start of the growing season. However, this early LU fits better to the observations in Reiffenhausen. For the adjusted simulations, GWR in 2012 is only zero for the Max1 parameterisation. For the year 2012 data of the matrix potential (tensiometer measurements) in 20, 60 and 120 cm soil depth are available. These data show a drought period in May 2012, where the tensiometers in 20 cm soil depth run out the measuring range; i.e. a matrix potential was lower than about −800 hPa. Starting from May 2012, the tensiometers in 60 cm and 120 cm soil depth indicated a consistent drying signal (not shown). Additionally, the poplar SRC in 2012 is younger and therefore less water demanding than the poplar SRC parameterised in the model, applied for these analyses. This indicates that GWR after May 2012 is very unlikely in these simulations.

The parameterisation using the adjusted course for LAI and Rsc, based on the measured course of LAI, in combination with the Max1 parameterisation for LU, calibrated at local observations, seems to be the most realistic model simulation (LAIadjusted Rsc40 Max1). By comparing all four model parameter combinations shown in Fig. 7, one can switch completely from GWR present in 2012 to absent.

Table 5 summarises the ETR and GWR for all simulations, averaged over all years for the period 1969–2013, as well as for the five driest and five wettest years of this period. GWR averaged over all years varies from 80 to 145 mm yr^{-1} depending on the approximation of the annual course of LAI and Rsc and the estimation of LU. The ratio of maximum and minimum of the all year averages of GWR for the different simulations is approx. 1.8. This factor is approx. 3 for the

five driest years and approx. 1.7 for the five wettest years, showing that especially the model results for dry years are sensitive to the parameterisations used.

4 Discussion

Not all necessary model parameters for WaSim could be measured in detail. One example is the implemented assumption on rooting depth which was measured in 2012, and was set to 1 m, which is comparable to the commonly used values presented in Raissi et al. (2009).

Measuring Rsc in the field is rather challenging. For hydrological modelling we are interested in more theoretically minimum values, indicating optimal transpiration. These conditions are hardly found in reality. In addition, the measurements are affected by soil water availability and rapidly changing atmospheric conditions. Breuer et al. (2003) summarised values for minimal stomatal resistance for various plants. Values for *Populus* clones (*Populus grandidenata, P. tremula* and *P. tremuloides*) range from 102 to 400 s m^{-1}. Our measured minimum of Rsc for poplar clone Max1 (*Populus nigra* × *Populus maximowiczii*) is lower: 80 s m^{-1}. Yet we needed to further reduce Rsc to 40 s m^{-1} for matching the observed soil water contents. On the one hand, the low observed minimum of 80 s m^{-1} shows that specific measurements of Rsc are helpful. On the other hand, the measurements of Rsc were still too high to produce plausible results with WaSim. One might interpret the reduction of Rsc from 80 to 40 s m^{-1} as a shift from the often reported iso-

Table 5. Precipitation (mm yr^{-1}), total evapotranspiration (ETR) and ground water recharge (GWR) for the period 1969–2013. Simulations are shown for the step-function simulation and the adjusted course for leaf area index and stomatal resistance, using the leaf unfolding (LU) according to the Max1 and IPG235 parameter set for simulating LU. The parameterisation for poplar is equal to that derived for 2013 for the whole period, i.e. the same vegetation hydrological modelled driven by different weather conditions. Values are summed up for all years (1969–2013) of the period and for the five driest and five wettest years, respectively.

	ETR (all years)	ETR (five driest)	ETR (five wettest)	GWR (all years)	GWR (five driest)	GWR (five wettest)
Precipitation	676.7	500.4	896.6	676.7	500.4	896.6
LAIstep Rsc40 Max1	527.8	487.1	533.6	79.7	23.2	140.2
LAIstep Rsc40 IPG235	488.5	463.0	482.6	105.0	37.9	170.4
LAIadjusted Rsc40adjusted Max1	484.1	460.8	477.9	107.1	41.3	173.9
LAIadjusted Rsc40adjusted IPG235	425.9	424.4	411.7	144.7	68.1	232.9

hydric behaviour of poplar clones (Tardieu and Simonneau, 1998) to a more anisohydric behaviour. But the diurnal or seasonal variations of leaf water potential that are characteristic for anisohydric plants are not expressed by the Rsc value in WaSim, which represents the minimal resistance for a state when plants are fully supplied with water. The reduction of transpiration in drought stress situations is done in a different way in WaSim. Furthermore, there are also more drought tolerant, anisohydric water use strategies reported from greenhouse experiments for poplar clones (Ceulemans et al., 1988; Larchevêque et al., 2011). Schmidt-Walter et al. (2014) reported also a poor stomatal control of water loss estimated from field measurements of a poplar SRC.

LAI measurements show a systematic difference between the two measurement devices, whereas LI191SA seems to overestimate LAI taking the destructive method as a reference. In situ measurements of LAI are helpful to determine the maximum value, but differences due to the different estimation methods including underlying assumptions should be considered. The annual development of LAI is indispensable information that is needed to adjust and improve the model parameterisation of annual course. The measured LAI development represents local conditions and is therefore valid for the measurement site and time period only. Approximations of seasonal course are advisable to enable the transferability to other sites and years. A crucial factor is LU, which determines the start of LAI increase. Determination of this date requires the definition of phenological stages. Various models are available to describe LU, some are based on air temperature, soil temperature, photoperiod, day length or radiation. All models have to be calibrated for the specific plant species. There is also evidence that local conditions like latitude or altitude of observations are influencing the calibration of the phenological model. Furthermore, the derivation of parameters for the phenological model will depend on the observed data, e.g. the detection of extremely early or late

LU as well as the climate data, which has to be appropriate for the observed site. For poplar clone Max1 we could not find parameter sets in literature, so we used for comparison the IPG235 parameters of Menzel (1997) for *Populus tremula*, which is better investigated. The period of parameter adjustment used by Menzel (1997) is 1959–1993 and it is based on several phenological stations, whereas our derived parameters are based on 2 years at one site. However, these 2 years show a wide variability in LU. The parameters from Menzel (1997) should be generally more valid, because of the higher number of observations. Yet the use of IPG235 parameters resulted in an underestimation of observed variability, compared to the observations for *Populus tremula* in Tharandt. Differences between IPG235 and Max1 also show the importance of parameterisations for local site conditions and specific species. Comparing the parameter sets presented in Table 1 these effects become evident.

Especially threshold temperatures for chilling and forcing units, T_0 and T_1, vary more widely between IPG235 and Max1. Due to this wider range, the model is able to describe extreme values and therefore a higher variability of LU, which was observed in our calibration years 2012 and 2013. We evaluated our parameter set to observations of the poplar SRC Reiffenhausen in 2014 and the poplar SRC Großfahner (2012–2014), which was planted with the same clone and in the same year like Reiffenhausen. The differences for LU between observations and the Max1 model set up are low and within the observed variability. The use of IPG235 parameters for the Max1 clone, which is a common procedure when specific values are missing, can result in large deviations as shown for GWR, especially in the year 2012 with the drought period in spring.

The source of temperature data also influences the parameters derived for phenological models as well as the results obtained by applying these parameter sets. We compared the estimated DOY of LU derived with local temperature measurements with estimated DOY derived with temperatures of the nearest DWD stations. For Reiffenhausen, with the nearest DWD station Göttingen, we additionally tested an altitude correction using the vertical temperature gradient of $-0.0065\,°C\,m^{-1}$ to account for 158 m altitude difference between Göttingen (167 m a.s.l.) and Reiffenhausen (325 m a.s.l.). Deviations in DOY of LU are small when using the DWD temperature instead of the local measurements. Interestingly, the altitude correction of temperature increases differences in DOY of LU compared to observations. The reason could be the often occurring thermal inversion, when the air temperature in Reiffenhausen is higher than in Göttingen, so that implemented altitude reduction of temperature increases the differences even more, due to that also the differences of the estimated DOY of LU increase. The effects of the altitude correction are larger for the Max1 than for the IPG235 parameter set, because our model is more sensitive to extreme values due to higher T_0 and lower T_1 temperatures. This shows the importance of applying the local temperatures, associated with the phenological observations, to calibrate and use the temperature-dependent phenological models. The use of local temperatures improves the estimation of LU and better represents interannual variability.

According to the criteria of Moriasi et al. (2007), the hydrological model results, using measured values of LAI and Rsc (start and development of LU is implemented), are satisfactory only for the simulation LI191SA with Rsc $= 80\,s\,m^{-1}$. The simulation using the LAI values from the LAI2000 does not satisfy the recommended criteria (Table 4). However, the model produces better agreement with observations when Rsc minima of $40\,s\,m^{-1}$ are used with any LAI data. The reduction of Rsc is a suitable way to simulate the observed soil water conditions. An increase of LAI could lead to lower soil water contents as well, but it is also affecting soil evaporation and interception evaporation. Additionally, larger values for LAI, necessary to minimise the model deviations to measurements, have to be unrealistically high for the poplar SRC investigated here. When using a Rsc minimum of $40\,s\,m^{-1}$ together with measured LAI the model evaluation is good for the year 2013, reaching NSC values of 0.87 and 0.89 for the LAI2000 and LI191SA simulations, respectively.

Data of such intense measurement campaigns are not available for all sites were hydrological modelling should be done. Therefore literature values, typically providing just maximum or minimum values for LAI and Rsc, are used and the annual course has to be modelled. The question of transferability of these values to different sites, years or even species has to be solved. The applied step function is the simplest approximation of the annual course for LAI and Rsc. These simulations also pass the recommended criteria for a satisfactory model performance (Table 4).

For the year 2013 the best model fit could be obtained by the adjusted annual courses for LAI and Rsc. They are based on the observed course and maximum values of LAI measurements.

The weather regime and therefore the development of soil water conditions are not suitable in 2013 to show the effects of different estimates for the start of LU in spring. Drought conditions started after July 2013; therefore, we performed scenario simulation by transferring the vegetation parameterisation for 2013 to the weather regime of 2012. This year was characterised by a drought period in spring. Consequently, the effects of different estimates of DOY of LU are pronounced. The adjusted simulations using the IPG235 parameters to estimate LU, i.e. later LU by approx. 30 days in 2012, show GWR in 2012. Due to the delayed start of the growing season, the drought stress in spring is not reproduced by the model, leading to wetter soil conditions which favours percolation and rewetting and finally enlarges GWR (Fig. 7). The tensiometer measurements available for 2012 suggest that GWR is rather unlikely for this year. The adjusted simulation using the Max1 parameters for LU and both step-function simulations (Max1 and IPG235) result in zero GWR for the year 2012. However, the strongest simplification of the course of LAI and Rsc, i.e. the step function, shows the lowest GWR for 2013 and for the long-term simulations (Table 5).

In Fig. 7 the effects of the different simulations on GWR are presented for the years 2012 and 2013, which are characterised by rather different weather regimes. Whereas a realistic description of LAI and Rsc seems to be less important in 2013, it is even more essential in 2012, showing the importance of distinct spatial and temporal characteristics for local modelling.

We performed a long-term simulation, by keeping the parameterisation for the vegetation constant for the period 1969–2013 to account for the effects of climate variability. This is a more theoretical scenario, because it accounts for changes in climate forcing only. In reality also the vegetation characteristics are changing over the years, as well as soil properties on a longer timescale, especially for SRC, whereas rotation cultivation is applied, e.g. harvesting and re-sprouting. Particularly the rotation cultivation can reduce extreme drought conditions, when dry years coincide with rotation stages that have a lower water demand. The vegetation parameterised here can be seen as fully developed in hydrological terms, characterised by a large water demand. The simulations here are rather artificial, especially by succeeding dry years when soil water storage is not refilled completely in winter and drought conditions are influencing the following growing season. Nevertheless, the effects caused by different descriptions of vegetation parameterisations are quiet large (Table 5). Especially on a local scale such differences can be important by evaluating effects of land use change, particularly in dry years.

Taking into account that the best model evaluation for 2013 is achieved with the adjusted course of LAI and Rsc, the adjusted simulation using the Max1 phenology parameters seems to be the most reasonable parameterisation. It gives the best fit to the evaluation in 2013.

5   Conclusions

In the context of hydrological analysis of sites with focus on land use change or climate change, an adequate parameterisation of the vegetation cover is important to determine processes like soil evaporation, interception evaporation and transpiration. Sources of model parameters for the vegetation cover are local measurements or scientific literature. The analysis shows simulation uncertainties evolving from the use of model parameters that are derived from (i) non-local measurements or (ii) some appropriate literature values.

Regarding the objective 1 of our study, we showed that LAI, Rsc as well as the beginning and length of growing season are very sensitive parameters when effects of an enhanced cultivation of SRC on local water budget are investigated. In particular, our analysis reveals that correct information about the beginning of the growing season is highly important to obtain correct and acceptable simulation results of evapotranspiration components and GWR. If the start of the growing season is inappropriate, such as shown for the different species as in the IPG235 and Max1 parameterisation (Table 1), the accuracy of other parameters (like LAI and Rsc) plays a minor role. Concerning GWR, LU is the most sensitive parameter. Its parameterisation is particularly important when interannual variations and hydrological extreme conditions are on focus.

The implementation of locally measured vegetation parameters for hydrological modelling has both advantages and drawbacks. Measurements are expensive, time-consuming and also not always feasible. In such cases the use of appropriate literature values and transposition of adjacent observations is necessary and common practice.

The present study displays that locally measured LAI are suitable information for hydrological simulations. The comparisons between locally measured and adjusted parameter sets reveal that simulation results are less affected by other model parameters, like Rsc or LU, when using adjusted parameters of LAI.

Opposite results appear for Rsc. Simulation results differ significantly when site-specific values of Rsc are available. However, for Rsc the benefit of direct use of local measurements is arguable: minimum has to be reduced within WaSim to produce model results comparable with soil water measurements. In consequence the implementation of Rsc values from literature for hydrological modelling without accompanying measurement data for model evaluation can produce very uncertain results. The analyses illustrate that the locally adjusted vegetation parameterisation gives the best model fit.

Additionally, the adjusted course of LAI and Rsc is less sensitive to different estimates for LU, due to a slower increase in spring compared to a step-functional annual course. However, the adjusted courses are also approximations and not a distinct measurement, and are therefore more generally valid for different sites and years, than a direct use of measured parameters.

For the land use poplar SRC there are certain years where the modelled GWR is reduced to zero, like in the year 2012 (Fig. 7). Different parameterisations for vegetation characteristics are influencing modelled GWR for those years producing a wide range from GWR present or completely absent.

Hydrological models are often used to analyse effects of climate and land use changes on spatial and temporal scale becoming smaller and smaller. Approximations in the description of vegetation, a lack of local information (also soil and climate description), the transfer of inappropriate parameters and deficiency in model formulation can cause large differences in simulation results. To account for small-scale and local effects of land use change, a more detailed descriptions of sites and processes are necessary to capture the spatial and temporal variability of effects. In particular, the extremes are often underestimated when the description of site and processes are insufficient.

Acknowledgements. The work was funded by the German Federal Ministry of Education and Research (BMBF) and is part of the BEST-Research Framework (http://www.best-forschung.de), we gratefully acknowledge this support.

We give our special thanks to D. Fellert and H. Kreilein for valuable advices and assistance in handling the meteorological equipment; to D. Böttger for his assistance in field work, to J. Sauer, H. Wendler and G. Kalberlah for their assistance in measurements. Thanks to A. Knohl for the fruitful discussions and for the help in solution of research problems. Special thanks to J. Schulla for advising in all WaSim questions arose during the study, solving problems and providing new model versions.

Edited by: N. Romano

References

Aronsson, P. G., Bergström, L. F., and Elowson, S. N. E.: Long-term influence of intensively cultured short-rotation Willow Coppice on nitrogen concentrations in groundwater, J. Environ. Manage., 58, 135–145, 2000.

Baum, S., Weih, M., and Bolte, A.: Stand age characteristics and soil properties affect species composition of vascular plants in short rotation coppice plantations, 7, 51–71, 2012.

Bernhofer, C.: Institut für Hydrologie und Meteorologie, Professur Meteorologie: Exkursions- und Praktikumsführer Tharandter

Wald Material zum "hydrologisch-meteorologischen Feldprak-tikum," Techn. Univ., Dresden, 2002.

Blume, H.-P., Brümmer, G. W., Horn, R., Kandeler, E., Kögel-Knabner, I., Kretzschmar, R., Stahr, K., Thiele-Bruhn, S., Welp, G., and Wilke, B.-M.: Lehrbuch der Bodenkunde, Springer, 552 pp., 2010.

Breda, N. J.: Ground-based measurements of leaf area index: a re-view of methods, instruments and current controversies, J. Exp. Bot., 54, 2403–2417, 2003.

Breuer, L., Eckhardt, K., and Frede, H.-G.: Plant parameter values for models in temperate climates, Ecol. Model., 169, 237–293, 2003.

Broeckx, L. S., Verlinden, M. S., Berhongaray, G., Zona, D., Fichot, R., and Ceulemans, R.: The effect of a dry spring on seasonal carbon allocation and vegetation dynamics in a poplar bioenergy plantation, GCB Bioenergy, 6, 473–487, 2013.

Cannell, M. G. R. and Smith, R. I.: Thermal time, chill days and pre-diction of budburst in Picea sitchensis, J. Appl. Ecol., 20, 951–963, 1983.

Ceulemans, R., Impens, I., and Imler, R.: Stomatal conductance and stomatal behavior in Populus clones and hybrids, Can. J. Bot., 66, 1404–1414, 1988.

Chmielewski, F.-M., Heider, S., Moryson, S., and Bruns, E.: In-ternational Phenological Observation Networks: Concept of IPG and GPM, in Phenology: An Integrative Environmental Science, 137–153, Springer, 2013.

Djomo, S. N., Kasmioui, O., and Ceulemans, R.: Energy and green-house gas balance of bioenergy production from poplar and wil-low: a review, GCB Bioenergy, 3, 181–197, 2011.

Don, A., Osborne, B., Hastings, A., Skiba, U., Carter, M. S., Drewer, J., Flessa, H., Freibauer, A., Hyvönen, N., Jones, M. B., Lanigan, G. J., Mander, Ü., Monti, A., Djomo, S. N., Valentine, J., Walter, K., Zegada-Lizarazu, W., and Zenone, T.: Land-use change to bioenergy production in Europe: implications for the greenhouse gas balance and soil carbon, GCB Bioenergy, 4, 372–391, 2012.

Hartmann, L., Richter, F., Busch, G., Ehret, M., Jansen, M., and Lamersdorf, N.: Etablierung von Kurzumtriebsplantagen im Rahmen des Verbundprojektes BEST in Süd-Niedersachsen und Mittel-Thüringen – Standorteigenschaften und anfängliche Erträge, Forstarchiv, 134–150, 2014.

International Phenological Gardens of Europe: IPG, Phasen, avail-able at: http://ipg.hu-berlin.de/ipg/faces/list_phases.xhtml (last access 9 September 2014), 2014.

Jonckheere, I., Fleck, S., Nackaerts, K., Muys, B., Coppin, P., Weiss, M., and Baret, F.: Review of methods for in situ leaf area index determination: Part I, Theories, sensors and hemispherical photography, Agric. For. Meteorol., 121, 19–35, 2004.

Kalberlah, G.: Wurzelverteilungsmuster einer Pappel-Kurzumtriebsplantage in Südniedersachsen, Master Thesis, Georg-August-Universität Göttingen, Göttingen, 16–22, 2013.

Larchevêque, M., Maurel, M., Desrochers, A., and Larocque, G. R.: How does drought tolerance compare between two improved hy-brids of balsam poplar and an unimproved native species?, Tree Physiol., 31, 240–249, 2011.

Lasch, P., Kollas, C., Rock, J., and Suckow, F.: Potentials and im-pacts of short-rotation coppice plantation with aspen in East-ern Germany under conditions of climate change, Reg. Environ. Change, 10, 83–94, 2010.

LGLN - Landesbetrieb Landesvermessung und Geobasisinforma-tion: Digitale Geländemodelle – DGM - ATKIS, 2013.

LI-CO179 pp., 1992.

Lorenz, K. and Müller, J.: Ergebnisse zur Nettoprimärproduktion von mit biologisch geklärtem Abwasser bewässerten Pappeln und Weiden im Kurzumtrieb, Landbauforsch., Appl. Agric. For. Res., 63, 307–320, 2013.

Menzel, A.: Phänologie von Waldbäumen unter sich ändernden Klimabedingungen: Auswertung der Beobachtungen in den in-ternationalen phänologischen Gärten und Möglichkeiten der Modellierung von Phänodaten, Frank, 147 pp., 1997.

Menzel, A.: Trends in phenological phases in Europe between 1951 and 1996, Int. J. Biometeorol., 44, 76–81, 2000.

Moriasi, D. N., Arnold, J. G., Van Liew, M. W., Bingner, R. L., Harmel, R. D., and Veith, T. L.: Model evaluation guidelines for systematic quantification of accuracy in watershed simulations, Trans ASABE, 50, 885–900, 2007.

Octave community: GNU Octave, available at: www.gnu.org/software/octave/ (last access: 30 July 2015), 2012.

Petzold, R., Wahren, A., and Feger, K.-H.: Steuerungsoptionen des Wasser- und Stoffhaushalts auf Landschaftsebene durch den Anbau von Kurzumtriebsplantagen – ein Forschungsansatz, in Bodenschutz in Europa – Ziele und Umsetzung, 6, 91–96, Marktredwitzer, available at: http://boku.forst.tu-dresden.de/pdf/petzold_wahren_feger_Steuerungsoptionen.pdf (last access: 30 July 2015), 2010.

Raissi, F., Müller, U., and Meesenburg, H.: Geofakten 9 Ermit-tlung der effektiven Durchwurzelungstiefe von Forststandorten, Red. Geofakten Landesamt Für Bergbau Energ. Geol. Hann., (4. Auflage), available at: http://www.lbeg.niedersachsen.de (last ac-cess: 30 July 2015), 2009.

R Development Core Team: RStudio for Windows, R Foundation for Statistical Computing, Vienna, Austria, available at: http://www.R-project.org (last access: 30 July 2015), 2011.

Rutter, A. J., Kershaw, K. A., Robins, P. C., and Morton, A. J.: A predictive model of rainfall interception in forests, 1. Derivation of the model from observations in a plantation of Corsican pine, Agric. Meteorol., 9, 367–384, 1971.

Schmidt-Walter, P. and Lamersdorf, N.: Biomass Production with Willow and Poplar Short Rotation Coppices on Sensitive Areas – the Impact on Nitrate Leaching and Groundwater Recharge in a Drinking Water Catchment near Hanover, Germany, BioEnergy Res., 5, 546–562, 2012.

Schmidt-Walter, P., Richter, F., Herbst, M., Schuldt, B., and Lamers-dorf, N. P.: Transpiration and water use strategies of a young and a full-grown short rotation coppice differing in canopy cover and leaf area, Agric. For. Meteorol., 195–196, 165–178, 2014.

Schulla, J.: Hydrologische Modellierung von Flussgebieten zur Abschätzung der Folgen von Klimaänderungen, ETH Zürich, doi:10.3929/ethz-a-001763261, 1997.

Schulla, J. and Jasper, K.: Model Description WaSiM-ETH, Inst. Atmospheric Clim. Sci. Swiss Fed. Inst. Technol. Zür, available at: http://www.wasim.ch/downloads/doku/wasim/wasim_2007_en.pdf (last access: 30 July 2015), 2013.

Spank, U., Schwärzel, K., Renner, M., Moderow, U., and Bernhofer, C.: Effects of measurement uncertainties of meteorological data on estimates of site water balance components, J. Hydrol., 492, 176–189, 2013.

Surendran Nair, S., Kang, S., Zhang, X., Miguez, F. E., Izaurralde,

R. C., Post, W. M., Dietze, M. C., Lynd, L. R., and Wullschleger, S. D.: Bioenergy crop models: descriptions, data requirements, and future challenges, GCB Bioenergy, 4, 620–633, 2012.

Tardieu, F. and Simonneau, T.: Variability among species of stomatal control under fluctuating soil water status and evaporative demand: modelling isohydric and anisohydric behaviours, J. Exp. Bot., 49, 419–432, 1998.

Van Genuchten, M. T.: A closed-form equation for predicting the hydraulic conductivity of unsaturated soils, Soil Sci. Soc. Am. J., 44, 892–898, 1980.

Volkert, E. and Schnelle, F.: Arboreta Phaenologica: Mitteilung der Arbeitsgemeinschaft Internationaler Phänologischer Gärten, Erläut. Zur Beob., Phänologischen Termins Blattentfaltung Bei Popolus Tremula, 7, 5 pp., 1966.

Watson, D. J.: Comparative physiological studies on the growth of field crops: I. Variation in net assimilation rate and leaf area between species and varieties, and within and between years, Ann. Bot., 11, 41–76, 1947.

Zambrano-Bigiarini, M.: hydroGOF: Goodness-of-fit functions for comparison of simulated and observed hydrological time series, available at: http://CRAN.R-project.org/package=hydroGOF (last access: 30 July 2015), 2014.

11

Simulation of semi-arid biomass plantations and irrigation using the WRF-NOAH model – a comparison with observations from Israel

O. Branch[1], K. Warrach-Sagi[1], V. Wulfmeyer[1], and S. Cohen[2]

[1]Institute of Physics and Meteorology, University of Hohenheim, Stuttgart, Germany
[2]Institute of Soil, Water and Environmental Sciences, Agricultural Research Organization, Volcani Center, Bet-Dagan, Israel

Correspondence to: O. Branch (oliver_branch@uni.hohenheim.de), K. Warrach-Sagi
(kirsten.warrach-sagi@uni-hohenheim.de), V. Wulfmeyer (volker.wulfmeyer@uni-hohenheim.de), and S. Cohen
(vwshep@volcani.agri.gov.il)

Abstract. A 10×10 km irrigated biomass plantation was simulated in an arid region of Israel to simulate diurnal energy balances during the summer of 2012 (JJA). The goal is to examine daytime horizontal flux gradients between plantation and desert. Simulations were carried out within the coupled WRF-NOAH atmosphere/land surface model. MODIS land surface data was adjusted by prescribing tailored land surface and soil/plant parameters, and by adding a controllable sub-surface irrigation scheme to NOAH. Two model cases studies were compared – *Impact* and *Control*. Impact simulates the irrigated plantation. Control simulates the existing land surface, where the predominant land surface is bare desert soil. Central to the study is parameter validation against land surface observations from a desert site and from a 400 ha *Simmondsia chinensis* (jojoba) plantation. Control was validated with desert observations, and Impact with Jojoba observations. Model evapotranspiration was validated with two Penman–Monteith estimates based on the observations.

Control simulates daytime desert conditions with a maximum deviation for surface 2 m air temperatures ($T2$) of $0.2\,°C$, vapour pressure deficit (VPD) of 0.25 hPa, wind speed (U) of $0.5\,\mathrm{m\,s^{-1}}$, surface radiation (R_n) of $25\,\mathrm{W\,m^{-2}}$, soil heat flux (G) of $30\,\mathrm{W\,m^{-2}}$ and 5 cm soil temperatures (ST5) of $1.5\,°C$. Impact simulates irrigated vegetation conditions with a maximum deviation for $T2$ of 1–$1.5\,°C$, VPD of 0.5 hPa, U of $0.5\,\mathrm{m\,s^{-1}}$, R_n of $50\,\mathrm{W\,m^{-5}}$, G of $40\,\mathrm{W\,m^{-2}}$ and ST5 of $2\,°C$. Latent heat curves in Impact correspond closely with Penman–Monteith estimates, and magnitudes of $160\,\mathrm{W\,m^{-2}}$ over the plantation are usual. Sensible heat fluxes, are around $450\,\mathrm{W\,m^{-2}}$ and are at least 100–$110\,\mathrm{W\,m^{-2}}$ higher than the surrounding desert. This surplus is driven by reduced albedo and high surface resistance, and demonstrates that high evaporation rates may not occur over Jojoba if irrigation is optimized. Furthermore, increased daytime $T2$ over plantations highlight the need for hourly as well as daily mean statistics. Daily mean statistics alone may imply an overall cooling effect due to surplus nocturnal cooling, when in fact a daytime warming effect is observed.

1 Introduction

The large-scale implementation of biomass plantations in arid regions is the subject of recent research due to the perceived potential for carbon sequestration, energy production, agricultural development and environmental services (Becker et al., 2013; Beringer et al., 2011). Such plantations are becoming feasible through modern desalination (Khawaji et al., 2008; Fritzmann et al., 2007), wastewater (Hamilton et al., 2007; Oron et al., 1999), and irrigation techniques (e.g., see Spreer et al. 2007). Valuable and hardy shrubs such as *Jatropha curcas* (jatropha) or *Simmondsia chinensis* (jojoba) can withstand heat and drought, and be irrigated with waste- or brackish water (Rajaona et al., 2012; Abou Kheira and Atta, 2009; Benzioni, 1995). These traits makes them more viable than many food crops and may reduce threats to food security if exclusive use of marginal land is adhered to (Becker et al., 2013).

Critical research is still missing however, on potential climatic impacts caused by significant land surface modifications in arid regions. Vital insights can be obtained using dynamically-downscaled simulations with coupled atmospheric/land surface models. Such models need careful calibration for regional arid conditions though, and validation to assess confidence in simulation results.

Large-scale agroforestry (AF) could modify the local and regional climate. Alpert and Mandel (1986), observed a reduction in amplitude and variance of wind speeds (U) and 2 m temperatures ($T2$) in Israel over three decades. They correlated changes with increases in irrigation since the 1960s, and attribute them to lower sensible heat fluxes (HFX) and changes in albedo and roughness. De Ridder and Gallée (1998) concurred with these trends. Increases in rainfall, especially around October were also found by Ben-Gai et al. (1998, 1994, 1993) and Otterman et al. (1990) (in Perlin and Alpert, 2001). This is likely due to the combination of autumn climatic conditions and the land surface perturbations. From De Ridder and Gallée (1998), Alpert and Mandel conclude that altered weather patterns are caused by lower HFX from irrigated cops, whereas Otterman cites increased HFX from non-irrigated shrubs. For the latter land use type, higher HFX and lower latent heat (LH) magnitudes would result from less water availability. Otterman (1989) also found that increased Saharan fringe vegetation increased daytime convection and atmospheric boundary layer (ABL) growth.

Given the likely dependence of flux partitioning on soil moisture, this presents some interesting questions to investigate: what partitioning of fluxes can be expected from a large arid irrigated plantation? How would these fluxes contrast with the surrounding desert surface?

The introduction of large vegetation patches into deserts is likely to induce significant horizontal flux gradients, increase surface roughness, moisten the ABL, modify turbulent flows, and induce pressure perturbations. These phenomena would influence ABL evolution and may cause convergences (Wulfmeyer et al., 2014) and mesoscale circulations (Hong, 1995; Mahfouf et al., 1987). Impacts could be dependent on the scale of the patches. Dalu et al. (1996) suggest flux gradients of the order of 1 to 10 km are sufficient to induce significant changes. Letzel and Raasch (2003) estimate scales of around 5 km from large eddy simulations (LES). Contiguous plantations on scales of this order could be feasible now for the reasons previously discussed.

Regarding fluxes, an expectation is that a freely transpiring canopy would result in low HFX and higher LH, in contrast to bare desert surfaces where LH is likely to be almost zero. However, it is not clear how plantation HFX magnitudes would compare with typically high desert HFX. Firstly, there is generally a greater surface net radiation (R_n) at canopy surfaces due to lower albedos. Secondly, leaves can be very efficient heat radiators, and have lower heat storage potential than most substrates (Warner, 2006). Finally, desert crops such as jojoba or jatropha may not transpire freely due to

their high water use efficiency and resistance to water stress (Silva et al., 2010; Abou Kheira and Atta, 2009; Benzioni and Dunstone, 1988) and also because efficient irrigation techniques such as partial root zone drying are used (Spreer et al., 2007). The final point is significant, because in arid regions a fine balance exists between maintaining yield and plant health, and the need to conserve water. Plantation evapotranspiration (ET) could therefore be limited, resulting in higher HFX magnitudes than a freely transpiring canopy implies. An example of large HFX magnitudes from drier vegetation is the Yatir pine forest in Israel, where Rotenberg and Yakir (2010) observed summer HFX magnitudes that were 1.3 times higher than over the Sahara and 1.6 and 2.4 times higher than tropical and temperate forests, respectively.

Relative HFX over plantation and desert will depend largely on the albedos and energy balance over the plantation. In turn, albedo generally depends on crop type, phenological stage, leaf area index (LAI), canopy cover, senescent material and so forth (e.g. Ingwersen et al., 2011; Zhang et al., 2013). ET and upward turbulent transport of moisture depends not only on available energy, water availability, soil characteristics and boundary layer conditions, but also on roughness and plantation/canopy/leaf homogeneity, geometry and scales (Bonan, 2008; Burt, 2002; Raupach and Finnegan, 1996). Specific plant characteristics and survival strategies also play a major role, such as modified reflectivity, photosynthesis pathways or stomatal closure. These characteristics are observed in many desert species (Warner, 2006), such as jojoba (Seventh International Conference on Jojoba and its uses: Proceedings, 1988) and *Jatropha curcas* (Silva et al., 2010).

In order to estimate impacts on atmospheric interactions, detailed simulations are carried out within coupled atmospheric and land surface models (LSM). This can be achieved by artificially modifying the land surface data used by the LSM to calculate surface exchanges. If the irrigated vegetation surface is correctly parameterized, we can then go on to assess the impacts on (a) diurnal fluxes, (b) feedbacks to and from the ABL, and (c) mesoscale impacts such as convection initiation. Furthermore, the effects of variables such as plantation size and location on these phenomena, can be explored.

The use of coupled 3-D models is preferable to the use of uncoupled models, where a land surface model is forced unidirectionally with atmospheric forcing data. This is because uncoupled models neglect the simultaneous feedbacks which occur between the surface, boundary layer and entrainment zone. These processes are central to ABL evolution (see, e.g. Van Heerwaarden et al., 2009), evaporation and therefore flux gradients over heterogeneous landscapes. The use of fine resolutions of e.g., less than 4 km, allows for detailed resolution of landscape features and can reduce systematic errors and biases in soil–cloud–precipitation feedbacks commonly seen in coarser models where convection is normally parameterized (Rotach et al., 2009, 2010; Wulfmeyer et al., 2008,

2011; Bauer et al., 2011; Weusthoff et al., 2010; Schwitalla et al., 2008).

Our ultimate goal is to use the WRF-NOAH model to conduct impact studies on meso-α scales. This study focuses on the parameterization and validation of the WRF (Skamarock et al., 2008) with its LSM NOAH (Chen and Dudhia, 2001) model for the region/vegetation/irrigation/soils, and also on the comparison of energy fluxes over the desert and plantation. Two model scenarios are set up – WRF Control and WRF Impact. WRF Control represents a baseline, using unmodified MODIS land surface type initialization data and the second is a simulation of a 10×10 km irrigated plantation (WRF Impact). WRF Control output is compared to observations from a desert surface and WRF Impact is compared with observations from the jojoba plantation. These observations were collected especially for the experiment. Specific objectives are

- to conduct an experimental study, to form the basis for a model configuration for later impact studies on large-scale arid plantations;

- to build and set up a WRF-NOAH model simulation for irrigated plantations in a semi-arid region;

- to verify the model for follow-up impact studies.

In Sect. 2, the study area in Israel is described, including climate, and specifics on the irrigated plantations and surrounding area. In Sect. 3, the field methodology and observation data from desert and plantations are presented. In Sect. 4, the methodology for the WRF-NOAH model simulations are covered, including configuration, domains, and the simulation of irrigated vegetation. In Sect. 5, WRF-NOAH simulations are compared with field observations, including calculated evapotranspiration. Finally the results are discussed in Sect. 6.

2 The study area and its climate

The impact of plantations is studied in the semi-arid region of Israel. In this area, long hot summers, clear skies and high radiation are the most common conditions. Synoptically, a pressure trough to the north generally runs from Turkey down to the Persian Gulf drawing north-north-western winds steadily in from the Mediterranean for most of the year. Until around October, the summer climate is dominated by Hadley subsidence and strong inversions, which inhibit convection. During autumn, these inversions tend to weaken, and the Mediterranean and its winter cyclones start to have more influence. Precipitation is usually convective when it occurs, either embedded in the passage of fronts or induced by local circulations (Perlin and Alpert, 2001).

This study focuses on the area of the northern edge of Israel's Negev Desert close to the city of Be'ér Sheva' (see Fig. 1, inset) around 50 km from the coast (31.24° N,

34.72° E). This lies roughly on the border of two climatic zones – a semi-arid one with crops and grasslands to the north, and an arid one to the south. In this desert area various plantations exist. Among them are a jatropha and a jojoba plantation, which are the subject of this investigation and as a control case, a dry desert area was chosen. All cases are located approximately 2 km to the west of Be'ér Sheva'.

The "Desert" case study is situated on bare, desert soil with no vegetation, marked (1) in Fig. 1. Some small plantations exist around 800 m upwind but it is assumed that moisture advection to the sensors would not be significant and that non-advected quantities such as surface radiation and soil temperature would be representative of a desert surface. The "Jatropha" case study (2) is a 2 ha irrigated *Jatropha curcas* plantation and the "Jojoba" case study (3) is a 400 ha plantation of irrigated jojoba with a canopy height of around 3–3.5 m. Both plantations are irrigated with secondary treated waste water from Be'ér Sheva', with low water salinity, i.e. the plantation's managers report that mean electrical conductivity (EC) of the irrigation water is ~ 1 dS m^{-1} (see Appendix B for more information about salinity). The experimental jatropha plantation is irrigated only from March to December and is heavily pruned during the winter. Because of this, the canopy cover is still only around 50–60 % at the beginning of June. This increased over the summer to nearly 100 % during July. This is likely to bias the Jatropha observations somewhat (compared to a fully mature jatropha canopy) due to gradual changes in wind speed, albedo, evaporation and so on. Due to the small size of the jatropha plantation and the changing canopy conditions, it was decided that only the Jojoba observations would be used for validation. The Jatropha observations are however, examined and compared with the Jojoba observations (see Sect. 3.2) and yield interesting information on differences in solar and thermal radiation components between the crops.

The jojoba plantation is fully mature and watered all year round. The shrubs are widely spaced for mechanized harvesting (4×2.5 m) producing canopy coverage of around 70 % for the mature sections. These factors, consequently, are likely to produce differences in albedo, wind flow, turbulence, evaporation, skin and air temperatures and other quantities when compared with a 100 % canopy closure. The soils within the plantation are mainly composed of silty to sandy loam, loess soils by local soil survey.

While the jatropha plantation is being tested with various sub-surface treatments, the jojoba plantation is fed by a sophisticated, sub-surface deficit irrigation system configured to maximize water use efficiency and yield. Water requirements are estimated by agronomists using meteorology data and standard methods. The irrigation flow rates, duration and dripper spacing are optimized to minimize losses to percolation, runoff and direct evaporation. Given that (a) there is little precipitation, (b) the irrigation is sub-surface and (c) the dosage is carefully calibrated, these losses are assumed to be negligible. Therefore, based on these assumptions, only the

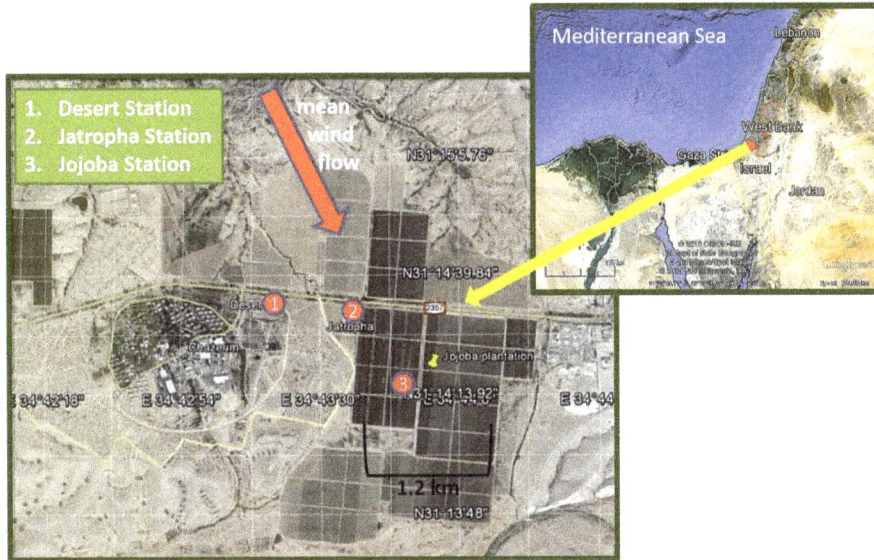

Fig. 1. The region of interest over Israel at the eastern edge of the Mediterranean Sea (inset) and the location of the three meteorological stations at Kibbutz Hatzerim in the centre of Israel, 40 km from the coast (the regional location is indicated by in the inset box). The location of the stations Desert (1), Jatropha (2) and Jojoba (3) are marked on the left-hand image. Mean wind flow is marked with an arrow.

Table 1. Measured quantities from Desert, Jatropha and Jojoba cases, sensor type and estimated measurement errors. Measurement errors are rated usually within a range of, or at a given temperature. Where this is applicable, the temperature range is indicated by and at sign.

Quantity	Sensor	Estimated error
2 m air temperature ($T2$)	Vaisala HMP155A	At $20\,°C \pm (0.055 + 0.0057 \times T)\,°C$
2 m relative humidity (RH)	Vaisala HMP155A	At $-20 + 40\,°C \pm (1.0 + 0.008 \times \text{reading})\,\%\,RH$
Short and long wave radiation (SW/LW)	Hukseflux NR01	$\pm 10\,\%$ for 12 h totals
6 m wind speed and direction (U and U_{dir})	Gill 2-D Windsonic	$U \pm 2\,\%$ U_{dir} 2–3°
Barometric surface pressure (BP)	Vaisala CS106	$\pm 0.6\,hPa$ at 0 to $+40\,°C$
Soil temperatures at 5 and 25 cm (ST5 and ST25)	CS 108 Thermopile	$\pm 0.3\,°C$ at -3 to $90\,°C$
Soil heat flux (G, two plates per station)	Hukseflux HFP01	within -15 to $+5\,\%$ for 12 h totals

potential and transpired evaporation terms would play a role within the NOAH evaporation equation (Appendix A). The plants are watered directly at the root ball (35 cm deep) in alternate crop rows. Soil moisture (Θ) is monitored by a sensor network, and constrained to a fraction of between 0.16 and 0.30 so that the plant is neither water-stressed nor over-watered. This means that the frequency of watering can be irregular depending on environmental conditions such as radiation magnitudes, phenological stage and so on. The stem spacing of the jojoba plantation also means that soil moisture is highly heterogeneous spatially. Irrigation information comes from the jojoba plantation agronomists.

3 Measurements

3.1 Site description and meteorological observations

The variables measured and the sensor array are described in Table 1. A scan rate of 5 s was used for measuring and the data was averaged over 10 min intervals. Other useful variables were then derived from these measurement data. Albedo was calculated from the individual observed radiation components as $SW_{\text{UP}}/SW_{\text{DOWN}}$ and net surface radiation (R_n) is calculated as $SW_{\text{DOWN}} - SW_{\text{UP}} + LW_{\text{DOWN}} - LW_{\text{UP}}$. The vapour pressure deficit (hPa) is derived as: $e_{\text{sat}} - e_{\text{actual}}$ where e_{sat} is the partial pressure of water vapour at saturation (hPa) and e_{actual} is the actual vapour pressure.

3.2 Analyses of observations

A summer time series of $T2$ and RH (Fig. 2) was examined from the three stations to assess the seasonal evolution of mean temperatures and relative humidity along with maxima and minima. The purpose of doing so was to reveal any major seasonal shifts, to assess the validity of examining seasonal diurnal curves, and to explain some of the hourly variance. Figure 2 indicates a general seasonal $T2$ pattern, peaking in

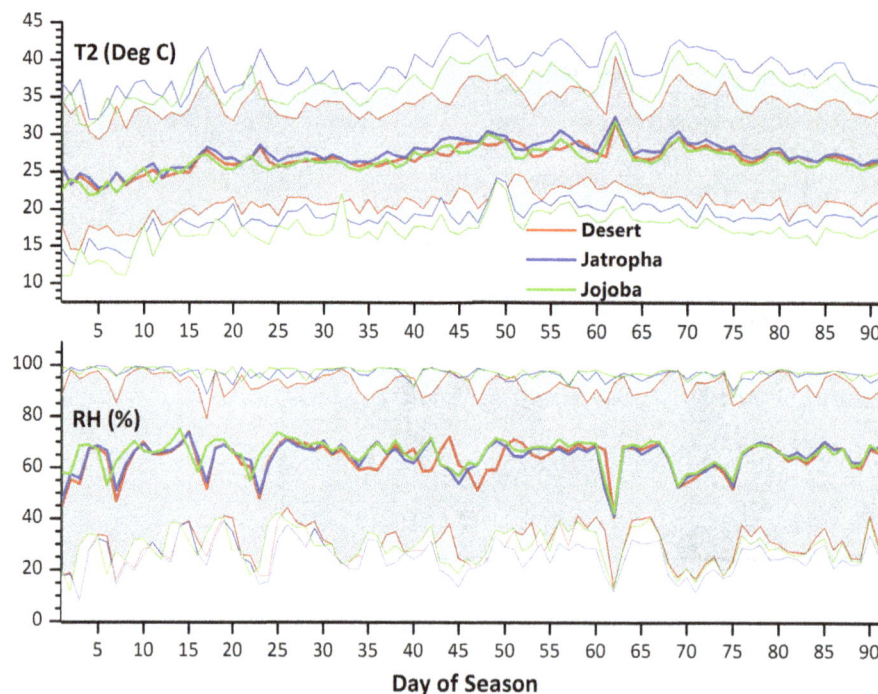

Fig. 2. Observed daily mean, maxima and minima of 2 m air temperatures and relative humidities for the Desert, Jatropha and Jojoba stations – summer 2012 (JJA). The thick curves at the centre of the shaded areas are the daily mean values. The upper and lower thin lines bounding the shaded areas are the daily maxima and minima.

July with daily means 2–3 °C warmer than in June. Desert is warmer at night but cooler during the daytime than the plantations with the Jojoba $T2$ lower in general than Jatropha. Relative humidities remain fairly constant indicating more humidity during July where higher July $T2$ signifies higher saturation values. The Jojoba and Jatropha RH curves match very closely but Desert exhibits lower maximum RH during the night, reflecting the higher $T2$ minimum.

Net radiation (R_n) values peak at the end of June and then decrease steadily over the season (see Fig. 3). Jojoba has a higher R_n than Jatropha especially over July, reflecting the lower albedo of Jojoba, and the Desert station has between 80 to 100 W m^{-2} less peak R_n than either of the plantations. Mean U is quite constant over Desert and Jojoba, but over Jatropha it decreases somewhat over the season (0.5 m s^{-1}) probably reflecting the Jatropha's canopy development and corresponding increase in drag. Winds speeds are in general a little higher over Jojoba than over Jatropha. There is a high variability in mean U direction and the 7 day mean indicates a slight shift of around 10° to the west through the season, for all stations. Surface wind speeds over Desert reach 5–6 m s^{-1} and are in general higher than over the plantations (3–4 m s^{-1}) and also very slightly more northerly, indicating the effect of drag and a tendency towards gradient flows over the canopy. Surface air pressures (P) in Fig. 3 tend to vary inversely with the seasonal temperatures, but with large

variations over periods of a few days, in accordance with changing large-scale pressure systems.

Considering mean diurnal statistics (Fig. 4), $T2$ values over the plantations exhibit larger amplitudes than over the desert being warmer in the daytime and cooler at night. Jatropha is warmer than the Jojoba during the day (+2 °C) and also at night (+1 or 2 °C). Day and nighttime RH values are fairly similar and reflect the differences in temperature over the day. Wind speeds over Desert are higher with a pronounced daytime peak late in the afternoon, 2.5 m s^{-1} higher than the plantations. Daytime peak R_n values are around 350 W m^{-2} higher over both plantations than over Desert, but with similar losses at night time. Daytime Jojoba albedo values are noticeably lower than the Jatropha and this is reflected in the greater Jojoba R_n values.

4 Model simulations

This analysis is carried out via high resolution model simulations with the WRF-NOAH model. In relation to subsequent impact studies, this study focuses on the configuration and validation of the model for an arid region. Additionally, as a first examination of vegetation impacts, the energy fluxes estimated by NOAH are investigated, since these express the interaction of the atmosphere and the land surface. The model domain spans 888×888 km and is centred over the state of Israel (see Fig. 5).

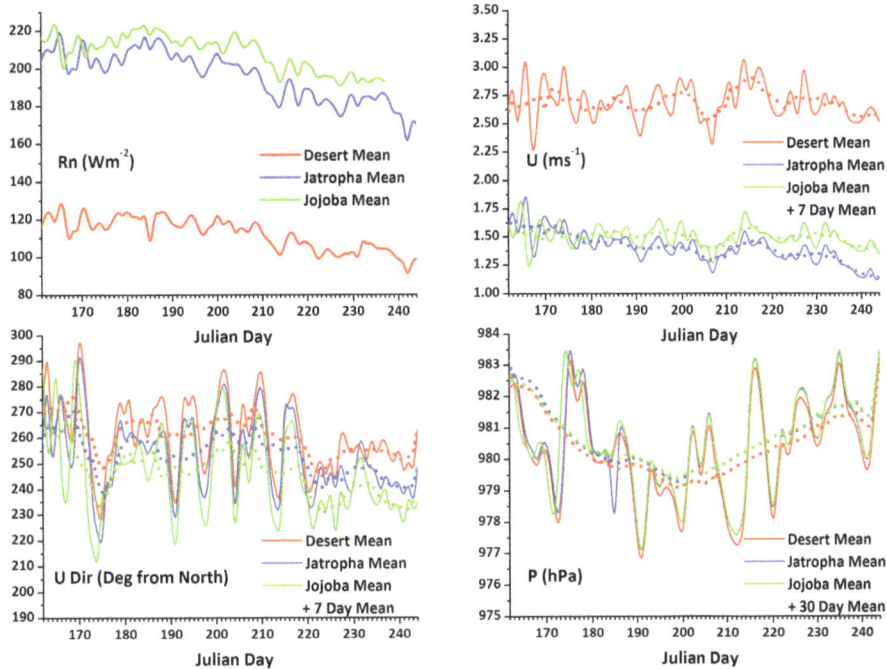

Fig. 3. Observed daily 24 h mean values of R_n, U, P, U and U direction. 7 or 30 day means are plotted for U, U_{Dir} and P based on peak analysis to highlight differences between the stations and the evolution of the summer climate (2012–JJA). Due to poor quality flags some Jojoba R_n data were rejected for the last 8 days of the season.

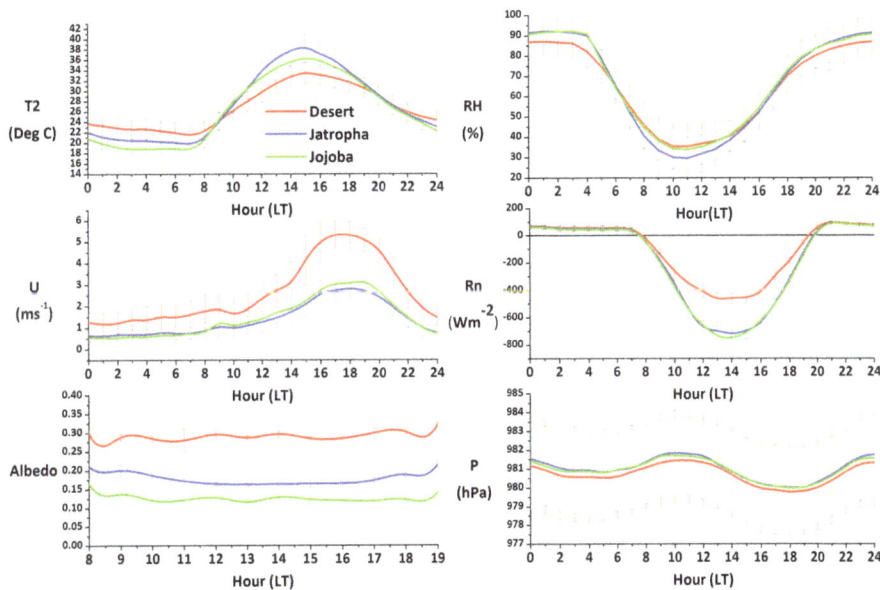

Fig. 4. Observed mean diurnal cycle of $T2$, RH, U (6 m), R_n, albedo and P – 2012 (JJA). The error bars represent temporal standard deviation.

To judge the model performance and configuration, two cases are assessed. The first is the baseline (WRF Control). The second is a simulation of a 10×10 km irrigated plantation (WRF Impact). WRF Control observations are compared to the Desert observations and WRF Impact observations are compared with observations from Jojoba and Jatropha. The ultimate goal, although not in this paper, will be to use the WRF-NOAH model to investigate the impact of horizontal flux gradients on ABL development and convection initiation or suppression. To ensure that these gradients are representative, the model should be able to reproduce the energy balance correctly over both surfaces. For that purpose, this

Table 2. Physics schemes used for the study within the WRF atmospheric model.

Physics	Scheme	References
Boundary layer	YSU (Yonsei University)	Hong et al. (2006)
Surface layer	MM5 Monin–Obhukov	Paulson (1970), Dyer and Hicks (1970), Webb (1970), Beljaars (1995)
Microphysics	Morrison 2-moment	Morrison and Gettelman (2008)
Shortwave radiation	RRTMG	Iacono et al. (2008)
Longwave radiation	RRTMG	Mlawer et al. (1997)

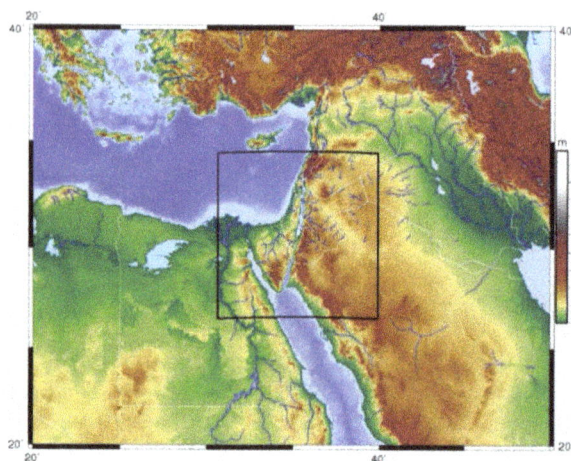

Fig. 5. Topographic map of the region of interest, at the eastern end of the Mediterranean. The model domain (approx. 888×888 km) is marked in the centre with a black line.

study examines the correct reproduction of the fluxes over desert and vegetated surfaces.

4.1 Modelling configuration

The Advanced Research WRF (ARW; WRF-ARW 3.4.1) non-hydrostatic atmospheric model, coupled with the NOAH land surface model (LSM) was configured with a 444×444 cell grid with 92 vertical levels and a 2 km grid horizontal increment. A single downscaled model domain was chosen with care to capture most large-scale features, such as the influx of sea air from the north-east, but to avoid orography and other strong features at the domain boundaries (see Fig. 5). The model was forced at the boundaries by ECMWF (European Centre for Medium Range Weather Forecasting) 6-hourly analysis data at $0.125°$ grid increments and with 6-hourly updated sea surface temperatures (SSTs). The model physics schemes used are shown in Table 2. Model physics schemes were chosen with consideration to the following:

- how relevant processes are dealt with and relevance to arid regions, land-surface/atmosphere feedbacks and convection;

- experience and sensitivity tests within the working group and within the WRF model community;

- which variables are explicitly calculated by, and are output from the scheme.

Additionally some schemes are designed to be paired (e.g. the SW and LW RRTMG schemes).

The YSU (ABL) and Morrison 2-moment (microphysics) schemes have been used for various publications relating to arid regions (Wulfmeyer et al., 2014; Becker et al., 2013) and temperate regions (Warrach-Sagi et al., 2013). YSU is the default WRF ABL scheme. It is non-local, explicitly handles entrainment, and is generally thought to perform well in unstable convective conditions (e.g. Shin and Hong, 2011; Hu et al., 2010), which is most relevant for examining the daytime fluxes. The MM5 surface layer scheme which computes surface exchange coefficients of heat, moisture, momentum using Monin–Obhukov stability functions, and is to be paired with the YSU (or WRF) scheme.

The Morrison 2-moment microphysics predicts total number concentration of ice species and may improve the representation of ice crystal aggregation and ice cloud radiation representation (Morrison and Gettelman, 2008). One study (Molthan and Colle, 2012) which used Morrison with WRF, cited that it gave the minimum difference between simulated and accumulated precipitation during a convective storm when compared to five other schemes. However, it is not known if this improvement in the representation of ice number concentrations would really improve simulations within our region of interest.

The land surface model was initialized using the International Geosphere Biosphere Programme (IGBP) MODIS 20-category land use and soil texture with the Food and Agriculture Organization (FAO) State Soil Geographic (STATSGO) 19-category soil data set. The initial soil moisture state and lower soil boundary temperatures come from the forcing data. The model duration was 92 days over JJA, 2012 and instantaneous values were generated every hour. Full observation data sets from Desert, Jatropha and Jojoba were available for this period.

In the absence of accurate, gridded, initial soil moisture (Θ) conditions, a spin-up period is needed to allow soil moisture within NOAH to approach equilibrium within the hydrological cycle. The optimal spin-up period for any particular application is uncertain and may depend on the, regional characteristics, accuracy of initial soil conditions, temporal and spatial resolution applied as well as other factors (Lim

Table 3. Modifications to model vegetation parameters, based on literature, sensitivity tests and local data.

Modifications	Default value	Prescribed value	Source
Roughness – Z_0 (m)	0.5 m	0.3 m	Literature, canopy height
Albedo	0.12	0.12	Observations
Veg. fraction – σ_f (%)	95 %	70 %	Local knowledge
Min. stom. resistance – $R_{C_{min}}$ ($s\,m^{-1}$)	120	250	7th International Conference on Jojoba

et al., 2012; Du et al., 2006). Du et al. (2006) simulate soil moisture in East Asia using the CLM model (Community Land Model) at 0.5° grid spacing, and proposes that the time interval between a precipitation perturbation and reaching an equilibrium is proportional to soil depth. He also says that surface soils (0–10 cm) may require a few months to reach equilibrium. Lim et al. (2012) ran five year comparisons in differing climates using NOAH at around a 0.1° spacing and concluded that arid soils may require considerable periods, possibly even years to reach equilibrium when compared with a monsoonal climate. Using such fine grid scales the required resources needed for such spin-up periods are simply prohibitive for many applications. In this case a one month spin up period was chosen as the longest period feasible, with the admission that for the control run, even one year might not be sufficient for a true equilibrium to be reached. However, within the simulated plantations, the sub-layers are in any case artificially moistened, and the target Θ level reached after approximately one day (see next section).

4.2 Irrigated plantations in NOAH

At the time of our simulations there were no official releases of WRF with irrigation schemes implemented in the accompanying NOAH or the newer NOAH-MP (multi-physics) land surface models. Schemes have been devised by others independently though and incorporated into WRF for impact studies (see Harding and Snyder, 2012; Sridhar, 2013). Sridhar for instance simulated two kinds of surface irrigation: flood and sprinkler systems. In our case, a controllable subsurface scheme was required, to reflect the sophisticated system used in Israel and so this was developed as a sub-routine and incorporated into NOAH for this study. The scheme was intended to mimic the actual jojoba plantation characteristics as closely as possible, therefore attention was given to the following factors: plantation location, size and shape; subsurface irrigation scheme/soil moisture and vegetation and soil parameters. A hypothetical plantation was introduced by modifying the land surface properties in the static land surface data used by the model (see Fig. 6). It should be emphasized that the intention is not to simulate detailed spatiotemporal phenomena over the actual $4\,km^2$ observed plantation, where a more explicit resolution could be more appropriate. Rather we seek a good statistical representation of the diurnal fluxes over a homogeneous plantation.

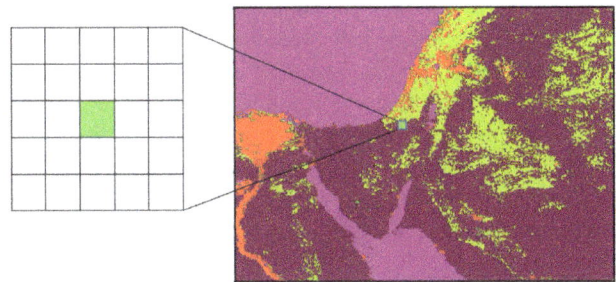

Fig. 6. Setup of the analysis of WRF output data. The image on the right is of the 20-category 30 arc second MODIS land use data set, a static data set for model initialization (all 25 cells are classed as desert/scrub in the MODIS data). A 25 cell grid box (left panel) was used, over which all variables values were averaged spatially, prior to the calculation of temporal statistics. The centre grid cell, marked in green corresponds geographically to the location of the three surface stations. The 25 cell box (10 × 10 km) was also used as a template for the simulated plantation.

A plantation size of 5 × 5 cell grid cells was used, representing dimensions of 10 km × 10 km. This is in fact larger than the actual jojoba plantation, which is closer to the size of one 2 km × 2 km grid cell. Pielke Sr. (2002) suggests a minimum of 4 grid cells to resolve any one feature. There is also a risk of introducing statistical anomalies due to clustering or artefacts and therefore using multiple cells allows for spatial averaging. Independent from the model simulations, the assumption could be made that surface quantities over the 2 km × 2 km jojoba plantation would acquire similar characteristics to those over a larger plantation of e.g., 10 km × 10 km across. This is uncertain though because, although advection effects are likely to be greatly reduced after a few hundred metres horizontally, differences in the scale of pressure perturbations and the mean wind field may well differ with the scale of the plantation. Another factor is that the wide spacing between the jojoba plants could lead to local heterogeneities and unusual turbulent characteristics above the canopy which may differ significantly with assumptions inherent in the model. Such effects are difficult to identify with point measurements. To check our assumptions on representativeness, we later compared fluxes from the 5 × 5 plantation with those from a 1 × 1 plantation (over one week), and the diurnal cycles and variability were not significantly different.

Grid cells for the plantation were first re-classified from *Desert/Scrub* to an *Evergreen Broadleaf* classification as a starting point for the configuration. Then, parameters such as canopy height, minimum stomatal resistance ($R_{C_{min}}$) and roughness were modified further based on literature on jatropha (Rajaona et al., 2012, 2013; Niu et al., 2012) and jojoba (Benzioni, 2010; Benzioni and Dunstone, 1988), sensitivity tests and site surveys (see Table 3).

Realistic simulation of the sophisticated irrigation system of the jojoba plantation using a soil moisture based system in NOAH is problematic, because the sub-grid heterogeneity of Θ cannot be reproduced at 2 km resolution. Nevertheless, if reasonable estimates of optimal Θ for irrigation can be made, and well-chosen plant soil and parameters are used, then we may expect a reasonable reproduction of the soil/plant water hydrology. We then make estimates of ET based on Penman–Monteith methods and observations, and compare them with the model results. These comparisons should indicate whether the irrigation scheme and parameterization produces ET of a realistic magnitude.

In order to calculate irrigation inputs, attention was paid to both the soil and plant properties. A method from Choudhury and DiGirolamo (1998) was used, who collated critical values of fractional root zone available water for various species from various publications. This value F_{AW} is the ratio of available water to maximum available water (Eq. 1):

$$F_{AW} = \frac{\Theta - \Theta_{WP}}{\Theta_{FC} - \Theta_{WP}} \qquad (1)$$

where F_{AW} is the critical value, Θ_{WP} is the soil wilting point and Θ_{FC} is the field capacity. If the soil conditions such as Θ and soil texture, are such that this ratio falls below the critical value then the plant is expected to experience stress. F_{AW} values for various species are quoted by Choudhury, but not for jatropha or jojoba. The variability in quoted F_{AW} values for plants of a similar biomass were not that varied – mostly between 0.3 and 0.4, with the only extremes being 0.25 for cotton and wheat and 0.50 for grasses. Sorghum, which like jojoba and jatropha requires a warm climate and is drought resistant, is accorded F_{AW} values of 0.37 and 0.35 by two separate studies, as reported by Choudhury. This represents the closest match in terms of climatic envelope as it can survive in semi-arid climates. Using the soil texture data a Θ value of around 0.39 was calculated for F_{AW} and rounded up to 0.4. This yields a value of 0.18 m³ m⁻³ which was used for the irrigation target moisture level. The results should therefore be interpreted under the assumption that 0.18 m³ m⁻³ is the minimum permissible water input for the plants. This also relies on the assumption that the deficit irrigation techniques minimize the water quantities need for the plants to thrive.

For soil moisture transport NOAH uses a layer discretized version of the Richards equation (Eq. 2) with four soil layers of thicknesses: 10, 30, 60 and 100 cm (from the surface layer

downwards). There is a free drainage scheme at the lower boundary.

$$d_{z_1} \frac{\partial \Theta_1}{\partial t} = -D\left(\frac{\partial \Theta}{\partial z}\right)_{z_1} - K_{z_1} + P_D - R - E_{dir} - E_{t_1}$$

$$d_{z_2} \frac{\partial \Theta_2}{\partial t} = D\left(\frac{\partial \Theta}{\partial z}\right)_{z_1} - D\left(\frac{\partial \Theta}{\partial z}\right)_{z_2} + K_{z_1} - K_{z_2} - E_{t_2} + I_\Theta$$

$$d_{z_3} \frac{\partial \Theta_3}{\partial t} = D\left(\frac{\partial \Theta}{\partial z}\right)_{z_2} - D\left(\frac{\partial \Theta}{\partial z}\right)_{z_3} + K_{z_2} - K_{z_3} - E_{t_3} + I_\Theta$$

$$d_{z_4} \frac{\partial \Theta_4}{\partial t} = D\left(\frac{\partial \Theta}{\partial z}\right)_{z_3} - D\left(\frac{\partial \Theta}{\partial z}\right)_{z_4} + K_{z_3} - K_{z_4} - E_{t_4} \qquad (2)$$

where D is hydraulic diffusivity, K is the soil hydraulic conductivity, P_D is precipitation, R is surface runoff and E_{t_i} is the layer root uptake.

The soil was irrigated by adding an extra irrigation term (I_Θ) to the appropriate soil layers. To assess which layers should be irrigated, a site inspection was made, and a lateral distribution radius of 30–40 cm was observed around the pipe. Therefore water was added to the second and third soil layers to approximate this depth and water distribution. The Θ level was replenished every 7 days to each sub-surface layer independently using the following logical statement: "IF $\Theta_{2,3} < 0.18$ THEN add water. IF $\Theta_{2,3} > 0.18$ THEN do not add water".

Adding water to only two of the sub-surface layers caused the model to become unstable due to the matrix method of solving the discretized equations. Therefore the water had to be added slowly (0.0004 mm/18 s timestep) and the constraining parameter in the water-balance error mechanism had to be adjusted.

Because the entire volume of each soil layer was wetted, the drainage over time is very slow, despite losses to uptake and deep percolation, and therefore after the first day of irrigation, Θ remains almost constant over the seven day intervals (0.18 ± 0.002). The reproduction of soil drainage characteristics over time still remains a problem therefore, because in reality the Θ fluctuation would be larger and more rapid from a smaller wetted volume. Therefore, the variability in canopy ET at short time scales of e.g., a few days, may not be well represented by the model because in reality the soil may dry before the sensors activate new irrigation and higher resistances are likely to occur briefly. In this model simulation therefore, ET is therefore still limited by Θ, but only in terms of the target level applied and not by varying levels of moisture due to soil moisture spatial heterogeneity, as may happen in reality. Well-reproduced daily variability therefore, might not always be expected but representative diurnal ET magnitudes based on target Θ levels and the environmental conditions are assumed.

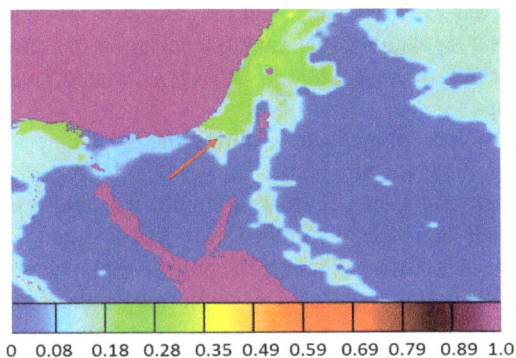

Fig. 7. ECMWF soil moisture initialization data for the second soil layer in NOAH (10–40 cm). The re-initialization of the soil moisture within the plantation can be seen on the image marked with a red arrow where there is a small patch which is much drier than the surroundings.

4.3 Soils within the plantations

The soils within the plantation are classified as clay loam by the FAO model soil data. Local soil survey data estimates that soils at the plantation are mainly composed of silty to sandy loam, loess soils. These were therefore reclassified to a sandy loam category both in Impact and Control. Parameters were then refined further using local survey data (see Table 4).

Soil moisture initialization values seemed to be unrealistic in the 2nd and 3rd soil layers (see Fig. 7) where Θ fractions of 0.2–0.28 were prevalent, particularly upwind of the plantations and in the desert. These Θ levels approach field capacity for sandy loam soils. Values closer to wilting point are likely to be more realistic during summer after a dry 2012 spring, even at 0.5 m. However, this could not be confirmed as quality data was not available. Nevertheless, if the sub-soil Θ data is unrepresentative, there could be implications for advection of moisture and perhaps the model spin-up time for the soils. The assumption was made that sub-soil Θ from unvegetated surfaces would not be a significant factor due to lack of a transport mechanism from sub-soils to the surface i.e. roots. Of course, there may still be an impact on the thermal diffusion and conductivity of the soil.

The 2nd and 3rd soil layers within the plantation boundary were re-initialized to wilting point (0.047) to ensure that initial levels are below the levels prescribed by the irrigation scheme (0.18 $m^3 m^{-3}$). Otherwise it may have taken some time for the soil moisture to decrease to that level which would increase the spin up time. Re-initializing a wider area of soil was considered, but a method for deciding the extent and Θ value was not found.

5 Validation

To judge the model performance and configuration, two cases are assessed. The first is a baseline run with unmodified

Table 4. Modifications to model soil parameters, based on literature and local soil data.

Modifications	Default value	Prescribed value	Source
Soil type	Clay Loam	Sandy Loam	Local soil survey
Sat Hyd. Cond K_s (m s^{-1})	2.45×10^{-6}	5.23×10^{-6}	Local soil survey
Porosity (m^3 m^{-3})	0.43	0.38	Local soil survey
Field capacity (m^3 m^{-3})	0.4	0.31	Local soil survey

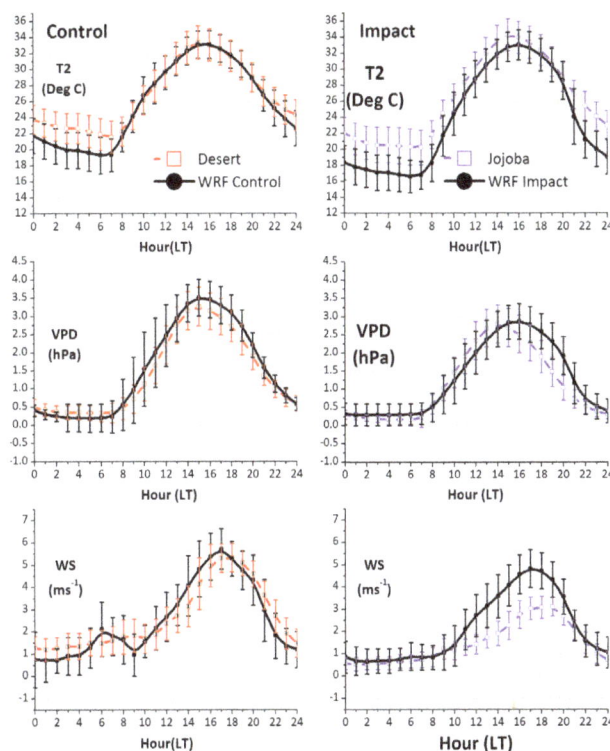

Fig. 8. Validation of WRF Control and Impact with observations for mean summer diurnal cycles of 2 m temperature ($T2$), 2 m vapour pressure deficit (VPD), wind speeds (U). Left hand panels show Control and the right panels, Impact. WRF variables were averaged over a 25 grid cell box centred at the geographical coordinates of the Desert, Jatropha and Jojoba sites. Note: The observations have been extrapolated from the sensor height of 6 to 10 m as calculated by WRF.

MODIS land surface data (WRF Control). The second is a simulation of a 10 km × 10 km irrigated plantation (WRF Impact). WRF Control is compared to observations from a desert surface. WRF Impact is compared with observations from jojoba and jatropha plantations.

5.1 Comparison with observed quantities

The validation of the Control model run against Desert observations, and Impact against Jojoba are shown in Figs. 8 and 9 as mean diurnal cycles with standard deviations as error bars. It is relevant to compare not only model against

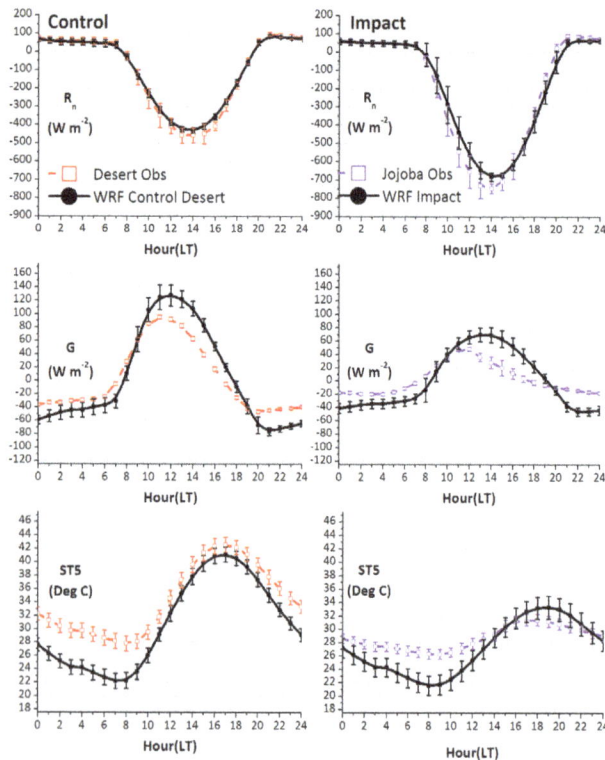

Fig. 9. Validation of WRF Control and Impact with observations for mean summer diurnal cycles of net surface radiation (R_n), ground flux (G) and 5 cm soil temperatures (ST5).

observations, but also how the quantities compare between the cases themselves.

$T2$ (model to observations) – during the daytime, Desert $T2$ values are reproduced extremely well by WRF Control with almost no deviation between 08:00 and 21:00 LT The variance is well reproduced throughout the day. After 21:00 LT, the model starts to diverge and there is a significant nighttime cold bias of around 2 °C. A similar overall pattern occurs with WRF Impact which shows a strong cold bias during the night time (up to 5 °C). Here though, the model is also a little too cool during the day (1–1.5 °C). WRF Impact $T2$ also accurately simulates the variability of Jojoba $T2$.

$T2$ (case comparison) – in reality, the observed daytime $T2$ over Jojoba is warmer than over Desert (1 °C) but up to 2 °C cooler during the night (Fig. 8, top panels). If we compare the model's representation of this phenomenon, WRF Impact correctly predicts cooler night time $T2$ than Control. However, this difference is larger than in reality by some margin (up to 2 °C). WRF Control and Impact have daytime $T2$ which are almost equivalent, with WRF Impact $T2$ around 0.5–1 °C cooler than WRF Control.

VPD (model to observations) – WRF Control models the VPD quite accurately throughout the 24 h period including variability, with a maximum bias of +0.2 hPa during the afternoon when temperatures are highest. WRF Impact also

models the VPD relatively well in terms of magnitudes, but exhibits a lag which could lead to a deviation in diurnal ET. The variability is also somewhat over estimated during the mornings.

VPD (case comparison) – during the middle of the day Desert VPD is 0.5 hPa higher than Jojoba. Disregarding the bias itself, WRF models this difference accurately in both magnitude and sign. At night Desert has a slightly higher VPD than Jojoba, reflecting the higher temperatures. Jojoba VPD approaches zero in the morning indicating near-saturated conditions.

U (model to observations) – WRF Control reproduces Desert U accurately, exhibiting biases of no more than $0.5 \, \text{m s}^{-1}$. In fact this bias appears to be one of phase rather than amplitude, with the WRF Control peak occurring later than Desert by an hour or so. There is also an unusual U peak ($0.5–1 \, \text{m s}^{-1}$) which occurs around 06:00–07:00 LT in WRF which is not reflected by the observations. The variability is overestimated somewhat (up to $0.5 \, \text{m s}^{-1}$). WRF Impact U was compared with height corrected U data from Jojoba to account for the difference in measuring height of 6 m and the model diagnostic height of 10 m (see Appendix C). The WRF Impact U peak matches the observations to within $0.5 \, \text{m s}^{-1}$. In contrast to Control, the WRF Impact peak occurs earlier than Jojoba by around 1 h but is similar in that the variability is overestimated.

U (case comparison) – Desert U is considerably more rapid than Jojoba during the middle part of the day (5.5 and $3 \, \text{m s}^{-1}$, respectively) as expected due to the difference in roughness. This is only reproduced partially by WRF – predicting Desert peak U to be only $1 \, \text{m s}^{-1}$ more rapid than Jojoba.

R_n (model to observations) – WRF Control R_n matches closely with Desert observations throughout the day and night with only a deficit of $25 \, \text{W m}^{-2}$ around midday. WRF Control underestimates the variability a little though during the daytime (around $15 \, \text{W m}^{-2}$). WRF Impact matches Jojoba somewhat less well with biases of up to $60–70 \, \text{W m}^{-2}$ (8–10 % of total magnitude), although the variability is simulated accurately.

R_n (case comparison) – during the day, observed Jojoba R_n at 01:00 LT ($740 \, \text{W m}^{-2}$) reaches $300 \, \text{W m}^{-2}$ higher than Desert. During the night, net losses for Desert and Jojoba are quite similar – between 60 and $100 \, \text{W m}^{-2}$. WRF models these relative characteristics very closely.

G (model to observations) – WRF Control overestimates G during the day from around 10:00 LT onward. This positive bias reaches $30 \, \text{W m}^{-2}$ at around 00:00–01:00 LT. The variability is also somewhat overestimated. During the night, G losses are overestimated – up to $25 \, \text{W m}^{-2}$ during the late evening. WRF Impact overestimates G by up to $20 \, \text{W m}^{-2}$ during the day and also overestimates the upward nighttime flux also up to $20 \, \text{W m}^{-2}$. In both model cases the morning G gradient has a slope that is too steep in comparison with the observations. Additionally in both model cases, the model appears to lag the observations (1 and 2 h for Control and Impact, respectively). If these anomalies were corrected, the biases would be considerably reduced over much of the day.

G (case comparison) – as expected, observed peak G is considerably higher in the Desert soil than in Jojoba – with a ratio of around 2 : 1. Observed nighttime G losses also hold to this ratio with Desert losses larger than Jojoba. Again disregarding the absolute values, WRF reflects this daytime ratio between the surfaces.

ST5 (model to observations) – WRF Control underestimates the 5 cm soil temperatures, although the variability is well simulated. Both cases exhibit a significant cold nighttime bias, reflecting the bias in $T2$. The nighttime biases approach 4 °C for both cases. During the day WRF Control converges significantly with Desert and is only 1 °C cooler around midday. WRF Impact ST5 exhibits a more damped amplitude than Jojoba and bisects Jojoba at around 3 pm, then underestimating the peak ST5 by up to 2 °C.

ST5 (case comparison) – Desert ST5 has a much greater amplitude (15 °C) than Jojoba (5 °C). WRF Control predicts an amplitude which is too large (18.5 °C) compared to Desert, likely due to the large nighttime $T2$ bias and WRF Impact also overestimates the amplitude by around 4 °C. ST5 drops to around 28 °C at night time whereas Joj LT for Desert and around 18:00 LT for Jojoba. These peaks are modelled very well by WRF in both cases.

5.2   Comparison with evapotranspiration estimates

Since measurements of vertical fluxes were not available, ET was calculated independently from WRF-NOAH by applying two formulas based on Penman–Monteith equations (see below) and the observed Jojoba meteorological data. Penman–Monteith methods were thought to be appropriate here because they are often used in conditions where water is not greatly limited such as with irrigated crops. Estimations in conditions where water stress is present are more problematic and other methods may be more suitable, e.g. parameterized ET sub-models based in hydrologic models (Sumner and Jacobs, 2005). Additional estimates were examined from Becker et al. (2013) and also from a United States Agency for International Development (USAID) report (Irrigation and Crop Management Plan, 2006) – both of which discuss ET estimates for a jatropha plantation (Luxor, Egypt) in a similar summer climate (though winter is warmer and drier). Jatropha ET was not estimated because of the small plantation size and the likelihood of biases from advection.

Two methods were used: (a) the combination Penman–Monteith equation (Penman R_a/R_s) and (b) a modified Penman–Monteith ASCE method (Penman FAO 56) (see Appendix B for descriptions). The first method, Penman R_a/R_s is based on the so called combination Penman–Monteith equation from Monteith (1965) which includes explicit surface and aerodynamic resistances. The second method, FAO-56 Penman–Monteith (Allen et al., 1998) was

Table 5. Mean diurnal summer evaporation over Jojoba based on calculations from Penman R_a/R_s and Penman FAO 56. The values highlighted in bold are the daily ET_C or crop and canopy fraction adjusted estimates. As mean summer diurnal values were used to calculate ET the monthly figures shown are the same. In reality there may be a little variability over the summer with changing temperatures and so on.

Jojoba	Variable	Mean summer value (mm d^{-1})
Penman R_a/R_s	ET for 100 % canopy	6.51
	ET$_c$ for 70 % canopy	**4.56**
Penman FAO 56	ET$_0$	8.83
	ET$_c$ ($K_c \cdot \sigma_f$)	**4.33**

developed by the Food and Agriculture Organization (FAO) – a standard analytic/empirical method, useful when stomatal resistance data are not available. It describes a potential or reference ET (ET$_0$) of a well-watered vegetated grass surface with canopy height of 0.12 m, a constant Rs of 70 s m^{-1} and an albedo of 0.23. This ET$_0$ value is then modified with a crop coefficient K_c associated with particular plant types (see Appendix B for a detailed description of both methods and for calculations).

Both methods generally assume a neutrally stable surface layer which, given the dry convective afternoon conditions in hot, arid climates, when thermal turbulence dominates, is often not the case. Methods have been devised to include stability functions. The MM5 surface layer scheme (see Table 1) selected for the model simulations does employ a stability correction factor (S) which is combined with wind speed to calculate evaporation (see Appendix A for a description). However, it is not completely clear whether the inclusion of stability correction affects ET calculations substantially or not. Mahrt and Ek (1984) discuss the significance of stability correction in a study based on the Wangara experiment. Otles and Gutowski (2005) also discuss this issue in relation to a modelling study carried out within a semi-arid climate. They tested methods with and without stability correction, and obtained fairly similar results which corresponded closely with lysimeter and flux observations. Bearing this in mind, and because the absence of profile data makes the stability regime hard to identify, no stability correction was used in this case for estimations.

The mean daytime Jojoba evaporation estimates (averaged for all summer months) from both Penman–Monteith methods are shown in Table 5. Both methods yield very similar mean daytime ET$_C$ values (Penman R_a/R_s 4.56 mm d^{-1} and Penman 56 FAO 4.33 mm d^{-1}). The quality of these estimations were assessed through comparison with annual data gathered for the Luxor jatropha plantation, from the USAID report (also based on the FAO 56 approach). In Luxor, mean jatropha ET$_C$ values of 4.86 mm d^{-1} are quoted for the summertime (see Fig. 10), which is a very close match.

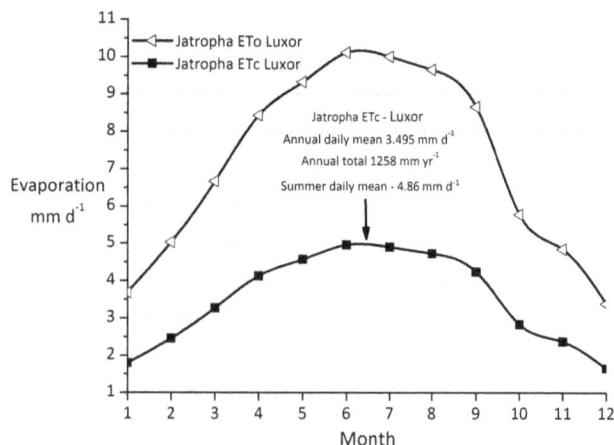

Fig. 10. ET_0 and ET_C values obtained for a jatropha plantation in Luxor as reported in the USAID report *Irrigation and Crop Management Plan*. The ET_C is calculated using the Penman–Monteith FAO 56 method and a crop coefficient K_C of 0.7 for jatropha. The annual total is calculated as 1258.61 mm yr^{-1}.

The annual total ET_C in Luxor is estimated to be around 1250 mm. In Israel on the other hand, agronomists quote an annual input of 700 mm for jojoba (650 mm for jatropha). The average 200 mm of winter rain in Be'ér Sheva' can be added to that. If the above annual irrigation inputs are accurate, there still remains around a 300 mm difference between the Israel and Egypt totals. This difference may be attributed to (a) cruder, surface irrigation in Luxor, where greater losses to direct evaporation and runoff could be assumed, and also (b) the cooler winter climate in Israel. Therefore, less water is needed in Israel during the winter and 900 mm yr^{-1} may therefore be a plausible water requirement. Additionally, three harvests are obtained every growing season in Luxor which necessitates more irrigation than if only two per year were taken. However, if we concentrate only on the summer months, where observations for Israel are available, then both Penman estimates (Table 5) match the Luxor ET rates (Fig. 10) to within 0.5 mm d^{-1}.

ET from the Penman–Monteith ET estimates and from the WRF-NOAH model, are compared in Fig. 11 (left panel), and are expressed in both W m^{-2} and mm d^{-1}. Given that we assume no losses to drainage or direct evaporation, the transpiration also represents an approximation of the plant water requirement. Therefore the JJA total requirement based on the WRF value of 4.42 mm d^{-1} would yield 406 mm over the 92 days of JJA (419 mm for Penman R_a/R_s and 398 mm for Penman FAO 56).

The remainder of the energy balance for these two methods was then estimated from the G and R_n observations. The WRF HFX fluxes were plotted against the plantation HFX values implied by the ET estimations, calculated as the residual of the energy balance (Eq. 3):

$$HFX = R_n(\text{Obs}) - G(\text{Obs}) - LH(\text{Estimate}).$$ (3)

R_n measurements should be fairly representative, but a good representation of G is difficult to obtain without many measurement points due to: soil heterogeneity, sharp temperature gradients and diurnal changes in shading caused by the partially open Jojoba canopy. Additionally, the heat storage needs to be accounted for, requiring good estimations of wet/dry soil thermal conductivities and Θ. In spite of these factors, during the middle part of the day, G magnitudes play a minor role in the energy balance (for Jojoba, 5% of R_n). Therefore, day time biases in G should not overly affect estimates of the other energy fluxes, based on $R_n - G$. During the night however, biases in G could play a larger role where R_n and G flux magnitudes approach each other.

The resulting Penman ET curves are quite similar in magnitude with a 1 mm d^{-1} or 28 W m^{-2} difference during the middle of the day. The Penman FAO 56 curve shows a slight lag of perhaps 1 h, when compared with Penman R_a/R_s, and has a less peaked shape. At night, Penman FAO 56 exhibits only a very small downward flux (perhaps 5 W m^{-2}), whereas Penman R_a/R_s shows a higher downward flux of 10 - 20 Wm^{-2}. WRF Impact ET matches well in magnitude with the Penman estimates, and during the day the WRF curve falls somewhere between the two Penman curves with a maximum latent heat value of 160–170 W m^{-2}. After sunrise, WRF Impact follows closely with the Penman R_a/R_s ET curve until midday when the model diverges a little. Around 14:00 LT, the Penman R_a/R_s curve drops sharply and bisects the WRF curve which predicts high ET for a longer period before dropping more smoothly downward between 14:00 and 16:00 LT. The peak time (of highest ET) in the model lies in between those from the Penman estimates, with all three curves being spaced around 30 min apart. During the latter part of the afternoon and evening, WRF matches more closely with Penman FAO 56. Both estimates vary more than WRF-NOAH especially during the morning up until midday. There could be many reasons for this. Biophysical factors not accounted for by WRF-NOAH are spatiotemporal heterogeneity in soil moisture and stomatal resistances. The influence of these factors are unquantifiable though without detailed hydrological measurements. In terms of atmospheric demand, for which we do have sufficient data, we know that R_n varies by 80 W m^{-2} during this time of day (Fig. 4) and decreases in mean daily values over the season by around 20 W m^{-2} (Fig. 3). From Fig. 9, G varies very little at this time (< 10 W m^{-2}) so is not likely to influence ET significantly. $T2$ in Jojoba also varies by 2 °C during these hours (Fig. 4) which is explained mostly by the seasonal peak in July (Fig. 3). U varies only by around 0.5 m s^{-1} diurnally and with little drift over the season, therefore aerodynamics do not seem to play a large role. VPD varies diurnally by around 0.4 hPa and from Fig. 3 we can surmise that there is a peak of humidity during July because it remains constant whilst $T2$ peaks at this time. Visually it appears that R_n is the most variable during the morning hours which corresponds most closely with ET variability. Given the size of the

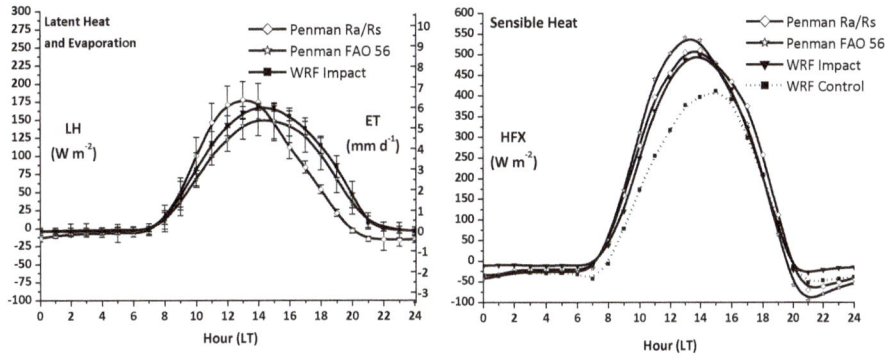

Fig. 11. Mean summer diurnal cycles of LH and HFX from WRF Impact (solid lines). Also indicated are the estimates from Penman–Monteith with R_a/R_s and FAO-56 Penman–Monteith (dashed lines). The left hand plot shows ET expressed in W m^{-2} (left y axis) and mm d^{-1} (right axis) along with standard deviation. The right hand plot shows the HFX fluxes from WRF Impact and also the implied HFX based on the Penman estimates [calculated as the residual of the energy balance $R_n(\text{Obs}) - G(\text{Obs}) - \text{LH}(\text{Estimation})$]. HFX from WRF Control is also plotted to assess the diurnal differences between HFX from desert and from irrigated plantations.

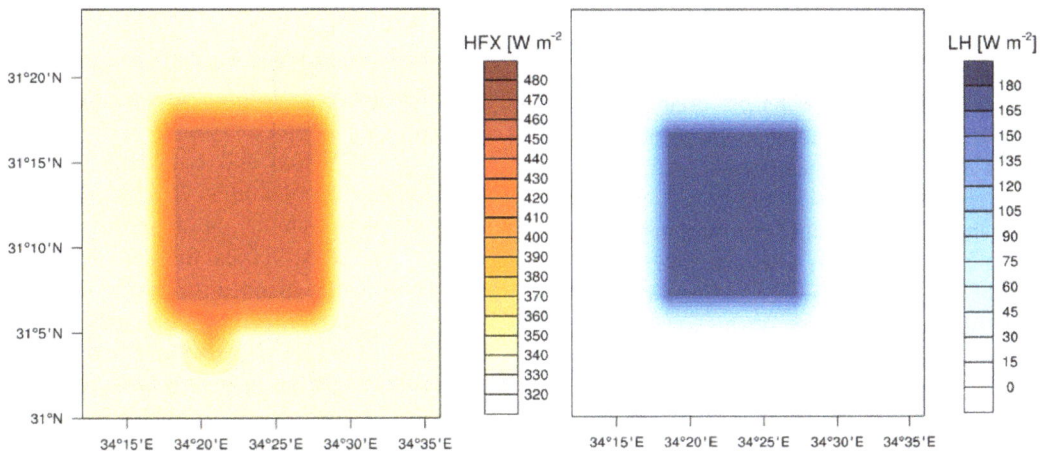

Fig. 12. Mean daily maximum of sensible and latent heat flux (JJA) in WRF Impact to show the spatial gradient between the plantation and the surrounding desert. Values of HFX over plantation and desert are around 460 and 330 W m^{-2} respectively (a 130 W m^{-2} gradient). Values of LH over plantation and desert are around 165 and 0 W m^{-2}, respectively (165 W m^{-2} gradient).

80 W m^{-2} fluctuation it is likely then that ET_0 variability is driven predominantly by that of R_n.

For the estimated HFX (Fig. 11, right panel), WRF Control HFX is also plotted alongside WRF Impact and the two Penman estimates. WRF Impact approaches most closely to Penman R_a/R_s in magnitude and shape. It is noticeable that HFX from both Penman estimates have higher peak magnitudes than WRF Impact, which seems contradictory to what the LH plot implies, where WRF LH falls in between the two estimates. This apparent anomaly can be explained by (a) the slightly lower WRF Impact R_n during the day and (b) the differences of alignment in peaks for observed and modelled R_n and G (Fig. 9). At night time, both Penman estimates exhibit large downward HFX (-50 to $-100\ \text{W m}^{-2}$) in the late evening which is not reflected by WRF Impact.

To examine the spatial gradient between simulated plantation and the surrounding desert directly, the mean daily max-

imum HFX and LH flux (JJA) were plotted from WRF Impact (Fig. 12). These statistics show a plantation surplus of roughly 130 W m^{-2} for HFX and 165 W m^{-2} for LH. There is an anomalous sensible heat flux high, extending to the south of the plantation, on a scale of around 1 grid cell. However it is not yet clear why this is so as there are no anomalies in soil or land use type at this location.

6 Discussion

The aim of the study was to simulate a sub-surface irrigated 100 km^2 plantation in WRF-NOAH and compare diurnal statistics of $T2$, $Q2$, U, R_n, G, ST5 and LH with desert and vegetation observations. Based on the results we can assess the model's ability to simulate desert and vegetation surfaces and examine the flux gradients over vegetation and desert.

6.1   Validation results

WRF Impact diagnoses cooler morning and midday temperatures over Jojoba in contrast to WRF Control and Desert (Fig. 8) but the reasons for this are not clear. $T2$ measurement error is likely to be negligible (see Table 1). Perhaps the Impact nighttime cold bias is extended into the convective ABL. Another possible cause is the lag in R_n, in WRF Impact during this period. Other possibilities are advection effects due to the disparity in simulated and real plantation sizes. This needs to be tested with varying plantation sizes.

The nighttime $T2$ cold bias is large and this is reflected in the 5 cm soil and skin temperatures and a correspondingly low upwelling long wave flux. This could be due to poor model simulation of the stable boundary layer and ABL transitions.

VPD is simulated quite well (deviation < 1 hPa) indicating a reasonable simulation of evaporative demand (see Appendix A for NOAH evaporation mechanism). How this relates in reality to diurnal surface resistances and ET_C is not so clear. Under constant light and VPD, the stomatal aperture of jojoba is controlled by the xylem water potential (ψ) of the plant. Furthermore, responses to changes in ψ are heavily dependent on soil and air temperatures and therefore highly non-linear (Benzioni and Dunstone, 1988).

U influences the energy balance partitioning greatly, through turbulent exchanges. U is well simulated over the desert and plantation. Variability is overestimated though by WRF Impact and underestimated by WRF Control. A possible source of bias is locally induced complexities in the turbulent wind field due to the open canopy. Finnegan et al. (2009) say that pressure gradients between the front and back of leaves and stems lead to unique turbulent characteristics, and suggest that Monin–Obhukov assumptions are not necessarily valid over canopies.

R_n is well modelled in Control and Impact over the day (Fig. 9). There is a slight underestimation for both cases around peak time (14:00 LT) of 30–50 W m^{-2}. This can be explained by a simulated atmosphere which is too dry, with a reduced downwelling long wave (LW) radiation. This was investigated and a deficit does exist which accounts for nearly all of the R_n bias. This represents only a small fraction of the R_n magnitude though so it should not compromise energy balance estimates much.

G is overestimated by around 30 % in WRF Control, but not by WRF Impact. In both cases the morning upward slope is too sharp in the model, especially in Control. This could indicate (a) a too large temperature gradient between skin and soil, (b) misparameterized thermal conductivity, dependent on Θ (Chen and Dudhia, 2001), or (c) misclassified soil texture/characteristics. Measurement error in the desert is also a possible factor. However, the very high correlation between the two desert flux plates (0.99), more or less rules out any relative error. The contribution of G to the energy balance is 5–6 % in the plantation, but 20 % in the desert and when

there is little ET, this G bias inevitably affects HFX exclusively. If the measurements are accurate, then WRF Control is underestimating HFX by around 30 W m^{-2} at peak time. This needs to be accounted for when comparing fluxes.

ST5 comparisons reflects the nighttime cold bias. However, the bias is strongest in Control. This can be partially explained by the greater upward G in Control (20 W m^{-2}) than in the plantation (10–12 W m^{-2}). During the daytime the model converges again with the observations. This can be explained by the steeper model slope which allows the model to reduce the deficit somewhat.

6.2   Diurnal energy fluxes

In terms of ET, the model matches closely with the observations, and lies within 20 W m^{-2} of both curves at peak time (Fig. 11, left panel). Both the shape and the magnitude of WRF Impact lies in the middle of Penman R_a/R_s and FAO 56. This lends confidence to the simulated peak LH of 160 W m^{-2}. Extrapolating these fluxes to HFX, a surplus of around 120–130 W m^{-2} (Fig. 11, right panel) is likely between the Jojoba and Desert surfaces (90–100 W m^{-2} if we adjust for the 30 W m^{-2} G bias in WRF Control).

Further calibration and sensitivity tests could improve NOAH for local conditions. One area that needs some improvement is the simulation of soil thermal transport, especially on bare soils. Vertical flux, profile and soil/plant measurements are also planned for Jojoba, for validation and for further calibration of the Penman methods. In summary, the simulation of irrigated plantations and corresponding land surface exchanges has been largely successful with only limited deviations in important variables. In particular, ET seems to be well simulated, when compared to Penman–Monteith estimates and the Luxor data. Therefore, conducting further impact studies, where flux gradients need to be correctly simulated, seems a reasonable prospect with this model.

In terms of the flux gradients, a prediction can be made that HFX over plantations will be higher than over desert surfaces, mainly due to the R_n surplus and a low ET from Jojoba. These predictions differ with conclusions from some regional irrigation studies (e.g. Qian et al., 2013 and Kueppers et al., 2007) which diagnose cooler (daily mean) $T2$ and high LH over irrigated plantations. This has significant implications for the impacts on local and regional climate if larger scale biomass plantations are planned.

Acknowledgements. Die Stiftung Energieforschung, Baden-Wüurttemberg is gratefully acknowledged for their generous financial backing of this research.

The authors would also like to thank their colleagues H. S. Bauer, J. Milovac, T. Schwitalla, H. D. Wizemann, A. Behrendt and Alex Geissler for continuing support and discussion. A special thank you to Dan Yakir at the Department of Environmental Sciences and Energy Research, Weizmann Institute of Science, Rehovot, Israel for his valuable scientific contributions and support.

With thanks to the personnel at Kibbutz Hatzerim and Netafim, Israel for their continuing technical support with our measurement campaign – especially Ami Charitan, Opher Silberberg, Ronen Rothschild, Oscar Lutenberg and Eli Matan.

A part of this work was supported by a grant from the Ministry of Science, Research and Arts of Baden-Württemberg (AZ Zu 33-721.3-2) and the Helmholtz Centre for Environmental Research – UFZ, Leipzig (WESS project) and a part was supported by the Project PAK 346/RU 1695 funded by DFG.

Many thanks for the support from the High Performance Computing Center Stuttgart (HLRS) of the University of Stuttgart, Germany, where the simulations and analysis were performed on the CRAY XE6 system.

Edited by: F. Tian

References

Abou Kheira, A. and Atta, N.: Response of Jatropha curcas L. to water deficits: Yield, water use efficiency and oilseed characteristics, Biomass Bioenerg., 33, 1343–1350, 2009.

Allen, R., Pereira, L., Raes, D., and Allen, M.: Crop evapotranspiration: Guidelines for computing crop water requirements, FAO Irrigation and Drainage Paper No. 56, retrieved from: http://scholar.google.co.uk/scholar?hl=en&as_sdt=0,5&cluster=8045698339079564915#0, Food and Agricultural Organization, Rome, 1998.

Alpert, P. and Mandel, M.: Wind Variability – An Indicator for a Mesoclimatic Change in Israel, J. Clim. Appl. Meteorol., 25, 1568–1576, 1986.

Ayers, R. S. and Westcot, D. W.: Water Quality For Agriculture, retrieved from: http://www.cabdirect.org/abstracts/19856755033.html;jsessionid=08B1B6FB48B32C38E2994EEEC7EFB14E?freeview=true (last access: 20 March 2013), 1985.

Bauer, H.-S., Weusthoff, T., Dorninger, M., Wulfmeyer, V., Schwitalla, T., Gorgas, T., Arpagaus, M., and Warrach-Sagi, K.: Predictive skill of a subset of models participating in D-PHASE in the COPS region, Q. J. Roy. Meteorol. Soc., 137, 287–305, doi:10.1002/qj.715, 2011.

Becker, K., Wulfmeyer, V., Berger, T., Gebel, J., and Münch, W.: Carbon farming in hot, dry coastal areas: an option for climate change mitigation, Earth Syst. Dynam., 4, 237–251, doi:10.5194/esd-4-237-2013, 2013.

Beljaars, A. C. M.: The parametrization of surface fluxes in large-scale models under free convection, Q. J. Roy. Meteorol. Soc., 121, 255–270, doi:10.1002/qj.49712152203, 1995.

Ben-Gai, T., Bitan, A., Manes, A., and Alpert, P., 1993. Long-term change in October rainfall patterns in southern Israel, Theor. Appl. Climatol., 46, 209–217, 1993.

Ben-Gai, T., Bitan, A., Manes, A., and Alpert, P.: Long-term changes in annual rainfall patterns in southern Israel, Theor. Appl. Climatol., 49, 59–67, doi:10.1007/BF00868190, 1994.

Ben-Gai, T., Bitan, A., Manes, A., Alpert, P., and Rubin, S.: Spatial and Temporal Changes in Rainfall Frequency Distribution Patterns in Israel, Theor. Appl. Climatol., 61, 177–190, doi:10.1007/s007040050062, 1998.

Benzioni, A.: Jojoba Domestication and Commercialization in Israel, in: Horticultural Reviews, John Wiley & Sons, Inc., 233–266, doi:10.1002/9780470650585.ch7, 1995.

Benzioni, A.: Jojoba Domestication and Commercialization in Israel, in: Horticultural Reviews, Volume 17, edited by: Janick, J., John Wiley & Sons, Inc., Oxford, UK, doi:10.1002/9780470650585.ch7, 2010.

Benzioni, A. and Dunstone, R. L.: Effect of air and soil temperature on water balance of jojoba growing under controlled conditions, Physiol. Plantarum, 74, 107–112, doi:10.1111/j.1399-3054.1988.tb04949.x, 1988.

Beringer, T., Lucht, W., and Schaphoff, S.: Bioenergy production potential of global biomass plantations under environmental and agricultural constraints, Global Change Biol. Bioenerg., 3, 299–312, doi:10.1111/j.1757-1707.2010.01088.x, 2011.

Bonan, G. B.: Ecological climatology: concepts and applications, Cambridge University Press, available at: http://cabdirect.org/abstracts/20093323860.html;jsessionid=CECED55A707E2DDD7D339E29DF836BE4 (last access: 25 March 2014, 2008.

Burt, P. J. A.: Introduction to micrometeorology, Q. J. Roy. Meteorol. Soc., 128, 1039–1040, doi:10.1256/0035900021643665, 2002.

Chen, F. and Dudhia, J.: Coupling an Advanced Land Surface–Hydrology Model with the Penn State – NCAR MM5 Modeling System, Part I: Model Implementation and Sensitivity, Mon. Weather Rev., 129, 569–585. doi:10.1175/1520-0493(2001)129<0587:CAALSH>2.0.CO;2, 2001.

Choudhury, B. J. and DiGirolamo, N. E.: A biophysical process-based estimate of global land surface evaporation using satellite and ancillary data I. Model description and comparison with observations, J. Hydrol., 205, 164–185, 1998.

Dalu, G. A., Pielke, R. A., Baldi, M., and Zeng, X.: Heat and Momentum Fluxes Induced by Thermal Inhomogeneities with and without Large-Scale Flow, J. Atmos. Sci., 53, 3286–3302, 1996.

De Ridder, K. and Gallée, H.: Land surface-induced regional climate change in southern Israel, J. Appl. Meteorol., 37, 1470–1485, doi:10.1175/1520-0450(1998)037<1470:LSIRCC>2.0.CO;2, 1998.

Du, C., Wu, W., Liu, X., and Gao, W.: Simulation of Soil Moisture and Its Variability in East Asia, edited by: Gao, W. and Ustin, S. L., Soc. Photo-Opt. Instru. (SPIE) Conference Series, San Diego, California, USA, 62982F–62982F-6, 2006.

Dyer, A. J. and Hicks, B. B.: Flux-gradient relationships in the constant flux layer, Q. J. Roy. Meteorol. Soc., 96, 715–721, doi:10.1002/qj.49709641012, 1970.

Finnigan, J. J., Shaw, R. H., and Patton, E. G.: Turbulence structure above a vegetation canopy, J. Fluid Mech., 637, 387–424, 2009.

Fritzmann, C., Löwenberg, J., Wintgens, T., and Melin, T.: State-of-the-art of reverse osmosis desalination, Desalination, 216, 1–76, doi:10.1016/j.desal.2006.12.00, 2007.

Hamilton, A. J., Stagnitti, F., Xiong, X., Kreidl, S. L., Benke, K. K., and Maher, P.: Wastewater Irrigation: The State of Play, Vadose Zone J., 6, 823–840, 2007.

Harding, K. J. and Snyder, P. K.: Modeling the Atmospheric Response to Irrigation in the Great Plains, Part I: General Impacts on Precipitation and the Energy Budget, J. Hydrometeorol., 13, 1667–1686, doi:10.1175/JHM-D-11-098.1, 2012.

Hong, S.-Y., Noh, Y., and Dudhia, J.: A New Vertical Diffusion Package with an Explicit Treatment of Entrainment Processes, Mon. Weather Rev., 134, 2318–2341, doi:10.1175/MWR3199.1, 2006.

Hong, X.: Role of vegetation in generation of mesoscale circulation, Atmos. Environ., 29, 2163–2176, 1995.

Hu, X.-M., Nielsen-Gammon, J. W., and Zhang, F.: Evaluation of Three Planetary Boundary Layer Schemes in the WRF Model, J. Appl. Meteorol. Clim., 49, 1831–1844, doi:10.1175/2010JAMC2432.1, 2010.

Hussain, G., Bashir, M. A., and Ahmad, M.: Brackish water impact on growth of jojoba (Simmondsia chinensis), J. Agric. Res, 49, 591–596, 2011.

Iacono, M. J., Delamere, J. S., Mlawer, E. J., Shephard, M. W., Clough, S. A., and Collins, W. D.: Radiative forcing by long-lived greenhouse gases: Calculations with the AER radiative transfer models, J. Geophys. Res., 113, D13103, doi:10.1029/2008JD009944, 2008.

Ingwersen, J., Steffens, K., Högy, P., Warrach-Sagi, K., Zhunusbayeva, D., Poltoradnev, M., Gäbler, R., Wizemann, H.-D., Fangmeier, A., Wulfmeyer, V., and Streck, T.: Comparison of Noah simulations with eddy covariance and soil water measurements at a winter wheat stand, Agr. Forest Meteorol., 151, 345–355, 2011.

Khawaji, A. D., Kutubkhanah, I. K., and Wie, J.-M.: Advances in seawater desalination technologies, Desalination, 221, 47–69, doi:10.1016/j.desal.2007.01.067, 2008.

Kueppers, L. M., Snyder, M. A., and Sloan, L. C.: Irrigation cooling effect: Regional climate forcing by land-use change, Geophys. Res. Lett., 34, L03703, doi:10.1029/2006GL028679, 2007.

Letzel, M. O. and Raasch, S.: Large Eddy Simulation of Thermally Induced Oscillations in the Convective Boundary Layer, J. Atmos. Sci., 60, 2328–2341, doi:10.1175/1520-0469(2003)060%3C2328%3ALESOTI%3E2.0.CO%3B2, 2003.

Lim, Y.-J., Hong, J., and Lee, T.-Y.: Spin-up behavior of soil moisture content over East Asia in a land surface model, Meteorol. Atmos. Phys., 118, 151–161, doi:10.1007/s00703-012-0212-x, 2012.

Mahfouf, J. F., Evelyne, R., and Mascart, P.: The Influence of Soil and Vegetation on the Development of Mesoscale Circulations, J. Appl. Meteorol., 26, 1483–1495, 1987.

Mahrt, L. and Ek, M.: The Influence of Atmospheric Stability on Potential Evaporation, J. Clim. Appl. Meteorol., 23, 222–234, 1984.

Mlawer, E. J., Taubman, S. J., Brown, P. D., Iacono, M. J. and Clough, S. A.: Radiative transfer for inhomogeneous atmospheres: RRTM, a validated correlated-k model for the longwave, J. Geophys. Res., 102, 16663–16682, 1997.

Molthan, A. L. and Colle, B. A.: Comparisons of Single- and Double-Moment Microphysics Schemes in the Simulation of a Synoptic-Scale Snowfall Event, Mon. Weather Rev., 140, 2982–3002, doi:10.1175/MWR-D-11-00292.1, 2012.

Monteith, J.: Evaporation and environment, Symp. Soc. Exp. Biol., 19, 205–234, 1965.

Morrison, H. and Gettelman, A.: A New Two-Moment Bulk Stratiform Cloud Microphysics Scheme in the Community Atmosphere Model, Version 3 (CAM3), Part I: Description and Numerical Tests, J. Climate, 21, 3642–3659, doi:10.1175/2008JCLI2105.1, 2008.

Niu, G., Rodriguez, D., Mendoza, M., Jifon, J., and Ganjegunte, G.: Responses of Jatropha curcas to Salt and Drought Stresses, Int. J. Agron., 2012, 1–7, doi:10.1155/2012/632026, 2012.

Oron, G., Campos, C., Gillerman, L., and Salgot, M.: Wastewater treatment, renovation and reuse for agricultural irrigation in small communities, Agr. Water Manage., 38, 223–234, doi:10.1016/S0378-3774(98)00066-3, 1999.

Otles, Z. and Gutowski, W. J.: Atmospheric Stability Effects on Penman-Monteith Evapotranspiration Estimates, Pure Appl. Geophys., 162, 2239–2254, doi:10.1007/s00024-005-2713-8, 2005.

Otterman, J.: Enhancement of surface-atmosphere fluxes by desert-fringe vegetation through reduction of surface albedo and of soil heat flux, Theor. Appl. Climatol., 40, 67–79, 1989.

Otterman, J., Manes, A., Rubin, S., Alpert, P., and Starr, D. O.: An increase of early rains in Southern Israel following land-use change? Bound.-Lay. Meteorol., 53, 333–351, doi:10.1007/BF02186093, 1990.

Paulson, C. A.: The Mathematical Representation of Wind Speed and Temperature Profiles in the Unstable Atmospheric Surface Layer, J. Appl. Meteorol., 9, 857–861, doi:10.1175/1520-0450(1970)009<0857:TMROWS>2.0.CO;2, 1970.

Penman, H. L.: Natural Evaporation from Open Water, Bare Soil and Grass, P. Roy. Soc. A, 193, 120–145, 1948.

Perlin, N. and Alpert, P.: Effects of land use modification on potential increase of convection: A numerical mesoscale study over south Israel, J. Geophys. Res., 106, 22621–22634, 2001.

Pielke Sr., R. A.: Mesoscale meteorological modeling, 2nd Edn., Academic Press, San Diego, CA, 676 pp., 2002.

Qian, Y., Huang, M., Yang, B., and Berg, L. K.: A Modeling Study of Irrigation Effects on Surface Fluxes and Land–Air–Cloud Interactions in the Southern Great Plains, J. Hydrometeorol., 14, 700–721, doi:10.1175/JHM-D-12-0134.1, 2013.

Rajaona, A. M., Brueck, H., Seckinger, C., and Asch, F.: Effect of salinity on canopy water vapor conductance of young and 3-year old Jatropha curcas L., J. Arid Environ., 87, 35–41, 2012.

Rajaona, A., Brueck, H., and Asch, F.: Leaf Gas Exchange Characteristics of Jatropha as Affected by Nitrogen Supply, Leaf Age and Atmospheric Vapour Pressure Deficit, J. Agron. Crop Sci., 199, 144–153, doi:10.1111/jac.12000, 2013.

Raupach, M. and Finnigan, J.: Coherent eddies and turbulence in vegetation canopies: the mixing-layer analogy, Bound.-Lay. Meteorol., 78, 351–382, doi:10.1007/BF00120941, 1996.

Rotach, M. W., Ambrosetti, P., Apenzeller, C., Arpagus, M., Fontannaz, L., Fundel, F., Germann, U., Hering, A., Liniger, M. A., Stoll, M., Walser, A., Bauer, H.-S., Behrendt, A., Wulfmeyer, V., Bouttier, F., Seity, Y., Buzzi, A., Davolio, S., Carazza, M., Denhard, M., Dorniger, M., Gorgas, T., Frick, J., Hegg, C., Zappa, M., Keil, C., Volkert, H., Marsigli, C., Montaini, A., McTaggart-Cowan, R., Mylne, K., Ranzi, R., Richard, E., Rossa, A., Santos-Muñoz, D., Schär, C., Staudinger, M., Wang, Y., and Werhahn, J.: MAP D-PHASE: Realtime Demonstration of Weather Forecast Quality in the Alpine Region, Bu. Am. Meteorol. Soc., 90, 1321–1336, 2009.

Rotach, M. W., Ambrosetti, P., Apenzeller, C., Arpagus, M., Fontannaz, L., Fundel, F., Germann, U., Hering, A., Liniger, M. A., Stoll, M., Walser, A., Bauer, H.-S., Behrendt, A., Wulfmeyer, V., Bouttier, F., Seity, Y., Buzzi, A., Davolio, S., Carazza, M., Denhard, M., Dorniger, M., Gorgas, T., Frick, J., Hegg, C., Zappa, M., Keil, C., Volkert, H., Marsigli, C., Montaini, A., McTaggart-Cowan, R., Mylne, K., Ranzi, R., Richard, E., Rossa, A., Santos-Muñoz, D., Schär, C., Staudinger, M., Wang, Y., and Werhahn,

J.: MAP D-PHASE: Real-Time Demonstration of Weather Forecast Quality in the Alpine Region, B. Am. Meteorol. Soc., 90, 1321–1336, doi:10.1175/2009BAMS2776.1, 2010.

Rotenberg, E. and Yakir, D.: Contribution of semi-arid forests to the climate system, Science, 327, 451–454, 2010.

Schwitalla, T., Bauer, H.-S., Wulfmeyer, V., and Zängl, G.: Systematic errors of QPF in low-mountain regions as revealed by MM5 simulations, Meteorol. Z., 17, 903–919, 2008.

Shin, H. H. and Hong, S.-Y.: Intercomparison of Planetary Boundary-Layer Parametrizations in the WRF Model for a Single Day from CASES-99, Bound.-Lay. Meteorol., 139, 261–281, doi:10.1007/s10546-010-9583-z, 2011.

Silva, E. N., Ribeiro, R. V., Ferreira-Silva, S. L., Viégas, R. A., and Silveira, J. A. G.: Comparative effects of salinity and water stress on photosynthesis, water relations and growth of Jatropha curcas plants, J. Arid Environ., 74, 1130–1137, 2010.

Spreer, W., Nagle, M., Neidhart, S., Carle, R., Ongprasert, S., Müller, J., and Muller, J.: Effect of regulated deficit irrigation and partial rootzone drying on the quality of mango fruits (Mangifera indica L., cv. "Chok Anan"), Agr. Water Manage., 88, 173–180, 2007.

Sridhar, V.: Tracking the Influence of Irrigation on Land Surface Fluxes and Boundary Layer Climatology, J. Contemp. Water Res. Educ., 152, 79–93, 2013.

Sumner, D. M. and Jacobs, J. M.: Utility of Penman–Monteith, Priestley–Taylor, reference evapotranspiration, and pan evaporation methods to estimate pasture evapotranspiration, J. Hydrol., 308, 81–104, doi:10.1016/j.jhydrol.2004.10.023, 2005.

Van Heerwaarden, C. C., Vilà-Guerau de Arellano, J., Moene, A. F., and Holtslag, A. A. M.: Interactions between dry-air entrainment, surface evaporation and convective boundary-layer development, Q. J. Roy. Meteorol. Soc., 135, 1277–1291, doi:10.1002/qj.431, 2009.

Warner, T.: Desert Meteorology, Int. J. Climatol., 26, 1737–1738, doi:10.1002/joc.1347, 2006.

Webb, E. K.: Profile relationships: The log-linear range, and extension to strong stability, Q. J. Roy. Meteorol. Soc., 96, 67–90, doi:10.1002/qj.49709640708, 1970.

Weusthoff, T., Ament, F., Arpagaus, M., and Rotach, M. W.: Assessing the Benefits of Convection-Permitting Models by Neighborhood Verification: Examples from MAP D-PHASE, Mon. Weather Rev., 138, 3418–3433, doi:10.1175/2010MWR3380.1, 2010.

Warrach-Sagi, K., Schwitalla, T., Wulfmeyer, V., and Bauer, H.-S.: Evaluation of a climate simulation in Europe based on the WRF-NOAH model system: precipitation in Germany, Clim. Dynam., 41, 755–774, doi:10.1007/s00382-013-1727-7, 2013.

Wulfmeyer, V., Behrendt, A., Bauer, H.-S., Kottmeier, C., Corsmeier, U., Blyth, A., Craig, G., Schumann, U., Hagen, M., Crewell, S., Di Girolamo, P., Flamant, C., Miller, M., Montani, A., Mobbs, S., Richard, E., Rotach, M. W., Arpagaus, M., Russchenberg, H., Schlüssel, P., König, M., Gärtner, V., Steinacker, R., Dorninger, M., Turner, D. D., Weckwerth, T., Hense, A., and Simmer, C.: The Convective and Orographically-induced Precipitation Study, B. Am. Meteorol. Soc., 89, 1477–1486, doi:10.1175/2008BAMS2367.1, 2008.

Wulfmeyer, V., Behrendt, A., Kottmeier, Ch., Corsmeier, U., Barthlott, C., Craig, G. C., Hagen, M., Althausen, D., Aoshima, F., Arpagaus, M., Bauer, H.-S., Bennett, L., Blyth, A., Brandau, C., Champollion, C., Crewell, S., Dick, G., Di Girolamo, P., Dorninger, M., Dufournet, Y., Eigenmann, R., Engelmann, R., Flamant, C., Foken, T., Gorgas, T., Grzeschik, M., Handwerker, J., Hauck, C., Höller, H., Junkermann, W., Kalthoff, N., Kiemle, C., Klink, S., König, M., Krauss, L., Long, C. N., Madonna, F., Mobbs, S., Neininger, B., Pal, S., Peters, G., Pigeon, G., Richard, E., Rotach, M. W., Russchenberg, H., Schwitalla, T., Smith, V., Steinacker, R., Trentmann, J., Turner, D. D., van Baelen, J., Vogt, S., Volkert, H., Weckwerth, T., Wernli, H., Wieser, A., and Wirth, M.: The Convective and Orographically Induced Precipitation Study (COPS): The Scientific Strategy, the Field Phase, and First Highlights, Q. J. Roy. Meteorol. Soc., 137, 3–30, doi:10.1002/qj.752, 2011.

Wulfmeyer, V., Branch, O., Warrach-Sagi, K., Bauer, H.-S., Schwitalla, T., and Becker, K.: The impact of plantations on weather and climate in coastal desert regions, J. Appl. Meteorol. Clim., doi:10.1175/JAMC-D-13-0208.1, in press, 2014.

Zhang, Y., Wang, X., Pan, Y., and Hu, R.: Diurnal and seasonal variations of surface albedo in a spring wheat field of arid lands of Northwestern China, Int. J. Biometeorol., 57, 67–73, 2013.

Appendix A

NOAH LSM ET$_C$

The following expression based on a Penman–Monteith formulation is used by NOAH to calculate ET$_C$ (see e.g. Chen and Dudhia, 2001):

$$
ET_C = \underbrace{\left(\frac{\Delta(R_n - G)}{L_v(\Delta + 1)} + \frac{\rho_a(e_s - e_a)}{S(1 + \Delta)} \right)}_{\text{Potential ET}} \left[\underbrace{(1 - \sigma_f) \frac{\Theta - \Theta_w}{\Theta_{FC} - \Theta_w}}_{\text{Direct ET}} \right.
$$
$$
\left. + \sigma_f \left(\underbrace{\left[\frac{w_C}{\mu} \right]^{0.5}}_{\text{Wet canopy ET}} + \underbrace{\left[1 - \left(\left[\frac{w_C}{\mu} \right]^{0.5} \right) B_C \right]}_{\text{Dry canopy ET}} \right) \right] \quad (A1)
$$

where $R_n - G$ is the available radiation (MJ d^{-1}), Δ is the slope of saturation vapour pressure against temperature, L_v is the latent heat of vaporization (J kg^{-1}), ρ_a is surface air density (kg m^{-3}), $e_s - e_a$ is the VPD (kPa), S is a stability coefficient and represents $C_q U$, where C_q is the turbulent exchange coefficient for water vapour, described by the Richardson number. Θ_{FC} is the field capacity and Θ_w is the wilting point. W_C is the intercepted canopy water content (kg m^{-2}), μ is the maximum canopy capacity (kg m^{-2}) and B_C is a modifier, analogous to K_C in Penman FAO 56, used to calculate ET$_C$ from ET$_0$:

$$
B_C = \frac{1 + \frac{\Delta}{R_r}}{1 + R_c C_q + \frac{\Delta}{R_r}}. \quad (A2)
$$

$R_r = f(U, T, P, C_h)$, and R_c is the canopy resistance. For R_c NOAH uses the Jarvis-type scheme, also described in Chen and Dudhia (2001), for calculating R_C:

$$
R_c = \frac{R_{c\,min}}{LAI_{eff}\, F1\, F2\, F3\, F4} \quad (A3)
$$

where $R_{c\,min}$ is an empirical constant and LAI$_{eff}$ is the effective leaf area index (generally 0.5 × LAI). $F1$, $F2$, $F3$, $F4$ are coefficients representing the effect of radiation, air humidity, air temperature and soil moisture on R_c respectively. $F3$, the temperature coefficient is given by $1 - 0.0016(T_{ref} - T)^2$ where T_{ref} is an optimum temperature for maximum photosynthesis and $F4$, the soil moisture factor is given by the layer discretized expression:

$$
F4 = \sum_{i=1}^{nroot} \frac{\Theta - \Theta_w}{\Theta_{FC} - \Theta_w} f_{root} \quad (A4)
$$

where "nroot" is the number of soil layers where roots are present and f_{root} is the layer's fraction of the root zone.

Appendix B

Field ET estimation methods

There are standard methods used to estimate ET from vegetated surfaces, including so called Penman–Monteith methods (Penman, 1948; Monteith, 1965). Since Penman developed a method to estimate ET from an open water surface, others included evaporation estimates from other surfaces like canopies by incorporating various resistance terms. Further research includes the effects of different stability regimes – for instance, Mahrt and Ek (1984) and Otles and Gutowski (2005). Different methods have been devised and are used depending on (a) data availability, (b) required interval for averaging, e.g. daily/hourly; and (c) what assumptions can be made, e.g. stability.

B1 Penman–Monteith R_a/R_s equation (Monteith, 1965)

The following expression is the Penman–Monteith (Penman R_a/R_s) equation formulated to account for explicit surfaces and aerodynamic resistances:

$$
ET_C = \frac{\Delta(R_n - G)}{L_v(\Delta + 1)} + \frac{\rho_a C_p \frac{(e_s - e_a)}{R_a}}{\Delta + \gamma\left(1 + \frac{R_s}{R_a}\right)} \quad (B1)
$$

where ET$_C$ is the crop ET (mm d^{-1}), C_p is the specific heat of air at constant pressure (J Kg^{-1} K^{-1}, γ is the psychrometric constant [kPa °K^{-1}], R_s and R_a are the surface and aerodynamic resistances respectively (s m^{-1}). The resistance terms are defined respectively as

$$
R_s = \frac{R_l}{LAI_{eff}} \quad (B2)
$$

$$
R_a = \frac{\ln\left(\frac{Z-d}{Z_{0m}}\right) \ln\left(\frac{Z-d}{Z_{0h}}\right)}{k^2 U_z} \quad (B3)
$$

where R_l represents the bulk stomatal resistance. In the second expression, Z is the standard measurement height, Z_{0m} is the roughness height for momentum, d is the displacement height, Z_{0h} is the roughness height for water vapour and k is von Kármán's constant. Estimations for d, Z_{0m}, Z_{0m} have been suggested (Allen et al., 1998), assuming that roughness heights for vapour and heat are equivalent:

$$
d = 2/3\,h \quad (B4)
$$
$$
Z_{om} = 0.123\,h \quad (B5)
$$
$$
Z_{oh} = 0.1\,Z_{om} \quad (B6)
$$

where h is the height of the canopy. In order to use this method, values for bulk stomatal resistances (R_l) are a prerequisite. Estimations for Jojoba resistances under different conditions were collated for the report of the Seventh Conference on Jojoba (1988) and are shown in Table B1.

Table B1. Measurements of the stomatal resistances of Jojoba using different methods, at different times of day; and under varying moisture and salinity conditions. Taken from the Seventh International Conference on Jojoba and Its Uses: Proceedings (1988). See the text for individual references.

Method	Conditions	Stomatal resistance $(s\,m^{-1})$
Heavy water scintillation	Well watered, low salt	333.3
Diffusive resistance	Well watered, low salt	625
Diffusive resistance	Well watered, high salt	1666
Diffusion porometer	Well watered, January	250
Diffusion porometer	Well watered, June	312.5
Continuous flow	Well watered, 07:00 LT	250
Continuous flow	Well watered, 11:00 LT	500
Unknown	Well watered, median	312

From these estimates it can be deduced that mean R_l values of between 300 and 650 seem feasible for well-watered plants in summertime. The difference in early morning resistance ($250\,s\,m^{-1}$) compared to 11:00 LT ($500\,s\,m^{-1}$) reflects a common desert plant strategy of closing the stomata during the hotter parts of the day. Another factor when choosing a suitable R_l value is that in reality, although the plants are watered adequately, only the minimum amount needed for plant health and optimized yields, are fed to the plants. In arid regions, even with slightly compromised yields, Jojoba production could still be optimal if the savings in water costs exceed the opportunity costs conceded due to lower yields. Lower inputs would indicate higher R_l values. In other words, under greater water stress it is reasonable to expect higher values of R_l. Regarding the effect of salinity on stomatal resistance, extremely high R_l values are estimated for salt-sensitive plants in saline soils, even exceeding $1000\,s\,m^{-1}$. Given the mean values of $1\,dS\,m^{-1}$ quoted for the irrigation water it is likely that only a minimum of leaching by winter rains may be needed to avoid salt accumulation. The FAO (Ayers and Westcott, 1985) quote one method for estimating the annual water requirement for leaching as $A_w = ET(1 - LR)$, where A_w is the annual water requirement including irrigation and ET is the annual irrigation applied. LR is a coefficient for the minimum leaching requirement needed to control salts within the tolerance of the crop. This is calculated as $LR = EC_w/(5\,EC_e - EC_w)$, where EC_w is the salinity of the applied irrigation water in $dS\,m^{-1}$. EC_e is the average soil salinity tolerated by the crop as measured on a soil saturation extract. The FAO recommends that the tolerance value used should represent a maximum of 90 % reduction of the potential yield, but 100 % for moderate to heavy salinity ($> 1.5\,dS\,m^{-1}$). Only sparse data on Jojoba salt tolerance is available and many factors could complicate estimates, such as plant varieties and age. Hussein et al. (2011) reported that young Jojoba plants can withstand up to $8\,dS\,m^{-1}$, with one variety "Siloh" tolerating $10\,dS\,m^{-1}$

with no reduction in flower production. However these are juvenile plants or seedlings. They also state that salt tolerance increases with the age and vigour of the plants. If we insert a value of $8\,dS\,m^{-1}$ and an annual ET of 700 mm for Jojoba, this yields a leaching requirement of 20 mm. If we are more conservative and choose a tolerance of $2\,dS\,m^{-1}$ (as estimated in Ayers and Westcott, 1985 for sorghum, grapefruit and orange trees), and assuming a maximum 10 % yield reduction this would require around 90 mm. Therefore, it is safe to assume that any accumulated salt is leached by the 200 mm of winter rain which is average for Be'ér Sheva'. Anecdotally, the managers report that there has been no significant soil degradation due to salinization, even dating back to 1948 when the first plantations were implemented. In spite of this, it is apparent that at least some salt is present in the soils during summer, and this is evident from small patches of salt accruing where water has occasionally breached the surface. Given the very high sensitivity of Jojoba stomatal resistance to salt stress (Table B1), it was thought to be safer to assume a small amount of salt stress. Therefore, a corresponding value of $800\,s\,m^{-1}$ was estimated for R_l. This higher value can also be justified by the deficit irrigation technique which is associated with higher resistances when compared with cruder methods, e.g. surface irrigation. In this regard, stomatal resistances could also exhibit a diurnal peak during the afternoon when soil water around the roots becomes depleted, and soil water is redistributed at night.

Mean seasonal hourly values calculated from this Penman method were multiplied by the vegetated fraction (σ_f) estimated at 70 %.

B2 FAO-56 Penman–Monteith Equation (Allen, 1998)

The following expression is the Penman–Monteith equation (FAO-56) using a crop coefficient K_c to modify ET_0 (Eq. B7):

$$ET_0 = \frac{0.408\,\Delta\,(R_n - G) + \gamma\left(\frac{900\,(K)}{T\,(^{\circ}C)+273.16}\right)\frac{u_z\,(m\,s^{-1})}{m\,s^{-1}}\,(e_s - e_a)}{\Delta + \gamma\,(0.34\,u_z\,(m\,s^{-1}) + 1\,m\,s^{-1})} \quad (B7)$$

where ET_0 is the reference evapotranspiration ($mm\,d^{-1}$) of a well-irrigated cut grass crop, $R_n - G$ is the net available radiation ($MJ\,d^{-1}$), T is temperature at standard height ($^{\circ}C$) and u_z is the wind speed at standard height ($m\,s^{-1}$). ET_c is modified using K_C i.e. $ET_c = ET_0 \times \sigma_f \times K_c$, where ET_c represents the actual crop ET estimate. In general terms the ET_0 can be thought of as the first-order climatic demand and K_C is a modifier. K_C accounts for species-specific physiological and physical factors, differentiating the crop from the reference vegetation and is intended to represent the effect of: crop type, albedo, stomatal resistance and direct soil evaporation. It is the ratio of ET_0 to the ET_c and is often < 1, but not in all cases. With closely-spaced, tall, freely-transpiring canopies, K_C can be as much as 15 to 20 % > 1. A dual K_C method can also be used by splitting the coefficient K_C into basal crop and soil evaporation components

$(K_{CB} + K_E)$. However, since we assume negligible soil evaporation with sub-surface irrigation, K_E would be negligible. We therefore concentrate on the basal effect only i.e. a single K_C value. Normally K_C values are available in lookup tables made available by the FAO but specific values for Jatropha and Jojoba are not given, so values were substituted from the nearest crop type in terms of height, biomass, geographical distribution and characteristics (oil seed crops and fruit tree categories). The validity of doing this is of course debateable. For nearly the whole year, a value of 0.7 is estimated by the FAO for these categories (this falls fractionally to 0.65 in the coolest winter months). Another source gives Jojoba a value of 0.5 all year round (Benzioni and Dunstone, 1998). A constant value is a reasonable assumption in arid regions, especially for the Israel Jojoba plants because they are fully mature perennials whose shape is maintained by periodic pruning. This contrasts with some crops such as annuals whose K_C varies widely over phenological stages.

It should be noted that when conditions are calm and humid, the aerodynamic factors of tall, dense canopies have less effect on the ET_0/ET_c ratio than the radiation which is the dominating driver of ET at low wind speeds. When relative humidities are lower than 45 % (assumed in the reference K_C estimates) the vapour pressure deficit (VPD) is higher, and the aerodynamics of taller crops has more effect. This can be better seen from the ratio $(e_s - e_a) R_a^{-1}$ in the numerator of Penman, R_a/R_s (Eq. B1). In arid conditions when the VPD is high, the ratio will be larger, which means that a significant change in R_a will have a large effect on the ET_0. Accordingly, the differences in ET_0 estimation would be amplified with tall crops experiencing higher wind speeds because the aerodynamic term is proportional to U_z and canopy height. Over the plantations, mean minimum daytime RH values are 29.6 % over the Jatropha and 30.0 % over the Jojoba which represents a large VPD. However the mean daytime wind speeds are low (1.71 m s^{-1} Jatropha and 1.88 m s^{-1} Jojoba) which would likely be a compensating factor. This was checked by using an adjustment modifier for mid-season K_C from Allen et al. (1998) (Eq. B8):

$$K_C = K_C(\text{Table}) + \left[0.04 \left(\frac{\overline{U_Z}(\text{m s}^{-1}) - 2\,\text{m s}^{-1}}{\text{m s}^{-1}} \right) \right.$$
$$\left. - 0.004 \left(\frac{\text{RH}_{\min} - 45\%}{\%} \right) \left(\frac{h}{3\,m} \right)^{0.3} \right] \qquad (B8)$$

where $\overline{u_Z}$ is the mean 2 m daytime wind speed (m^{-1}) and RH is the minimum daytime relative humidity (%). Using this correction and the observation data, the K_C value remains virtually unchanged because the low RH is offset by low wind speeds. Therefore the K_C values used were not changed for the calculations.

There are other factors which may affect ET, such as crop and leaf geometry to name just two examples. Regarding geometry, in Jojoba the leaf orientation is almost vertical which,

as well as affecting the turbulent wind characteristics, would also reduce incident solar radiation on the leaves when the solar zenith angle is low and temperatures are at their highest. Transpiration and photosynthesis rates tend to be correlated with solar radiation intensity. This would therefore minimize evaporation and heat loading at midday and optimize photosynthesis, when heat and water potential losses are low (Seventh International Conference on Jojoba and Its Uses: Proceedings, 1988). Factors such as these are difficult to take into account within a general K_C coefficient however but further research to improve species-specific estimates are out of the scope of this study.

Appendix C

Wind observation height extrapolation

The wind data (6 m) was extrapolated to a standard 10 m height using the following standard neutral stability log profile law (WMO):

$$\frac{\overline{U_1}}{U_{\text{ref}}} = \ln \left(\frac{Z_1 - d}{Z_{0m}} \right) / \ln \left(\frac{Z_{\text{ref}}}{Z_{0m}} \right) \qquad (C1)$$

where $\overline{U_1}$ is the wind speed at measurement height, U_{ref} is the wind speed at the standard height of 10 m, Z_1 is the height of the wind sensor, Z_{ref} is the standard height (10 m), d is the displacement height (Eq. B4) and Z_{0m} is the roughness length for momentum (Eq. B5). For Desert and Jojoba, values of 0 m and 2 m (2/3 h) were used for d and 0.0005 m and 0.369 m (0.123 h) for Z_{0m}, respectively.

12

Principal components of thermal regimes in mountain river networks

Daniel J. Isaak, Charles H. Luce, Gwynne L. Chandler, Dona L. Horan, and Sherry P. Wollrab

U.S. Forest Service, Rocky Mountain Research Station, Aquatic Sciences Lab, Boise, ID 83702, USA

Correspondence: Daniel J. Isaak (disaak@fs.fed.us)

Abstract. Description of thermal regimes in flowing waters is key to understanding physical processes, enhancing predictive abilities, and improving bioassessments. Spatially and temporally sparse data sets, especially in logistically challenging mountain environments, have limited studies on thermal regimes, but inexpensive sensors coupled with crowd-sourced data collection efforts provide efficient means of developing large data sets for robust analyses. Here, thermal regimes are assessed using annual monitoring records compiled from several natural resource agencies in the northwestern United States that spanned a 5-year period (2011–2015) at 226 sites across several contiguous montane river networks. Regimes were summarized with 28 metrics and principal component analysis (PCA) was used to determine those metrics which best explained thermal variation on a reduced set of orthogonal axes. Four principal components (PC) accounted for 93.4 % of the variation in the temperature metrics, with the first PC (49 % of variance) associated with metrics that represented magnitude and variability and the second PC (29 % of variance) associated with metrics representing the length and intensity of the winter season. Another variant of PCA, T-mode analysis, was applied to daily temperature values and revealed two distinct phases of spatial variability – a homogeneous phase during winter when daily temperatures at all sites were $< 3\,°C$ and a heterogeneous phase throughout the year's remainder when variation among sites was more pronounced. Phase transitions occurred in March and November, and coincided with the abatement and onset of subzero air temperatures across the study area. S-mode PCA was conducted on the same matrix of daily temperature values after transposition and indicated that two PCs accounted for 98 % of the temporal variation among sites. The first S-mode PC was responsible for 96.7 % of that variance and correlated with air temperature variation ($r = 0.92$), whereas the second PC accounted for 1.3 % of residual variance and was correlated with discharge ($r = 0.84$). Thermal regimes in these mountain river networks were relatively simple and responded coherently to external forcing factors, so sparse monitoring arrays and small sets of summary metrics may be adequate for their description. PCA provided a computationally efficient means of extracting key information elements from the temperature data set used here and could be applied broadly to facilitate comparisons among more diverse stream types and develop classification schemes for thermal regimes.

1 Introduction

Temperatures of flowing waters control many physicochemical processes (Likens and Likens, 1977; Gordon et al., 1991; Ducharne, 2008) and affect the ecology of aquatic organisms and communities (Isaak et al., 2017b; Neuheimer and Taggart, 2007; Woodward et al., 2010). Knowledge of thermal regimes, characterized as the annual sequence of temperature conditions specific to locations within river networks (Caissie, 2006), is key to understanding natural conditions and diagnosing anthropogenic impairments. Seminal work by Poff and colleagues (Poff and Ward, 1989; Poff et al., 1997) created a robust framework for describing flow regimes based on metric descriptions of magnitude, frequency, timing, duration, and variability that are largely transferrable to thermal regimes (Poole et al., 2004; Olden and Naiman, 2010). Recent studies have contributed useful derivations of temperature metrics (Arismendi et al., 2013; Chu et al., 2010; Rivers-Moore et al., 2013; Steel et al., 2016) or classification schemes based on a small number of pre-selected metrics (Maheu et al., 2016), but the limited availability of annual temperature records (Orr et al., 2015; Isaak et al., 2018a) has slowed broad development and adoption of thermal regime concepts. Data inadequacies are often compounded for montane riverscapes that are difficult to sample (Brown and Hannah, 2008; Isaak et al., 2013), a shortfall that needs to be overcome given the importance of these areas as climate refugia for cold-water biodiversity (Brown et al., 2009; Isaak et al., 2016b; Quaglietta et al., 2018) and as the focus of costly regional conservation strategies (Roni et al., 2002; Rieman et al., 2015).

Despite existing limitations, the importance of temperature to stream biota is well recognized and inculcated to regulatory standards based on metrics used within threshold-based approaches (Poole et al., 2004; Todd et al., 2008). Most often, those metrics represent some aspect of conditions during warm summer months when temperature sensitive species or life stages are thought to be most vulnerable (Ice et al., 2004; McCullough, 2010), which contributes to the preponderance of short monitoring records spanning only these months (Isaak et al., 2017a). However, thermally mediated ecological processes occur throughout the year (Neuheimer and Taggart, 2007; Olden and Naiman, 2010), so adequate understanding requires broader characterization of thermal conditions from annual data sets. While that may bring additional complexity, most warm season metrics are strongly correlated and therefore redundant (Isaak and Hubert, 2001; Dunham et al., 2005; Steel et al., 2016). If redundancy is also the norm among a broader array of annual temperature metrics, then multivariate data reduction techniques might be useful for identifying a few key aspects of thermal regimes.

Supporting that idea, Rivers-Moore et al. (2013) used principal component analysis (PCA) to describe covariation among 39 temperature metrics calculated for 82 South African stream sites and found that two PCs accounted for 75 % of the total variation among metrics. Similarly in the field of hydrology, Olden and Poff (2003) examined 171 flow metrics calculated from 420 gage sites across the United States (U.S.) and found that two to four PCs accounted for 76 %–97 % of variation in the data set. In addition to metric-based PCA that is commonly used in the hydrological sciences, several other PCA variants are standard analytical tools in the field of climatology and may be relevant for characterizing the dynamics of thermal regimes (Richman, 1986; Demsar et al., 2013). Most notably, PCA can be done on repeated measurements of a single variable to identify common spatial or temporal behavior among monitoring stations. In the climatology literature, for example, empirical orthogonal function analysis (S-mode PCA in the taxonomy of Richman, 1986) is used to determine which sites covary temporally as a means of developing regionalization schemes for precipitation, air temperatures, or wind speeds (Piechota et al., 1997; Jiménez et al., 2008; Martins et al., 2012). If com-

mon temporal patterns are identified, it suggests potential redundancy in the monitoring network and the information can be used to refine future sampling designs. The closely allied T-mode PCA identifies dominant spatial patterns in data sets and the times when these phases occur (Richman, 1986; Gallacher et al., 2017). A single dominant spatial pattern suggests the spatial distribution of a variable is temporally consistent, whereas more than one spatial phase suggests change points and different states.

The advent of inexpensive sensors, combined with regulatory requirements and concerns about climate change, have led to the recent expansion in temperature monitoring networks for rivers and streams (Isaak et al., 2010; Rivers-Moore et al., 2013; Hilderbrand et al., 2014; Luce et al., 2014b; Trumbo et al., 2014; Hannah and Garner, 2015; Jackson et al., 2016; Molinero et al., 2016; Daigle et al., 2016; Mauger et al., 2016; Steel et al., 2016). What was once a data dearth is becoming a deluge and opportunities exist to study thermal regimes with robust data sets. Here, we use annual temperature records compiled from several natural resource agencies for 226 monitoring sites in a mountainous landscape to conduct an initial assessment of thermal regimes. We limit the geographic scope of our effort to several adjacent river basins in the northwestern U.S. that are geologically and topographically similar but which have particularly dense monitoring networks to maximize analytical flexibility. Our objectives were to (1) provide a basic description of the annual thermal characteristics in mountain rivers and streams because these are rare within the literature, (2) develop metrics to describe thermal regime characteristics based on magnitude, frequency, timing, duration, and variability, and (3) explore spatiotemporal variation among those metrics and temperature dynamics in relation to basin morphology and hydroclimatic conditions to better discern the principal components of thermal regimes and their regulating factors.

2 Study area

The study area encompasses 79 500 km^2 of mountainous, topographically complex terrain that spans a broad elevation range of 200–3600 m at a latitude of 45° N in the northwestern United States (Fig. 1). Climate is characterized by cold, wet winters with moderate to heavy snow accumulations at high elevations and hot, dry summers. Hydrographs are typical of snowmelt runoff systems, with high flows during spring and early summer and low flows during late summer, fall, and winter (Fig. 2). Vegetation is dominated by conifer forests except at low elevations and south facing aspects where grasses and shrubs predominate. Wildfires are common within the landscape and burned 8 % of the area from 2011 to 2015 (Morgan et al., 2014). Parent geology consists mostly of resistant granites of the Idaho Batholith and a smaller easterly portion of intrusive volcanics (Bond and Wood, 1978; Meyer et al., 2001). Both geologies are

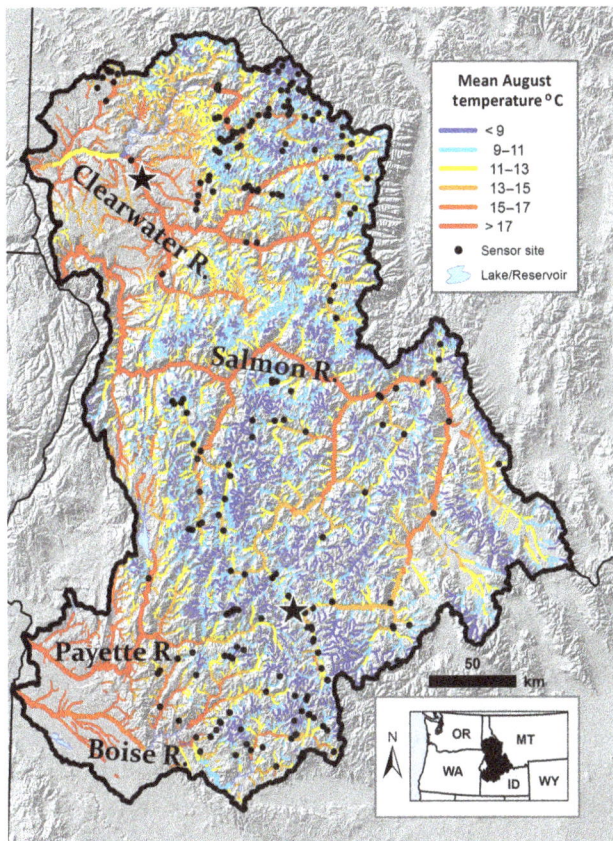

Figure 1. Locations of 226 monitoring sites overlaid on an August stream temperature scenario for the 29 600 km network in the study area. Stars denote where air temperature and stream discharge data were obtained from a low-elevation site (294 m, northern station) and a high-elevation site (1850 m, southern station).

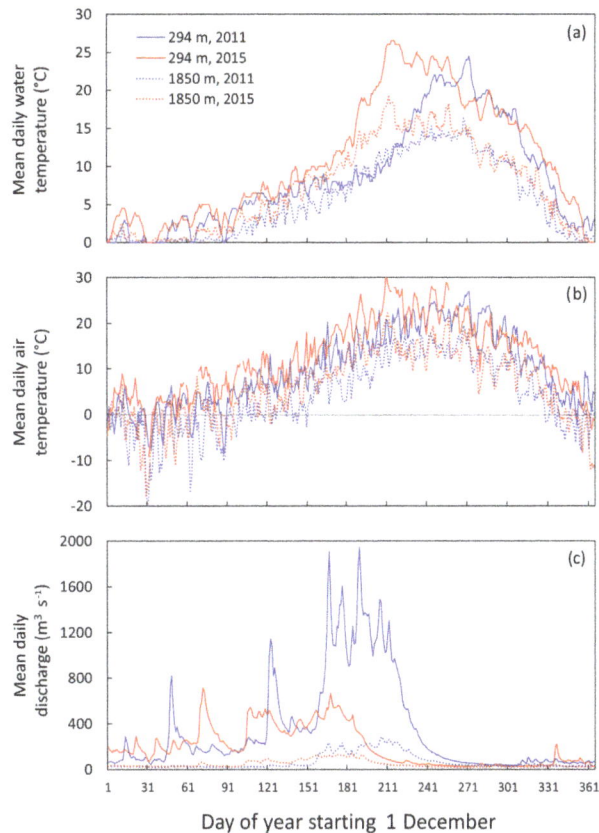

Figure 2. Annual cycle of mean daily water temperatures (**a**), air temperatures (**b**), and discharge (**c**) at a high-elevation site and a low-elevation site during 2 contrasting climate years. Discharge values at the high-elevation site are multiplied by 10 for better visibility.

heavily dissected and stream valleys are V-shaped except for some alpine valleys at the highest elevations that were once glaciated. Human population densities are low except along wider segments of river valleys where fertile floodplains and easy access to water accommodate small amounts of agriculture and ranching. Most of the study area is publicly owned (81 %) and federally administered by the United States National Forest Service and Bureau of Land Management for a variety of land-use, recreational, and conservation purposes. Unpaved road networks have been developed in some drainages for timber harvest, but many drainages are protected in large wilderness areas with minimal anthropogenic effects or roads (Swanson, 2015).

River networks and temperature data set

Rivers and streams within the study area were delineated using the 1 : 100 000-scale National Hydrography Dataset (NHD; http://www.horizon-systems.com/NHDPlus/index.php, last access: 2 December 2018; McKay et al., 2012), which was attributed with mean annual flow val-

ues from data at the Western U.S. Stream Flow Metrics website (http://www.fs.fed.us/rm/boise/AWAE/projects/modeled_stream_flow_metrics.shtml, last access: 2 December 2018; Wenger et al., 2010). To highlight the perennial subset of the network where temperature monitoring occurred, reaches with annual flows less than $0.03 \, \mathrm{m^3 \, s^{-1}}$ were removed from the network, as were reaches with channel slopes > 15 %, and those coded as intermittent in the NHD (Fcode = 46003). Filtering reduced the original network extent from 58 000 to 29 600 km with streams flowing at elevations of 221–3105 m. To visualize thermal heterogeneity in the network, a scenario representing mean August temperatures for a baseline climate period of 1993–2011 was downloaded from the Northwestern Stream Temperature website (NorWeST: https://www.fs.fed.us/rm/boise/AWAE/projects/NorWeST.html, last access: 2 December 2018; Isaak et al., 2016a) and linked to the NHD reaches (Fig. 1). Several large rivers drain the area in a generally westerly direction, the largest of which is the Salmon River with a mean annual discharge of $315 \, \mathrm{m^3 \, s^{-1}}$ and a basin that comprised 44 % of the study area. Six large dams and reservoirs are in downstream

Table 1. Descriptive statistics for spatial attributes of the study network and 226 monitoring sites with annual temperature data in the northwestern United States.

Network reaches	Mean	Median	SD	Minimum	Maximum
Elevation (m)	1493	1533	536	221	3105
Drainage area (km^2)	915	17.7	4359	0.005	34 865
Mean annual flow ($m^3 s^{-1}$)	9.73	0.229	43.2	0.0253	379
Reach slope ($m m^{-1}$)	0.0584	0.0519	0.0429	0	0.150
Monitoring sites					
Elevation (m)	1392	1407	464	280	2369
Drainage area (km^2)	687	47.3	3011	2.18	34 865
Mean annual flow ($m^3 s^{-1}$)	7.37	0.692	26.4	0.0253	281
Reach slope ($m m^{-1}$)	0.0389	0.0273	0.0403	0	0.150

portions of the network (three in the Boise River basin, two in the Payette River basin, and one in the Clearwater River basin), but these affect thermal conditions in less than 300 km of river and no temperature data were used from these sections. Spatial attributes and environmental characteristics of the study area network are summarized in Table 1.

To obtain a water temperature data set for analysis, we intersected the filtered network with the NorWeST database of daily temperature summaries (Chandler et al., 2016) and extracted data for sites that had mean daily temperature values on at least 70 % of the days from 1 December 2010 to 30 November 2015. We started the thermal year on 1 December because temperatures usually reach their annual lows by this date and the 3-month period thereafter constituted a logical winter season (i.e., December, January, February). Subsequent 3-month periods were considered to be spring (March, April, May), summer (June, July, August), and fall (September, October, November) seasons. NorWeST temperature records were supplemented with additional data solicited from hydrologists and fisheries biologists employed by the Idaho Department of Fish and Game and the U.S. Forest Service, and we also downloaded data from online databases maintained by the Columbia Habitat Monitoring Program (https://www.champmonitoring.org/Home/Index, last access: 2 December 2018) and the NOAA Northwest Fisheries Science Center (https://www.webapps.nwfsc.noaa.gov/WaterQuality/, last access: 2 December 2018). Geographic gaps in monitoring were identified using geospatial analysis (e.g., Jackson et al., 2016) and additional sensors were strategically deployed where needed (Isaak et al., 2010, 2013). Data from the different sources were recorded at a variety of sub-daily intervals, so records were summarized to mean daily temperatures for standardization. Data were collected using different sensor models (TidbiT, Stowaway, and Pendant models from Onset Computer Corporation, Pocasset, Massachusetts, USA; Temp101a model from

MadgeTech, Warner, New Hampshire, USA), which had measurement accuracies of ±0.2 to ±0.5 °C and resolutions of 0.02 to 0.14 °C based on manufacturer specifications and calibration tests we performed. Sensors were deployed using underwater epoxy or steel cables for connection to large boulders and other immobile channel structures and were shielded from direct sunlight (Isaak et al., 2013; Stamp et al., 2014). Temperature records were subject to standard quality assurance–quality control measures as described elsewhere (Chandler et al., 2016).

The stream temperature data set consisted of records from 226 sites across a range of elevations, stream sizes, and reach slopes (Fig. 1; Table 1). Although we set the minimum threshold for record completeness at 70 % during the 5-year period, the average completeness of records was higher at 88 %. Missing daily values were imputed using the Miss-MDA package (Missing Values with Multivariate Data Analysis; Josse and Husson, 2016) in R (R Development Core Team, 2014) because temporal covariation among proximate stream temperature sites is usually strong. That was confirmed in our data set by the high correlations between observed daily temperatures and predictions from the imputation technique, which ranged from $r = 0.98$ to 0.99. All temperature records at the 226 sites were complete after imputation and consisted of 1826 mean daily temperatures from 1 December 2010 to 30 November 2015. Climatological variation during the same period was described using discharge data downloaded from the National Water Information System database (https://waterdata.usgs.gov/usa/nwis/nwis, last access: 2 December 2018) for a high-elevation gage site at 1850 m and a low-elevation gage site at 294 m and air temperature data from monitoring stations in the Cooperative Observer Network (https://www.ncdc.noaa.gov/data-access, last access: 2 December 2018) that were near the gage sites (Fig. 1).

Table 2. Temperature metrics used to describe thermal regimes of mountain rivers and streams.

Category	Thermal metric	Definition
Magnitude	M1. Mean annual temperature	Average of mean daily temperatures during a year
	M2. Mean winter temperature	Average of mean daily temperatures during December, January, and February
	M3. Mean spring temperature	Average of mean daily temperatures during March, April, and May
	M4. Mean summer temperature	Average of mean daily temperatures during June, July, and August
	M5. Mean August temperature	Average of mean daily temperatures during August
	M6. Mean fall temperature	Average of mean daily temperatures during September, October, and November
	M7. Minimum daily temperature	Lowest mean daily temperature during a year
	M8. Minimum weekly average temperature	Lowest 7-day running average of mean daily temperature during a year
	M9. Maximum daily temperature	Highest mean daily temperature during a year
	M10. Maximum weekly average temperature	Highest 7-day running average of mean daily temperature during a year
	M11. Annual degree days	Cumulative total of degree days during a year ($1\,°C$ for $24\,h = 1°$ day)
Variability	V1. Annual standard deviation	Standard deviation of mean daily temperature during a year
	V2. Winter standard deviation	Standard deviation of mean daily temperature during winter months
	V3. Spring standard deviation	Standard deviation of mean daily temperature during spring months
	V4. Summer standard deviation	Standard deviation of mean daily temperature during summer months
	V5. Fall standard deviation	Standard deviation of mean daily temperature during fall months
	V6. Range in extreme daily temperatures	Difference between minimum and maximum mean daily temperatures during a year (M9 minus M7)
	V7. Range in extreme weekly temperatures	Difference between minimum and maximum weekly average temperatures during a year (M10 minus M8)
Frequency	F1. Frequency of hot days	Number of days with mean daily temperatures $> 20\,°C$
	F2. Frequency of cold days	Number of days with mean daily temperatures $< 2\,°C$
Timing	T1. Date of 5 % of degree days	Number of days from 1 December until 5 % of degree days are accumulated
	T2. Date of 25 % of degree days	Number of days from 1 December until 25 % of degree days are accumulated
	T3. Date of 50 % of degree days	Number of days from 1 December until 50 % of degree days are accumulated
	T4. Date of 75 % of degree days	Number of days from 1 December until 75 % of degree days are accumulated
	T5. Date of 95 % of degree days	Number of days from 1 December until 95 % of degree days are accumulated
Duration	D1. Growing season length	Number of days between the 95 % and 5 % of degree days (T5 minus T1)
	D2. Duration of hot days	Longest number of consecutive days with mean daily temperatures $> 20\,°C$
	D3. Duration of cold days	Longest number of consecutive days with mean daily temperatures $< 2\,°C$

3 Data analysis

3.1 PCA of thermal metrics

Prior to calculating metrics for thermal characteristics, mean daily temperatures for 365 days were calculated from the 5 years of data at each site to provide representative values. Twenty-eight temperature metrics were then calculated to describe aspects of those annual records based on five categories associated with magnitude, variability, frequency, tim-ing, and duration (Tables 2 and 3). Metrics were similar to those used in previous studies of thermal regimes (Arismendi et al., 2013; Chu et al., 2010; Rivers-Moore et al., 2013; Steel et al., 2016) and in studies assessing the effects of peak summer temperatures on the distribution and abundance of aquatic organisms (Dunham et al., 2003; Huff et al., 2005; Isaak et al., 2017b). A wide range of variability occurred among sites where mean annual temperatures ranged from 3.1 to 10.3 °C and annual standard deviations ranged from 2.51 to 7.40 °C (Table 3). Relationships among the thermal

Table 3. Descriptive statistics for temperature metrics used to describe thermal regimes at 226 monitoring sites in mountain river networks. Statistics were calculated from the imputed time series and mean daily values for the period 2011–2015.

	Mean (°C)	Median (°C)	SD (°C)	Minimum (°C)	Maximum (°C)
M1. Mean annual temperature	5.36	5.10	1.44	3.10	10.34
M2. Mean winter temperature	0.75	0.63	0.60	−0.10	4.03
M3. Mean spring temperature	3.67	3.47	1.61	1.14	9.38
M4. Mean summer temperature	11.2	10.9	2.68	6.55	19.1
M5. Mean August temperature	12.5	12.1	2.78	7.78	22.5
M6. Mean fall temperature	5.71	5.50	1.53	3.04	11.5
M7. Minimum daily temperature	0.21	0.14	0.35	−0.45	2.18
M8. Minimum weekly average temperature	0.31	0.23	0.40	−0.42	2.69
M9. Maximum daily temperature	13.5	13.0	3.00	8.26	23.5
M10. Maximum weekly average temperature	13.2	12.7	2.99	7.96	23.2
M11. Annual degree days	1956	1863	527	1132	3775
V1. Annual standard deviation	4.43	4.27	1.05	2.51	7.40
V2. Winter standard deviation	0.30	0.29	0.16	0.00	0.87
V3. Spring standard deviation	1.62	1.57	0.72	0.33	5.36
V4. Summer standard deviation	1.99	1.88	0.61	0.61	4.45
V5. Fall standard deviation	3.43	3.34	0.73	2.13	6.05
V6. Range in extreme daily temperatures	13.3	12.8	3.06	7.50	23.3
V7. Range in extreme weekly temperatures	12.9	12.3	3.06	6.99	22.9
F1. Frequency of hot days	0.81	0	5.82	0	61
F2. Frequency of cold days	131	132	35.6	0	212
T1. Date of 5 % of degree days	109	113	25.5	44	168
T2. Date of 25 % of degree days	193	194	10.9	148	217
T3. Date of 50 % of degree days	237	238	5.01	215	251
T4. Date of 75 % of degree days	276	276	2.99	264	288
T5. Date of 95 % of degree days	323	323	4.78	309	340
D1. Growing season length	214	210	29.7	141	296
D2. Duration of hot days	0.691	0	5.61	0	61
D3. Duration of cold days	124	124	39.0	0	207

metrics were described by conducting PCA on a data matrix in which columns represented the 28 metrics and rows were the 226 monitoring sites. Linear combinations of the data were estimated with coefficients equal to the eigenvectors of their correlation matrix, which were the principal components (PCs; Pearson, 1901; Sergeant et al., 2016). The first principal component accounted for the largest possible variance in the data set and succeeding components accounted for the largest portions of the remaining variance while being orthogonal (i.e., uncorrelated) to the preceding components. Correlations, or loadings, between each metric and the PCs were also calculated to assist in subsequent interpretations. The Princomp procedure in SAS (SAS Institute Inc., 2015) was used to conduct the PCA. To describe geographical relationships, PC scores were mapped to the 226 temperature sites and bivariate correlations were calculated with descriptors of network conditions such as elevation, reach slope, and discharge summarized in Table 1.

3.2   PCA of daily water temperatures

To assess the consistency of spatial temperature patterns

among monitoring sites, a T-mode PCA (Richman, 1986) was done on a data matrix of mean daily temperatures in which the columns were the 365 days starting on 1 December and the rows were the 226 monitoring sites. In this analysis, the number of principal components explaining significant variation indicates the number of distinct spatial phases that occur throughout the year (Gallacher et al., 2017). Eigenvector loadings on the dominant PCs were plotted for each day of the year to describe when each phase occurred, and mean daily temperatures were mapped during these periods for visualization.

To assess temporal covariance among sites, an S-mode PCA (Richman, 1986) was done by transposing the T-mode data matrix so that monitoring sites were columns and the time-ordered daily mean temperatures were rows. Because hydroclimatic conditions among years could have affected the results, the S-mode PCA was done not only for the 5-year averages of daily water temperatures, but also on the disag-

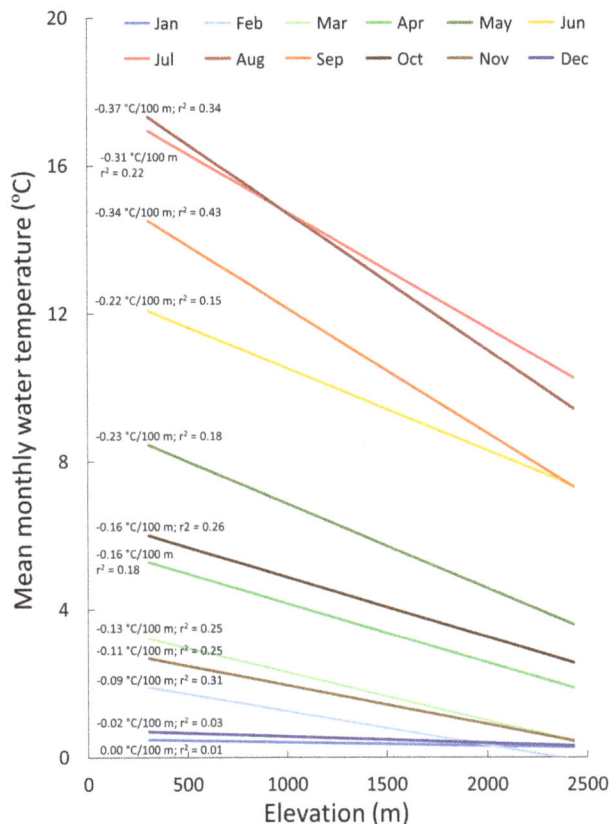

Figure 3. Linear regression trends between elevation and mean monthly temperatures at 226 river and stream sites during 2013 (data values are not shown for clarity). Values next to the trend lines are regression slopes and r^2 values from the regressions.

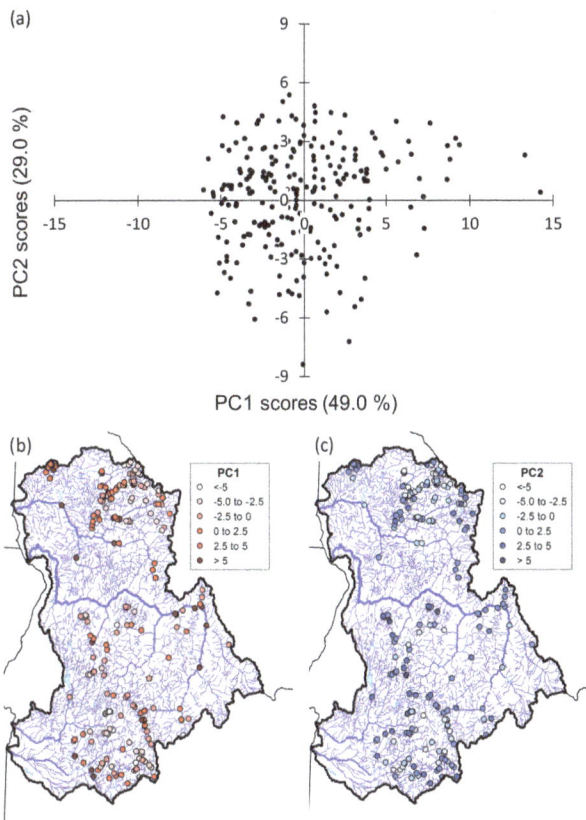

Figure 4. Ordination plot that shows principal component scores of the first two axes derived from water temperature data measured at 226 sites and summarized with 28 thermal metrics (**a**). (**b**) and (**c**) show principal component scores mapped to network locations.

gregated time series of 1826 daily values at the 226 monitoring sites. Concordance between the S-mode PC scores, air temperature, and discharge was examined post hoc by plotting standardized time series and calculating bivariate correlations.

4 Results

Water temperatures within the study area network exhibited spatial and temporal variation that reflected the local topography and annual hydroclimatic cycle. The annual temperature cycle is illustrated in Fig. 3 by the slopes of linear regressions between mean monthly temperatures and elevation at the 226 monitoring sites throughout the course of the year in 2013. No elevation trend occurred during cold winter months when many sites had water temperatures at or near 0 °C and were frequently exposed to subzero air temperatures. As temperatures warmed during the spring a small elevation trend appeared, which became most pronounced (approximately −0.37 °C/100 m) during peak temperatures in the months of July and August. Examples of inter-annual variation are shown in Fig. 2, which contrasts the extreme conditions ob-

served in 2011 and 2015. The former year was relatively cool with a large winter snow accumulation and spring runoff, whereas 2015 had below average snowfall, low runoff, and particularly warm early summer air temperatures. As a result, the median discharge date occurred 1–2 months earlier in 2015 than in 2011 and peak water temperatures were 4–5 °C warmer.

Four PCs accounted for 93.4 % of the variation in the 28 temperature metrics (Table 4). The first PC explained 49 % of the variation and was strongly correlated with metrics that represented magnitude and variability during most seasonal periods. Correlations between PC1 scores and elevation ($r = -0.59$) and mean flow ($r = 0.58$) suggested gradients in these network characteristics were important controls on this component of thermal regimes (Table 5). PC2 explained 29 % of thermal variation and represented the length and intensity of the winter period, with strong loadings for mean winter temperature, minimum temperature, and timing metrics that determined growing season length. PC3 accounted for 9.8 % of total variation and was associated with summer temperature variability and two timing metrics, whereas PC4 accounted for 5.6 % of thermal variance. An

Table 4. Loadings of 28 temperature metrics on the first four principal components in a PCA of annual temperature records from mountain river networks in the northwestern United States.

Temperature metric	PC1	PC2	PC3	PC4
M1. Mean annual temperature	0.99	−0.07	−0.05	−0.03
M2. Mean winter temperature	0.26	−0.92	0.14	0.00
M3. Mean spring temperature	0.91	−0.19	−0.25	0.04
M4. Mean summer temperature	0.97	0.21	−0.06	−0.05
M5. Mean August temperature	0.95	0.22	0.16	−0.10
M6. Mean fall temperature	0.96	−0.18	0.14	−0.08
M7. Minimum daily temperature	−0.02	−0.86	0.08	−0.02
M8. Minimum weekly average temperature	−0.03	−0.90	0.08	0.00
M9. Maximum daily temperature	0.95	0.26	0.09	−0.08
M10. Maximum weekly average temperature	0.95	0.25	0.09	−0.07
M11. Annual degree days	0.99	−0.07	−0.05	−0.03
V1. Annual standard deviation	0.90	0.41	0.01	−0.07
V2. Winter standard deviation	0.69	−0.54	0.16	0.00
V3. Spring standard deviation	0.71	0.30	−0.55	0.04
V4. Summer standard deviation	0.42	0.32	0.78	−0.14
V5. Fall standard deviation	0.87	0.39	0.19	−0.12
V6. Range in extreme daily temperatures	0.93	0.33	0.08	−0.07
V7. Range in extreme weekly temperatures	0.93	0.33	0.08	−0.07
F1. Frequency of hot days	0.47	−0.01	0.30	0.82
F2. Frequency of cold days	−0.70	0.61	0.09	0.11
T1. Date of 5 % of degree days	0.02	0.96	−0.10	0.01
T2. Date of 25 % of degree days	−0.43	0.74	0.46	−0.08
T3. Date of 50 % of degree days	−0.45	0.37	0.79	−0.16
T4. Date of 75 % of degree days	−0.19	−0.51	0.72	−0.19
T5. Date of 95 % of degree days	0.30	−0.88	0.12	−0.09
D1. Growing season length	0.03	−0.97	0.11	−0.03
D2. Duration of hot days	0.44	−0.03	0.32	0.84
D3. Duration of cold days	−0.64	0.66	0.07	0.11
Variance explained (%)	49.0 %	29.0 %	9.8 %	5.6 %
Cumulative variance (%)	49.0 %	78.0 %	87.8 %	93.4 %
Eigenvalue	13.73	8.12	2.74	1.56

Table 5. Correlations among stream temperature principal components and spatial attributes of 226 monitoring sites with annual data from river networks in the northwestern United States.

	Elevation	Mean flow	Reach slope	PC1	PC2	PC3	PC4
Elevation	1						
Mean flow	−0.34	1					
Reach slope	−0.10	−0.23	1				
PC1	−0.59	0.58	−0.34	1			
PC2	0.27	−0.06	−0.49	0.00	1		
PC3	−0.23	0.35	0.13	0.00	0.00	1	
PC4	0.12	0.54	−0.02	0.00	0.00	0.00	1

ordination plot of scores from the two dominant PCs showed a symmetrical distribution except for several sites with large positive scores on the first axis that were from large rivers at low elevations and had the warmest temperatures (Fig. 4a). A map of PC1 scores indicated that the spatial pattern in magnitude and variability (Fig. 4b) was congruent with the network scenario of mean August temperatures as would be expected (Fig. 1). In fact, the correlation between PC1 scores and the NorWeST August scenario predictions at the 226 monitoring sites was strong at $r = 0.86$. The PC2 map showed several clusters of stream sites with high scores scattered throughout the study area (Fig. 4c), which tended to be associated with lower reach slopes (Table 5).

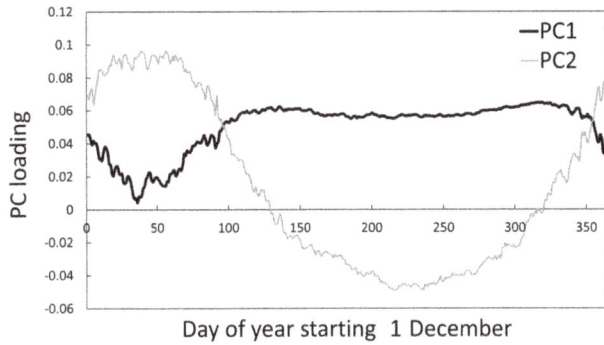

Figure 5. T-mode PCA results showing times when dominant spatial phases occurred in water temperatures at 226 sites based on principal component eigenvector loadings during an average year.

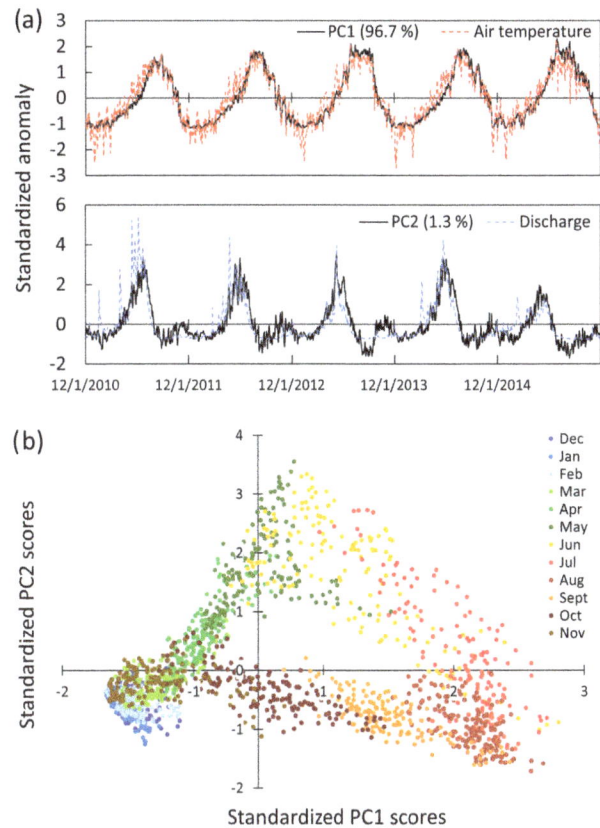

Figure 7. S-mode PCA results showing principal component scores that describe temporal patterns in mean daily water temperatures for 226 stream sites during 5 years (**a**). Average daily air temperatures and discharge values from two monitoring stations are aligned with the principal component scores for comparative purposes. A plot of PC1 versus PC2 reveals that variation along the two axes differs by monthly and seasonal periods (**b**).

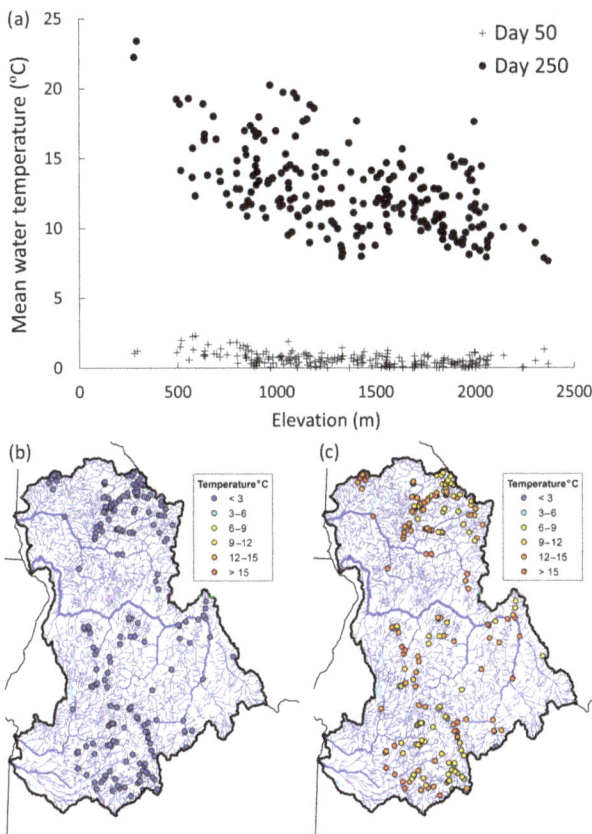

Figure 6. Thermal patterns during two periods with distinct spatial phases based on T-mode PCA results (**a**). Day 50 occurs in mid-January and represents the homogenous winter period (**b**), whereas day 250 occurs in late July and represents the heterogeneous period (**c**).

In the T-mode analysis, the first two PCs explained 88 % of the total variation in mean daily temperatures. A plot of the daily eigenvector loadings indicated that one distinct spatial phase occurred in the winter and a second phase spanned the year's remainder (Fig. 5). Phase transitions occurred around days 100 and 350, which closely aligned with the abatement and onset of subzero air temperatures in the study area (Fig. 2). Figure 6 illustrates the spatial patterns characteristic of the two phases by mapping mean daily water temperatures at the monitoring sites on days 50 and 250, which occurred in mid-January and late July, respectively. Temperatures during the winter phase were spatially homogenous and exhibited a narrow range from 0 to 2.5 °C, whereas the non-winter phase was heterogeneous and had a broader temperature range from 7.6 to 23.4 °C.

In the S-mode analysis, the first PC accounted for 98 % of the variation when applied to the average year of 365 daily temperatures at the 226 monitoring sites. Nearly an identical result was obtained when the analysis was repeated on the disaggregated time series of 1826 daily temperatures, as PC1 then explained 96.7 % of total variation (Fig. 7a). The correlation between PC1 scores and mean daily air temperatures in the disaggregated series was strong ($r = 0.92$), suggesting that water temperatures were responding coherently to vari-

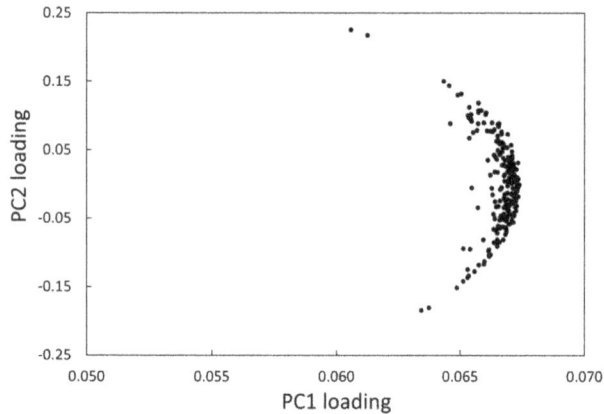

Figure 8. Plot of S-mode eigenvector loadings from 226 stream sites on PC1 and PC2. Note that the range of variation in the PC1 loadings is small relative to the loadings along PC2, which indicates that most of the differences among sites were associated with the second principal component.

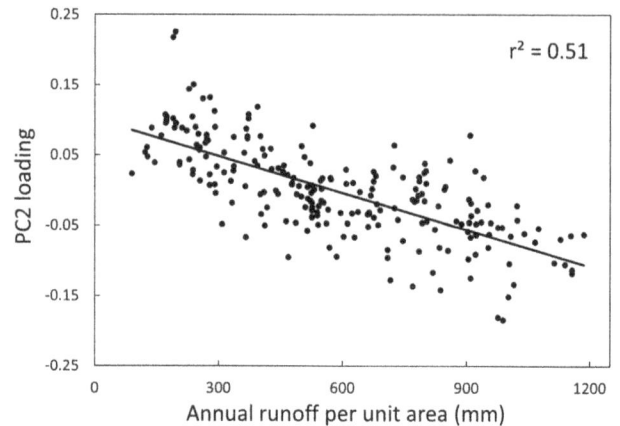

Figure 10. Relationship between the S-mode eigenvector loadings from PC2 and the annual unit-area runoff in basins upstream of 226 water temperature sites.

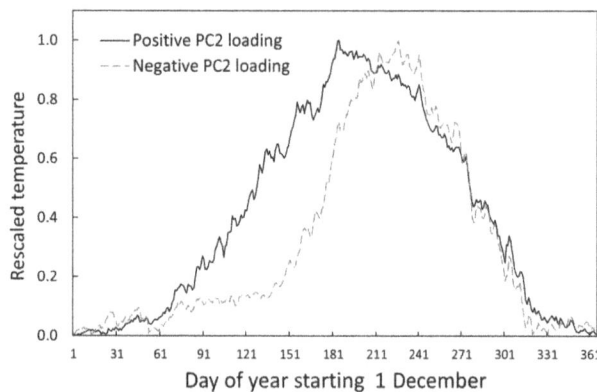

Figure 9. Annual water temperature timing patterns reconstructed from S-mode PCs using the mean eigenvector loading value for PC1 and ±0.16 for PC2 to demonstrate the effects of strong negative loadings and positive loadings on PC2.

ability in air temperatures across the study area. A second PC accounted for 1.3 % of water temperature variation in the disaggregated series and was strongly correlated with variation in mean daily discharge ($r = 0.84$). A plot of PC1 versus PC2 indicated that variation along these axes corresponded to monthly and seasonal periods (Fig. 7b). As was expected, little variation occurred during the cold winter months, but during spring and early summer, variation was observed along both axes as air temperatures warmed and snowmelt runoff created a large discharge pulse. Once discharge returned to baseflow conditions in late summer, variability along PC1 was the primary signal until air temperatures cooled significantly in late fall and the homothermous period began.

Although PC1 and PC2 are linearly uncorrelated, the loop structure of Fig. 7b indicates there was some mutual information and that one driver of temperature variation was out of

phase with the other. Examining this more closely by plotting the site loading values on each component from the S-mode analysis, we see little variability among the loadings for PC1 relative to the much larger range of loading values for PC2 (Fig. 8). This confirms that PC1 represented the common behavior among all stream sites and that deviations in timing of water temperature increases and decreases were dictated by PC2. As a result, when annual temperature signals were reconstructed for two sites from the PCs based on the mean loading value for PC1 and ±0.16 for PC2 to represent strong negative and positive loadings, the expected timing shift was apparent (Fig. 9). Notably, the site with the -0.16 PC2 loading had a later, sharper rise in water temperature that peaked in late summer approximately 1 month after the site with the positive loading. The correspondence of PC2 to stream discharge in Fig. 7a suggests the timing shift could be related to runoff patterns. And indeed, the annual unit-area runoff for the basins associated with the 226 sites was a strong predictor of the PC2 loadings in a linear regression ($r^2 = 0.51$; Fig. 10). Site elevation provides some indication of rainfall–snowfall fraction that may help explain timing shifts, but this covariate added little predictive capacity beyond annual runoff when examined across all sites ($r^2 = 0.54$). However, when sites with basin sizes less than 50 km^2 were examined (because site elevation relates more strongly to mean basin elevation in smaller basins), elevation accounted for a large increase in the explainable variance of PC2 loadings beyond that attributable to annual runoff ($r^2 = 0.69$). Although orographic enhancement of precipitation is evident in the study area, there is enough difference in circulation patterns across the north–south extent of the area that elevation and annual runoff were only weakly correlated in the small basins ($r = -0.2$), so the elevation effect was largely independent of annual precipitation. As a result, both factors appeared to contribute to the PC2 loadings such that either wetter or colder locations had more negative loadings and later rises in water temperatures.

5 Discussion

5.1 Thermal regimes in mountain settings

Thermal regimes in the mountain river networks we studied were simple and responded relatively coherently to climatic variability across a geomorphically consistent area with few reservoirs. Strong seasonal patterns in water temperatures characteristic of temperate latitudes were apparent in response to the primary signal set by the annual air temperature cycle and accompanying changes in solar radiation. Not surprisingly given the pronounced elevational gradients in the study landscape, the dominant regime aspect represented by PC1 in the metric-based PCA was associated with magnitude. Less expected was that many of the variability metrics also loaded heavily on the first PC because variation has been treated as a distinct element of thermal regimes (e.g., Steel et al., 2012; Kovach et al., 2018). The concurrence of magnitude and variability metrics probably also relates to elevation and changes in the importance of groundwater buffering, which both cools streams and dampens diurnal and seasonal variations (Caissie and Luce, 2017). For example, the coldest streams at the highest elevations are usually strongly buffered by groundwater inputs derived from large annual snowpacks in mountain environments and often show limited thermal variability (Luce et al., 2014b; Isaak et al., 2016b). Downstream from the headwaters, the proportional inputs of groundwater decrease and streams are more coupled to climatic variability even as their average temperatures increase due to solar gains over longer flow distances (Caissie, 2006). In contrast to the metrics associated with PC1, metrics that described the winter period and the extent of the growing season largely defined PC2. This "winter" PC is probably common to stream thermal regimes in mountain landscapes where subzero air temperatures are frequent and result in prolonged periods with water temperatures near 0 °C. The orthogonal nature of PC1 and PC2 suggests that streams with otherwise similar magnitude and variance structures will sometimes differ substantially with regards to their winter and growing seasons – a distinction that could have important implications for biological communities or stream physicochemical processes.

Our results also suggest that important local nuances in water temperature dynamics like the differences in timing of spring warming and peak temperatures may emerge from the interactions among annual climate cycles, basin geomorphology, and hydrology. Because precipitation, air temperatures, snowpack, runoff volume, and runoff timing are all evolving in response to climate change in mountain environments across the study region (Mote et al., 2005; Luce et al., 2013) and globally (Stewart, 2009), better understanding of these

connections is needed. In particular, more insight into the relationship of water temperatures with annual unit-area runoff and whether the underlying mechanisms relate to changes in snowpack accumulation (Luce et al., 2014a; Lute and Luce, 2017), snowmelt timing and rate (Musselman et al., 2017), the volume of water stored in groundwater (e.g., Tague et al., 2007), or the outcomes of extreme low flows (e.g., Kormos et al., 2016; Luce and Holden, 2009) could lead to better predictions about water temperatures and the evolution of thermal regimes in response to expected changes in air temperatures and precipitation.

5.2 Implications for modeling and monitoring

Water temperature models are often developed for use in ecological assessments and to understand how habitat degradation or restoration efforts may affect thermal regimes (Benyahya et al., 2007; Gallice et al., 2015; Dugdale et al., 2017). Our results, like several previous studies that have compared multiple temperature metrics (Isaak and Hubert, 2001; Rivers-Moore et al., 2013; Steel et al., 2016), confirm that numerous metrics are strongly correlated and provide redundant information. The specific choice of a metric, therefore, may not be critical as long as it represents an important aspect of a thermal regime and is suited to the goals of a study. Metrics associated with temperature magnitude and variability, which have been the focus of most modeling efforts, are good choices because they represent significant portions of the information about thermal regimes and have been shown on many occasions to be important determinants of ecological attributes such as species distributions and abundance or physical processes in streams and rivers (Isaak et al., 2017b; Webb et al., 2008). Our preferred metrics in previous research have been mean August or mean summer temperatures because the data records for their calculation are typically available at the largest number of sites in mountain environments, which maximizes sample sizes and minimizes the distances over which interpolations are made when developing and applying network-scale temperature models (e.g., Detenbeck et al., 2016; Isaak et al., 2017a). Metrics based on longer-term means rather than short-term daily or weekly maxima are also more stable and easier to predict (Isaak et al., 2010; Turschwell et al., 2016), although a focus on the latter metrics is often mandated within regulatory environments and may negate these considerations (Todd et al., 2008; McCullough, 2010).

Comparatively little effort has gone towards modeling temperature metrics associated with growing season length or the dates of spring and winter season onset, despite the significant information these metrics provide about thermal regimes and their relevance to the phenology and life histories of organisms that constitute aquatic communities (Huryn and Wallace, 2000; Neuheimer and Taggart, 2007). These aspects of thermal regimes, as well as magnitude and variability characteristics, are also likely to be evolving in response to climate change, so new models are needed to provide forecasting abilities about changes later this century. Rather than focusing on individual metrics, researchers could also instead use PCA to efficiently summarize multiple temperature metrics and then model the eigenvector loadings that define one or more of the principal components. This approach would maximize the amount of thermal information represented by a response metric, but would yield results that were more ambiguous to interpret.

The growth of new stream and river temperature monitoring and data collection activities has been remarkable in recent years. Although optimization of those efforts ultimately depends on local considerations, some general guidelines emerge from this work that may be applicable to other areas. For example, the coherent behavior we observed among temperatures at many sites suggests that a limited number of monitoring stations will often be sufficient to represent the temporal dynamics of thermal regimes. Those stations would need to be spread geographically and along major environmental gradients and replicated to mitigate against sensor losses, but 20–30 stations might prove sufficient at scales comparable to our study area. Given low sensor costs and the availability of standardized data collection protocols (Isaak et al., 2013; Stamp et al., 2014), monitoring arrays could also be crowd-sourced effectively if site locations were coordinated and chosen strategically using geospatial analyses to describe and stratify networks for sample allocation (Jackson et al., 2016). Monitoring networks might also be supplemented by incorporating data from sites established for other purposes such as documenting thermal responses to habitat restoration efforts (Nichols and Ketcheson, 2013) or disturbances associated with land management, wildfires, or livestock grazing (Mahlum et al., 2011; Nusslé et al., 2015). In fact, those factors motivated collection of many of the data sets compiled for this analysis, although supplementation with additional sites was needed to ensure adequate coverage within the study area.

If one of the goals of temperature data collection efforts is to develop accurate prediction maps that show spatial variation in one or more thermal metrics (e.g., Isaak et al., 2017a; Steel et al., 2016), monitoring sites may need to be established more densely than the temporal considerations discussed above otherwise suggest. Spatial autocorrelation in temperature metric values is minimal in mountain river networks beyond distances of 10–100 km (Isaak et al., 2010; Zimmerman and Ver Hoef, 2017), so this level of sensor spacing would be required to generate the most accurate maps. Given the extent of many river networks, that could translate into a large number of sites, but most of these could be monitored for short periods while temporal dynamics were represented by a subset of long-term sites because temporal covariance among sites would be strong. Costs associated with numerous sensor deployments could be prohibitive, so aggregation of existing data sets from multiple natural resource agencies into a centralized database often becomes

an attractive option. Moreover, if those central databases are made publicly accessible, professionals from the contributing agencies may begin to coordinate data collection activities more consistently and effectively across larger areas (e.g., Isaak et al., 2018b).

As new data collection and database development efforts proceed, it is commonly the case that temperature records have inconsistent period lengths or missing values. Usually it is desirable to have complete records for analysis, so missing values are sometime imputed based on the correlations between two monitoring site records that strongly covary (e.g., Rivers-Moore et al., 2013). However, the process can be tedious if required at more than a few sites, so an efficient improvement is offered by the imputation technique described by Josse and Husson (2012) that is easily used in the MissMDA software package (Josse and Husson, 2016) for the R statistical program (R Development Core Team, 2014). This technique examines and uses correlations among multiple site records simultaneously to estimate missing values by first applying standard PCA to the incomplete data set where missing values are replaced with the respective record mean. Data are then reconstructed from the PCs, and the initial analysis step is repeated but with missing values replaced using estimates from the reconstructed data. The process is repeated until convergence, and the missing values in the original data records are ultimately replaced with estimates from the last PCA reconstruction (Josse and Husson, 2012). Care should be taken against overreliance on the technique to impute particularly sparse records, but the MissMDA package provides a useful tool for addressing gaps when working with large temperature data sets or time series of other measurements common to hydrology such as gage discharge records (e.g., Isaak et al., 2018a).

6   Conclusions

Our analysis of thermal regimes follows previous work that has proven fundamental to advancing the understanding of hydrologic regimes (Poff et al., 1997; Olden and Poff, 2003) but also adds novel applications of PCA variants from the field of climatology that hold utility for stream temperature research and monitoring design. Insights from those applications indicate that thermal conditions in the mountain river networks studied here were strongly coherent through time, exhibited two distinct spatial phases, could be adequately described by a few principal components or allied metrics, and reflected landscape geomorphology and hydroclimatic conditions. A logical next step involves application of PCA techniques to larger stream and river temperature data sets at regional, continental, or intercontinental scales to encompass greater heterogeneity and discern the geographic domains over which distinct thermal regimes are operable. Across sufficiently diverse landscapes, we might expect to find classes of thermal regimes that, at a minimum, mimicked previously described classes of hydrologic regimes (e.g., rainfall, snowmelt, spring groundwater, and regulated), but possible divergences from, or additions to, these categories would be useful to ascertain. In a national-scale assessment for the United States, Maheu et al. (2016) classified stream thermal regimes into six categories, but the 135 temperature stations that supported the analysis were limited in comparison to a drainage network comprised of millions of kilometers. Subsequent iterations on that effort could document additional, undescribed thermal classes and might also prove beneficial by developing detailed maps of classification schemes to aid in assessments of ecological conditions or anthropogenic effects on stream thermal regimes. As research on the topic of thermal regimes matures, syntheses with flow regime concepts and databases could also be sought to more fully describe the hydroclimatic conditions of flowing waters.

Author contributions. DJI and CHL conceived the study, conducted the analysis, and co-wrote the manuscript. DLH, SPW, and DJI collected water temperature data. GLC and SPW developed the temperature database. DLH developed map figures.

Competing interests. The authors declare that they have no conflict of interest.

Acknowledgements. We thank Dave Schoen, Bart Gamett, Dan Garcia, Scott Vuono, Caleb Zurstadt, and Clayton Nalder with the U.S. Forest Service, Tim Copeland, Eric Stark, and Ron Roberts with the Idaho Department of Fish and Game, Eric Archer and Jeff Ojala with the Pacfish-Infish Biological Opinion monitoring program, and Boyd Bowes and Chris Jordan with the CHaMP monitoring program that contributed water temperature data to enable this research. Comments from Nicholas Rivers-Moore and one anonymous reviewer improved the quality of the final manuscript. The authors of this work were supported by the U.S. Forest Service Rocky Mountain Research Station.

Edited by: Jim Freer

References

Arismendi, I., Johnson, S. L., Dunham, J. B., and Haggerty, R.: Descriptors of natural thermal regimes in streams and their responsiveness to change in the Pacific Northwest of North America, Freshwater Biol., 58, 880–894, 2013.

Benyahya, L., Caissie, D., St-Hilaire, A., Ouarde, T., and Bobee, B.: A review of statistical water temperature models, Can. Water Resour. J., 32, 179–192, 2007.

Bond, J. G. and Wood, C. H.: Geologic map of Idaho, 1 : 500 000 scale, Idaho Department of Lands, Bureau of Mines and Geology, Moscow, Idaho, 1978.

Brown, L. E. and Hannah, D. M.: Spatial heterogeneity of water temperature across an alpine river basin, Hydrol. Process., 22, 954–967, 2008.

Brown, L. E., Cereghino, R., and Compin, A.: Endemic freshwater invertebrates from southern France: diversity, distribution

and conservation implications, Biol. Conserv., 142, 2613–2619, 2009.

Caissie, D.: The thermal regime of rivers: a review, Freshwater Biol., 51, 1389–1406, 2006.

Caissie, D. and Luce, C. H.: Quantifying streambed advection and conduction heat fluxes, Water Resour. Res., 53, 1595–1624, https://doi.org/10.1002/2016WR019813, 2017.

Chandler, G. L., Wollrab, S. P., Horan, D. L., Nagel, D. E., Parkes, S. L., Isaak, D. J., Wenger, S. J., Peterson, E. E., Ver Hoef, J. M., Hostetler, S. W., Luce, C. H., Dunham, J. B., Kershner, J. L., and Roper, B. B.: NorWeST stream temperature data summaries for the western U.S., Forest Service Research Data Archive, Fort Collins, CO, https://doi.org/10.2737/RDS-2016-0032, 2016.

Chu, C., Jones, N. E., and Allin, L.: Linking the thermal regimes of streams in the Great Lakes Basin, Ontario, to landscape and climate variables, River Res. Appl., 26, 221–241, 2010.

Daigle, A., Caudron, A., Vigier, L., and Pella, H.: Optimization methodology for a river temperature monitoring network for the characterization of fish thermal habitat, Hydrol. Sci. J., 62, 483–497, 2016.

Demsar, U., Harris, P., Brunsdon, C., Fotheringham, A. S., and McLoone, S.: Principal components analysis on spatial data: An overview, Ann. Assoc. Am. Geogr., 103, 106–128, 2013.

Detenbeck, N. E., Morrison, A., Abele, R. W., and Kopp, D.: Spatial statistical network models for stream and river temperature in New England, USA, Water Resour. Res., 52, 6018–6040, 2016.

Ducharne, A.: Importance of stream temperature to climate change impact on water quality, Hydrol. Earth Syst. Sci., 12, 797–810, https://doi.org/10.5194/hess-12-797-2008, 2008.

Dugdale, S. J., Hannah, D. M., and Malcolm, I. A.: River temperature modelling: A review of process-based approaches and future directions, Earth-Sci. Rev., 175, 97–113, 2017.

Dunham, J., Rieman, B., and Chandler, G.: Influences of temperature and environmental variables on the distribution of bull trout within streams at the southern margin of its range, N. Am. J. Fish. Manage., 23, 894–904, 2003.

Dunham, J. B., Chandler, G., Rieman, B. E., and Martin, D.: Measuring stream temperature with digital dataloggers: a user's guide, U.S. Forest Service General Technical Report, Rocky Mountain Research Station, 150WWW, Fort Collins, Colorado, USA, 2005.

Gallacher, K., Miller, C., Scott, E. M., Willows, R., Pope, L., and Douglass, J.: Flow-directed PCA for monitoring networks, Environmetrics, 28, e2434, https://doi.org/10.1002/env.2434, 2017.

Gallice, A., Schaefli, B., Lehning, M., Parlange, M. B., and Huwald, H.: Stream temperature prediction in ungauged basins: review of recent approaches and description of a new physics-derived statistical model, Hydrol. Earth Syst. Sci., 19, 3727–3753, https://doi.org/10.5194/hess-19-3727-2015, 2015.

Gordon, N. D., McMahon, T. A., and Finlayson, B. L.: Stream hydrology: an introduction for ecologists, Stream hydrology: an introduction for ecologists, John Wiley and Sons, Chichester, UK, 1991.

Hannah, D. and Garner, G.: River water temperature in the United Kingdom: Changes over the 20th century and possible changes over the 21st century, Prog. Phys. Geogr., 39, 68–92, 2015.

Hilderbrand, R. H., Kashiwagi, M. T., and Prochaska, A. P.: Regional and local scale modeling of stream temperatures and spatio-temporal variation in thermal sensitivities, Environ. Manage., 54, 14–22, 2014.

Huff, D. D., Hubler, S. L., and Borisenko, A. N.: Using field data to estimate the realized thermal niche of aquatic vertebrates, N. Am. J. Fish. Manage., 25, 346–360, 2005.

Huryn, A. D. and Wallace, J. B.: Life history and production of stream insects, Annu. Rev. Entomol., 45, 83–110, 2000.

Ice, G. G., Light, J., and Reiter, M.: Use of natural temperature patterns to identify achievable stream temperature criteria for forest streams, West. J. Appl. For., 19, 252–259, 2004.

Isaak, D. J. and Hubert, W. A.: A hypothesis about factors that affect maximum summer stream temperatures across montane landscapes, J. Am. Water Resour. As., 37, 351–366, 2001.

Isaak, D. J., Luce, C. H., Rieman, B. E., Nagel, D. E., Peterson, E. E., Horan, D. L., Parkes, S., and Chandler, G. L.: Effects of climate change and wildfire on stream temperatures and salmonid thermal habitat in a mountain river network, Ecol. Appl., 20, 1350–1371, 2010.

Isaak, D. J., Horan, D. L., and Wollrab, S. P.: A simple protocol using underwater epoxy to install annual temperature monitoring sites in rivers and streams, U.S. Forest Service General Technical Report, Rocky Mountain Research Station, 314, Fort Collins, Colorado, USA, 2013.

Isaak, D. J., Wenger, S. J., Peterson, E. E., Ver Hoef, J. M., Nagel, D. E., Luce, C. H., Hostetler, S. W., Dunham, J. B., Roper, B. B., Wollrab, S., Chandler, G., Parkes, S., and Horan, D.: NorWeST modeled summer stream temperature scenarios for the western United States. U.S. Forest Service, Rocky Mountain Research Station Research Data Archive, Fort Collins, CO, https://doi.org/10.2737/RDS-2016-0033, 2016a.

Isaak, D. J., Young, M. K., Luce, C. H., Hostetler, S., Wenger, S., Peterson, E. E., Ver Hoef, J. M., Groce, M., Horan, D. L., and Nagel, D.: Slow climate velocities of mountain streams portend their role as refugia for cold-water biodiversity, P. Natl. Acad. Sci. USA, 113, 4374–4379, 2016b.

Isaak, D. J., Wenger, S. J., Peterson, E. E., Ver Hoef, J. M., Nagel, D. E., Luce, C. H., Hostetler, S. W., Dunham, J. B., Roper, B. B., Wollrab, S., Chandler, G., Parkes, S., and Horan, D.: The NorWeST summer stream temperature model and scenarios: A crowd-sourced database and new geospatial tools foster a user community and predict broad climate warming of rivers and streams in the western United State, Water Resour. Res., 53, 9181–9205, 2017a.

Isaak, D. J., Wenger, S. J., and Young, M. K.: Big biology meets microclimatology: Defining thermal niches of aquatic ectotherms at landscape scales for conservation planning, Ecol. Appl., 27, 977–990, 2017b.

Isaak, D. J., Luce, C. H., Horan, D. L., Chandler, G. L., Wollrab, S. P., and Nagel, D.: Global warming of salmon and trout rivers in the northwestern United States: Road to ruin or path through purgatory?, T. Am. Fish. Soc., 147, 566–587, 2018a.

Isaak, D. J., Young, M. K., McConnell, C., Roper, B. B., Archer, E. K., Staab, B., Hirsch, C., Nagel, D. E., Schwartz, M. K., and Chandler, G. L.: Crowd-sourced databases as essential elements for Forest Service partnerships and aquatic resource conservation, Fisheries, 43, 423–430, 2018b.

Jackson, F. L., Malcolm, I. A., and Hannah, D. M.: A novel approach for designing large-scale river temperature monitoring networks, Hydrol. Res., 47, 569–590, 2016.

Jiménez, P. A., García-Bustamante, E., González-Rouco, J. F., Valero, F., Montávez, J. P., and Navarro, J.: Surface wind regionalization in complex terrain, J. Appl. Meteorol. Clim., 47, 308–325, 2008.

Josse, J. and Husson, F.: Handling missing values in exploratory multivariate data analysis methods, Journal of the Société Francaise de Statistique, 153, 79–99, 2012.

Josse, J. and Husson, F.: MissMDA: a package for handling missing values in multivariate data analysis, J. Stat. Softw., 70, 1–31, 2016.

Kormos, P., Luce, C., Wenger, S. J., and Berghuijs, W. R.: Trends and Sensitivities of Low Streamflow Extremes to discharge Timing and Magnitude in Pacific Northwest Mountain Streams, Water Resour. Res., 52, 4990–5007, https://doi.org/10.1002/2015WR018125, 2016.

Kovach, R. P., Muhlfeld, C. C., Al-Chokhachy, R., Ojala, J. V., and Archer, E. K.: Effects of land use on summer thermal regimes in critical salmonid habitats of the Pacific Northwest, Can. J. Fish. Aquat. Sci., 76, https://doi.org/10.1139/cjfas-2018-0165, 2018.

Likens, G. E. and Likens, G. E.: Biogeochemistry of a forested ecosystem, Springer-Verlag, New York, 1977.

Luce, C. H. and Holden, Z. A.: Declining annual streamflow distributions in the Pacific Northwest United States, 1948–2006, Geophys. Res. Lett., 36, L16401, https://doi.org/10.1029/2009GL039407, 2009.

Luce, C. H., Abatzoglou, J. T., and Holden, Z. A.: The Missing Mountain Water: Slower Westerlies Decrease Orographic Enhancement in the Pacific Northwest USA, Science, 342, 1360–1364, 2013.

Luce, C. H., Lopez-Burgos, V., and Holden, Z.: Sensitivity of snowpack storage to precipitation and temperature using spatial and temporal analog models, Water Resour. Res., 50, 9447–9462, 2014a.

Luce, C. H., Staab, B., Kramer, M., Wenger, S., Isaak, D., and McConnell, C.: Sensitivity of summer stream temperatures to climate variability in the Pacific Northwest, Water Resour. Res., 50, 3428–3443, 2014b.

Lute, A. C. and Luce, C. H.: Are model transferability and complexity antithetical? Insights from validation of a variable-complexity snow model in space and time, Water Resour. Res., 53, 8825–8850, 2017.

Maheu, A., Poff, N. L., and St-Hilaire, A.: A classification of stream water temperature regimes in the conterminous USA, River Res. Appl., 32, 896–906, 2016.

Mahlum, S. K., Eby, L. A., Young, M. K., Clancy, C. G., and Jakober, M.: Effects of wildfire on stream temperatures in the Bitterroot River Basin, Montana, Int. J. Wildland Fire, 20, 240–247, 2011.

Martins, D. S., Raziei, T., Paulo, A. A., and Pereira, L. S.: Spatial and temporal variability of precipitation and drought in Portugal, Nat. Hazards Earth Syst. Sci., 12, 1493–1501, https://doi.org/10.5194/nhess-12-1493-2012, 2012.

Mauger, S., Shaftel, R., Leppi, J. C., and Rinella, D. J.: Summer temperature regimes in southcentral Alaska streams: watershed drivers of variation and potential implications for Pacific salmon, Can. J. Fish. Aquat. Sci., 74, 702–715, 2016.

McCullough, D. A.: Are coldwater fish populations of the United States actually being protected by temperature standards?, Freshwater Reviews, 3, 147–199, 2010.

McKay, L., Bondelid, T., Dewald, T., Johnston, J., Moore, R., and Reah, A.: NHDPlus Version 2: User Guide, available at: ftp://ftp.horizon-systems.com/NHDPlus/NHDPlusV21/Documentation/NHDPlusV2_User_Guide.pdf (last access: 2 December 2018), 2012.

Meyer, G. A., Pierce, J. L., Wood, S. H., and Jull, A. J. T.: Fire, storms, and erosional events in the Idaho batholith, Hydrol. Process., 15, 3025–3038, 2001.

Molinero, J., Larrañaga, A., Pérez, J., Martínez, A., and Pozo, J.: Stream temperature in the Basque Mountains during winter: thermal regimes and sensitivity to air warming, Clim. Change, 134, 593–604, 2016.

Morgan, P., Heyerdahl, E., Miller, C., Wilson, A., and Gibson, C.: Northern Rockies pyrogeography: an example of fire atlas utility, Fire Ecology, 10, 14–30, 2014.

Mote, P. W., Hamlet, A. F., Clark, M. P., and Lettenmaier, D. P.: Declining mountain snowpack in western North America, B. Am. Meteorol. Soc., 86, 39–49, 2005.

Musselman, K. N., Clark, M. P., Liu, C., Ikeda, K., and Rasmussen, R.: Slower snowmelt in a warmer world, Nat. Clim. Change, 7, 214–219, 2017.

Neuheimer, A. B. and Taggart, C. T.: The growing degree-day and fish size-at-age: the overlooked metric, Can. J. Fish. Aquat. Sci., 64, 375–385, 2007.

Nichols, R. A. and Ketcheson, G. L.: A two-decade-watershed approach to stream restoration log jam design and stream recovery monitoring: Finney Creek, Washington, J. Am. Water Resour. As., 49, 1367–1384, 2013.

Nusslé, S., Matthews, K. R., and Carlson, S. M.: Mediating water temperature increases due to livestock and global change in high elevation meadow streams of the Golden Trout Wilderness, PloS ONE, 10, e0142426, https://doi.org/10.1371/journal.pone.0142426, 2015.

Olden, J. D. and Naiman, R. J.: Incorporating thermal regimes into environmental flows assessments: modifying dam operations to restore freshwater ecosystem integrity, Freshwater Biol., 55, 86–107, 2010.

Olden, J. D. and Poff, N. L.: Redundancy and the choice of hydrologic indices for characterizing streamflow regimes, River Res. Appl., 19, 101–121, 2003.

Orr, H. G., Johnson, M. F., Wilby, R. L., Hatton-Ellis, T., and Broadmeadow, S.: What else do managers need to know about warming rivers? A United Kingdom perspective, Wiley Interdisciplinary Reviews Water, 2, 55–64, 2015.

Pearson, K.: On Lines and Planes of Closest Fit to Systems of Points in Space, Philos. Mag., 6, 559–572, 1901.

Piechota, T. C., Dracup, J. A., and Fovell, R. G.: Western US streamflow and atmospheric circulation patterns during El Nino-Southern Oscillation, J. Hydrol., 201, 249–271, 1997.

Poff, N. L. and Ward, J. V.: Implications of streamflow variability and predictability for lotic community structure: a regional analysis of streamflow patterns, Can. J. Fish. Aquat. Sci., 46, 1805–1818, 1989.

Poff, N. L., Allan, J. D., Bain, M. B., Karr, J. R., Prestegaard, K. L., Richter, B. D., Sparks, R. E., and Stromberg, J. C.: The natural flow regime, BioScience, 47, 769–784, 1997.

Poole, G. C., Dunham, J. B., Keenan, D. M., Sauter, S. T., McCullough, D. A., Mebane, C., Lockwood, J. C., Essig, D. A., Hicks, M. P., Sturdevant, D. J., and Materna, E. J.: The case for regime-based water quality standards, BioScience, 54, 155–161, 2004.

Quaglietta, L., Paupério, J., Martins, F. M. S., Alves, P. C., and Beja, P.: Recent range contractions in the globally threatened Pyrenean desman highlight the importance of stream headwater refugia, Anim. Conserv., 21, https://doi.org/10.1111/acv.12422, 2018.

R Development Core Team, R: A language and environment for statistical computing, R Foundation for Statistical Computing, Vienna, Austria, 2014.

Richman, M. B.: Rotation of principal components, Int. J. Climatol., 6, 293–335, 1986.

Rieman, B. E., Smith, C. L., Naiman, R. J., Ruggeronee, G. T., Wood, C. C., Huntly, N., Merrill, E. N., Alldredge, J. R., Bisson, P. A., Congleton, J., Fausch, K. D., Levings, C., Pearcy, W., Scarnecchia, D., and Smouse, P.: A comprehensive approach for habitat restoration in the Columbia Basin, Fisheries, 40, 124–135, 2015.

Rivers-Moore, N. A., Dallas, H. F., and Morris, C.: Towards setting environmental water temperature guidelines: A South African example, J. Environ. Manage., 128, 380–392, 2013.

Roni, P., Beechie, T. J., Bilby, R. E., Leonetti, F. E., Pollock, M. M., and Pess, G. R.: A review of stream restoration techniques and a hierarchical strategy for prioritizing restoration in Pacific Northwest watersheds, N. Am. J. Fish. Manage., 22, 1–20, 2002.

SAS Institute Inc.: SAS/STAT 14.1 User's Guide Cary, NC, USA, 2015.

Sergeant, C. J., Starkey, E. N., Bartz, K. K., Wilson, M. H., and Mueter, F. J.: A practitioner's guide for exploring water quality patterns using principal components analysis and Procrustes, Environ. Monit. Assess., 188, 249, https://doi.org/10.1007/s10661-016-5253-z, 2016.

Stamp, J., Hamilton, A., Craddock, M., Parker, L., Roy, A., Isaak, D., Holden, Z., Passmore, M., and Bierwagen, B.: Best practices for continuous monitoring of temperature and flow in wadeable streams, Global Change Research Program, National Center for Environmental Assessment, Washington, D.C., EPA/600/R-13/170F, 2014.

Steel, E. A., Tillotson, A., Larsen, D. A., Fullerton, A. H., Denton, K. P., and Beckman, B. R.: Beyond the mean: the role of variability in predicting ecological effects of stream temperature on salmon, Ecosphere, 3, 1–11, 2012.

Steel, E. A., Sowder, C., and Peterson, E. E.: Spatial and temporal variation of water temperature regimes on the Snoqualmie River network, J. Am. Water Resour. As., 52, 769–787, 2016.

Stewart, I. T.: Changes in snowpack and snowmelt runoff for key mountain regions, Hydrol. Process., 23, 78–94, 2009.

Swanson, F.: Where roads will never reach: Wilderness and its visionaries in the Northern Rockies, University of Utah Press, Salt Lake City, Utah, 2015.

Tague, C., Farrell, M., Grant, G., Lewis, S., and Rey, S.: Hydrogeologic controls on summer stream temperatures in the McKenzie River Basin, Oregon, Hydrol. Process., 21, 3288–3300, 2007.

Todd, A. S., Coleman, M. A., Konowal, A. M., May, M. K., Johnson, S., Vieira, N. K. M., and Saunders, J. F.: Development of new water temperature criteria to protect Colorado's fisheries, Fisheries, 33, 433–443, 2008.

Trumbo, B. A., Nislow, K. H., Stallings, J., Hudy, M., Smith, E. P., Kim, D., Wiggins, B., and Dolloff, C. A.: Ranking site vulnerability to increasing temperatures in southern appalachian brook trout streams in Virginia: an exposure-sensitivity approach, T. Am. Fish. Soc., 143, 173–187, 2014.

Turschwell, M. P., Peterson, E. E., Balcombe, S. R., and Sheldon, F.: To aggregate or not? Capturing the spatio-temporal complexity of the thermal regime, Ecol. Indic., 67, 39–48, 2016.

Webb, B. W., Hannah, D. M., Moore, R. D., Brown, L. E., and Nobilis, F.: Recent advances in stream and river temperature research, Hydrol. Process., 22, 902–918, 2008.

Wenger, S. J., Luce, C. H., Hamlet, A. F., Isaak, D. J., and Neville, H. M.: Macroscale hydrologic modeling of ecologically relevant flow metrics, Water Resour. Res., 46, W09513, https://doi.org/10.1029/2009WR008839, 2010.

Woodward, G., Perkins, D. M., and Brown, L. E.: Climate change and freshwater ecosystems: impacts across multiple levels of organization, Philos. T. R. Soc. B, 365, 2093–2106, 2010.

Zimmerman, D. L. and Ver Hoef, J. M.: The Torgegram for fluvial variography: characterizing spatial dependence on stream networks, J. Comput. Graph. Stat., 26, 253–264, 2017.

13

Can mussels be used as sentinel organisms for characterization of pollution in urban water systems?

Elke S. Reichwaldt and Anas Ghadouani

Aquatic Ecology and Ecosystem Studies, School of Civil, Environmental and Mining Engineering, M015, The University of Western Australia, 35 Stirling Highway, Crawley, Western Australia 6009, Australia

Correspondence to: Elke S. Reichwaldt (elke.reichwaldt@uwa.edu.au)

Abstract. Urbanization strongly impacts aquatic ecosystems by decreasing water quality and altering water cycles. Today, much effort is put towards the restoration and conservation of urban waterbodies to enhance ecosystem service provision, leading to liveable and sustainable cities. To enable a sustainable management of waterbodies, the quantification of the temporal and spatial variability of pollution levels and biogeochemical processes is essential. Stable isotopes have widely been used to identify sources of pollution in ecosystems. For example, increased nitrogen levels in waterbodies are often accompanied with a higher nitrogen stable isotope signature (δ^{15}N), which can then be detected in higher trophic levels such as mussels. The main aim of this study was to assess the suitability of nitrogen stable isotopes as measured in mussels (*Mytilus edulis*), as an indicator able to resolve spatial and temporal variability of nitrogen pollution in an urban, tidally influenced estuary (Swan River estuary in Western Australia). Nitrogen concentrations were generally low and nitrogen stable isotope values of nitrate throughout the estuary were well within natural values of uncontaminated groundwater, organic nitrate from soils, or marine-derived sources, indicating groundwater inflow rather than pollution by human activity was responsible for differences between sites. The δ^{15}N signature in mussels was very stable over time within each site which indicated that mussels can be used as time-integrated sentinel organisms in urban systems. In addition, our study shows that the nature of the relationship between δ^{15}N in the mussels and the nitrate in the water can provide insights into site-specific biogeochemical transformation of nutrients. We suggest that mussels and other sentinel organisms can become a robust tool for the detection and characterization of the dynamics of a number of

emerging anthropogenic pollutants of concern in urban water systems.

1 Introduction

Humans exert a growing impact on the environment supporting them. Today, more than 50 % of the world's population is living in cities and this percentage is projected to further increase to up to 80 % by 2050 (Pickett et al., 2011; United Nations, 2013). Impervious surfaces in cities lead to less rainfall infiltrating the soil. Instead, stormwater runoff is directly transported to waterbodies, polluting them with nutrients, heavy metals, and bacteria (Makepeace et al., 1995; Brezonik and Stadelmann, 2002). Urbanization has resulted in increased eutrophication of waterbodies leading to deteriorated ecosystems worldwide, reducing natural biodiversity and ecosystem services (Heathwaite, 2010). Environmental management is often hampered by a limited understanding of the temporal and spatial variability of pollution levels, the sources of contamination and the processes within systems that affect the recovery of a system (Kooistra et al., 2001; Scheffer et al., 2001; Lahr and Kooistra, 2010). In addition, the traditional hierarchical water management practices that are still in use around the world have been criticized as being ineffective and leaving little scope for adaptation to changes (Pahl-Wostl, 2007; van de Meene et al., 2011). The current trend to decentralize such urban water management might allow for more local management of water resources, indicating the need for improving our understanding of the variability of pollution levels in a range of urban waterbodies with greater emphasis on local processes.

Many urban estuaries are highly impacted by human activity due to direct input of pollutants, such as nitrogen from urban, agriculture, and industry areas (e.g. Oczkowski et al., 2008), which can lead to eutrophication. In urban estuaries, tributaries often transport high amounts of nitrogen from the watershed into the estuary, causing water-quality problems including toxic bloom development (Hamilton, 2000; Atkins et al., 2001). Nitrogen concentration gradients might develop with higher upstream and lower downstream values, where nutrients are diluted by seawater (Dähnke et al., 2010; Fry et al., 2011). This can lead to a spatial variability of nitrogen concentration within estuaries. Nitrogen pollution can also be highly variable in time with higher nitrogen concentrations in estuaries found during times of high water input by tributaries. Smaller-scale variability in temporal and spatial nitrogen concentrations can additionally stem from local differences in hydrological processes (Linderfelt and Turner, 2001), variations in fertilizer use in agricultural areas or temporal failure of septic tank systems causing leakage of sewage, leading to localized places of concern for water management.

Anthropogenic nitrogen and organic pollution of water systems, including the interaction between surface and groundwater, have been successfully investigated using a range of stable isotopes (Sikdar and Sahu, 2009; Yang et al., 2012; Lutz et al., 2013). In addition, stable isotopes have been widely used in purely hydrological studies focused on flow paths, hydraulic residence time, and other hydrological dynamics (Clay et al., 2004; Rodgers et al., 2005; Volkmann and Weiler, 2014). Stable isotopes of nitrogen (N), carbon (C), sulfur (S), and oxygen (O) in water and biota have also been applied as an integrated measure of ecosystem processes (Robinson, 2001; Chaves et al., 2003; Pace et al., 2004). Furthermore, the analysis of the nitrogen signature has proven to be an especially powerful tool as an indicator of anthropogenic contamination (Lake et al., 2001; McKinney et al., 2002; Fry and Allen, 2003; Xu and Zhang, 2012) and land use (Harrington et al., 1998; Broderius, 2013; Carvalho et al., 2015), bearing on the fact that the sources of contamination such as animal manure, sewage, septic waste, and some fertilizers carry higher nitrogen signatures values and consequently a higher $\delta^{15}N$ (Heaton, 1986; Cabana and Rasmussen, 1996; Kellman, 2005; Choi et al., 2007). This signal is then passed on to higher trophic levels up the food chain (Cabana and Rasmussen, 1994; Carvalho et al., 2015): elevated $\delta^{15}N$ signals in nitrate have been shown to lead to elevated $\delta^{15}N$ signals in organisms that directly take up nitrate from the water, such as phytoplankton and microbes (Harrington et al., 1998). These organisms form an important part of particulate organic matter (POM), which serves as food for filter feeders (e.g. mussels). Mussels that ingest POM with an elevated $\delta^{15}N$ signal will then also show a higher $\delta^{15}N$ signal.

Assessing anthropogenic pollution of a system by directly measuring the isotopic signature of nitrogen containing nutrients (e.g. nitrate, ammonium) or of aquatic short-lived organisms with fast tissue turnover times, such as phytoplankton, may significantly under- or overestimate the average level of pollution, as the result strongly depends on the time of measurement. Bivalves, on the other hand, which include the blue mussel, are primary consumers with limited movement, and have been suggested as suitable site-specific bioindicators of time-averaged persistence of nutrient pollutants, because their isotopic signature fluctuates less than that of their food sources due to longer tissue turnover rates (Raikow and Hamilton, 2001; Post, 2002; Fukumori et al., 2008; Fertig et al., 2010; Wang et al., 2013). The blue mussel, *Mytilus edulis*, is a common sessile bivalve in estuarine and marine environments that is able to adapt to a wide range of environmental conditions, such as food concentration, temperature, and salinity (e.g. Thompson and Bayne, 1974; Widdows et al., 1979; Zandee et al., 1980; Almadavillela, 1984), and that shows low sensitivity to anthropogenic pressures (Mainwaring et al., 2014). As such, this species is able to thrive at different pollution levels and has therefore been used as an indicator species for pollution (Phillips, 1976) and as a model organism for physiological, genetic, and toxicological studies (Luedeking and Koehler, 2004) for some time. Earlier studies in polluted freshwater and marine systems found positive relationships between the concentration of nitrogen and the isotopic signature of nitrogen in mussels, and between the isotopic signature of nitrate-N and that of mussels (Cabana and Rasmussen, 1996; McClelland et al., 1997; Costanzo et al., 2001; Anderson and Cabana, 2005; Gustafson et al., 2007; Wen et al., 2010), suggesting that bivalves are suitable indicators of changes in the nutrient pollution load from agriculture and wastewater to waterbodies. However, very little information exists on the use of these stable isotopic signatures in urban systems.

The main aim of this study was to identify the variability of nitrogen concentration in an urban estuary over time and space and to ascertain the suitability of the isotopic signature ($\delta^{15}N$) of blue mussel (*Mytlius edulis*) tissue as an indicator of nitrogen pollution in urban water systems. Specifically, we anticipated that (1) a higher input of nitrogen-rich waters upstream would lead to a higher isotopic signatures of nitrate, (2) spatial differences in the level of nitrates in the water would lead to spatial differences in mussel isotopic signature, and (3) the increased distance from the estuary mouth would lead to elevated ^{15}N values in mussels due to elevated ^{15}N inputs from nitrogen-rich waters upstream.

2   Materials and methods

2.1   Study sites

The study was performed in the lower reaches of the heavily urbanized Swan River estuary that flows through Perth, Western Australia (Fig. 1) (Atkins and Klemm, 1987). The catchment of this estuary is approximately $121\,000\,\text{km}^2$ (Peters and Donohue, 2001) and encompasses urban, rural, agricultural and forested areas. In the urban area, drains contain areas with and without sewers (Peters and Donohue, 2001). The Swan River estuary experienced a major toxic cyanobacterial bloom in 2000, when a large rainfall event increased nutrient concentrations and decreased salinity within the estuary (Hamilton, 2000; Atkins et al., 2001), indicating that this estuary is prone to nutrient pollution from the watershed. The Swan River estuary is influenced by mostly diurnal tides with a mean tidal range at the mouth of the estuary of 0.8 m. At the same time, the estuary is seasonally forced with a large discharge of freshwater from the tributaries during the wetter winter months (May–September), and little freshwater discharge during dry summers. This leads to fresh-to-brackish water in parts of the estuary in winter with a freshwater lens overlying saltwater, and an inland progression of the saltwater wedge, making the estuary a saltwater habitat during drier months (Stephens and Imberger, 1996). The Swan River estuary is permanently open to the ocean and has two major freshwater tributaries, the Swan River and the Canning River (Fig. 1). While there are also several short stormwater drains leading into the lower Swan River estuary that could potentially provide nutrient input into the Swan River estuary from the adjacent land; these drains did not flow during the study.

Seven sites within the lower Swan River estuary were sampled six times for blue mussels and nine times for nutrients, particulate organic matter, chlorophyll a, temperature, salinity, pH, and oxygen during the wetter season (April–November 2010). The sites were jetties at Point Walter (WP) (32°0′39.23″ S, 115°47′15.11″ E), Minim Cove (MC) (32°1′21.23″ S, 115°45′57.38″ E), Swan River Canoe Club (SCC) (32°0′27.31″ S, 115°46′18.73″ E), Claremont (Cl) (31°59′23.80″ S, 115°46′52.97″ E), Broadway (BRD) (31°59′25.55″ S, 115°49′5.49″ E), Applecross (AC) (32°0″ 17.59″ S, 115°49′58.29″ E), and Como Beach (CB) (31°59″ 37.46″ S, 115°51′10.33″ E) (Fig. 1). While MC and SCC are situated at the deeper part of the estuary (depth < 17 m), all other sites are located in the shallower part (depth < 10 m) (Stephens and Imberger, 1996). The jetty at Cl is situated in a shallow bay (depth approximately 2 m) with established seagrass meadows and abundant macroalgae and macrophytes (Department of Water, 2010). Additionally, a one-time marine reference measurement was performed towards the end of the study outside the estuary at Woodman Point Jetty (WO; 32°7′26.97″ S, 115°45′32.10″ E) (Fig. 1).

2.2 Sampling and analyses

On each date, sampling was performed 0.5–1 h prior to high and low tide at each site, respectively. While mussels were sampled only once per day, all other parameters were sampled at high and low tide. Salinity, pH, water temperature, and oxygen were measured at 20 cm depth with hand-held probes (WP-81; TPS-DO$_2$). At each site, water samples for

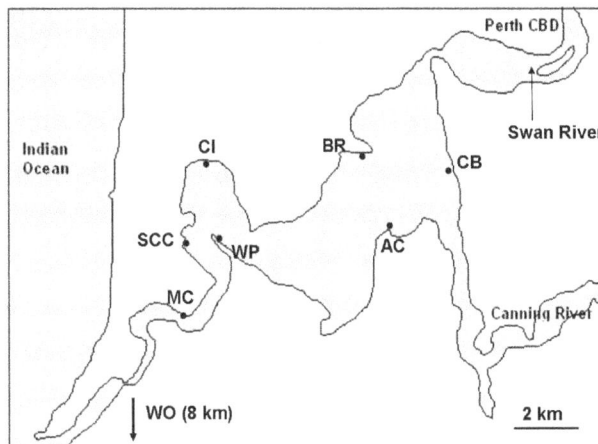

Figure 1. Map indicating the seven sampling sites (jetties) within the lower Swan River estuary, Perth, Western Australia. AC: Applecross, BRD: Broadway, CB: Como Beach, Cl: Claremont (Freshwater Bay), MC: Minim Cove, SCC: Swan River Canoe Club, WP: Point Walter; the ocean reference site was located 8 km south of the estuary mouth (WO: Woodman Jetty).

quantification of nutrient concentration (TP is total phosphorous; NO$_x$ is nitrate (NO$_3$) plus nitrite (NO$_2$); NH$_4^+$ is ammonium), phytoplankton biomass (as chlorophyll a), and stable isotope analysis of NO$_3$ (δ^{15}N, δ^{18}O) and particulate organic matter (POM; δ^{15}N) were taken from 10 to 20 cm below the surface and brought back to the laboratory in glass bottles that were stored on ice. Nine blue mussels per site were randomly taken from the pylons of the jetties at each site from between 20 and 40 cm depth and brought into the laboratory on ice in bags containing water from the respective site. There were no mussels at WP in November.

In the laboratory, total phytoplankton concentration at each site was measured with a bench top version of the FluoroProbe (bbe Moldaenke, Germany) as μg chl $a\,\text{L}^{-1}$ (Beutler et al., 2002; Ghadouani and Smith, 2005). Water for quantification of NO$_x$ (LOQ = 0.14 μM) and NH$_4^+$ (LOQ = 0.21 μM) concentrations was filtered through 0.45 μm syringe filters (Ht Tuffryn, Pall, Australia) and kept frozen until analysis at the Marine and Freshwater Research Laboratory (Murdoch University, Western Australia) using a Lachat QuikChem Flow Injection Analyser. Water for analysis of nitrate δ^{15}N and δ^{18}O was filtered through 0.2 μm syringe filters (Ht Tuffryn, Pall, Australia) and kept frozen until analysis at the UC Davis Stable Isotope Facility (Davis, California, USA) using a ThermoFinnigan GasBench plus PreCon trace gas concentration system interfaced to a ThermoScientific Delta V Plus isotope-ratio mass spectrometer (Bremen, Germany), with the bacteria denitrification method (Sigman et al., 2001). All values are reported in per mill (‰) with respect to the international standards (δ^{15}N: air; δ^{18}O: Vienna Standard Mean Ocean Water, VSMOW). The limit of quantification for this analysis was 0.71 μM NO$_3$-N and the external errors of analysis were 0.4‰ for nitrate $\delta^{15}N$ and 0.8‰ for nitrate $\delta^{18}O$. Raw water was used for quantification of TP with the ascorbic acid method (APHA, 1998).

To determine the isotopic composition of nitrogen in particulate organic matter (POM), which is the food source for mussels that presents the direct link between nitrate and the mussels, 0.7–2.5 L of water was filtered onto pre-combusted 25 mm GF/C filters (Whatman), which were then dried for 24 h at 60 °C and stored in a desiccator until analysis. Harvested mussels were measured and dissected to obtain the foot tissue for stable isotope analysis. The feet of three individuals per site were combined, dried at 60 °C for at least 24 h, fully homogenized with mortar/pestle, and stored in a desiccator until a subsample was analysed for mussel $\delta^{15}N$ and C : N ratio. As nine mussels per site were collected, this resulted in three replicates for stable isotope analysis per site, with each replicate comprised of the feet of three mussels. This method was adopted from Lancaster and Waldron (2001), as the minimum detectable difference between two populations was negatively associated with the number of replicate samples and the number of individual animals combined in each replicate. Therefore, this method is preferred when only small differences in the stable isotope signatures are expected. We used foot tissue for the analysis, because it is easy to identify and obtain, and because its $\delta^{15}N$ value presents a time-averaged value of $\delta^{15}N$ of the food source. Stable isotope analysis of mussel feet tissue and POM was performed at the West Australian Biogeochemistry Centre (University of Western Australia, Australia) with a continuous flow Delta V Plus mass spectrometer (connected with a Thermo Flush 1112 via Conflo IV) (ThermoFinnigan, Germany). All values are reported in per mill (‰) with respect to the international standard (air). The external errors of analysis were 0.10‰ for $\delta^{15}N$. To check whether the size of mussels was correlated with their $\delta^{15}N$, 13 mussels with shell lengths between 30 and 54 mm were sampled from MC in July.

2.3 Data processing and statistical analyses

Relationships between parameters (i.e. nutrient concentrations, physical parameters, chlorophyll a, stable isotope values), and distance to the estuary mouth were analysed with linear regressions. Differences between sites were analysed with one-way ANOVA or Kruskal–Wallis one-way ANOVA, in cases where the normality test failed (Sokal and Rohlf, 1995). If significant, the parametric Tukey (equal variances) or the non-parametric Games–Howell (non-equal variances) post hoc tests were used to identify which sites were different. The Mann–Whitney U test was used to compare chlorophyll a concentrations between high and low tide. All analyses were done with IBM® SPSS® Statistics 20 or Sigma Plot® Statistics 11.0, and significance level was set to $P < 0.05$ unless stated otherwise.

3 Results

3.1 Physicochemical parameters

Rainfall was below average in 2010, with 421 mm for the entire sampling period, while the average for this period was 690 mm in the previous 17 years (1993–2009; Bureau of Meteorology, 2016). This resulted in a lower-than-usual discharge from the tributaries into the estuary with a mean discharge from the Swan River of 7.5×10^5 m³ d⁻¹ in 2010 compared to an average discharge of 8.4×10^6 m³ d⁻¹ for the period of 1993–2009 for the same season (min–max: 1.99×10^6 m³ d⁻¹ (2002) – 2.21×10^7 m³ d⁻¹ (1996) (Department of Water, 2016)). This might have contributed to higher salinities throughout the entire estuary during this study than previously reported (Stephens and Imberger, 1997) and no relationship between salinity and distance to the estuary mouth was detected. During high tide, the salinity at all sites was between 24.2 and 32.4 and there was no difference in salinity between sites which can be considered brackish to saline (salinity of seawater is 35). Although salinity was not different between sites at low tide either, sites further away from the ocean (AC, CB, BRD) were entirely freshwater between April and June, while saline (mean ± SE; 27.4 ± 0.4) conditions prevailed at all sites between July and November. There were no differences between sites in temperature (temporal range 12.5–23 °C; Kruskal–Wallis $H = 0.584$, df $= 6$), dissolved oxygen (temporal range 6.4–11.6 mg L⁻¹, one-way ANOVA $F_{6,84} = 0.764$; 63–124 % $sat.$, one-way ANOVA $F_{6,84} = 0.515$), and pH (temporal range 6.7–8.4; one-way ANOVA $F_{6,112} = 0.163$). Total chlorophyll a concentration was between 1.4 and 9.5 µg L⁻¹ with a mean of 3.9 µg L⁻¹ (coefficient of variation $= 0.18$). Total chlorophyll a concentration was similar between sites (ANOVA; $F_{6,70} = 1.45$), and did not differ between low and high tide at any site (Mann–Whitney U test).

3.2 Nutrient concentrations

Overall, NO_x and NH_4^+ concentrations were low in the Swan River estuary. The concentration of NO_x ranged between below quantifiable limits (LOQ $= 0.14$ µM) and 15.0 µM (median 0.29; mean ± SD 0.72 ± 1.7), and differed significantly between sites (Kruskal–Wallis one-way ANOVA, $H = 50.03$, df $= 6$) (Fig. 2). The concentration of NH_4^+ ranged between the limit of quantification (LOQ $= 0.21$ µM) and 2.6 µM (median 0.78; mean ± SD $= 0.85 \pm 0.58$) and did not differ between sites (Kruskal–Wallis one-way ANOVA, $H = 7.9$, df $= 6$). On average, NO_x was the dominant N source at MC, SCC, and WO, while nitrogen from NH_4^+ was greater at all other sites (Fig. 2) (Kruskal–Wallis one-way ANOVA, $H = 59.0$, df $= 6$). Total phosphorous was below or just above the limit of quantification (LOQ $= 0.32$ µM) throughout the study and did not show any spatial or temporal trend.

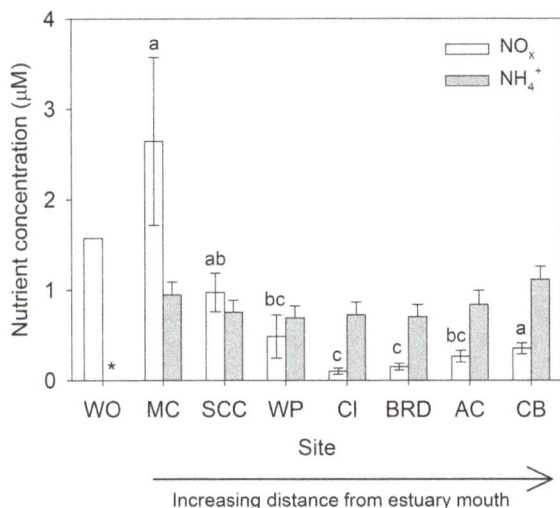

Figure 2. Mean concentration of NO_x and NH_4^+ (μM) at each site. Letters indicate differences between sites for NO_x concentrations, with sites sharing the same letter being not significantly different. Error bars represent one standard error ($N = 17$). Asterisk at WO indicates that mean value of NH_4 was below the limit of quantification.

The concentrations of total dissolved inorganic nitrogen (TDIN = NO_x + NH_4) (μM) and NO_x (μM) were higher towards the estuary mouth (Fig. 2), although these relationships were weak (TDIN: $r^2 = 0.113$, $y = -0.186x + 3.69$, $F_{1,117} = 14.86$; NO_x: $r^2 = 0.153$, $y = -0.196x + 2.98$, $F_{1,117} = 21.16$) and were driven by site MC only. Ammonium concentrations were not correlated with the distance from the estuary mouth ($F_{1,117} = 0.41$).

The TN : TP ratio (weight) of particulate organic matter was between 0 and 6.5 with 84 % of the samples being below 2.2 in our study, indicating a high possibility of nitrogen limitation in this system (Redfield, 1958; Geider and La Roche, 2002).

3.3　Stable isotope values of NO_3

Analysis of the stable isotope composition of NO_3 was limited to a total of 25 samples that fulfilled nutrient concentration requirements for the analysis (0.71 μM NO_3-N). Of these, 9 were from MC, 10 from SCC, 2 from AC, 3 from CB, and 1 from WP. Nitrate $\delta^{15}N$ values varied between -1.3 and 10.4 ‰, while nitrate $\delta^{18}O$ values ranged between 18.4 and 72.9 ‰. Nitrate $\delta^{15}N$ increased exponentially with increasing NO_x concentration ($F_{1,23} = 10.50$) (Fig. 3) and differed between sites (one-way ANOVA; $F_{4,25} = 5.94$). A post hoc test (Games–Howell) indicated that nitrate at MC was ^{15}N enriched (mean ± SD; 7.92 ‰ ±2.55; $n = 12$) compared to SCC (2.71 ‰ ±1.02; $n = 10$) and AC (-0.19 ‰ ±1.51; $n = 2$). There was no temporal trend in nitrate $\delta^{15}N$ at sites MC and SCC, respectively, which were the only two sites for which sufficient data for such an analysis were available.

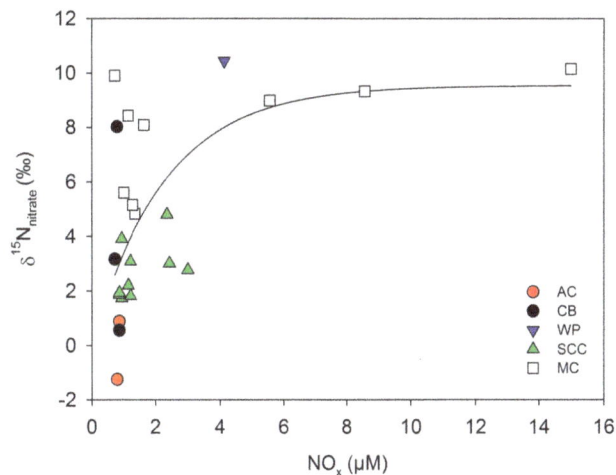

Figure 3. Relationship between nitrate $\delta^{15}N$ (‰) and the concentration of NO_x (μM) ($r^2 = 0.313$, $y = 9.54(1 - e^{-0.44x})$, $P < 0.05$).

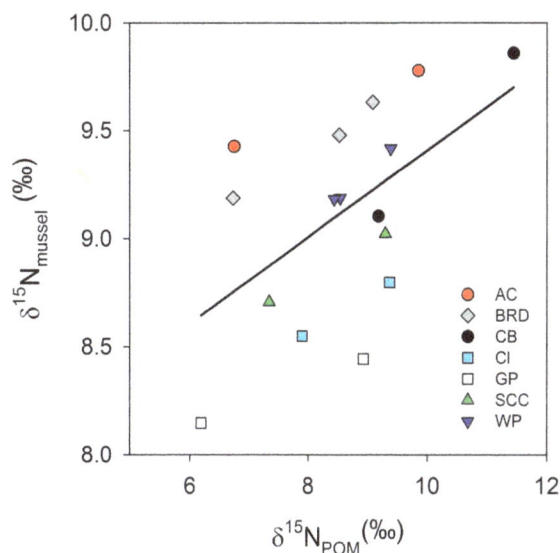

Figure 4. Relationship between mean $\delta^{15}N$ of POM and mussel (‰). Linear regression is calculated using all data points ($r^2 = 0.303$, $y = 0.20x + 7.40$, $F_{1,14} = 6.08$, $P < 0.05$).

Nitrate $\delta^{18}O$ was not significantly different between sites ($F_{4,25} = 0.059$).

3.4　Particulate organic matter (POM) $\delta^{15}N$ values

POM $\delta^{15}N$ values were between 6.2 and 9.9 ‰ with no significant difference between sites ($F_{6,25} = 1.327$). A weak but significant negative relationship between POM $\delta^{15}N$ values and TDIN concentration was detected ($r^2 = 0.163$, $y = -0.044x + 9.37$, $F_{1,28} = 5.44$), while a significant positive relationship between nitrogen stable isotope signatures of POM and mussels was found ($r^2 = 0.303$, $y = 0.20x + 7.40$, $F_{1,14} = 6.08$) (Fig. 4). The relationship be-

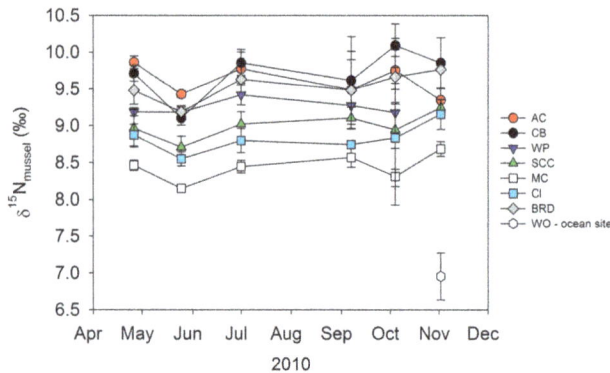

Figure 5. Mean δ^{15}N mussel signature (‰) at each site over time. Error bars represent standard deviations of $N = 3$ for April–July and WO, and $N = 2$ for September–November 2010.

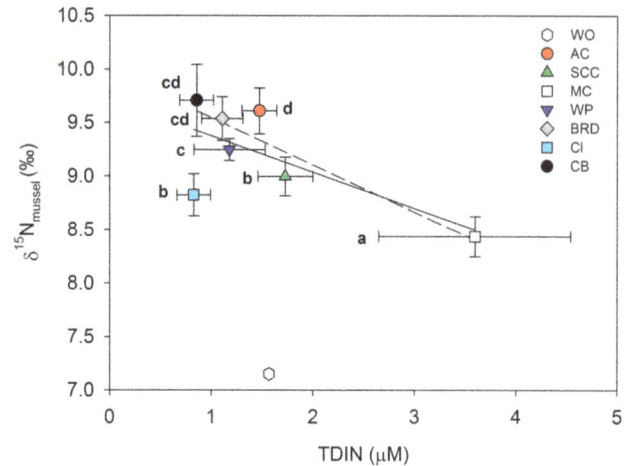

Figure 6. Relationship between mean mussel δ^{15}N (‰) and total dissolved inorganic nitrogen (TDIN) (µM). Error bars represent standard deviation for mussels ($N = 6$ for all sites except for WP where $N = 5$) and standard error of for TDIN ($n = 17$). The solid line represents the relationship calculated for all sites ($r^2 = 0.486$, $y = -0.338x + 9.71$), the dashed line when site Cl is omitted ($r^2 = 0.838$, $y = -0.440x + 9.98$). Letters indicate differences in δ^{15}N$_{mussels}$ (ANOVA with Games–Howell post hoc test), with sites sharing the same letter being not significantly different. WO was not included in the regressions.

tween δ^{15}N of POM and nitrate was not significant; however, as this calculation was based on only five data points where simultaneous measurements of the two δ^{15}N values were available, the value of this result is uncertain.

3.5 Mussel δ^{15}N values

Values of δ^{15}N of mussels varied between 6.8 and 10.3 ‰ and the range was therefore smaller than the range seen in nitrate δ^{15}N (-1.3 and 10.4 ‰). No significant relationship between mussel length and mussel δ^{15}N (linear regression; $F_{1,13} = 2.235$) and no temporal trend in mussel δ^{15}N was detected (Fig. 5). Mussel δ^{15}N was significantly different between sites (one-way ANOVA; δ^{15}N: $F_{6,98} = 42.53$) and was negatively correlated with the concentration of total dissolved inorganic nitrogen ($r^2 = 0.486$, $F_{1,5} = 4.73$, $P < 0.1$) (Fig. 6). When site Cl was omitted, the strength of the relationship increased ($r^2 = 0.838$, $F_{1,4} = 20.69$, $P < 0.05$), while the relationship was not significant with an r^2 of 0.009 only when site MC was omitted (Fig. 6). Mussel δ^{15}N increased significantly with distance from the estuary mouth ($r^2 = 0.563$, $y = 0.12x + 7.74$, $F_{1,110} = 141.65$) (Fig. 7) and showed a significant negative relationship between the δ^{15}N values of mussel and nitrate ($r^2 = 0.711$, $F_{2,10} = 24.65$) (Fig. 8).

4 Discussion

Urban development poses a major threat to aquatic ecosystems, resulting in a range of systems with different impact levels. The management of these waterbodies, whether they are historical, hybrid, or novel (Hobbs et al., 2014), requires a detailed knowledge of the complex interactions of processes in these systems. The limited understanding of spatial and temporal variabilities of pollutants is often the major limitation to successful and long-lasting restoration and protection efforts (Kooistra et al., 2001; Lahr and Kooistra, 2010). As

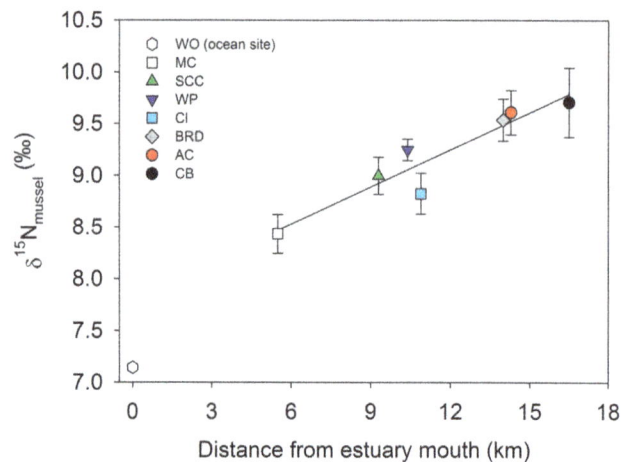

Figure 7. Relationship between mean δ^{15}N of mussels (‰) and distance of sites from estuary mouth ($r^2 = 0.563$, $y = 0.12x + 7.74$). Error bars represent standard deviation of $N = 6$ for all sites except for WP where $N = 5$.

such, it is essential to develop in-depth knowledge of local processes and pollution levels that will allow a decentralized management approach adapted to local issues (van de Meene et al., 2011).

Our study supports this notion by showing that the concentration of nitrates and the nitrogen stable isotope signatures of nitrate and of mussels were different between sites in the Swan River estuary. Site-specific differences in nutrient

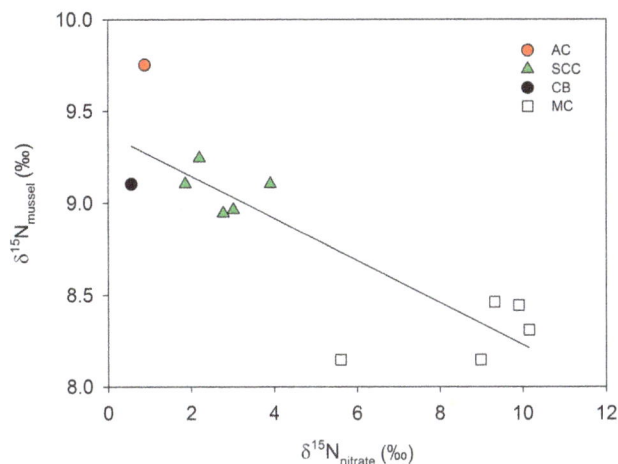

Figure 8. Relationship between nitrogen stable isotope signature of mussel and nitrate in the water ($r^2 = 0.711$, $y = -0.114x + 9.37$).

concentrations can be caused by local input of nutrients or by spatial differences in nutrient cycling caused by physicochemical conditions or biological factors (Michener and Lajtha, 2007). Additionally, nutrient input from the watershed often leads to higher nutrient concentrations upstream. During our study, freshwater input into the estuary was weak, leading to the estuary being mainly influenced by ocean water. This might have been the reason that nutrient concentrations did not increase upstream in our study and that nitrogen concentrations were low in general, leading to the conclusion that the Swan River estuary does not represent a highly impacted urban estuary. However, differences in NO_x and TDIN concentrations between sites suggested a significant site-specific input of nutrients into the Swan River estuary. This is supported by the fact that mean nitrogen concentrations at the site closest to the ocean (MC) were higher than the concentrations in the ocean (WO), pointing towards a local input of non-marine NO_x at MC.

Earlier studies indicated that the nitrogen stable isotope ratio of dissolved inorganic nitrogen was often higher at sites with high anthropogenic nitrogen pollution (Heaton, 1986; Cabana and Rasmussen, 1996). In the Swan River estuary, NO_3 was enriched and there was a positive relationship between nitrate $\delta^{15}N$ and the concentration of NO_x throughout the estuary, although this was strongly driven by site MC. Because the isotopic signatures of nitrates were well in the range of values reported for surface water (ca. -4 to $+9‰$; Xue et al., 2009), uncontaminated groundwater (ca. -1 to $+8‰$; Xue et al., 2009), organic nitrate from soils (0 to $+10‰$; Heaton, 1986), pristine streams ($+1.8$ to $+2.2‰$; Harrington et al., 1998), or naturally available marine-derived dissolved inorganic nitrogen (ca. 6–8‰; Dudley and Shima, 2010), our study does not suggest differences in the level of human impact between sites. Additionally, nitrate $\delta^{18}O$ values in our study are similar to values in-

dicative of the atmospheric source ($+20$ to $+80‰$; Kendall, 1998; Xue et al., 2009), suggesting that the higher concentration and enriched signature of NO_x at site MC is unlikely to result from anthropogenic pollution, but might rather be due to addition of NO_x by groundwater inflow, potentially in combination with different productivity or biochemical processes at this site compared to any of the other sites. Overall, results from the stable isotope analysis in combination with nitrogen concentrations indicate that anthropogenic nutrient pollution in the Swan River estuary is low.

The fraction of NO_x of the TDIN pool (%) was significantly different between sites (data not shown; $y = 0.15x - 6.9$, $r^2 = 0.215$, $F_{1,23} = 6.30$, $P < 0.05$), with site MC having a higher mean fraction (mean $= 62.5\%$) compared to all other sites, except for SCC. An earlier study by Sugimoto et al. (2009) also found a positive relationship between nitrate $\delta^{15}N$ values and the nitrate fraction in TDIN, which they explained by in situ isotopic effects during nitrification. However, whether higher $\delta^{15}N$ values of nitrate at MC are related to site-specific nitrification rates in our estuary needs further investigation, as the $\delta^{18}O$ and $\delta^{15}N$ values of nitrate are rather representative of atmospheric NO_3 deposition values (Durka et al., 1994; Fang et al., 2011) and nitrification is likely to play a minor role at ammonium concentrations $< 5\,\mu M$ (Day et al., 1989) that prevail in the Swan River estuary.

Earlier studies found that nitrogen $\delta^{15}N$ values are reflected in higher trophic levels in a predictable way, with a positive relationship between $\delta^{15}N$ of nitrate, primary producer and primary consumer (e.g. mussels) (Cabana et al., 1994; Cabana and Rasmussen, 1996; Harrington et al., 1998; Oczkowski et al., 2008; Carvalho et al., 2015). In addition and identical to our study, the range of $\delta^{15}N$ values for nitrate and POM has been shown to be wider than the range for primary producers, indicating a time-averaging effect in mussels (Gustafson et al., 2007; Wang et al., 2013). Previous studies reported mussel $\delta^{15}N$ values between $+6.6$ and $+16.7‰$ in densely populated areas (Cabana and Rasmussen, 1996), polluted inland waterbodies (Wen et al., 2010; Wang et al., 2013), and a eutrophic estuary (Fry et al., 2011). Our values are at the lower end of this range, with mussel $\delta^{15}N$ values in our study being between 6.8 and 10.3‰, indicating the estuary is not highly polluted by wastewater, agriculture or fertilizers. We also found a positive relationship between food (POM) and mussel $\delta^{15}N$, but a negative relationship between nitrate $\delta^{15}N$ and consumers (mussels), which was strongly affected by site MC. Such negative relationships were previously found in systems with very high nitrogen concentrations (DIN $> 40\,\mu M$) (Oczkowski et al., 2008), because in these systems primary producers can be choosy and will preferentially uptake lighter NO_x, leading to a higher fractionation at higher concentrations (Lake et al., 2001; Oczkowski et al., 2008). Therefore, the residual NO_x in those waters retains more ^{15}N-enriched material, leading to a positive relationship between nitrogen concentration and nitrate $\delta^{15}N$, while consumers which incorporate primary producers will have a lighter signature. Because such fractionation is unlikely at TDIN concentrations below 1 µM (Oczkowski et al., 2008), this mechanism is unlikely for most of our sites where mean TDIN concentration was < 1.5 µM. This is also supported by the lack of relationship between mussel $\delta^{15}N$ and TDIN concentration when omitting MC. However, we cannot rule out that this process contributed partially to the low mussel $\delta^{15}N$ values detected at MC, as TDIN concentrations were higher at this site with a mean of 3.6 µM. An alternative explanation would be that POM could originate upstream where nitrate might have had higher $\delta^{15}N$ values (not quantified in this study). Upon entering the estuary, POM mixes with estuarine POM, uncoupling the within-estuary $\delta^{15}N$ nitrate and POM $\delta^{15}N$ values. This could also explain the strong relationship between $\delta^{15}N$ in mussels and the distance from the estuary mouth found in our study. Such a strong relationship can be expected in estuaries with low pollution levels due to the aforementioned mixing, while little spatial variability in $\delta^{15}N$ values of primary consumers can be expected in heavily polluted estuaries due to the dominance of upstream POM, as was shown by Oczkowski et al. (2008).

The relationship between mussel $\delta^{15}N$ and TDIN concentration within the estuary was much stronger when omitting site Cl and not significant when omitting site MC. Site Cl was the shallowest site with a high density of macroalgae and seagrass. These benthic primary producers are known to incorporate nutrients from the groundwater and pore water (Pennifold and Davis, 2001). As pore water in the Swan River estuary contains a high concentration of ammonium (Linderfelt and Turner, 2001), this is taken up by the benthic primary producers, and, when recycled, nitrogen with a different $\delta^{15}N$ value is released into the water column. Therefore, nitrogen $\delta^{15}N$ in the water column at this site is likely to differ from that of all other sites, which could explain why mussel $\delta^{15}N$ values at Cl do not fit the general negative relationship. Due to constantly low nitrate concentration at this site, the stable isotope composition of nitrate could not be tested in our study. Site MC was closest to the ocean, was one of the deepest sites, and had a higher TDIN concentration compared to all other sites, which in turn did not show differences in TDIN concentrations between them. This emphasizes that the differences in mussel $\delta^{15}N$ between sites detected in our estuary might rather reflect site-specific nutrient cycling processes than nitrogen pollution itself.

Fluctuation of mussel $\delta^{15}N$ at each site over time was low compared to the differences between sites, indicating that observed differences between sites prevailed and were not obscured by time effects. This is important for assessing site-specific processes and source inputs and highlights that mussels can be used as time-integrated sentinel organisms in urban systems. The limited temporal variation likely reflected the physiochemical state of the system during the study period; in our study, the estuary was dominated by marine in-

fluences due to reduced river discharge. This might have further resulted in a dampening effect of possible fluctuations of the nitrate $\delta^{15}N$ value caused by changes in watershed input. Our results therefore highlight that while high seasonal variations of stable isotope signature in mussels can be connected to seasonal changes in watershed input and chemistry in large rivers (Fry and Allen, 2003), this is less pronounced in tidally influenced estuaries or during drier conditions with low freshwater input.

5   Conclusion

The findings of our study corroborate that stable isotope analysis is a valuable tool for identifying spatial variability of nutrient pollution and local processes in an urban, tidally influenced estuary. As such, stable isotope analysis of a model organism, such as the blue mussel, can deliver essential information for future decentralized water management practices that are focused on local process understanding. We propose to further investigate its use for assessing the pollution by co-occurring, non-nutrient pollutants, such as oils and heavy metals, which are entering waterbodies simultaneously with nutrients during stormwater events and which are critical in urban systems.

Based on nutrient concentrations and stable isotope analysis, our data provide detailed evidence that the lower Swan River estuary did not present a highly impacted urban estuary during our study. The nitrate stable isotope signature in the water suggested that the higher concentrations of nitrate at two sites (MC, SCC) were due to natural input of nitrate uncontaminated groundwater (Xue et al., 2009) rather than human pollution. The stable spatial differences in mussel $\delta^{15}N$ values over time highlight the value of this organism as a bioindicator of spatial water quality assessment. The negative trends between mussel $\delta^{15}N$ values and nitrate concentration or nitrate $\delta^{15}N$ values emphasize that mussels might not be good indicators for NO_3 sources in systems with low pollution levels. Instead, the small differences in mussel stable isotope signatures might reflect differences in site-specific nutrient cycling caused by physicochemical conditions or biological factors rather than nitrogen pollution. This is important information for local management, but would have gone undetected at high pollution levels as the larger deviations of nitrogen stable isotope values would have made such small differences in mussel values invisible. In addition, we advocate future studies in similarly (low) polluted systems that include stable isotope analysis of other food web end-members and nutrients of the groundwater, to develop baselines of spatial natural isotopic variability in urban aquatic systems which will help identifying the importance of local biogeochemical processes for pollution control.

In conclusion, this work shows the value of using stable isotope analysis as an integrative tool to establish an understanding of local processes and pollution levels in aquatic systems. With an increasing importance of managing urban aquatic systems sustainably, our work presents an important proof-of-concept study in this context. In addition, we propose that it could help to define divisions in tidal estuaries based on natural characteristics and the human dimension that are meaningful for monitoring and management and for which reference conditions have to be identified (Ferreira et al., 2006).

Acknowledgements. This study was supported by a Research Development Award (2009) from The University of Western Australia to Elke S. Reichwaldt and by an Australian Research Council Linkage Project (LP0776571) and the Water Corporation of Western Australia. The authors would like to thank S. C. Sinang, H. Song, and L. X. Coggins for help in the field and in the laboratory; L. X. Coggins for editing an earlier version of the manuscript; and C. Harrod for valuable help during the preparation of the manuscript. The permit for sampling mussels was obtained from the Department of Environment and Conservation, Western Australia (Licence no. SF007464). Discharge data were courtesy of the Department of Water, Western Australia.

Edited by: A. Guadagnini

References

Almadavillela, P. C.: The effects of reduced salinity on the shell growth of small *Mytilus Edulis*, J. Mar. Biol. Assoc. UK, 64, 171–182, 1984.

Anderson, C. and Cabana, G.: $\delta^{15}N$ in riverine food webs: effects of N inputs from agricultural watersheds, Can. J. Fish. Aquat. Sci., 62, 333–340, 2005.

APHA: Standard methods for the examination of water and wastewater, 20th Edn., edited by: Greenberg, A. E., American Public Health Association, Washington, DC, 1998.

Atkins, R., Rose, T., Brown, R. S., and Robb, M.: The *Microcystis* cyanobacteria bloom in the Swan River – February 2000, Water Sci. Technol., 43, 107-114, 2001.

Atkins, R. P. and Klemm, V. V.: The effect of discharges, effluent and urbanisation on the Swan River, in: Swan River Estuary: Ecology and Management, edited by: John, J., Curtin University of Technology, Perth, Australia, Environmental Studies Group, Report no. 1, 296–313, 1987.

Beutler, M., Wiltshire, K. H., Meyer, B., Moldaenke, C., Lüring, C., Meyerhöfer, M., Hansen, U. P., and Dau, H.: A fluorometric method for the differentiation of algal populations in vivo and in situ, Photosynth. Res., 72, 39–53, 2002.

Brezonik, P. L. and Stadelmann, T. H.: Analysis and predictive models of stormwater runoff volumes, loads, and pollutant concentrations from watersheds in the Twin Cities metropolitan area, Minnesota, USA, Water Res., 36, 1743–1757, doi:10.1016/s0043-1354(01)00375-x, 2002.

Broderius, C.: Anthropogenically altered land and its effect on $\delta^{15}N$ values in periphyton on a fourth order stream in Utah's Cache Valley, Nat. Resour. Env. Iss., 18, 61–69, 2013.

Bureau of Meteorology: Climate statistics for Australian locations Monthly climate statistics, Australian Government, http://www.

bom.gov.au/climate/averages/tables/cw_009034.shtml (last access: 3 March 2016), 2016.

Cabana, G. and Rasmussen, J. B.: Modeling food-chain structure and contaminant bioaccumulation using stable nitrogen isotopes, Nature, 372, 255–257, 1994.

Cabana, G. and Rasmussen, J. B.: Comparison of aquatic food chains using nitrogen isotopes, P. Natl. Acad. Sci. USA, 93, 10844–10847, 1996.

Cabana, G., Tremblay, A., Kalff, J., and Rasmussen, J. B.: Pelagic food-chain structure in Ontario Lakes – a determinant of mercury levels in Lake Trout (*Salvelinus-Namaycush*), Can. J. Fish. Aquat. Sci., 51, 381–389, 1994.

Carvalho, D. R., Castro, D., Callisto, M., Moreira, M. Z., and Pompeu, P. S.: Isotopic variation in five species of stream fishes under the influence of different land uses, J. Fish Biol., 87, 559–578, 10.1111/jfb.12734, 2015.

Chaves, M. M., Maroco, J. P., and Pereira, J. S.: Understanding plant responses to drought – from genes to the whole plant, Funct. Plant Biol., 30, 239–264, doi:10.1071/fp02076, 2003.

Choi, W. J., Han, G. H., Lee, S. M., Lee, G. T., Yoon, K. S., Choi, S. M., and Ro, H. M.: Impact of land-use types on nitrate concentration and $\delta^{15}N$ in unconfined groundwater in rural areas of Korea, Agr. Ecosyst. Environ., 120, 259–268, doi:10.1016/j.agee.2006.10.002, 2007.

Clay, A., Bradley, C., Gerrard, A. J., and Leng, M. J.: Using stable isotopes of water to infer wetland hydrological dynamics, Hydrol. Earth Syst. Sci., 8, 1164–1173, doi:10.5194/hess-8-1164-2004, 2004.

Costanzo, S. D., O'Donohue, M. J., Dennison, W. C., Loneragan, N. R., and Thomas, M.: A new approach for detecting and mapping sewage impacts, 42, 149–156, 2001.

Dähnke, K., Emeis, K., Johannsen, A., and Nagel, B.: Stable isotope composition and turnover of nitrate in the German Bight, Mar. Ecol.-Prog. Ser., 408, 7-U26, doi:10.3354/Meps08558, 2010.

Day, J. W. J., Hall, C. A. S., Kemp, W. M., and Yanez-Arancibia, A.: Estuarine chemistry, in: Estuarine Ecology, edited by: Day, J. W. J., Hall, C. A. S., Kemp, W. M., and Yanez-Arancibia, A., John Wiley & Sons, New York, 1989.

Department of Water: Macrophytes and macroalgae in the Swan-Canning Estuary (Volume 20), Department of Water, Perth, Australia, Perth, Australia, 2010.

Department of Water: River monitoring stations, Government of Western Australia, available at: http://kumina.water.wa.gov.au/waterinformation/telem/stage.cfm (last access: 4 March 2016), 2016.

Dudley, B. D. and Shima, J. S.: Algal and invertebrate bioindicators detect sewage effluent along the coast of Titahi Bay, Wellington, New Zealand, New Zeal. J. Mar. Fresh., 44, 39–51, doi:10.1080/00288331003641687, 2010.

Durka, W., Schulze, E. D., Gebauer, G., and Voerkelius, S.: Effects of forest decline on uptake and leaching of deposited nitrate determined from ^{15}N and ^{18}O measurements, Nature, 372, 765–767, doi:10.1038/372765a0, 1994.

Fang, Y. T., Koba, K., Wang, X. M., Wen, D. Z., Li, J., Takebayashi, Y., Liu, X. Y., and Yoh, M.: Anthropogenic imprints on nitrogen and oxygen isotopic composition of precipitation nitrate in a nitrogen-polluted city in southern China, Atmos. Chem. Phys., 11, 1313–1325, doi:10.5194/acp-11-1313-2011, 2011.

Ferreira, J. G., Nobre, A. M., Sirnas, T. C., Silva, M. C., Newton, A., Bricker, S. B., Wolff, W. J., Stacey, P. E., and Sequeira, A.: A methodology for defining homogeneous water bodies in estuaries – Application to the transitional systems of the EU Water Framework Directive, Estuar. Coast. Shelf S., 66, 468–482, doi:10.1016/j.ecss.2005.09.016, 2006.

Fertig, B., Carruthers, T. J. B., Dennison, W. C., Fertig, E. J., and Altabet, M. A.: Eastern oyster (*Crassostrea virginica*) delta(15)N as a bioindicator of nitrogen sources: Observations and modeling, Mar. Pollut. Bull., 60, 1288–1298, doi:10.1016/j.marpolbul.2010.03.013, 2010.

Fry, B. and Allen, Y. C.: Stable isotopes in zebra mussels as bioindicators of river-watershed linkages, River Res. Appl., 19, 683–696, 2003.

Fry, B., Rogers, K., Barry, B., Barr, N., and Dudley, B.: Eutrophication indicators in the Hutt River Estuary, New Zeal. J. Mar. Fresh., 45, 665–677, doi:10.1080/00288330.2011.578652, 2011.

Fukumori, K., Oi, M., Doi, H., Takahashi, D., Okuda, N., Miller, T. W., Kuwae, M., Miyasaka, H., Genkai-Kato, M., Koizumi, Y., Omori, K., and Takeoka, H.: Bivalve tissue as a carbon and nitrogen isotope baseline indicator in coastal ecosystems, Estuar. Coast. Shelf'S., 79, 45–50, 2008.

Geider, R. J., and La Roche, J.: Redfield revisited: variability of $C:N:P$ in marine microalgae and its biochemical basis, Eur. J. Phycol., 37, 1–17, doi:10.1017/s0967026201003456, 2002.

Ghadouani, A. and Smith, R. E. H.: Phytoplankton distribution in Lake Erie as assessed by a new in situ spectrofluorometric technique, J. Great Lakes Res., 31, 154–167, 2005.

Gustafson, L., Showers, W., Kwak, T., Levine, J., and Stoskopf, M.: Temporal and spatial variability in stable isotope compositions of a freshwater mussel: implications for biomonitoring and ecological studies, Oecologia, 152, 140–150, 2007.

Hamilton, D. P.: Record summer rainfall induced first recorded major cyanobacterial bloom in the Swan River, The Environmental Engineer, 1, 25 pp., 2000.

Harrington, R. R., Kennedy, B. P., Chamberlain, C. P., Blum, J. D., and Folt, C. L.: ^{15}N enrichment in agricultural catchments: field patterns and applications to tracking Atlantic salmon (*Salmo salar*), Chem. Geol., 147, 281–294, doi:10.1016/S0009-2541(98)00018-7, 1998.

Heathwaite, A. L.: Multiple stressors on water availability at global to catchment scales: understanding human impact on nutrient cycles to protect water quality and water availability in the long term, Freshwater Biol., 55, 241–257, 2010.

Heaton, T. H. E.: Isotopic studies of nitrogen pollution in the hydrosphere and atmosphere: A review, Chem. Geol., 59, 87–102, 1986.

Hobbs, R. J., Higgs, E., Hall, C. M., Bridgewater, P., Chapin, F. S., Ellis, E. C., Ewel, J. J., Hallett, L. M., Harris, J., Hulvey, K. B., Jackson, S. T., Kennedy, P. L., Kueffer, C., Lach, L., Lantz, T. C., Lugo, A. E., Mascaro, J., Murphy, S. D., Nelson, C. R., Perring, M. P., Richardson, D. M., Seastedt, T. R., Standish, R. J., Starzomski, B. M., Suding, K. N., Tognetti, P. M., Yakob, L., and Yung, L.: Managing the whole landscape: historical, hybrid, and novel ecosystems, Front. Ecol. Environ., 12, 557–564, doi:10.1890/130300, 2014.

Kellman, L. M.: A study of tile drain nitrate – δ^{15}N values as a tool for assessing nitrate sources in an agricultural region, Nutr.

Cycl. Agroecosys., 71, 131–137, doi:10.1007/s10705-004-1925-0, 2005.

Kendall, C.: Tracing nitrogen sources and cycling in catchments, in: Isotope tracers in catchment hydrology, edited by: Kendall, C. and McDonnell, J. J., Elsevier Science B.V., Amsterdam, 1998.

Kooistra, L., Leuven, R. S. E. W., Nienhuis, P. H., Wehrens, R., and Buydens, L. M. C.: A procedure for incorporating spatial variability in ecological risk assessment of Dutch River floodplains, Environ. Manage., 28, 359–373, doi:10.1007/S0026702433, 2001.

Lahr, J. and Kooistra, L.: Environmental risk mapping of pollutants: State of the art and communication aspects, Sci. Total Environ., 408, 3899–3907, doi:10.1016/j.scitotenv.2009.10.045, 2010.

Lake, J. L., McKinney, R. A., Osterman, F. A., Pruell, R. J., Kiddon, J., Ryba, S. A., and Libby, A. D.: Stable nitrogen isotopes as indicators of anthropogenic activities in small freshwater systems, Can. J. Fish. Aquat. Sci., 58, 870–878, 2001.

Lancaster, J. and Waldron, S.: Stable isotope values of lotic invertebrates: Sources of variation, experimental design, and statistical interpretation, Limnol. Oceanogr., 46, 723–730, 2001.

Linderfelt, W. R. and Turner, J. V.: Interaction between shallow groundwater, saline surface water and nutrient discharge in a seasonal estuary: the Swan-Canning system, Hydrol. Process., 15, 2631–2653, 2001.

Luedeking, A. and Koehler, A.: Regulation of expression of multi-xenobiotic resistance (MXR) genes by environmental factors in the blue mussel *Mytilus edulis*, Aquat. Toxicol., 69, 1–10, 2004.

Lutz, S. R., van Meerveld, H. J., Waterloo, M. J., Broers, H. P., and van Breukelen, B. M.: A model-based assessment of the potential use of compound-specific stable isotope analysis in river monitoring of diffuse pesticide pollution, Hydrol. Earth Syst. Sci., 17, 4505–4524, doi:10.5194/hess-17-4505-2013, 2013.

Mainwaring, K., Tillin, H., and Tyler-Walters, H.: Assessing the sensitivity of blue mussels (*Mytilus edulis*) to pressures associated with human activities, JNCC Report No 506, Joint Nature Conservation Committee, Petersborough, UK, 2014.

Makepeace, D. K., Smith, D. W., and Stanley, S. J.: Urban stormwater quality – summary of contaminant data, Crit. Rev. Env. Sci. Tec., 25, 93–139, 1995.

McClelland, J. W., Valiela, I., and Michener, R. H.: Nitrogen-stable isotope signatures in estuarine food webs: A record of increasing urbanization in coastal watersheds, Limnol. Oceanogr., 42, 930–937, 1997.

McKinney, R. A., Lake, J. L., Charpentier, M. A., and Ryba, S.: Using mussel isotope ratios to assess anthropogenic nitrogen inputs to freshwater ecosystems, Environ. Monit. Assess., 74, 167–192, 2002.

Michener, R. and Lajtha, K.: Stable isotopes in ecology and environmental science, 2nd Edn., Blackwell Publishing Ldt., Malden, USA, 2007.

Oczkowski, A., Nixon, S., Henry, K., DiMilla, P., Pilson, M., Granger, S., Buckley, B., Thornber, C., McKinney, R., and Chaves, J.: Distribution and trophic importance of anthropogenic nitrogen in Narragansett Bay: An assessment using stable isotopes, Estuar. Coast, 31, 53–69, doi:10.1007/s12237-007-9029-0, 2008.

Pace, M. L., Cole, J. J., Carpenter, S. R., Kitchell, J. F., Hodgson, J. R., Van de Bogart, M. C., Bade, D. L., Kritzberg, E. S., and

Bastviken, D.: Whole-lake carbon-13 additions reveal terrestrial support of aquatic food webs, Nature, 427, 240–243, 2004.

Pahl-Wostl, C.: Transitions towards adaptive management of water facing climate and global change, Water Resour. Manag., 21, 49–62, doi:10.1007/s11269-006-9040-4, 2007.

Pennifold, M. and Davis, J.: Macrofauna and nutrient cycling in the Swan River Estuary, Western Australia: experimental results, Hydrol. Process., 15, 2537–2553, 2001.

Peters, N. E. and Donohue, R.: Nutrient transport to the Swan-Canning Estuary, Western Australia, Hydrol. Process., 15, 2555–2577, 2001.

Phillips, D. J. H.: Common mussel Mytilus edulis as and indicator of pollution by zinc, cadmium, lead and copper. 1. Effects of environmental variables on uptake of metals, Mar. Biol., 38, 59–69, 1976.

Pickett, S. T. A., Cadenasso, M. L., Grove, J. M., Boone, C. G., Groffman, P. M., Irwin, E., Kaushal, S. S., Marshall, V., McGrath, B. P., Nilon, C. H., Pouyat, R. V., Szlavecz, K., Troy, A., and Warren, P.: Urban ecological systems: Scientific foundations and a decade of progress, J. Environ. Manage., 92, 331–362, doi:10.1016/j.jenvman.2010.08.022, 2011.

Post, D. M.: Using stable isotopes to estimate trophic position: Models, methods, and assumptions, Ecology, 83, 703–718, 2002.

Raikow, D. F. and Hamilton, S. K.: Bivalve diets in a midwestern U.S. stream: A stable isotope enrichment study, Limnol. Oceanogr., 46, 514–522, 2001.

Redfield, A. C.: The biological control of chemical factors in the environment, Am. Sci., 46, 205–221, 1958.

Robinson, D.: δ15N as an integrator of the nitrogen cycle, Trends Ecol. Evol., 16, 153–162, doi:10.1016/s0169-5347(00)02098-x, 2001.

Rodgers, P., Soulsby, C., Waldron, S., and Tetzlaff, D.: Using stable isotope tracers to assess hydrological flow paths, residence times and landscape influences in a nested mesoscale catchment, Hydrol. Earth Syst. Sci., 9, 139–155, doi:10.5194/hess-9-139-2005, 2005.

Scheffer, M., Carpenter, S., Foley, J. A., Folke, C., and Walker, B.: Catastrophic shifts in ecosystems, Nature, 413, 591–596, 2001.

Sigman, D. M., Casciotti, K. L., Andreani, M., Barford, C., Galanter, M., and Bohlke, J. K.: A bacterial method for the nitrogen isotopic analysis of nitrate in seawater and freshwater, Anal. Chem., 73, 4145–4153, 2001.

Sikdar, P. K. and Sahu, P.: Understanding wetland sub-surface hydrology using geologic and isotopic signatures, Hydrol. Earth Syst. Sci., 13, 1313–1323, doi:10.5194/hess-13-1313-2009, 2009.

Sokal, R. R. and Rohlf, F. J.: Biometry: The principles and practices of statistics in biological research, 3rd Edn., W. H. Freeman, New York, 1995.

Stephens, R. and Imberger, J.: Dynamics of the Swan River estuary: The seasonal variability, Mar. Freshwater Res., 47, 517–529, 1996.

Stephens, R. and Imberger, J.: Intertidal motions within deep basin of Swan River estuary, J. Hydraul. Eng.-ASCE, 123, 863-873, 1997.

Sugimoto, R., Kasai, A., Miyajima, T., and Fujita, K.: Controlling factors of seasonal variation in the nitrogen isotope ratio of nitrate in a eutrophic coastal environment, Estuar. Coast. Shelf S., 85, 231–240, 2009.

Thompson, R. J. and Bayne, B. L.: Some relationships between growth, metabolism and food in mussel Mytilus edulis, Mar. Biol., 27, 317–326, 1974.

United Nations: World Economic and Social Survey 2013 – Sustainable Development Challenges, New York, USA, 2013.

van de Meene, S. J., Brown, R. R., and Farrelly, M. A.: Towards understanding governance for sustainable urban water management, Global Environ. Chang., 21, 1117–1127, doi:10.1016/j.gloenvcha.2011.04.003, 2011.

Volkmann, T. H. M. and Weiler, M.: Continual in situ monitoring of pore water stable isotopes in the subsurface, Hydrol. Earth Syst. Sci., 18, 1819–1833, doi:10.5194/hess-18-1819-2014, 2014.

Wang, Y., Yu, X., Zhang, L., and Lei, G.: Seasonal variability in baseline delta N-15 and usage as a nutrient indicator in Lake Poyang, China, J. Freshwater Ecol., 28, 365–373, doi:10.1080/02705060.2013.763296, 2013.

Wen, Z. R., Xie, P., and Xu, J.: Mussel isotope signature as indicator of nutrient pollution in a freshwater eutrophic lake: species, spatial, and seasonal variability, Environ. Monit. Assess., 163, 139–147, doi:10.1007/s10661-009-0823-y, 2010.

Xu, J. and Zhang, M.: Primary consumers as bioindicator of nitrogen pollution in lake planktonic and benthic food webs, Ecol. Indic., 14, 189–196, doi:10.1016/j.ecolind.2011.02.012, 2012.

Xue, D., Botte, J., De Baets, B., Accoe, F., Nestler, A., Taylor, P., Van Cleemput, O., Berglund, M., and Boeckx, P.: Present limitations and future prospects of stable isotope methods for nitrate source identification in surface- and groundwater, Water Res., 43, 1159–1170, doi:10.1016/j.watres.2008.12.048, 2009.

Yang, L., Song, X., Zhang, Y., Han, D., Zhang, B., and Long, D.: Characterizing interactions between surface water and groundwater in the Jialu River basin using major ion chemistry and stable isotopes, Hydrol. Earth Syst. Sci., 16, 4265–4277, doi:10.5194/hess-16-4265-2012, 2012.

Widdows, J., Fieth, P., and Worrall, C. M.: Relationships between seston, available food and feeding activity in the common mussel Mytilus edulis, Mar. Biol., 50, 195-207, 1979.

Zandee, D. I., Kluytmans, J. H., Zurburg, W., and Pieters, H.: Seasonal variations in biochemical composition of Mytilus edulis with reference to energy metabolism and gametogenesis, Neth. J. Sea Res., 14, 1–29, 1980.

14

Local nutrient regimes determine site-specific environmental triggers of cyanobacterial and microcystin variability in urban lakes

S. C. Sinang[1,*], E. S. Reichwaldt[1], and A. Ghadouani[1]

[1] Aquatic Ecology and Ecosystem Studies, School of Civil, Environmental and Mining Engineering, The University of Western Australia, 35 Stirling Highway, M015, Crawley, WA 6009, Western Australia, Australia

[*] present address: Faculty of Science and Mathematics, Sultan Idris Education University, 35900 Tanjong Malim, Perak, Malaysia

Correspondence to: A. Ghadouani (anas.ghadouani@uwa.edu.au)

Abstract. Toxic cyanobacterial blooms in urban lakes present serious health hazards to humans and animals and require effective management strategies. Managing such blooms requires a sufficient understanding of the controlling environmental factors. A range of them has been proposed in the literature as potential triggers for cyanobacterial biomass development and cyanotoxin (e.g. microcystin) production in freshwater systems. However, the environmental triggers of cyanobacteria and microcystin variability remain a subject of debate due to contrasting findings. This issue has raised the question of whether the relevance of environmental triggers may depend on site-specific combinations of environmental factors. In this study, we investigated the site-specificity of environmental triggers for cyanobacterial bloom and microcystin dynamics in three urban lakes in Western Australia. Our study suggests that cyanobacterial biomass, cyanobacterial dominance and cyanobacterial microcystin content variability were significantly correlated to phosphorus and iron concentrations. However, the correlations were different between lakes, thus suggesting a site-specific effect of these environmental factors. The discrepancies in the correlations could be explained by differences in local nutrient concentration. For instance, we found no correlation between cyanobacterial fraction and total phosphorous (TP) in the lake with the highest TP concentration, while correlations were significant and negative in the other two lakes. In addition, our study indicates that the difference of the correlation between total iron (TFe) and the cyanobacterial fraction between lakes might have been a consequence of differences in the cyanobacterial community structure, specifically the presence or absence of nitrogen-fixing species. In conclusion, our study suggests that identification of significant environmental factors under site-specific conditions is an important strategy to enhance successful outcomes in cyanobacterial bloom control measures.

1 Introduction

Urban lakes often serve as recreational spaces for communities and habitats for wildlife (Yan et al., 2012; Liu, 2014). To date, many urban lakes continue to deteriorate due to increased anthropogenic activities and often face water quality problems including toxic cyanobacteria blooms (Pineda-Mendoza et al., 2012; Reichwaldt and Ghadouani, 2012; Lei et al., 2014; Sun et al., 2014; Zhang et al., 2014). This issue has received great attention from water authorities worldwide as it presents health hazards to humans and animals who either directly or indirectly received services provided by urban lakes (O'Bannon et al., 2014; Rastogi et al., 2014; Waajen et al., 2014). The management of toxic cyanobacterial blooms is often challenging due to the variability in cyanobacteria biomass and microcystins (Rolland et al., 2013; Carey et al., 2014). In addition, microcystin production by cyanobacteria is a complex issue that might depend on their competition with other phytoplankton (e.g. Huisman and Hulot, 2005; Jang et al., 2006). From these earlier studies it can be concluded that the toxin concentration produced by a certain cyanobacterial biomass level might differ, depending on the level of competition (i.e. cyanobacterial fraction) indicating that management should consider biomass and cyanobacterial fractions concurrently.

Cyanobacterial biomass and the amount of microcystins being produced during toxic cyanobacterial blooms can vary significantly on a spatial basis within and between lakes (Reichwaldt et al., 2013; Sinang et al., 2013; Thi Thuy et al., 2014; Waajen et al., 2014). Past studies have found large variations in the percentage of potentially toxic cyanobacteria and in the microcystin concentration between spatially isolated phytoplankton communities (Sitoki et al., 2012; Li et al., 2014). Furthermore, it was reported that the variability of cyanobacterial biomass in lakes only explained a small fraction of the variability in microcystin concentration (Sinang et al., 2013; Eva and Lindsay, 2014). These findings highlight the importance of fully understanding the roles of environmental factors controlling both the cyanobacteria and the microcystin variability.

It has been suggested that cyanobacterial biomass and microcystin variability largely depends upon physical, chemical and biological properties of the water bodies (Engström-Öst et al., 2013; Lehman et al., 2013; Paerl and Otten, 2013; Ruiz et al., 2013). A range of environmental factors, including nitrogen and phosphorus concentrations (Schindler, 2012; Srivastava et al., 2012; Chaffin and Bridgeman, 2014; Van de Waal et al., 2014), TN:TP ratio (Smith, 1983; Wang et al., 2010b; Van de Waal et al., 2014), temperature (Davis et al., 2009; Rolland et al., 2013), salinity (Tonk et al., 2007), and iron concentration (Ame and Wunderlin, 2005; Nagai et al., 2007; Wang et al., 2010a) have been shown to have pronounced effects on cyanobacterial biomass, cyanobacterial dominance and microcystin production. Nevertheless, the results between studies differ, and there is no clear understanding of the roles of these environmental factors as the triggers of cyanobacterial bloom development and microcystin production. Furthermore, the occurrence of cyanobacterial toxins in a system is the result of a complex interaction between abiotic and biotic factors, including the competition with other phytoplankton. It therefore remains an important challenge for bloom management to fully understand the mechanisms behind toxic cyanobacterial bloom development and the drivers for biomass development, cyanobacterial dominance (fraction) and toxin production. For instance, regardless of the fact that many studies suggest the important role of phosphorus, reduction of internal and external phosphorus concentration is not always successful in preventing the occurrence of toxic cyanobacterial blooms in water bodies (Lewis and Wurtsbaugh, 2008; Amano et al., 2010; Koreiviene et al., 2014).

By taking into account the contrasting findings of earlier studies, including inconsistent outcomes of nutrient reduction strategies, we suggest that the main environmental triggers of cyanobacterial and microcystin variability may vary between water bodies due to the complex, lake-specific interplay of environmental conditions. Therefore, the main objective of this study was to investigate the site-specificity of en-

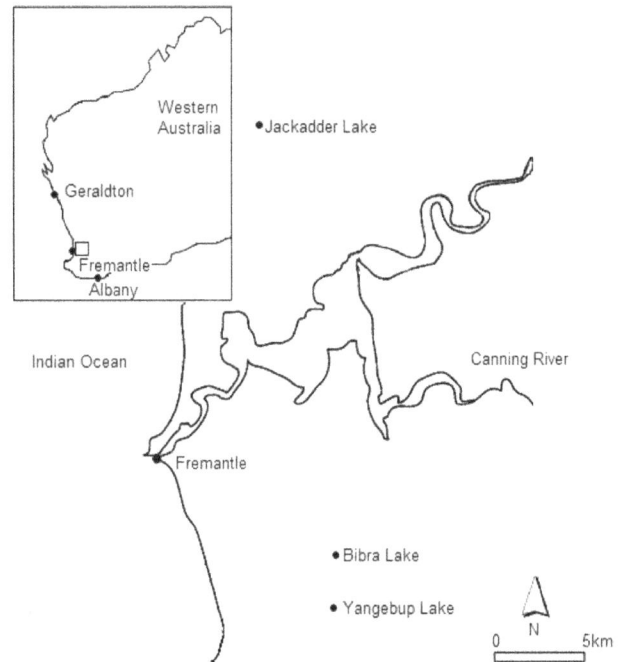

Figure 1. The locations of three studied lakes on Swan Coastal Plain.

vironmental triggers for cyanobacterial biomass and microcystin variability in a local urban lake system. More specifically, the objectives were to (1) determine the variability of cyanobacterial biomass and microcystin concentration in a set of local urban lakes, and (2) identify the site-specific relationships between environmental factors and cyanobacterial or microcystin dynamics.

2 Material and methods

2.1 Study lakes

This study was carried out in Jackadder Lake ($31°54'30$ S, $115°47'36$ E), Bibra Lake ($32°5'25$ S, $115°49'16$ E) and Yangebup Lake ($32°6'56$ S, $115°49'33$ E) located on the Swan Coastal Plain, Western Australia (Fig. 1). Sampling was carried out between January and March 2010. These lakes are shallow with mean depth of 2.1, 1.1, and 2.5 m for Jackadder Lake, Bibra Lake and Yangebup Lake, respectively. Jackadder Lake and Yangebup Lake are permanent lakes while Bibra Lake is subjected to seasonal drying due to progressive decline in groundwater levels over the Jandakot Mound. Jackadder Lake has an area of 7.18 ha, is surrounded by 6.6 ha of parkland and is draining a 152 ha catchment area (Arnold, 1990; Woodward, 2008). Water levels in Jackadder Lake are maintained by the input of surface runoff via 10 drain inlets (Rajah, 1991, as cited in Kemp, 2009). Jackadder Lake receives water from the Herdsman Lake catchment area and Osborne Park main drain during dry summers (Department of Planning, 2010). Bibra Lake has a size of 135 ha with an open water area of approximately 100 ha (Strategen, 2009) and is located within a 250 ha catchment are. This lake is surrounded by urban areas and a golf course and serves as habitat for many species of water birds (Kemp, 2009). Water enters Bibra Lake via direct rainfall recharge onto the lake surface or from surface runoff from the surrounding catchment (Strategen, 2009). Yangebup Lake has a total area of 90.5 ha with an open water area of approximately 68 ha, and is surrounded by residential, agriculture and industrial areas. Yangebup Lake is a groundwater through-flow wetland that accepts groundwater from the east and discharges groundwater to the west (Dunlop, 2008). Yangebup Lake receives urban runoff from three storm-water drains and additionally serves as a compensation basin for the South Jandakot drainage system with an approximate area of $200 \, km^2$. This includes receiving water from neighbouring Thomson Lake when it reaches its maximum water level. Once Yangebup Lake reaches its maximum allowable water level, water is pumped into nearby Cockburn Sound (Environmental Protection Authority, 1989). The hydrology of Jackadder, Bibra and Yangebup lakes is mainly affected by the strong seasonal rainfall pattern due to the Mediterranean climate. The region's mean annual rainfall is reported as 771.5 mm and monthly mean rainfall is 35.1, 156.3, 433.3 and 144.2 mm during summer, autumn, winter and spring, respectively (Bureau of Meteorology, 2014). In response, the maximum water levels in all lakes occur in September and October, and the minimum water levels occur in March and April at the end of summer months (Davis et al., 1993). The region's mean maximum annual temperature is 24.5 °C and monthly maximum temperatures are 30.9, 25.4, 18.0 and 22.6 °C during summer, autumn, winter and spring, respectively (Bureau of Meteorology, 2014). Prolonged stable thermal stratification is usually prevented in these lakes during summer due to continuous or intermittent wind mixing that creates a homogeneous environment throughout the water column (Davis et al., 1993; Arnold and Oldham, 1997).

These lakes were selected due to differences reported on physicochemical properties, levels of cyanobacterial biomass and microcystin concentration. Based on an earlier study conducted between November 2008 and July 2009 (Sinang et al., 2013), these lakes represent systems with low, medium and high cyanobacterial biomass and microcystin concentration. In this earlier study, the highest cyanobacterial biomass was reported as 28, 108, and 80 μg chl $a \, L^{-1}$ in Jackadder, Bibra and Yangebup Lake, respectively. The highest cellular microcystin concentrations (mg g^{-1} cyanobacterial dry mass) was 4.8 mg g^{-1} in Jackadder Lake, 35 mg g^{-1} in Bibra Lake and 1.7 mg g^{-1} in Yangebup Lake (Sinang et al., 2013).

2.2 Sampling and analyses

The lakes were sampled twice a month between January and March 2010, leading to 6 sampling days. Three samples were

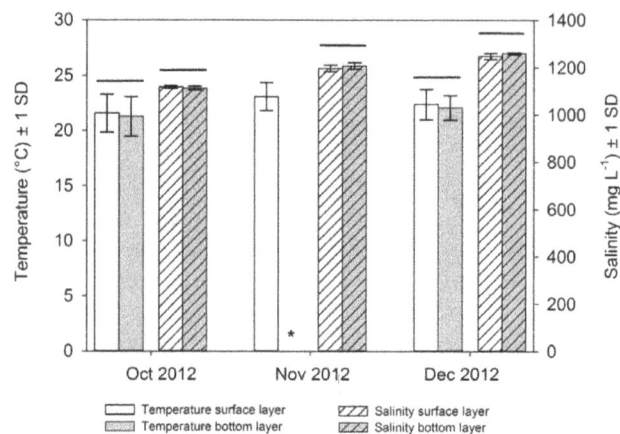

Figure 2. Temperature (°C) and salinity (ppm) in the surface and bottom layers measured at seven sites over three months in Lake Yangebup during a previous study in 2012. "*" indicates missing data; horizontal line indicates that no significant difference between layers were detected (*t*-test) (from Song et al., 2015).

collected from the same three points at each lake on every sampling occasion. As Bibra Lake dried up in late February no samples were taken from this lake in March, leading to only 4 sampling days. On-site measurements and samples were taken from shore sites at a water depth of 0.6 to 1 m. Temperature (Temp), pH and Salinity (Sal) were measured on-site with a WP-81 probe (TPS Pty Ltd) at a depth of 0.6 m. Grab water samples for cyanobacteria, microcystin and total phosphorus quantification were taken from approximately 0.15 m below the surface to avoid surface scum. Although there was a slight difference in the depth from which the samples for the physicochemical and water samples were taken, this is not expected to influence the interpretation of the results, as earlier studies in these lakes indicated that the water bodies at these shallow shore sites are well mixed with respect to physicochemical conditions (Arnold and Oldham, 1997; Song et al., 2015) (Fig. 2). Water samples were stored immediately in glass bottles in the dark on ice. Variables analysed from these samples were total phosphorus (TP), total dissolved phosphorus (TDP), total iron (TFe), total dissolved iron (TDFe), total nitrogen (TN), total dissolved nitrogen (TDN), ammonium (NH_4^+), cyanobacterial biomass, total phytoplankton biomass, intracellular and extracellular microcystin fractions. Samples for dissolved nutrient analyses were pre-filtered with a 0.45 μm syringe filter (Acrodisc, HT Tuffryn) before freezing at -20 °C.

Surface water temperatures were between 19.9 and 28.7 °C during the study period. However, the on-site measurements of surface water temperatures were dependent on the time of sampling and varied by up to 3.9 °C over the course of a day. Therefore, maximum air temperature on each sampling day recorded by weather stations located nearest to the studied lakes was used as a substitute for surface water temperature in all analyses (Yen et al., 2007).

2.2.1 Nutrients and phytoplankton biomass

TP and TDP concentrations were analysed using the ascorbic acid method, while TFe and TDFe concentrations were analysed with the phenanthroline method, according to standard methods (APHA, 1998). TN, TDN, and NH_4^+ were analysed at the South Coast Nutrients Analysis Laboratory, Albany, Western Australia with the standard colorimetric methods on a segmented flow auto-analyser (Alpkem, Wilsonville, OR, USA). Cyanobacterial and total phytoplankton chlorophyll a were measured with a top-bench version of a FluoroProbe (bbe Moldaenke, Germany).The FluoroProbe measures chl a fluorescence and differentiates four groups of phytoplankton (chlorophytes, cryptophytes, diatoms and cyanobacteria) by their specific fluorescence emission spectrum (Beutler et al., 2002). The fluorescence is used to calculate total biomass of each phytoplankton group that is expressed as chl a concentration equivalents (μg chl a L^{-1}) (Beutler et al., 2002; Ghadouani and Smith, 2005). FluoroProbe chl a measurements were validated against chl a data of samples extracted according to standard methods (APHA, 1998) (linear regression analysis: $R^2 = 0.94$, $N = 32$, $p < 0.05$). In our study, chl a fluorescence as measured by FluoroProbe was used as a proxy for cyanobacterial biomass (Geis et al., 2000; Eisentraeger et al., 2003).

For quantification of cyanobacterial biomass and to separate the intracellular from the dissolved microcystin fraction, water samples were filtered through pre-combusted and pre-weighed 47 mm GF/C filter papers. Filter papers containing particulate organic matter were dried for 24 h at 60 °C and re-weighed to obtain total dry weight (Harada et al., 1999). These filter papers were then moistened with Milli-Q water and kept frozen (at -20 °C) until intracellular microcystin extraction. As we were interested in the microcystin concentration per unit cyanobacterial dry mass, cyanobacterial dry mass was calculated from the total dry mass (from the filters) by adjusting it to the percentage of cyanobacteria measured with the FluoroProbe. Cyanobacterial dry mass was only used for microcystin quantification.

Water samples collected for cyanobacterial identification and enumeration were preserved with acidic Lugol's iodine solution (5 g I_2 + 10 g KI, 20 mL distilled water and 50 mL of 10 % acetic acid) and cyanobacteria were identified to the genus level using phytoplankton taxonomic guideline (Komarek and Hauer, 2011). The relative abundance of each cyanobacterial genera (cells or colonies mL^{-1}) was determined from 10–50 mL of sample using an inverse microscope (Utermöhl, 1958) and converted into biovolume per mL (μm^3 mL^{-1}) by multiplying the mean cell or colony biovolume (μm^3) with the total cells or colonies per millilitre (cells or colonies mL^{-1}). Mean cell or colony biovolume for each cyanobacterial genus was calculated by finding the geometric figure that best approximated the shape of each genera, and by measuring the dimension of 20 individual cells or colonies (Hillebrand et al., 1999). A minimum of 200 cells or colonies of the most abundant cyanobacteria were counted for each sample. Different cyanobacterial species within each genus can vary in size by several orders of magnitude. However, as we measured the mean biovolume of each cyanobacterial genus, differences in sizes between species are evened out as a larger mean is expected, if larger species are more abundant and vice versa. The calculated mean biovolume of each cyanobacterial genus was used to compute the dominant cyanobacteria genera in the studied lakes.

2.2.2 Microcystin extraction and quantification

Filters were freeze-thawed twice to break the cells prior to methanol extraction (Lawton et al., 1994). Filters were placed into centrifuge tubes and 5 mL of 75 % methanol–water (v/v) was added. Filters were sonicated on ice for 25 min, followed by gentle shaking for another 25 min. The extracts were then centrifuged at 3273 g (Beckman and Coulter, Allegra X-12 Series) for 10 min at room temperature. Extracts were carefully transferred into conical flasks, and two more extractions were done per filter. All three extracts were pooled and diluted with Milli-Q to 20 % methanol (v/v).

Intracellular microcystin extracts and the pre-filtered water containing dissolved (extracellular) microcystin were subjected to solid-phase extraction (SPE) (Waters Oasis HLB) for clean-up and concentration with a loading speed of < 10 mL min^{-1}. SPE cartridges were then rinsed with 10 mL of 10, 20 and 30 % methanol–water (v/v), before microcystin was eluted with acidified methanol (0.1 % v/v trifluoroacetic acid (TFA)) and evaporated with nitrogen gas at 40 °C. Finally, samples were re-dissolved in 30 % acetonitrile and analysed with high-performance liquid chromatography (HPLC) by using the Alliance 2695 (Waters, Australia) with a PDA detector (1.2 nm resolution) and an Atlantis T3 3 μm column (4.6 × 150 mm i.d.). Mobile phases used were acidified acetonitrile (0.05 % v/v TFA) and acidified Milli-Q water (0.05 % v/v TFA). Microcystin peaks were separated using a linear gradient as described in Lawton et al. (1994) but with a maximum acetonitrile concentration of 100 % and a run time of 37 min. Column temperature was maintained at 37.5 ± 2.5 °C. The limit of detection per microcystin peak was 1.12 ng. Microcystin variants were identified based upon their typical absorption spectrum detected by PDA detector at 238 nm (Meriluoto and Codd, 2005). Commercially available microcystin-LR standard (Sapphire Bioscience, Australia; purity ≥ 95 %) was used to quantify microcystin concentrations. Throughout this paper we refer to the total concentration of microcystin variants per sample as microcystin concentration.

In this study, cellular (intracellular) microcystin concentration was expressed as μg microcystin-LR mass equivalents per gram cyanobacterial dry mass to illustrate cyanobacterial microcystin content. Extracellular microcystin was expressed as the fraction of extracellular microcystin concentration per total microcystin concentration to allow the quantification of the proportion of microcystin released into the water column in comparison to the total microcystin being produced.

2.3   Data processing and statistical analyses

Differences in physicochemical factors, cyanobacterial biomass and microcystin between lakes were analysed with one-way ANOVA (SPSS 17.0) with post hoc test (Least Significance Difference; LSD) as all assumptions for an ANOVA were met (homogeneity of variances, normality). For the descriptive phase, bivariate correlation analysis (Pearson's) was carried out to identify the environmental variables which significantly correlate with cyanobacterial fraction, cyanobacterial biomass, cellular microcystin concentration and extracellular microcystin fraction (SPSS 21.0). We used linear mixed models to identify correlations between environmental variables and cyanobacterial fraction, cyanobacterial biomass, cellular microcystin concentration and extracellular microcystin fraction in each lake using sampling site and sampling date as random factors, and for all lakes combined adding lake as random factor (SPSS 21.0). All dependent variables were ln transformed. As extracellular microcystins were only detected in 5 out of 12 samples in Bibra Lake, this resulted in only 5 data points for this dependent variable in Bibra Lake, making the calculation of linear mixed models for this explanatory variable impossible. Two redundancy analyses (RDA) were calculated to identify the best combination of variables to explain the variability of intracellular microcystin concentration, extracellular microcystin fraction and either cyanobacterial fraction or cyanobacterial biomass (R version 2.15.1) for each lake. Canonical ordination (999 permutations) with forward selection was computed with standardised explanatory and response variables. All data was ln transformed to meet the assumption of normality. RDA analysis on Bibra Lake was conducted without the inclusion of pH and temperature due to an inadequate number of data points (residual d.f. < 0). In all analyses, results were considered significant at $p < 0.05$, unless stated differently.

3   Results

3.1   Physical and chemical characteristics of studied lakes

On the sampling days, mean pH fluctuated between 8.2 and 9.2 (Fig. 3a) and mean air temperature (Fig. 3b) ranged from 27 to 43 °C in all lakes. Salinity in Jackadder and Yangebup was mostly below 1.0 ppk and much lower than in Bibra Lake (Fig. 3c). The sharp increase in salinity in Bibra Lake was probably due to the decreasing water level as the lake dried up by end of February. Nutrient concentrations varied on a temporal basis within lakes and spatially between lakes. Phosphorus concentrations were higher

in Bibra Lake than in Jackadder and Yangebup lakes throughout the sampling period. Mean TP concentrations (Fig. 3d) ranged from 22 to 92, from 230 to > 1000, from and 28 to > 150 μg L^{-1} in Jackadder, Bibra and Yangebup lakes, respectively. Meanwhile, mean TDP concentrations (Fig. 3e) ranged from 12 to 24, from 17 to 142, and from 14 to 37 μg L^{-1} in Jackadder, Bibra and Yangebup lakes, respectively. Temporal variation of macronutrient concentrations in Yangebup and Jackadder lakes were much smaller than in Bibra Lake. The large increase of TP, TDP, TN and TDN in Bibra Lake might again have been a concentration effect due to the lake drying up. Mean TFe and TDFe concentrations were higher in Bibra Lake during the earlier three sampling dates. Mean TFe (Fig. 3f) ranged from 77 to 247, from 147 to 220, and from 51 to 110 μg L^{-1} in Jackadder, Bibra and Yangebup lakes, respectively. Mean TDFe (Fig. 3g) ranged from 24 to 174, from 61 to 117, and from 21 to 89 μg L^{-1} in Jackadder, Bibra and Yangebup lakes, respectively. TN (Fig. 3h) and TDN (Fig. 3i) concentrations were up to one order of magnitude higher in Bibra lakes compared to concentrations in Jackadder and Yangebup lakes. In contrast, mean TN:TP in Bibra Lake were lower than the ratios in Jackadder and Yangebup lakes (Fig. 3j). Mean TN:TP ranged from 18 to 60, 16 to 38, and 29 to 115 in Jackadder, Bibra and Yangebup lakes, respectively. NH$_4^+$ decreased over time in Jackadder and Yangebup lakes (Fig. 3k) and mean concentrations ranged from 43 to 170, from 157 to 239, and from 40 to 143 μg L^{-1} in Jackadder, Bibra and Yangebup lakes, respectively.

The three lakes were significantly different in salinity, phosphorus, nitrogen and iron, either as total or dissolved forms (except TDFe) (ANOVA; Table 1), but did not show a significant difference in pH, air temperature and TDFe. The post hoc tests (LSD) indicated that Jackadder and Yangebup Lake did not differ in TP, TDP and NH$_4^+$; however, both lakes were different to Bibra Lake. Furthermore, all lakes were different in salinity, TN, TDN, and TFe. Jackadder and Yangebup lakes can be classified as eutrophic, while Bibra Lake can be classified as hypereutrophic, based on the mean TP concentrations (Carlson, 1977). Nitrogen-limited conditions in a lake are usually defined when the TN:TP weight ratios are less than 10 (Graham et al., 2004). As our results indicate that TN:TP ratios below 10 were rare, the studied lakes were not associated with persistent nitrogen limitation.

3.2   Variability of cyanobacterial biomass and microcystin concentration

Cyanobacterial communities in all lakes contained potentially toxin-producing cyanobacteria including *Microcystis* spp., *Planktothrix* spp., *Anabaenopsis* spp., *Anabaena* spp. and *Nodularia* spp. (Fig. 4) with *Microcystis* spp. being the most abundant cyanobacterial genera in all lakes. Mean total cyanobacterial biomass was 5.41, 29.60, 15.14 μg L^{-1} in Jackadder, Bibra and Yangebup Lake, re-

Table 1. Physical and chemical properties of the three lakes throughout the sample period (January–March 2010), including analysis of differences between lakes (one-way ANOVA).

Factors	Jackadder Lake ($N = 18$)		Bibra Lake ($N = 12$)		Yangebup Lake ($N = 18$)		Differences between lakes
	Mean $\pm$ SD	Range	Mean $\pm$ SD	Range	Mean $\pm$ SD	Range	(one-way ANOVA)
pH	8.7 ± 0.3	8.1–9.0	8.9 ± 0.2	8.5–9.2	8.9 ± 0.4	7.5–9.3	$F_{(2,45)} = 2.16$
Air Temp	33.0 ± 4.9	27.4–42.7	35.7 ± 4.7	30.8–43.0	34.7 ± 4.1	30.8–43.0	$F_{(2,45)} = 1.31$
Sal (ppk)	0.4 ± 0.04	0.3–0.4	2.9 ± 1.0	1.7–4.1	0.9 ± 0.1	0.8–1.1	$F_{(2,45)} = 99.08^*$
TP ($\mu g\,L^{-1}$)	44.0 ± 28.0	20.0–131.6	598.1 ± 362.0	214.7–1145.9	64.8 ± 44.2	24.0–168.0	$F_{(2,45)} = 40.28^*$
TDP ($\mu g\,L^{-1}$)	17.6 ± 4.8	12.0–26.7	67.9 ± 51.3	16.0–180.0	23.2 ± 7.6	13.3–40.7	$F_{(2,45)} = 15.27^*$
TFe ($\mu g\,L^{-1}$)	123.3 ± 66.2	63.6–261.8	192.1 ± 43.4	138.2–289.3	81.5 ± 24.1	48.4–122.9	$F_{(2,45)} = 18.91^*$
TDFe ($\mu g\,L^{-1}$)	69.2 ± 66.3	20.0–200.0	89.1 ± 30.4	38.6–154.1	52.9 ± 28.9	11.2–92.6	$F_{(2,45)} = 2.15$
NH$_4$ ($\mu g\,L^{-1}$)	100.8 ± 54.9	30.0–180.0	191.5 ± 33.8	150.0–250.3	86.3 ± 45.6	30.0–160.0	$F_{(2,45)} = 20.04^*$
TN ($mg\,L^{-1}$)	1.3 ± 0.4	0.7–2.2	11.7 ± 5.2	4.9–17.3	3.5 ± 0.8	1.9–5.2	$F_{(2,45)} = 59.38^*$
TDN ($mg\,L^{-1}$)	0.8 ± 0.2	0.4–1.1	8.7 ± 3.0	4.9–14.0	2.4 ± 0.3	1.9–2.8	$F_{(2,45)} = 104.98^*$
TN : TP	35.6 ± 14.9	11.1–76.1	23.1 ± 10.0	10.3–41.1	68.6 ± 29.9	25.0–124.1	$F_{(2,45)} = 19.51^*$

N = number of samples; SD = standard deviation; * = $p < 0.05$.

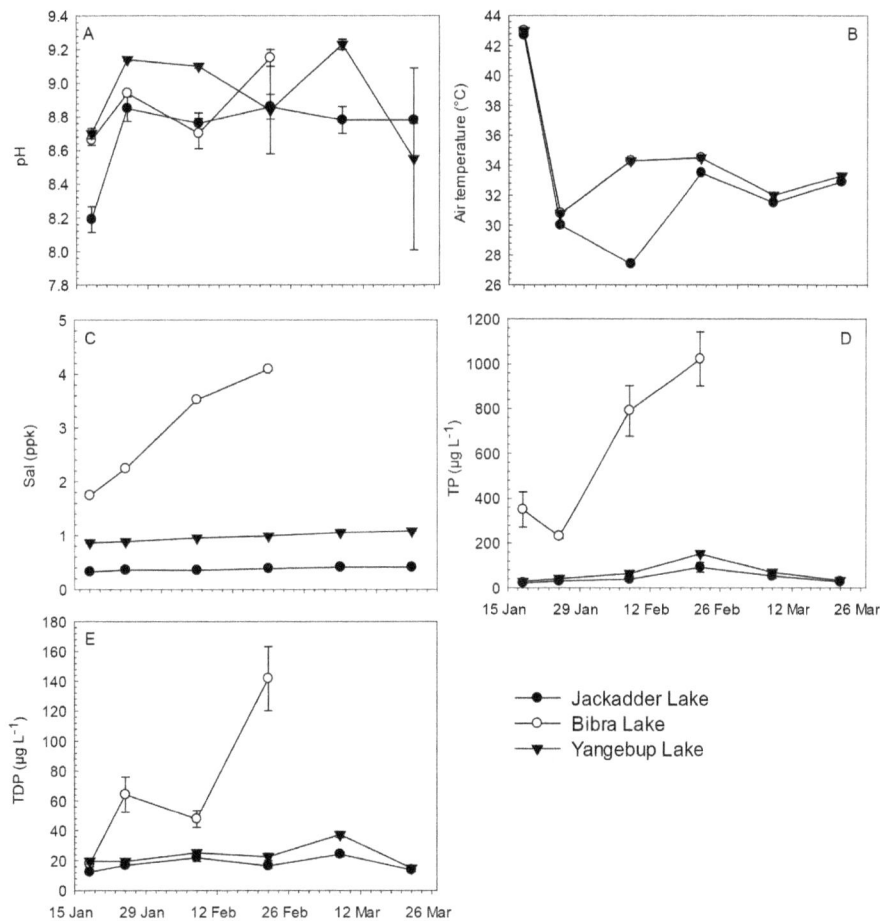

Figure 3.

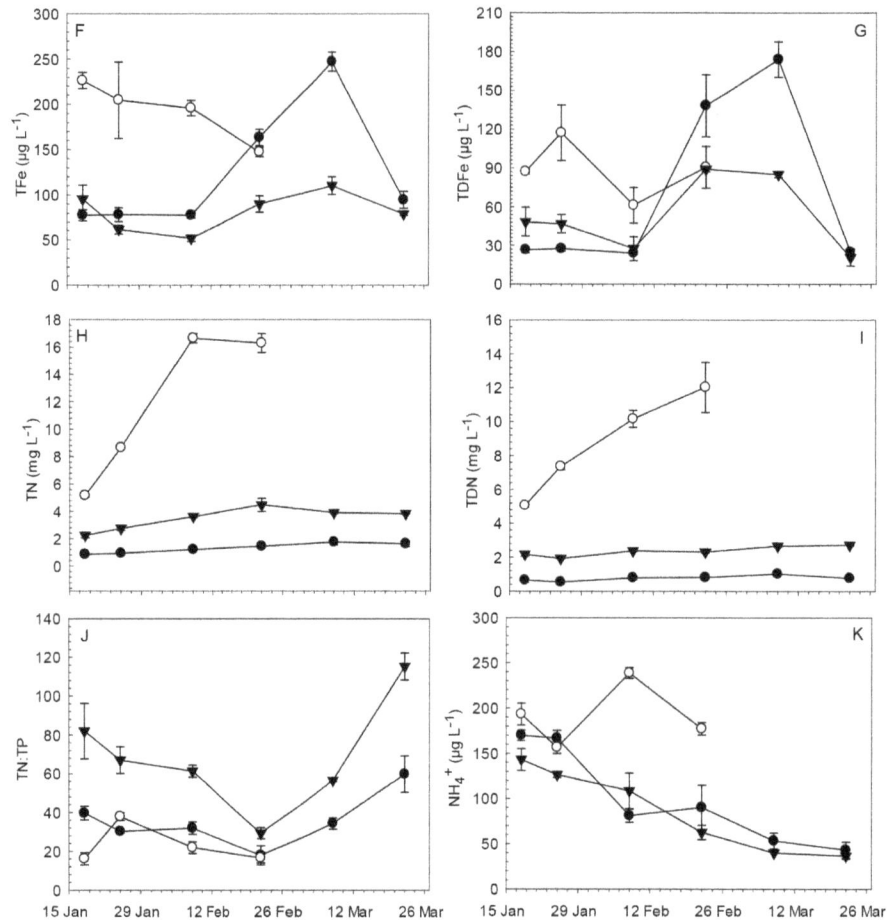

Figure 3. Mean values ($\pm$ one standard error) of physicochemical variables over time (A = pH; B = Air Temp; C = Sal; D = TP; E = TDP; F = TFe; G = TDFe; H = TN; I = TDN; J = TN : TP; K = NH_4^+) in Jackadder, Bibra and Yangebup lakes from January to March 2010. The mean is calculated from the three locations per lakes.

spectively (Fig. 5a). Cyanobacterial biomass varied within an order of magnitude on a temporal basis in Bibra and Jackadder lakes (Jackadder: $1-12 \mu g\, L^{-1}$, Bibra: $5-83\ \mu g\, L^{-1}$, Yangebup: $8-32 \mu g\, L^{-1}$). Although cyanobacterial biomass was significantly higher in Bibra Lake compared to the other two lakes ($F_{(2,45)} = 7.62$, $p < 0.05$), the cyanobacterial fraction (the ratio of cyanobacterial chlorophyll a to total phytoplankton chlorophyll a) in this lake was significantly lower than in Jackadder and Yangebup Lake ($F_{(2,45)} = 3.59$, $p < 0.05$) (Fig. 5b). Cyanobacterial fraction ranged between 0.05 to 0.71 in Jackadder Lake, 0.16 to 0.68 in Yangebup Lake, and 0.11 to 0.51 in Bibra Lake. The post hoc tests indicated that Jackadder and Yangebup lakes did not differ in cyanobacterial biomass and cyanobacterial fraction, but both lakes were different to Bibra Lake.

Cellular microcystin concentration ($mg\, g^{-1}$ cyanobacterial dry mass) varied over 3 orders of magnitude in Jackadder Lake, and 2 orders of magnitude in both Bibra Lake and Yangebup Lake (Fig. 5c) throughout the sampling events. Mean cellular microcystin concentrations were

$0.407\, mg\, g^{-1}$ in Jackadder Lake, $0.233\, mg\, g^{-1}$ in Bibra Lake, and $0.150\, mg\, g^{-1}$ in Yangebup Lake. Cellular microcystin concentration was not significantly different between lakes ($F_{(2,45)} = 2.07$, $p > 0.05$). Mean extracellular microcystin fraction was 0.18 in Jackadder Lake, 0.04 in Bibra Lake, and 0.26 in Yangebup Lake (Fig. 5d). The post hoc tests indicated that Bibra Lake was the only lake that had a significantly different extracellular microcystin fraction when compared to other lakes ($F_{(2,45)} = 6.49$, $p < 0.05$).

3.3 Relationship between environmental factors and cyanobacterial fraction, cyanobacterial biomass, or microcystin concentration

Most analysed nutrients were weakly but significantly correlated with cyanobacterial fraction, biomass and microcystin concentrations when data from all lakes were combined (Tables 2 and 3). The correlations presented in Tables 2 and 3 suggest that, in general, cyanobacterial dominance in the phytoplankton community was favoured at relatively lower nutrient concentrations as it was negatively correlated to TP,

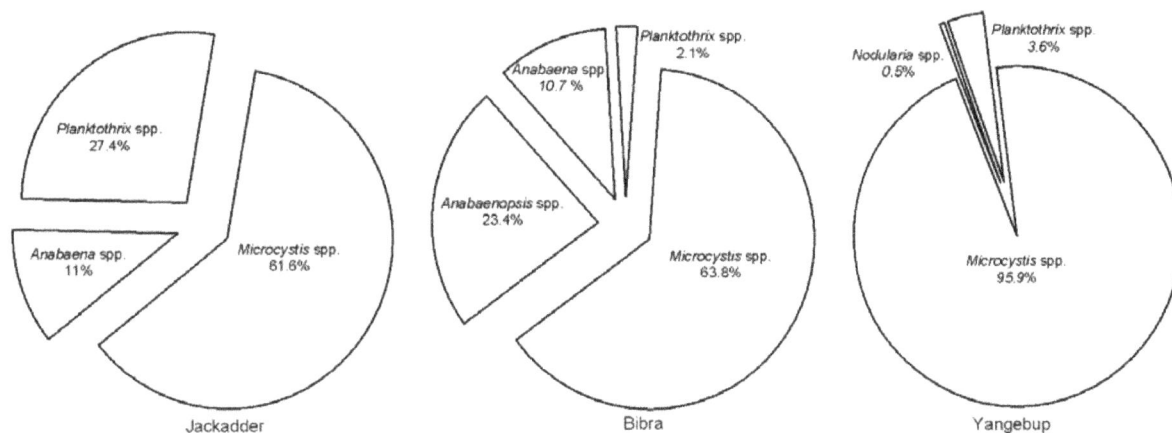

Figure 4. Mean biomass (μm^3 mL^{-1}) proportions of potentially toxic cyanobacterial genera in Jackadder, Bibra and Yangebup lakes during the study period.

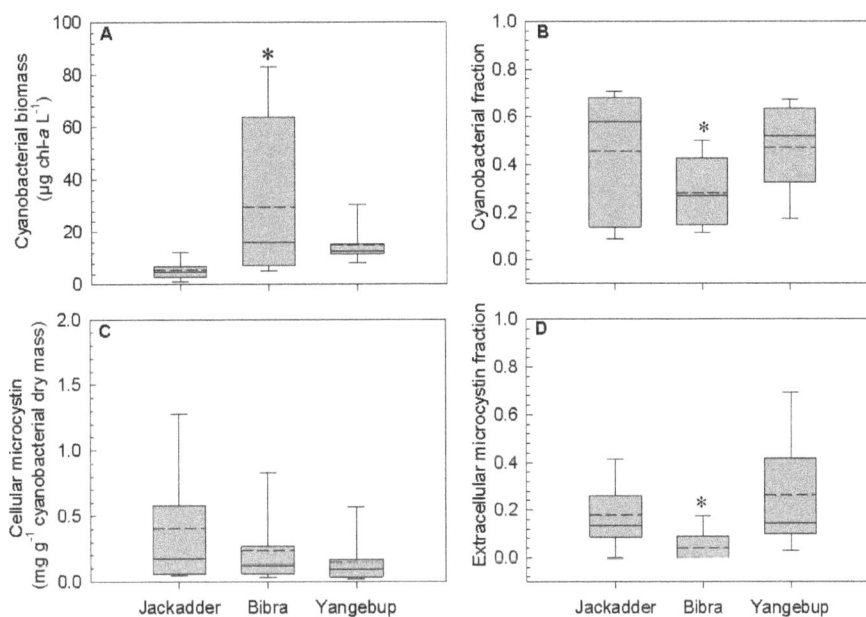

Figure 5. The variability of (**a**) cyanobacterial biomass (μg chl a L^{-1}), (**b**) cyanobacterial fraction (cyanobacterial biomass to total biomass), (**c**) cellular microcystin concentration (mg g^{-1} cyanobacterial dry mass) and (**d**) extracellular microcystin fraction over time for each lake. Boxes represent 25th to 75th percentiles; straight lines within the boxes mark the median, short dashed lines the mean; whiskers below and above the boxes indicate 10th and 90th percentiles. Asterisks (*) indicate lakes that are significantly ($p < 0.05$) different from other lakes.

TDP, TFe, and TDFe. In contrast, cyanobacterial fraction was positively correlated with TN : TP ratio, potentially due to relatively lower TP concentrations in comparison to TN concentrations. Cyanobacterial biomass on the other hand was positively correlated to salinity, TN, TDN and NH_4^+, but negatively correlated with TDFe. Cellular microcystin concentration was positively correlated with phosphorus and iron, but not with nitrogen. TDFe showed the strongest positive correlation with cellular microcystin concentration, followed by TP, TFe and TDP. Cellular microcystin was also negatively correlated with TN : TP ratio (Table 3). In contrast to

cellular microcystin, extracellular microcystin fraction was negatively correlated with salinity, TP, TDP, TN, TDN, and positively correlated with TN : TP ratio (Table 3). Correlating environmental variables with cyanobacteria or microcystin for each lake separately, the correlations that were significant (Pearson's) were different between lakes (Tables 2 and 3).

Using data from all lakes combined in linear mixed models, cyanobacterial fraction was negatively correlated to TP, TDP, TFe, TDFe (Fig. 6a–d), and positively to TN : TP (Fig. 6e). However, within each lake, the correlations with cyanobacterial fraction were significant only for TP, TDP

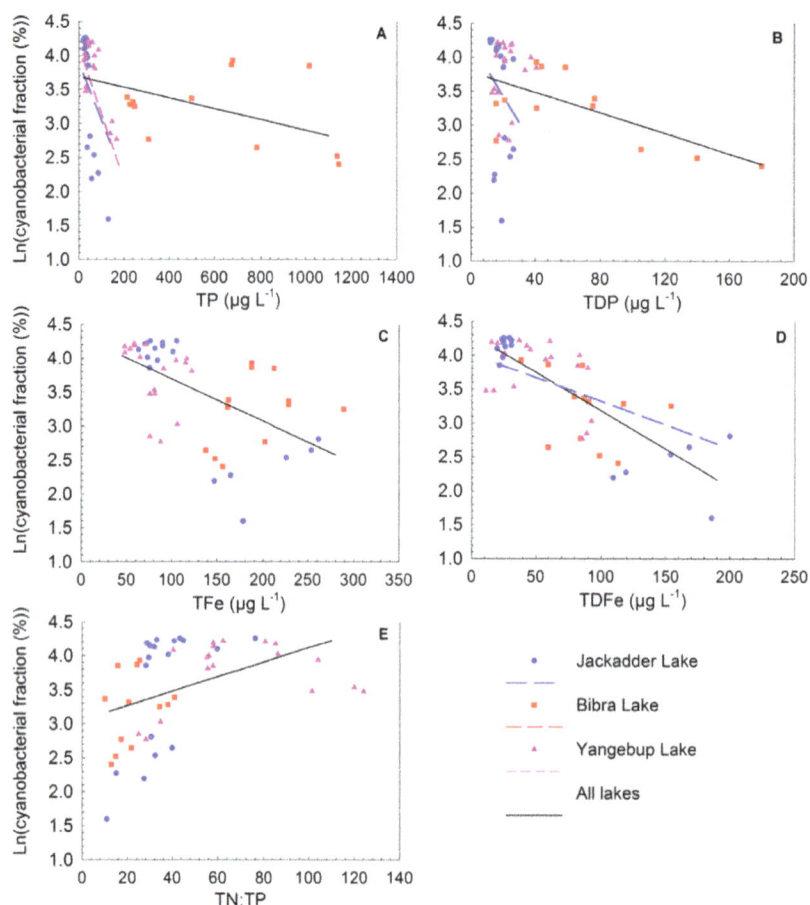

Figure 6. The correlations between cyanobacterial fraction and (**a**) TP, (**b**) TDP, (**c**) TFe, (**d**) TDFe, (**e**) TN : TP in Jackadder, Bibra and Yangebup lakes during the study period. Regression curves for each individual lake were calculated by linear mixed models with site and date as random factors on data from each lake (broken lines) while all data points were combined for the overall regression using a linear mixed model adding lake as random factor (solid line). Only significant ($p < 0.05$) regressions are shown.

Table 2. Pearson's correlation coefficients (R) between the environmental factors and cyanobacterial fraction (%) or cyanobacterial biomass (μg chl a L^{-1}) analysed for each lake and for all lakes combined using bivariate correlation analysis. The dependent variables are ln transformed.

Factor	Cyanobacterial fraction (%)				Cyanobacterial biomass (μg chl a L^{-1})			
	All lakes $N = 48$	Jackadder $N = 18$	Bibra $N = 12$	Yangebup $N = 18$	All lakes $N = 48$	Jackadder $N = 18$	Bibra $N = 12$	Yangebup $N = 18$
pH	−0.108	−0.363	**−0.653**	0.225	0.087	−0.181	**−0.671**	0.287
Air Temp	0.018	0.119	−0.112	0.016	0.138	0.002	0.080	0.043
Salinity	−0.250	−0.423	−0.204	−0.460	**0.454**	−0.063	−0.038	−0.236
TP	**−0.337**	**−0.873**	−0.272	**−0.742**	0.282	**−0.808**	−0.090	0.092
TDP	**−0.357**	−0397	**−0.641**	0.147	0.060	−0.320	−0.574	0.406
TFe	**−0.570**	**−0.789**	0.389	−0.304	−0.040	**−0.577**	0.340	−0.326
TDFe	**−0.777**	**−0.903**	−0.355	−0.432	**−0.339**	**−0.727**	−0.424	−0.113
NH$_4$	0.105	0.375	0.576	**0.543**	0.345	0.042	**0.721**	0.222
TN	−0.236	**−0.487**	0.035	**−0.628**	0.477	−0.185	0.197	0.025
TDN	−0.265	**−0.534**	−0.219	−0.305	0.430	−0.314	−0.078	−0.084
TN : TP	**0.423**	**0.570**	0.299	0.264	0.164	**0.741**	0.145	−0.339

Significant ($p < 0.05$) correlations are highlighted in bold.

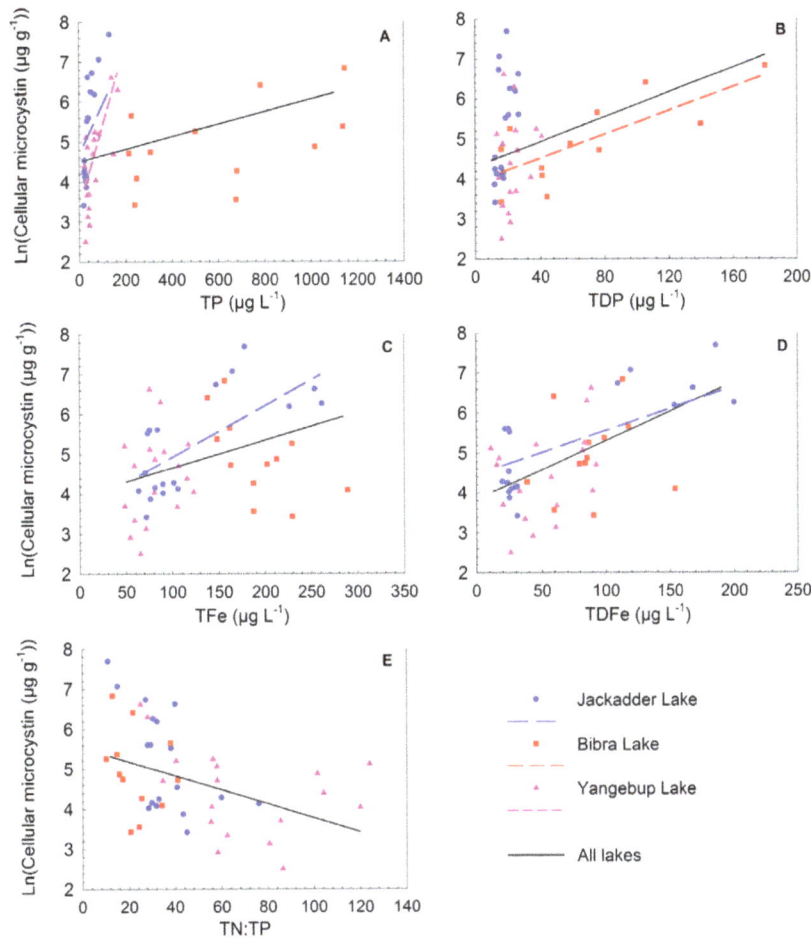

Figure 7. The correlations between cellular microcystin concentration and (**a**) TP, (**b**) TDP, (**c**) TFe, (**d**) TDFe, (**e**) TN : TP in Jackadder, Bibra and Yangebup lakes during the study period. Regression curves for each individual lake were calculated by linear mixed models with site and date as random factors on data from each lake (broken lines) while all data points were combined for the overall regression using a linear mixed model adding lake as random factor (solid line). All regression shown are $p < 0.05$, except for the regression calculated for all lakes combined in (**a**), which is $p < 0.1$.

and TDFe in Jackadder Lake and TP in Yangebup Lake. Cellular microcystin concentration was on the other hand positively correlated to TP, TDP, TFe and TDFe (Fig. 7a–d) and negatively to TN : TP (Fig. 7e). Within each lake, these correlations were only significant for TP, TFe, TDFe in Jackadder Lake (Fig. 7a, c and d), for TDP in Bibra Lake (Fig. 7b) and for TP in Yangebup Lake (Fig. 7a). When combining all lakes, extracellular microcystin fraction was negatively correlated to salinity (linear mixed model; $p < 0.1$), TP and TDP, but positively to TN : TP (Fig. 8a–d). Jackadder Lake was the only lake showing significant correlations between extracellular microcystin fraction and salinity (positive, Fig. 8a) and TP (negative, Fig. 8b). Using linear mixed models, cyanobacterial biomass was only significantly correlated to TDP and TDFe when combining all lakes (Fig. 8e and f), with Bibra Lake showing a significant negative correlation to TDFe (Fig. 8f). The 95 % confidence intervals of the slopes of the correlations between TP and cyanobacte-

rial fraction or extracellular microcystin fraction in Jackadder Lake and in all lakes combined (Figs. 6a and 8b) or between salinity and extracellular microcystin fraction in Jackadder Lake and in all lakes combined (Fig. 8a) did not overlap, providing a conservative estimate that the slopes were significantly different (Payton et al., 2003).

3.4 Multivariate analysis of site-specific environmental factors and the variability of cyanobacteria and microcystin concentration

The first RDA analysis showed significant relationships ($p < 0.05$) between the measured environmental factors and the combined variability of cyanobacterial fraction, cellular microcystin concentration and extracellular microcystin fraction for each lake. The canonical ordination indicated that 75 % (Jackadder Lake; $R^2_{adj.} = 0.75$; $F = 5.726$), 80 % (Bibra Lake; $R^2_{adj.} = 0.80$; $F = 5.888$) and 75 % (Yange-

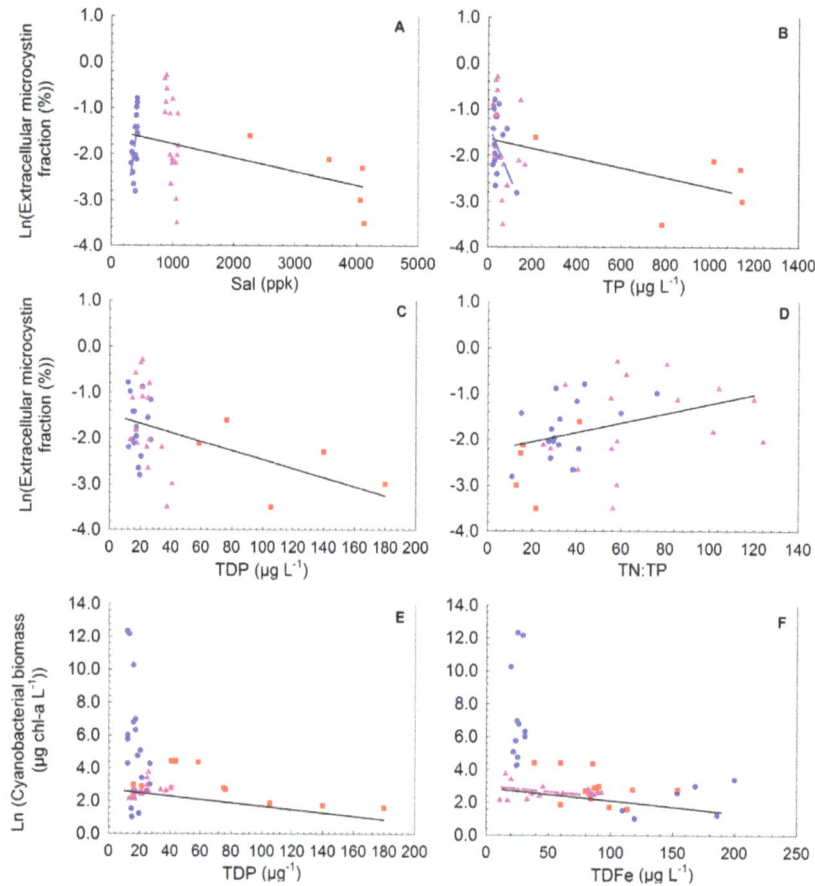

Figure 8. The correlations between extracellular microcystin fraction and **(a)** Sal, **(b)** TP, **(c)** TDP, **(d)** TN : TP, and between cyanobacterial biomass and **(e)** TDP, **(f)** TDFe in Jackadder, Bibra and Yangebup lakes during the study period. Regression curves for each individual lake were calculated by linear mixed models with site and date as random factors on data from each lake (broken lines) while all data points were combined for the overall regression using a linear mixed model adding lake as random factor (solid line). All regression shown are $p < 0.05$, except for the regression calculated for all lakes combined in **(a)**, which is $p < 0.1$. Symbols and lines are explained in Fig. 6.

Table 3. Pearson's correlation coefficients (R) between the environmental variables and cellular microcystin concentration ($\mu g\,g^{-1}$) or extracellular microcystin fraction (%) analysed for each lake and for all lakes combined using bivariate correlation analysis. The dependent variables are ln transformed. Extracellular microcystin fraction was zero in seven cases, leading to an $N = 5$ only.

Factor	Cellular microcystin concentration ($\mu g\,g^{-1}$)				Extracellular microcystin fraction (%)			
	All lakes $N = 48$	Jackadder $N = 18$	Bibra $N = 12$	Yangebup $N = 18$	All lakes $N = 38$	Jackadder $N = 18$	Bibra $N = 5$	Yangebup $N = 18$
pH	0.227	0.426	**0.762**	0.190	−0.297	0.155	−0.714	−0.360
Air Temp	−0.246	−0.288	−0.185	−0.160	0.077	0.138	−0.686	0.130
Salinity	0.067	0.330	0.448	**0.587**	**−0.375**	**0.570**	−0.775	**−0.659**
TP	**0.399**	**0.826**	0.489	**0.696**	**−0.392**	−0.303	−0.441	−0.295
TDP	**0.296**	**0.553**	**0.764**	0.225	**−0.428**	−0.088	−0.498	**−0.587**
TFe	**0.343**	**0.715**	**−0.605**	0.230	−0.037	0.380	0.499	−0.245
TDFe	**0.590**	**0.811**	0.135	0.400	−0.063	0.166	0.162	−0.252
NH	−0.267	−0.433	−0.338	**−0.579**	−0.115	−0.382	0.013	**0.530**
TN	0.085	0.441	0.268	**0.613**	**−0.376**	0.420	−0.633	−0.417
TDN	0.095	**0.482**	0.533	**0.479**	**−0.400**	0.324	**−0.921**	**−0.633**
TN : TP	**−0.446**	**−0.593**	−0.257	−0.382	**0.386**	0.492	0.514	0.239

Significant ($p < 0.05$) correlations are highlighted in bold.

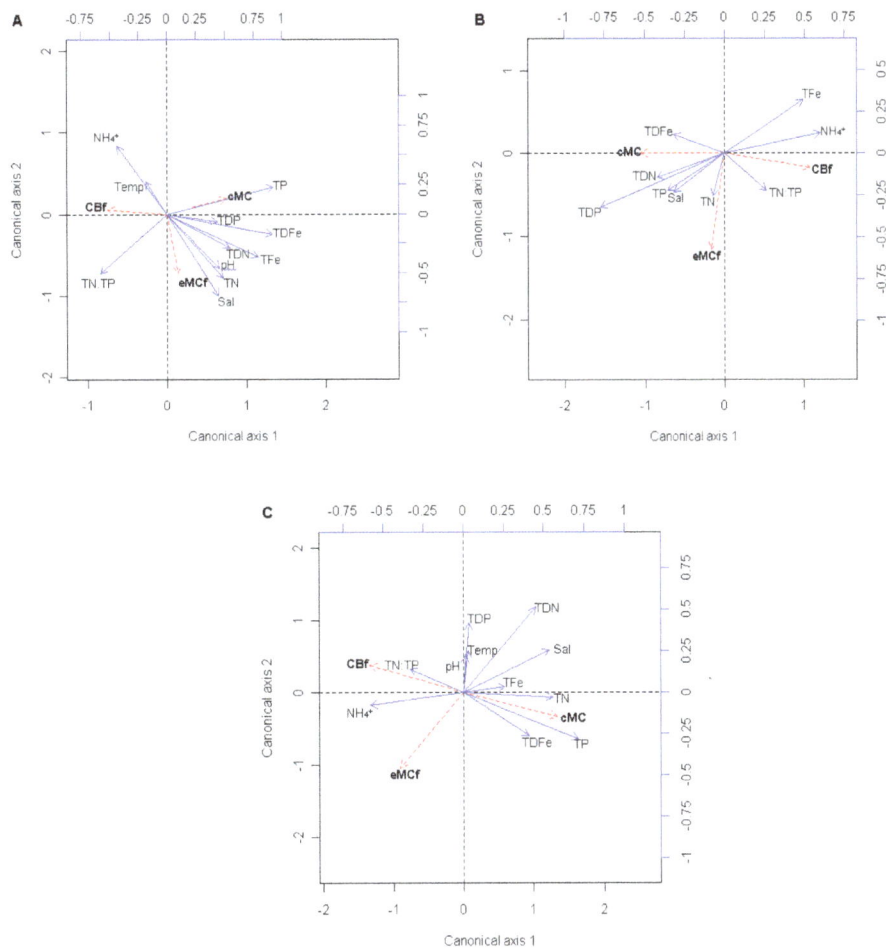

Figure 9. RDA biplots for the environmental variables and the cyanobacterial fraction (CBf), cellular microcystin (cMC) and extracellular microcystin fraction (eMCf) in **(a)** Jackadder Lake, **(b)** Bibra Lake, **(c)** Yangebup Lake; solid arrows indicate environmental variables; short dashed arrows indicate response variables. Canonical axes 1 and 2 represent a linear combination of the environmental variables, and axes are scaled by the square root of their eigenvalues.

bup Lake; $R^2_{adj.} = 0.75$; $F = 5.804$) of the combined variability of cyanobacterial fraction, cellular microcystin concentration and extracellular microcystin fraction can be explained by the measured environmental factors (Fig. 9a–c). The second RDA analysis, which sought to find relationships between environmental factors and absolute cyanobacterial biomass, cellular microcystin concentration and extracellular microcystin fraction for each lake found that 71 % (Jackadder Lake; $R^2_{adj.} = 0.71$; $F = 4.725$), 80 % (Bibra Lake; $R^2_{adj.} = 0.80$; $F = 5.806$) and 66 % (Yangebup Lake; $R^2_{adj.} = 0.66$; $F = 3.953$) of the combined variability of absolute cyanobacterial biomass, cellular microcystin concentration and extracellular microcystin fraction can be explained by the measured environmental factors (Fig. 10a–c).

In both sets of analyses, many of the environmental factors that were closely correlated to cyanobacteria and microcystins were slightly different between lakes. TDP was only correlated to either cyanobacteria fraction or cellular micro-

cystin concentration in Bibra and Jackadder lakes (Fig. 9a and b) but not in Yangebup Lake (Fig. 9c). Additionally, TFe was positively correlated to cyanobacteria only in Bibra Lake (Figs. 9b and 10b) but not in the other two lakes (Figs. 9a, c and 10a, c). In comparison to the other factors, TDFe was always negatively correlated to cyanobacterial fraction and biomass and positively correlated to cellular microcystin concentration variability (Figs. 9 and 10).

4 Discussion

The relationships between the environmental factors and cyanobacterial and microcystin variability were different between lakes. This is an indication that the relevance of factors that drive cyanobacteria and their toxin production depends on their site-specific combinations. Our results suggest that the site-specificity of environmental triggers may be related to spatial heterogeneity of the respective environmental fac-

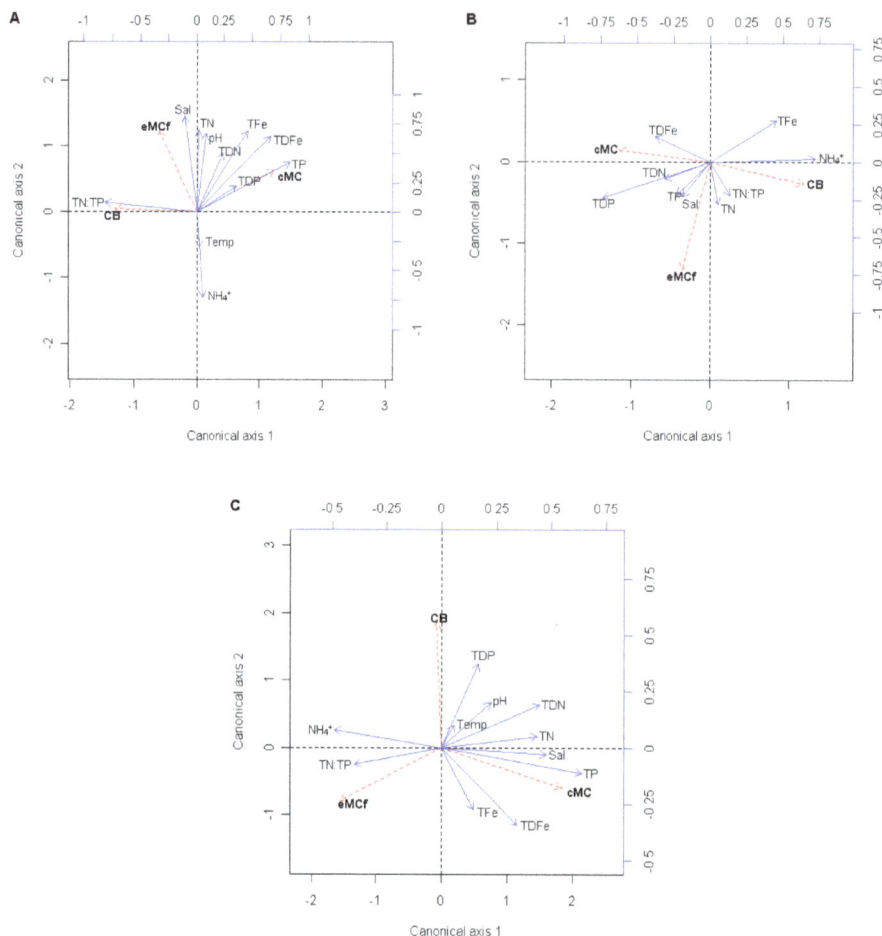

Figure 10. RDA biplots for the of environmental variables and the absolute cyanobacteria biomass (CB), cellular microcystin (cMC) and extracellular microcystin fraction (eMCf) in (**a**) Jackadder Lake, (**b**) Bibra Lake, (**c**) Yangebup Lake; solid arrows indicate environmental variables; short dashed arrows indicate response variables. Canonical axes 1 and 2 represent a linear combination of the environmental variables, and axes are scaled by the square root of their eigenvalues.

tor, as each factor can be present at different concentration regimes in each lake. Graham et al. (2004) and Dolman et al. (2012) have suggested that the correlations between the environmental factors and cyanobacterial biomass and microcystin concentration could change when the concentrations of the respective environmental factors increase from low to high in systems. Our results support these previous findings as the relationships between cyanobacterial fraction, cyanobacterial biomass and cellular microcystin concentration with TFe and TDFe were closely related to the concentration levels of TFe and TDFe in each lake. Mean TFe concentration in Bibra Lake was 1 order of magnitude higher than in Jackadder and Yangebup lakes, while mean TDFe concentrations in all lakes ranged within the same order of magnitude (Table 1). This could explain why the relationship between cyanobacterial fraction or cellular microcystin and TFe was different for between lakes, while TDFe was not. Further, the correlation between cyanobacterial fraction and TP was only significant in Yangebup and Jackadder lakes,

which both had lower TP concentrations than Bibra Lake, in which no significant correlation was found. Meanwhile, the correlation between cellular microcystin concentration and TFe was negative only in Bibra Lake, where TFe was present at significantly higher concentrations compared to the other two lakes. This indicates that the effect of environmental factors on cyanobacterial and microcystin variability may depend on site-specific factors such as concentration regimes, even in non-nutrient-limited lakes. Therefore, a generalisation by only using concentrations of nutrients might not be sufficient for future management of lakes.

The site-specificity of the environmental triggers of cyanobacterial and microcystin variability may also be a consequence of the variation of cyanobacterial communities between the systems. TFe was negatively correlated to cyanobacterial fraction in Jackadder and Yangebup Lake, and positively in Bibra Lake. The cyanobacterial community in Jackadder Lake was composed of only one nitrogen-fixing cyanobacterial genera (Fig. 4). In contrast, multiple

Local nutrient regimes determine site-specific environmental triggers of cyanobacterial and microcystin variability...

225

nitrogen-fixing cyanobacterial genera were present in Bibra Lake. Nitrogen-fixing cyanobacteria are known to utilise more iron in comparison to non-nitrogen-fixers (Wilhelm, 1995). Therefore, the site-specific correlation between TFe and cyanobacterial fraction may be explained through a greater iron requirement of the cyanobacterial community in Bibra Lake, in comparison to the cyanobacterial community in Jackadder Lake.

Currently, in the absence of lake-specific information, cyanobacterial management strategies are based on knowledge derived from general trends of the relationship between environmental factors and cyanobacteria or their toxins. Our study clearly indicates that the environmental variables explaining the variability in cyanobacteria and their toxins might be lake-specific and, more importantly, that these lake-specific correlations might also be different to the correlation derived from combining all data (e.g. Figs. 6a and 8a and b). This strongly supports the conclusion that site-specific conditions have to be taken into account for managing lakes with cyanobacterial blooms. Due to the site-specific environmental triggers of cyanobacterial and microcystin variability, the results presented in this study are important for the management of these lakes or lakes with similar physical, chemical and biological characteristics. In this study, the cyanobacterial fraction was negatively related with TP, TDP, TFe, TDFe, and positively correlated with TN:TP ratio. These relationships illustrate that in our study, cyanobacteria may dominate under lower phosphorus availability (Amano et al., 2010). Although the lakes in our study were not limited in phosphorus per se, the differences in phosphorus levels could have been responsible for the differences in the phytoplankton communities between lakes. At high concentration, phosphorus had been shown to potentially limit the ability of cyanobacteria to become dominant in the phytoplankton community, even though cyanobacteria as a group can dominate under a wide range of conditions (Chorus and Bartram, 1999; Reynolds et al., 2006). One reason for that is the higher growth rate of other phytoplankton groups compared to cyanobacteria, and, as such, their ability to utilise nutrients faster under high nutrient conditions. This can explain the negative correlation between cyanobacterial fraction and phosphorus concentration found in our study, and, maybe as a consequence of this, a positive correlation with TN:TP. In terms of iron, low availability was correlated to high cyanobacterial fraction in these lakes. This result indicated that cyanobacteria pose a competitive advantage to dominate the phytoplankton community under low iron availability. Cyanobacteria are capable to alter their cellular iron requirements, and increase the ability to utilise iron at a low concentration, through the presence of siderophores (Boyer et al., 1987; Lee et al., 2011). As reported in Nagai et al. (2007), cyanobacteria including *Microcystis* spp. and *Planktothrix* spp., can produce siderophores and become a superior competitor under iron-limited conditions. These results indicate that phosphorus and iron reduction in water

bodies might not be a sufficient remedial strategy against the occurrence of toxic cyanobacterial bloom.

In contrast to cyanobacterial fraction, cellular microcystin concentration was positively related to TP, TDP, TFe, TDFe and negatively correlated to TN:TP in all lakes. High availability of phosphorus relative to other nutrients is required for energy and material supply in microcystin biosynthesis as microcystin production in cyanobacterial cells is an energy-intensive process (Vezie et al., 2002). This is further supported through the observed negative relationship between cellular microcystin and TN:TP ratio, as low microcystin production is expected under conditions where phosphorus is present at lower concentrations in relation to other nutrients. In addition, the positive correlation between iron and cellular microcystin concentration is in agreement with earlier studies which suggested that iron plays an essential role in many metabolic pathways including microcystin biosynthesis in cyanobacteria (Jiang et al., 2008; Wang et al., 2010a). Our results illustrate that reducing phosphorus and iron concentrations in water bodies could potentially reduce the overall toxicity of cyanobacterial bloom, even though it might not completely prevent the occurrence of cyanobacterial bloom.

Environmental conditions influencing the release of microcystin into the environment, besides cells lyses, are not well understood (Rohrlack and Hyenstrand, 2007; Barrington et al., 2013). Our results showed that correlations exist between extracellular microcystin fraction and nutrients; however, the correlations could be direct or indirect ones. If they are direct, our results suggest that regardless of the potentially low microcystin production, cyanobacteria may release microcystins at lower nitrogen and phosphorus concentrations. This would support by the hypothesis that microcystin is involved in nutrient competition in the phytoplankton community (Huisman and Hulot, 2005).

Based on the RDA results, the measured environmental factors were able to better predict the variability of cyanobacterial fraction than the variability of absolute cyanobacterial biomass in two out of three lakes (Yangebup and Jackadder lakes). Both descriptors are important indicators for management. The competition with other phytoplankton, described by the cyanobacterial fraction in this study can affect the toxin production within a cell through allelopathy (Huisman and Hulot, 2005). Therefore, understanding the importance of site-specific drivers of both biomass and the cyanobacterial fraction is of highest importance to develop successful and sustainable management strategies.

5 Conclusions

The current approach to water body restoration and the prevention of toxic cyanobacterial blooms relies on reducing nutrient loading into water bodies and limiting the availability of nutrients in the water column. This approach might not always be successful in preventing the occurrence of cyanobacterial blooms, due to the roles of physicochemical factors on cyanobacteria and microcystin variability being dependent on the site-specific combination of environmental factors. Our study clearly highlights the importance of taking between-lake heterogeneity in the management of toxic cyanobacterial blooms into account. Site-specific studies may be required to determine the factors causing cyanobacterial dominance and microcystin production in different systems with different characteristics such as the hydrology, land use and water chemistry.

In our study, the dominance of cyanobacteria in the phytoplankton community is correlated to lower phosphorus and iron concentrations in the systems. In contrast, cyanobacteria required higher phosphorus and iron concentrations in the water column to produce a high amount of microcystin. Therefore, reducing phosphorus and iron concentration in the water column might not be a sufficient remedial strategy against the occurrence of toxic cyanobacterial bloom, if these nutrients are still available in sufficient amount to support the growth of highly competitive cyanobacteria. However, reducing phosphorus and iron could reduce the amount of microcystin being produced within cyanobacterial cells.

Acknowledgements. This project was funded by the Australian Research Council's Linkage Project funding scheme (LP0776571) and the Water Corporation of Western Australia. We wish to thank Pierre Legendre, Laura Firth and Kevin Murray for their valuable statistical advice, and Liah Coggins for her help in the editing of the paper. During the study, S. C. Sinang, was supported by a scholarship from Universiti Pendidikan Sultan Idris (UPSI) and Malaysian Government.

Edited by: C. Stamm

References

Amano, Y., Sakai, Y., Sekiya, T., Takeya, K., Taki, K., and Machida, M.: Effect of phosphorus fluctuation caused by river water dilution in eutrophic lake on competition between blue-green alga *Microcystis aeruginosa* and diatom *Cyclotella sp*, J. Environ. Sci.-China, 22, 1666–1673, 2010.

Ame, M. V. and Wunderlin, D. A.: Effects of iron, ammonium and temperature on microcystin content by a natural concentrated *Microcystis aeruginosa* population, Water Air Soil Poll., 168, 235–248, 2005.

APHA: Standard methods for the examination of water and wastewater, 20th Edn., edited by: Clesceri, L. S., Greenberg, A. E., and Eaton, A. D., American Public Health Association, Washington DC, 1998.

Arnold, J.: Perth Wetlands Resource Book, Environmental Protection Authority, Perth, 1990.

Arnold, T. N. and Oldham, C. E.: Trace-element contamination of a shallow wetland in Western Australia, Mar. Freshwater Res., 48, 531–539, 1997.

Barrington, D. J., Ghadouani, A., and Ivey, G. N.: Cyanobacterial and microcystins dynamics following the application of hydrogen peroxide to waste stabilisation ponds, Hydrol. Earth Syst. Sci., 17, 2097–2105, doi:10.5194/hess-17-2097-2013, 2013.

Beutler, M., Wiltshire, K. H., Meyer, B., Moldaenke, C., Luring, C., Meyerhofer, M., Hansen, U. P., and Dau, H.: A fluorometric method for the differentiation of algal populations *in vivo* and *in situ*, Photosynth. Res., 72, 39-53, 2002.

Boyer, G. L., Gillam, A. H., and Trick, C.: Iron chelation and uptake, in: The Cyanobacteria, edited by: Fay, P., and Baalen, C. V., Elsevier Science Publishers, the Netherlands, 415–431, 1987.

Bureau of Meteorology: Climate Data Online, http://www.bom.gov.au/climate/data/?ref=ftr (last access: 1 Sepember 2014), 2014.

Carey, C. C., Weathers, K. C., Ewing, H. A., Greer, M. L., and Cottingham, K. L.: Spatial and temporal variability in recruitment of the cyanobacterium *Gloeotrichia echinulata* in an oligotrophic lake, Freshwater Sci., 33, 577–592, doi:10.1086/675734, 2014.

Carlson, R. E.: A trophic state index for lakes, Limnol. Oceanogr., 22, 361–369, 1977.

Chaffin, J. D. and Bridgeman, T. B.: Organic and inorganic nitrogen utilization by nitrogen-stressed cyanobacteria during bloom conditions, J. Appl. Phycol., 26, 299–309, doi:10.1007/s10811-013-0118-0, 2014.

Chorus, I. and Bartram, J.: Toxic cyanobacteria in water: A guide to their public health consequences, monitoring and management, E & FN Spon, London and New York, 1999.

Davis, J. A., Rosich, R. S., Bradley, J. S., Growns, J. E., Schmidt, L. G., and Cheal, F.: Wetland classification on the basis of water quality and invertebrate community data, R/N:0730952487, Water Authority of Western Australia, Perth, 1993.

Davis, T. W., Berry, D. L., Boyer, G. L., and Gobler, C. J.: The effects of temperature and nutrients on the growth and dynamics of toxic and non-toxic strains of *Microcystis* during cyanobacteria blooms, Harmful Algae, 8, 715–725, 2009.

Department of Planning: Stirling City Centre District Water Management Strateg, Stirling City Centre District Water Management Strategy, Perth, 2010.

Dolman, A., Rucker, J., Pick, F., Fastner, J., Rohrlack, T., Mischke, U., and Wiedner, C.: Cyanobacteria and cyanotoxins: The influence of nitrogen versus phosphorus, PLoS ONE, 7, e38757, doi:10.1371/journal.pone.0038757, 2012.

Dunlop, M.: Yangebup lake environmental management study, Perth, Prepared for City of Cockburn, ENV Australia Pty Ltd, Perth, 2008.

Eisentraeger, A., Dott, W., Klein, J., and Hahn, S.: Comparative studies on algal toxicity testing using fluorometric microplate and Erlenmeyer flask growth-inhibition assays, Ecotox. Environ. Safe, 54, 346–354, 2003.

Engström-Öst, J., Repka, S., Brutemark, A., and Nieminen, A.: Clay- and algae-induced effects on biomass, cell size and toxin concentration of a brackish-water cyanobacterium, Hydrobiologia, 714, 85–92, doi:10.1007/s10750-013-1523-8, 2013.

Environmental Protection Authority: Drainage Management in South Jandakot and Beeliar Wetlands, EPA Bulletin, 371, 1989.

Eva, P. and Lindsay, B.: Microcystin and algal chlorophyll in relation to nearshore nutrient concentrations in Lake Winnipeg, Canada, Environ. Pollut., 3, 36–47, 2014.

Geis, S. W., Fleming, K. L., Korthals, E. T., Searle, G., Reynolds, L., and Karner, D. A.: Modifications to the algal growth inhibition test for use as a regulatory assay, Environ. Toxicol. Chem., 19, 36–41, 2000.

Ghadouani, A. and Smith, R. E. H.: Phytoplankton distribution in Lake Erie as assessed by a new in situ spectrofluorometric technique, J. Great Lakes Res., 31, 154–167, 2005.

Graham, J. L., Jones, J. R., Jones, S. B., Downing, J. A., and Clevenger, T. E.: Environmental factors influencing microcystin distribution and concentration in the Midwestern United States, Water Res., 38, 4395–4404, 2004.

Harada, K., Kondo, F., and Lawton, L. A.: Laboratory analysis of cyanotoxins, in: Toxic cyanobacteria in water: A guide to their public health consequences, monitoring and management, edited by: Chorus, I. and Bartram, J., E & FN Spon, London, New York, 363–367, 1999.

Hillebrand, H., Durselen, C., Kirschtel, D., Pollingher, U., and Zohary, T.: Biovolume calculation for pelagic and benthic microalgae, J. Phycol., 35, 403–424, 1999.

Huisman, J. and Hulot, F. D.: Population dynamic of harmful cyanobacteria, in: Harmful cyanobacteria, edited by: Huisman, J., Matthijs, H. C. P., and Visser, P. M., Springer, the Netherlands, 143–176, 2005.

Jang, M. H., Ha, K., Jung, J. M., Lee, Y. J., and Takamura, N.: Increased microcystin production of Microcystis aeruginosa by indirect exposure of nontoxic cyanobacteria: Potential role in the development of Microcystis bloom, B. Environ. Contam. Tox., 76, 957–962, 2006.

Jiang, Y., Ji, B., Wong, R. N. S., and Wong, M. H.: Statistical study on the effects of environmental factors on the growth and microcystins production of bloom-forming cyanobacterium Microcystis aeruginosa, Harmful Algae, 7, 127–136, 2008.

Kemp, A. S.: Freshwater cyanoprokaryota blooms in the Swan Coastal plain wetlands: Ecology, taxonomy and toxicology, PhD thesis, Department of Environmental Biology, Curtin University of Technology, Perth, 2009.

Komarek, J. and Hauer, T.: On-line database of cyanobacterial genera, http://www.cyanodb.cz (last access: 14 August 2011), 2011.

Koreiviene, J., Anne, O., Kasperoviciene, J., and Burskyte, V.: Cyanotoxin management and human health risk mitigation in recreational waters, Environ. Monit. Assess., 186, 4443–4459, doi:10.1007/s10661-014-3710-0, 2014.

Lawton, L. A., Edwards, C., and Codd, G. A.: Extraction and high-performance liquid chromatographic method for the determination of microcystins in raw and treated waters, Analyst, 119, 1525–1530, 1994.

Lee, W., van Baalen, M., and Jansen, V. A. A.: An evolutionary mechanism for diversity in siderophore-producing bacteria, Ecol. Lett., 15, 119–125, 2011.

Lehman, P. W., Marr, K., Boyer, G. L., Acuna, S., and Teh, S. J.: Long-term trends and causal factors associated with Microcystis abundance and toxicity in San Francisco Estuary and implications for climate change impacts, Hydrobiologia, 718, 141–158, doi:10.1007/s10750-013-1612-8, 2013.

Lei, L., Peng, L., Huang, X., and Han, B.-P.: Occurrence and dominance of Cylindrospermopsis raciborskii and dissolved cylindrospermopsin in urban reservoirs used for drinking water supply, South China, Environ. Monit. Assess., 186, 3079–3090, doi:10.1007/s10661-013-3602-8, 2014.

Lewis, W. M. and Wurtsbaugh, W. A.: Control of lacustrine phytoplankton by nutrients: Erosion of the phosphorus paradigm,

Int. Rev. Hydrobiol., 93, 446–465, doi:10.1002/iroh.200811065, 2008.

Li, D., Yu, Y., Yang, Z., Kong, F., Zhang, T., and Tang, S.: The dynamics of toxic and nontoxic Microcystis during bloom in the large shallow lake, Lake Taihu, China, Environ. Monit. Assess., 186, 3053–3062, doi:10.1007/s10661-013-3600-x, 2014.

Liu, Y.: Dynamic evaluation on ecosystem service values of urban rivers and lakes: A case study of Nanchang City, China, Aquat. Ecosyst. Health, 17, 161–170, doi:10.1080/14634988.2014.907223, 2014.

Meriluoto, J, and Codd, G.: Toxic–cyanobacterial monitoring and cyanotoxin analysis, in: Acta Academiae Aboensis Ser. B, Mathematica et physica, edited by: Högnäs, G., Åbo Akademi University Press, Åbo, 2005.

Nagai, T., Imai, A., Matsushige, K., and Fukushima, T.: Growth characteristics and growth modeling of Microcystis aeruginosa and Planktothrix agardhii under iron limitation, Limnology, 8, 261–270, 2007.

O'Bannon, C., Carr, J., Seekell, D. A., and D'Odorico, P.: Globalization of agricultural pollution due to international trade, Hydrol. Earth Syst. Sc., 18, 503–510, doi:10.5194/hess-18-503-2014, 2014.

Paerl, H. W. and Otten, T. G.: Harmful cyanobacterial blooms: Causes, consequences, and controls, Microb. Ecol., 65, 995–1010, doi:10.1007/s00248-012-0159-y, 2013.

Payton, M. E., Greenstone, M. H., and Schenker, N.: Overlapping confidence intervals or standard error intervals: What do they mean in terms of statistical significance?, , 3, 1–6, 2003.

Pineda-Mendoza, R. M., Olvera-Ramirez, R., and Martinez-Jeronimo, F.: Microcystins produced by filamentous cyanobacteria in urban lakes. A case study in Mexico City, Hidrobiologica, 22, 290–298, 2012.

Rastogi, R. P., Sinha, R. P., and Incharoensakdi, A.: The cyanotoxin-microcystins: current overview, Rev. Environ. Sci. Bio-Technol., 13, 215–249, doi:10.1007/s11157-014-9334-6, 2014.

Reichwaldt, E. S., Song, H., and Ghadouani, A.: Effects of the distribution of a toxic Microcystis bloom on the small scale patchiness of zooplankton, PLoS ONE, 8, 66674, doi:10.1371/journal.pone.0066674, 2013.

Reichwaldt, E. S. and Ghadouani, A.: Effects of rainfall patterns on toxic cyanobacterial blooms in a changing climate: Between simplistic scenarios and complex dynamics, Water Res., 46, 1372–1393, doi:10.1016/j.watres.2011.11.052, 2012.

Reynolds, C. S., Usher, M., Saunders, D., Dobson, A., Peet, R., Adam, P., Birks, H. J. B., Gustafssor, L., McNelly, J., Paine, R. T., and Richardson, D.: Growth and replication of phytoplankton, in: The ecology of phytoplankton, Cambridge University Press, 178–238, 2006.

Rohrlack, T. and Hyenstrand, P.: Fate of intracellular microcystins in the cyanobacterium Microcystis aeruginosa (Chroococcales, Cyanophyceae), Phycologia, 46, 277–283, 2007.

Rolland, D. C., Bourget, S., Warren, A., Laurion, I., and Vincent, W. F.: Extreme variability of cyanobacterial blooms in an urban drinking water supply, J. Plankton Res., 35, 744–758, doi:10.1093/plankt/fbt042, 2013.

Ruiz, M., Galanti, L., Laura Ruibal, A., Ines Rodriguez, M., and Alberto Wunderlin, D.: First report of microcystins and anatoxin-a co-occurrence in San Roque Reservoir (Cordoba, Argentina), Water Air Soil Poll., 224, 1593–1593, 2013.

Schindler, D.: The dilemma of controlling cultural eutrophication of lakes, P. Roy. Soc. B, 279, 4322–4333, 2012.

Sinang, S., Reichwaldt, E., and Ghadouani, A.: Spatial and temporal variability in the relationship between cyanobacterial biomass and microcystins, Environ. Monit. Assess., 185, 6379–6395, 2013.

Sitoki, L., Kurmayer, R., and Rott, E.: Spatial variation of phytoplankton composition, biovolume, and resulting microcystin concentrations in the Nyanza Gulf (Lake Victoria, Kenya), Hydrobiologia, 691, 109–122, 2012.

Smith, V. H.: Low nitrogen to phosphorus ratios favor dominace by blue-green algae in lake phytoplankton, Science, 221, 669–671, 1983.

Song, H., Coggins, L. X., Reichwaldt, E. S., and Ghadouani, A.: The importance of lake sediments as a pathway for microcystin dynamics in shallow eutrophic lakes, Toxins, 7, 900–918, 2015.

Srivastava, A., Choi, G.-G., Ahn, C.-Y., Oh, H.-M., Ravi, A., and Asthana, R.: Dynamics of microcystin production and quantification of potentially toxigenic Microcystis sp. using real-time PCR, Water Res., 46, 817–827, 2012.

Strategen: Bibra Lake: Landscape, recreational and environmental management plan, Perth, Prepared for City of Cockburn, Glenwood Nomineed Pty Ltd, Perth, 2009.

Sun, F., Yang, Z., and Huang, Z.: Challenges and solutions of urban hydrology in Beijing, Water Resour. Manage., 28, 3377–3389, doi:10.1007/s11269-014-0697-9, 2014.

Thi Thuy, D., Jaehnichen, S., Thi Phuong Quynh, L., Cuong Tu, H., Trung Kien, H., Trung Kien, N., Thi Nguyet, V., and Dinh Kim, D.: The occurrence of cyanobacteria and microcystins in the Hoan Kiem Lake and the Nui Coc reservoir (North Vietnam), Environ. Earth Sci., 71, 2419–2427, doi:10.1007/s12665-013-2642-2, 2014.

Tonk, L., Bosch, K., Visser, P. M., and Huisman, J.: Salt tolerance of the harmful cyanobacterium Microcystis aeruginosa, Aquat. Microb. Ecol., 46, 117–123, 2007.

Utermöhl, H.: Zur vervollkommnung der quantitativen phytoplankton-methodik, Mitt. Int. Ver. Theor. Angew. Limnol., 9, 1–38, 1958.

Van de Waal, D. B., Smith, V. H., Declerck, S. A. J., Stam, E. C. M., and Elser, J. J.: Stoichiometric regulation of phytoplankton toxins, Ecol. Lett., 17, 736–742, doi:10.1111/ele.12280, 2014.

Vezie, C., Rapala, J., Vaitomaa, J., Seitsonen, J., and Sivonen, K.: Effect of nitrogen and phosphorus on growth of toxic and nontoxic Microcystis strains and on intracellular microcystin concentrations, Microb. Ecol., 43, 443–454, 2002.

Waajen, G. W. A. M., Faassen, E. J., and Lürling, M.: Eutrophic urban ponds suffer from cyanobacterial blooms:Dutch examples, Environ. Sci. Pollut. Res., 21, 9983–9994, 2014.

Wang, C., Kong, H.-N., Wang, X.-Z., Wu, H.-D., Lin, Y., and He, S.-B.: Effects of iron on growth and intracellular chemical contents of Microcystis aeruginosa, Biomed. Environ. Sci., 23, 48–52, 2010a.

Wang, Q., Niu, Y. A., Xie, P., Chen, J., Ma, Z. M., Tao, M., Qi, M., Wu, L. Y., and Guo, L. G.: Factors affecting temporal and spatial variations of microcystins in Gonghu Bay of Lake Taihu, with potential risk of microcystin contamination to human health, Thes Scient. World J., 10, 1795–1809, 2010b.

Wilhelm, S.: Ecology of iron-limited cyanobacteria: A review of physiological responses and implications for aquatic systems, Aquat. Microb. Ecol., 9, 295–303, 1995.

Woodward, B.: Literature and Interview Project: Constructed Lakes in the Perth Metropolitan and South West Region, Perth, Prepared for Department of Water, Western Australian Local Government Association, Perth, 2008.

Yan, D. H., Wang, G., Wang, H., and Qin, T. L.: Assessing ecological land use and water demand of river systems: a case study in Luanhe River, North China, Hydrol. Earth Syst. Sci., 16, 2469–2483, doi:10.5194/hess-16-2469-2012, 2012.

Yen, H., Lin, T., Tseng, I., Tung, S., and Hsu, M.: Correlating 2-MIB and microcystin concentrations with environmental parameters in two reservoirs in South Taiwan, Water Sci. Technol., 55, 33–41, 2007.

Zhang, T., Zeng, W. H., Wang, S. R., and Ni, Z. K.: Temporal and spatial changes of water quality and management strategies of Dianchi Lake in southwest China, Hydrol. Earth Syst. Sci., 18, 1493–1502, doi:10.5194/hess-18-1493-2014, 2014.

Permissions

The contributors of this book come from diverse backgrounds, making this book a truly international effort. This book will bring forth new frontiers with its revolutionizing research information and detailed analysis of the nascent developments around the world.

We would like to thank all the contributing authors for lending their expertise to make the book truly unique. They have played a crucial role in the development of this book. Without their invaluable contributions this book wouldn't have been possible. They have made vital efforts to compile up to date information on the varied aspects of this subject to make this book a valuable addition to the collection of many professionals and students.

This book was conceptualized with the vision of imparting up-to-date information and advanced data in this field. To ensure the same, a matchless editorial board was set up. Every individual on the board went through rigorous rounds of assessment to prove their worth. After which they invested a large part of their time researching and compiling the most relevant data for our readers.

The editorial board has been involved in producing this book since its inception. They have spent rigorous hours researching and exploring the diverse topics which have resulted in the successful publishing of this book. They have passed on their knowledge of decades through this book. To expedite this challenging task, the publisher supported the team at every step. A small team of assistant editors was also appointed to further simplify the editing procedure and attain best results for the readers.

Apart from the editorial board, the designing team has also invested a significant amount of their time in understanding the subject and creating the most relevant covers. They scrutinized every image to scout for the most suitable representation of the subject and create an appropriate cover for the book.

The publishing team has been an ardent support to the editorial, designing and production team. Their endless efforts to recruit the best for this project, has resulted in the accomplishment of this book. They are a veteran in the field of academics and their pool of knowledge is as vast as their experience in printing. Their expertise and guidance has proved useful at every step. Their uncompromising quality standards have made this book an exceptional effort. Their encouragement from time to time has been an inspiration for everyone.

The publisher and the editorial board hope that this book will prove to be a valuable piece of knowledge for researchers, students, practitioners and scholars across the globe.

List of Contributors

G. Tang
Division of Earth and Ecosystem Sciences, Desert Research Institute, Reno, NV, USA

T. Hwang
Institute for the Environment, University of North Carolina, Chapel Hill, NC, USA

S. M. Pradhanang
Institute for Sustainable Cities, City University of New York, New York, NY, USA

Thomas B. Hasper and Johan Uddling
Department of Biological and Environmental Sciences, University of Gothenburg, 40530 Gothenburg, Sweden

Fernando Jaramillo
Department of Biological and Environmental Sciences, University of Gothenburg, 40530 Gothenburg, Sweden
Department of Physical Geography, Stockholm University, 106 91, Stockholm, Sweden
Stockholm Resilience Center, Stockholm University, 106 91, Stockholm, Sweden

Neil Cory
Department of Forest Resource Management; Division of Forest Resource Data, Swedish University of Agricultural Sciences, Umeå, Sweden

Berit Arheimer
Swedish Meteorological and Hydrological Institute, 601 76 Norrköping, Sweden

Hjalmar Laudon
Department of Forest Ecology and Management, Swedish University of Agricultural Sciences, 750 07 Umeå, Sweden

Ype van der Velde
Faculty of Earth and Life Sciences, University of Amsterdam, 1081 HV, Amsterdam, the Netherlands

Claudia Teutschbein
Department of Earth Sciences, Uppsala University, 75236, Uppsala, Sweden

N. Martínez-Carreras, C. E. Wetzel, J. Frentress, L. Ector, L. Hoffmann and L. Pfister
Luxembourg Institute of Science and Technology, Department Environmental Research and Innovation, Belvaux, Luxembourg

J. J. McDonnell
Global Institute for Water Security, University of Saskatchewan, Saskatoon, Canada
School of Geosciences, University of Aberdeen, Aberdeen, Scotland, UK

Harald Biester
IGÖ, Umweltgeochemie, TU Braunschweig, Langer Kamp 19c, 38106 Braunschweig, Germany

Klaus-Holger Knorr
ILÖK, Hydrologie, WWU Münster, Heisenbergstr. 2, 48149 Münster, Germany

Tanja Broder
IGÖ, Umweltgeochemie, TU Braunschweig, Langer Kamp 19c, 38106 Braunschweig, Germany
ILÖK, Hydrologie, WWU Münster, Heisenbergstr. 2, 48149 Münster, Germany

Rui Rivaes, José M. Santos and Teresa Ferreira
Forest Research Centre, Instituto Superior de Agronomia, Universidade de Lisboa, Tapada da Ajuda 1349-017 Lisbon, Portugal

Isabel Boavida and António N. Pinheiro
CERIS, Civil Engineering Research Innovation and Sustainability Centre, Instituto Superior Técnico, Universidade de Lisboa, Av. Rovisco Pais, 1049-001 Lisbon, Portugal

V. Srinivasan, K. Madhyastha, K. Jeremiah and S. Lele
Ashoka Trust for Research in Ecology and the Environment, Royal Enclave Sriramapura, Jakkur Post, Bangalore, Karnataka, India

S. Thompson and G. Penny
Department of Civil and Environmental Engineering, University of California, Berkeley, Berkeley, California, USA

Gonzalo Sapriza-Azuri and Pablo Gamazo
Departamento del Agua, Centro Universitario Regional Litoral Norte, Universidad de la República, Salto, Uruguay

Saman Razavi and Howard S. Wheater
Global Institute for Water Security, University of Saskatchewan, Saskatoon, SK, Canada
School of Environment and Sustainability, University of Saskatchewan, Saskatoon, SK, Canada
Department of Civil and Geological Engineering, University of Saskatchewan, Saskatoon, SK, Canada

Zhentao Cong, Qinshu Li, Kangle Mo, Lexin Zhang and Hong Shen
Department of Hydraulic Engineering, Tsinghua University, Beijing, 100084, China
State Key Laboratory of Hydroscience and Engineering, Beijing, 100084, China

Magdalena Uber
Univ. Grenoble Alpes, CNRS, IRD, Grenoble-INP, IGE Grenoble, 38000, France
Institute of Earth and Environmental Science, University of Potsdam, Potsdam, 14476, Germany
Univ. Grenoble Alpes, CNRS, IRD, Grenoble-INP, IGE Grenoble, 38000, France

Jean-Pierre Vandervaere, Isabella Zin, Cédric Legoût, Gilles Molinié and Guillaume Nord
ILÖK, Hydrologie, WWU Münster, Heisenbergstr. 2, 48149 Münster, Germany

Maik Heistermann
Institute of Earth and Environmental Science, University of Potsdam, Potsdam, 14476, Germany

Isabelle Braud
Irstea, UR RiverLy, Lyon-Villeurbanne Centre, Villeurbanne, 69625, France

F. Richter, C. Döring and M. Jansen
Department of Soil Science of Temperate Ecosystems, Georg-August-Universität Göttingen, Büsgenweg 2, 37077 Göttingen, Germany

O. Panferov
Department of Bioclimatology, Georg-August-Universität Göttingen, Büsgenweg 2, 37077 Göttingen, Germany
Institute of Climatology and Climate Protection, University of Applied Sciences, Bingen am Rhein, Berlinstr. 109, 55411 Bingen am Rhein, Germany

U. Spank and C. Bernhofer
Institute of Hydrology and Meteorology, Technische Universität Dresden, Pienner Str. 23, 01737 Tharandt, Germany

O. Branch, K. Warrach-Sagi and V. Wulfmeyer
Institute of Physics and Meteorology, University of Hohenheim, Stuttgart, Germany

S. Cohen
Institute of Soil, Water and Environmental Sciences, Agricultural Research Organization, Volcani Center, Bet-Dagan, Israel

Daniel J. Isaak, Charles H. Luce, Gwynne L. Chandler, Dona L. Horan and Sherry P. Wollrab
U.S. Forest Service, Rocky Mountain Research Station, Aquatic Sciences Lab, Boise, ID 83702, USA

Elke S. Reichwaldt and Anas Ghadouani
Aquatic Ecology and Ecosystem Studies, School of Civil, Environmental and Mining Engineering, M015, The University of Western Australia, 35 Stirling Highway, Crawley, Western Australia 6009, Australia

S. C. Sinang, E. S. Reichwaldt and A. Ghadouani
Aquatic Ecology and Ecosystem Studies, School of Civil, Environmental and Mining Engineering, The University of Western Australia, 35 Stirling Highway, M015, Crawley, WA 6009, Western Australia, Australia
Faculty of Science and Mathematics, Sultan Idris Education University, 35900 Tanjong Malim, Perak, Malaysia

Index